NAPOLEON'S GENERALS
The Waterloo Campaign

Tony Linck

Portraits by Keith Rocco

EMPEROR'S PRESS

NAPOLEON'S GENERALS

My sincere thanks to the following individuals:
Fanny and François François who many years ago found a copy
of Georges Six's great work that started me off on this project.

AUTHOR'S ACKNOWLEDGMENTS
Sincere thanks are due to the following individuals and
organizations who were helpful in providing prints
for this book:
Kevin Krause for the use of Knötel prints from his collection,
and to Waxtel & Hasenauer for the use of Detaille pictures from
their book, *L'Armée Française*

DEDICATION
This book is dedicated to my best friend and long suffering wife
Maria-Louise for having to share her life with myself and
seventy of Napoleon's Generals.

PREFACE

When I started to research the lives of Napoleon's generals and later decided to put pen to paper I never realized the daunting task I had taken on as their lives encompass such a wide field. Over four hundred men held the rank of général de division or its equivalent during the Napoleonic and Revolutionary Wars. While nearly another thousand held that of général de brigade. My problem was where to begin and who to include.

I have used Waterloo as a starting point. The work covers in strict alphabetical sequence the lives and military careers of those generals who attained the rank of général de division and served with the Army of the North during Napoleon's final campaign. There was a temptation to exclude the marshals - Grouchy, Ney and Soult. They have been covered extensively by others and I felt I could not do them justice in light of comparatively recent biographies by Horrocks and Hayman on the lives of Ney and Soult. As the title maréchal was a dignity rather than a military rank these men were still générals de division the most senior rank within the French Army at the time. For completeness their inclusion seemed justified.

This book is not another analysis of Waterloo, but brings to light the often ignored careers of the men, who were largely responsible for the implementation of Napoleon's style of warfare. Books and articles written covering Napoleon's life and his campaigns probably number more than his army at Waterloo. However the dominant personalities of Napoleon and to a lesser degree those of his Marshals have pushed the achievements of his generals to the background. There is no doubt they achieved a great deal. This is largely due to the fostering of the Napoleonic Legend by the great man himself, and then by biographers and politicians for differing motives in the middle years of the nineteenth century. It has as a result unintentionally diminished the greatness of the men who served under him.

To cover the subject is like walking in a minefield. It is difficult not to be biased as one has preconceived ideas as to whom are the heroes and who are the villains. I acknowledge the stance taken is generally sympathetic towards these men. However I do not believe balance is lost and where they may have failed, any reasons given are justified. Conversely I hope that my admiration is not too effusive where they have triumphed.

There is a gap in the chronicles of the Revolutionary and Napoleonic Wars covering the achievements of his commander outside the immediate circle of the marshalate. These men were the implementors of a new style of warfare that burst on to the European stage. They were cogs in the machine that welded the French Army into such a formidable fighting force. Attention is focused to their careers, their backgrounds, early careers, strengths, weaknesses and achievements as leaders. Also covered are their lives after that fateful day on 18 June 1815 when after years of fame, glory and adoration, their careers came to a sudden and dramatic end. For those, who survived, they all had to adjust. Some faced trial and the prospect of execution, whilst others suffered persecution, exile and impoverishment in the years that followed. Most reconciled themselves to the new order and continued to have illustrious and varied careers.

Their comparative youth at the time of Waterloo, which on average was a little over forty, gives recognition to the boundless ambition and energy these men had. Further analysis shows that on average most had held the rank of général de division for only three to four years. With a few exceptions like Vandamme, Friant, Durutte and Duhesme they were predominantly from the later generation of younger aggressive commanders, who when they cast their lot with the returned Emperor were anxious to win new titles and glory in place of the marshals who had deserted him. Many lived well into the second half of the century, Flahaut the last survivor till 1870, rich in memories of past glories, but forgotten by all but a few.

Numerous sources have been referred to: the closing pages contain a more detailed bibliography. I pay tribute to Georges Six who compiled the Dictionaire des Admiraux et Generaux de la Revolution et l'Empire, which gave me the idea to delve further into the careers of these men. It is this publication that forms the bedrock around which my work is based. Without the benefit of Georges Six's painstaking research this Napoleonic writer would never have got past first base.

There is little new unpublished material to be gleaned from this work. It is a compilation of facts, figures and anecdotes from great works past and present on the Wars from 1793 to 1815. Phipps's five volumes of The Generals of the Revolution and the Rise of Napoleon's Marshals have been of great assistance in charting the movements of the generals during their earlier years. Unsurpassed is Oman in covering events in the Peninsular in his seven volumes of A History of the Peninsula War. Petre's five solid works cover the middle and declining years of Napoleon's campaigns. David Chandler in his tome The Campaigns of Napoleon is a great chronicle of the period whose occasional gems and perceptive comments have helped in my endeavors.

The new generation of Napoleonic writers and researchers from the United States who have largely replaced the Old Guard of British writers in recent years need mentioning. Nafziger's recent work, Napoleon's Invasion of Russia is the best covering the 1812 Campaign in Russia. That has been followed by Lutzen and Bautzen, a detailed account of the

Spring Campaign of 1813 in Germany that gives much detail on happenings away from Napoleon's immediate center of operations. Bowden's works the Armies of the Danube 1809, Napoleon's Grande Armée of 1813 and the Armies at Waterloo for detailed orders of battle and the composition of the French armies are invaluable and masterpieces in their own right.

The books on the Waterloo Campaign itself are endless. Lachouque's, Waterloo - The End of a World stirs up such l'esprit within the French Army and covers the events of the campaign with such dashing style one is left breathless wondering how Napoleon ever managed to lose. Becke's Napoleon at Waterloo is excellent in analyzing events and apportioning blame to everyone except Napoleon himself.

I have tried to cater for all tastes and interests in this work. After much thought decided to include two detailed Orders of Battle gathered from archive sources showing the strengths and organization of the Army of the North. The first gives the strength of formations at the beginning of the campaign whilst corresponding figures were compiled from returns gathered as the Army rallied around Laon. They help to illustrate the debilitating effect of the campaign as the Army streamed back into Northern France after Waterloo.

Another question arose whether to provide a detailed index. Little benefit would be gained for a work which is in effect an alphabetical dictionary of a group of French generals. To partly overcome the problem of purists all Waterloo commanders are asterisked for easy cross reference. There are also detailed charts containing the service record of each commander in addition to a summary of all their major campaigns.

Finally this work is directed at every aspiring wargamer, military historian and armchair general who has at sometime has debated the merits of the commanders at Waterloo. To stimulate discussion and in particular help wargamers wishing to engage in Napoleonic games all French generals have been given a one to ten rating based on given criteria. They are the judgement of the author, who like any judge is not infallible. Wargamer or not, use your own judgement having read these pages or not and compare them. Many preconceived ideas I held changed during my research.

I hope this book succeeds in recognizing the influence these men had on the events of the day and pays tribute to their strength and courage. Finally I hope I have been fair to their memory and in the pages that follow may have aroused some interest or even debate.

Anthony Linck
Castletown, Isle of Man, November 1993.

NAPOLEON LEAVING ELBA, 1 MARCH 1815

INTRODUCTION

The generals that marched into Belgium in June 1815 chose to serve Napoleon for a variety of motives. Their stances invariably changed as the events of the Hundred Days unfolded. When Napoleon landed near Antibes on 1 March 1815 there was the group of die hard Bonapartists accompanying him, plus many in France both in and out of the army who possessed a fervent belief in him and lusted for past glories. Others of his former generals who were serving under the Bourbons respected and admired Napoleon but had little enthusiasm for the coming conflict they thought was inevitable. They had everything to lose. The Napoleonic Wars may have brought them fame and fortune but they were weary to put their lives on the line for what appeared at first a reckless gamble.

As Napoleon dodged the first columns sent out to intercept him and disappeared into the mountainous depths of the Isére, far northern hotheads like Exelmans and the Lallemand brothers supported by the region's commander comte d'Erlon prematurely tried to raise the northern garrisons and march on Paris.

At the same time many of Napoleon's former commanders who were now with the Bourbon armies actively opposed his return, in particular Ney, who promised the King to bring Napoleon back in an iron cage. Others, the vast majority, would have preferred that he never made an appearance at all, but began to change their positions with news of the first defections. From outright opposition they took the middle road, dragged their feet when orders came to march against him and in the end did little to hinder or halt the triumphant march to Paris.

When Napoleon was installed in the Tuileries they all with a few exceptions had the option to serve or not. Most took the former course not out of love or loyalty but out of sheer necessity. Failure to rally to the new order meant dishonor, dismissal and disgrace. Conversely there was the group of unemployed officers, which included such notables as Davout and Vandamme who had no ties of loyalty to the King.

There were only eight active serving marshals who chose to follow Napoleon and of these only three were to serve with the Army of the North. Soult as Chief of Staff was a most controversial choice as he had only recently resigned as Bourbon Minister for War. A "notetaker" of the Berthier mold would have been preferable and as a result Soult's considerable talents were misused. As commander of the army's left wing he would have been the man to engage the Anglo-Dutch forces. He had the greatest experience of all the marshals in facing Wellington. Bailly de Monthion, Berthier's understudy for so many years would have been a more suitable choice as chief of staff. Napoleon for considerations of rank and etiquette desired a Marshal of France for the post.

Ney, Soult's arch enemy, took charge of the army's left wing. His appointment was even more controversial, and had it not been made till after the army was on the move in all likelihood it would have sparked a rash of resignations by senior officers serving under him. It was a misfortune for France. The energy he had shown during the retreat from Moscow was simply not there. He was tired of war and slow to react, but this only showed once he was in the saddle. He brought with him a war weary lassitude that seeped down his command structure. His subordinates could only shake their heads in disbelief, and it certainly made them uneasy and suspicious.

Why did Napoleon add to the confusion by delaying Ney's appointment to the last moment? It gave the marshal no time to familiarize himself with his command. Possibly he was waiting for the faithful Mortier, ill but nominally in command of the Guard, to recover and lead the wing. Ney had been worsted in the Peninsula so often before by Wellington that it affected his performance. So overawed by the martial abilities of his opponent it accounts for the lassitude and indecision he so often displayed coupled with the sheer terror of having to face Napoleon if he failed. With Napoleon at his side he was a different man. Unfortunately for Napoleon and France, caution was thrown to the winds as the "Bravest of the Brave" turned into a man possessed, with a reckless bravery that made him about as effective as a cavalry lieutenant. Courage he did not lack. He would have been better suited to have been under Napoleon's direct control at the head of the Guard where the unfortunate Drouot had the dual responsibility of being its chief of staff, artillery expert and senior field commander.

Grouchy the most recently promoted marshal, was at first to head the Cavalry Reserve. It was the most logical and sensible choice, for after the disgraced Murat he was the most experienced man in France to lead large mounted formations. That he later took charge of the right wing once the campaign was underway was another misfortune. In the days preceding his non-appearance at Waterloo he began to show the indecision displayed by a man not confident in handling a large force of mixed arms led by some very strong willed commanders. That he showed such resolution after Waterloo and not before was an equal misfortune for France.

Who could have led the right wing in place of Grouchy? The logical choice was Davout. He knew how to deal with the Prussians, his greatest triumph Auerstadt had been against them. He was the one man who the hard headed Vandamme and the equally strong willed Gérard respected. His grasp of strategy and natural instinct to be at the right place at the right time would never have allowed him to have faced Grouchy's predicament on the day of Waterloo. Napoleon's biggest mistake of all was to leave his Minister for War at Paris where he felt he needed a strong man to watch the city. He totally missed the point that if he remained successful in the field, any opposition at home was of no consequence. It was however imperative that to guarantee victory quickly and decisively he field his best available team.

The choice of the army corps commanders cannot really be faulted. D'Erlon and Reille had shown their loyalty. Both had campaigned extensively against Wellington in the Peninsula and Southern France. That d'Erlon showed such indecision and tactical ineptitude in the campaign was generally out of keeping with his past form. Vandamme often so close to earning a marshal's baton but for his own failings, nurtured a deep hatred of the Prussians. Captured at Kulm in 1813 and a prisoner of the Allies, along with Lobau who joined him after the fall of Dresden, they both thirsted for revenge They had missed the humiliations of the previous year when the Allies trampled over the soil of France. Devoted to Napoleon, tenacious and tough, both were natural choices. Gérard the youngest of the corps commanders had the most to prove. He was the classic example of the younger generation of commander ready and prepared to give his all.

The cavalry corps leaders whilst not lacking in courage when called to do their duty showed indifferent form. Kellermann the most talented of them, was a late choice. He accepted a command not out of loyalty to Napoleon, but more as a duty to his country. Exelmans was a wild Bonapartist, who was as reckless as he was overrated. Pajol a steady professional was a commander of few faults. The regicide Milhaud had ability, but possessed loyalty to himself and no one else.

The appointment of commanders at divisional level and below was Davout's responsibility as Minister for War. He was not infallible, any more than Napoleon. Several choices he made were unfortunate, others pitiful. Fine commanders willing to serve were missed out. Molitor was a division commander at thirty-two in 1809 whose qualities suited him to a field command rather than the National Guard formation he received. Treilhard a magnificent dragoon general who performed with distinction in France the previous year, took a remote posting at Belle Isle en Mer. Sebastiani an indifferent corps commander, certainly knew how to handle a cavalry division and was willing to take a step down to lead one. He to his rage found himself sent to Corsica as military governor.

D'Erlon's corps was particularly plagued by second rate men. Marcognet for several years in Italy had shown a lassitude that made him well past his best. Donzelot had lost his sharpness during the seven years he had spent as governor of Corfu. Durutte showed little enthusiasm yet when it came to the crunch, he above all did his duty.

In light of the above, there was a need for Napoleon's aide Flahaut to examine the files and query Davout's appointments before they were submitted for final approval. Overworked Napoleon scanned his minister's and aide's proposals without looking at them too closely. It was Flahaut's at times tactless manner in carrying out his mission rather than his being there that caused a problem. He should have known better, Davout was a difficult man and Flahaut was after all an accomplished diplomat.

Amongst those who held commands before Napoleon's return discord was rife. They without being ardent royalists would have preferred to end their careers peacefully under the Bourbons. Instead many scowled at those who rallied to Napoleon before his arrival in Paris so plunging the country into a perilous venture that was to provoke a terrifying war. These men were in turn suspected by others and denounced as officers without energy, lukewarm patriots or even shameless royalists. Those who first rallied to the Emperor's standard were amazed to see men like Soult, Kellermann, Durutte and Bourmont still in the Army. Certainly few generals would have served Ney had he not taken up his command till after the campaign started. What they overlooked was Ney's common touch with the rank and file, in addition to him being a fine political prize. Things came to such a pass that generals were fighting duels. Bonnet, a fine Peninsular veteran without a command, incensed because he felt General Ornano had prejudiced him against the Emperor, shot him through the lungs. The Army and in particular the Guard lost a fine cavalry leader.

As senior officers they saw matters as they really were, there being a sense of cynicism and uncertainty amongst them. The rank and file on the other hand approached the campaign with enthusiasm and confidence. They still dreamt of glory and believed Napoleon invincible. If the senior officers lacked a will, the rank and file were filled with an eagerness to fight, a resolution to conquer, a hatred of foreigners and an idolatry of the Emperor unsurpassed. It was a dangerous cocktail with which to wage war and in the end proved so fragile.

Conversely when the generals stood to be counted, all with but one exception, Bourmont, did their duty. In many cases they may not have been prepared to die for Napoleon but faced by the prospect of Allied armies once more marching over the soil of France they were prepared to make the ultimate sacrifice. It was a high price even for Napoleonic battles. An appalling statistic was that of the remaining sixty-four generals, a further twenty-six were struck down with wounds of varying severity. This showed in spite of all their misgivings, that French generals still led from the front and were prepared to give their all for their country.

INTRODUCTION

AUTHOR'S NOTE

For the convenience of readers who are not familiar with the English equivalent of certain military ranks when these are given in French, here is an explanatory table:-

French Rank	English - American Equivalent
Soldat	Private
Brigadier or Caporal	Corporal
Sergent or maréchal des logis	Sergeant or corporal of Horse
Sergent-major	Sergeant major
Adjudant	Company sergeant major
Adjudant-chef or adjudant sous officer	Warrant officer first class
Sous lieutenant	Subaltern or second lieutenant
Lieutenant	Lieutenant
Capitaine	Captain
Chef de bataillon	Major infantry
Chef d'escadron	Major cavalry
Lieutenant colonel	Lieutenant colonel
Colonel	Colonel
Général de brigade or maréchal de camp	Brigadier general
Général de division	Major general
Maréchal	Field Marshal

Staff Ranks

Adjudant major	Staff Officer (usually grade of major)
Chef de brigade	Brigade major
Adjudant général or adjudant commandant	Staff Officer (First Class), colonel or lieutenant colonel
Major général or général en chef	Army Chief of Staff

NAPOLEON WATCHING THE BATTLE OF LIGNY

BACHELU - GILBERT DÉSIRÉ JOSEPH, BARON (1777 - 1849)

BACKGROUND AND EARLY CAREER

A native of Eastern France, Bachelu was born in the market town of Dole (Jura) on 9 February 1777. From a solid middle class background, his father having served for some years as a senior councillor in the Dole Parlement, it was hoped he would enter his father's legal practise. Instead, he chose the military and after attending the Metz Military Academy he was commissioned an engineer *sous lieutenant* on 2 February 1794.

From 1795-1797 he served with the Army of the Rhine. He made steady progress gaining promotion to *lieutenant* on 21 March and then *capitaine* on 19 June 1795. His daring and initiative came to the fore during Moreau's retreat to the river in 1796 and General Bonaparte in need of promising engineer officers recruited him for the Army of the Orient in May 1798. Three very active years followed in Egypt, the first being spent from late August 1798 with Desaix during his journey up the Nile and in Upper Egypt in pursuit of Murad Bey. He was then posted to Kosseir on the Red Sea to help strengthen the crude fortifications and distinguished himself during its defence when the Royal Navy attacked the port on 14-16 August 1799. He was in Cairo when the revolt engulfed the city in April 1800 and took part in the reduction of the citadel when at the head of a detachment of engineers he blew a hole in the walls. Promoted acting *chef de bataillon* by Kléber on 1 May 1800 a Consular Decree later confirmed his rank on 10 October 1801 after his return to France.

Appointed a Subdirector of Fortifications on 21 November 1801, the following month he joined the expedition to Santo Domingo. Arriving on the island in March 1802 he played a lead role in the bloody struggle that led to the reduction of Fort Crête à Pierrot in April. His exploits caught the eye of General Leclerc who promoted him acting *chef de brigade du genie* on 12 July 1802, and made him one of his aides. When Leclerc died of yellow fever in November 1802 he had the fortune to accompany the widowed Pauline Bonaparte back to France. Rumors of an affair with Pauline while on Santo Domingo abounded, as did her affairs with many with others but did not seem to affect his prospects. On 18 January 1803 he received confirmation of his rank soon after his arrival and delivery of the distraught Pauline to the Bonaparte family.

In December 1803 he took the post of Director of Fortifications and Chief of Staff of the Engineers under Soult at the Camp of Boulogne. His time in Santo Domingo had given him a taste for infantry life. Anxious to secure an infantry command, he left Boulogne on 1 February 1805 when he secured a posting as *colonel* of the 11th Line with Marmont at the Camp of Utrecht.

AUSTRIA AND DALMATIA 1805-1809

During the Austrian Campaign his regiment was with Boudet's division of Marmont's II Corps. The association was a good one, as Bachelu had served briefly under Boudet while on Santo Domingo. On 21 October 1805 he was present at the capitulation at Ulm, and with II Corps' advance into the Tyrol he fought at Weyer on 3 November and was at the capture of Leoben on 12 November 1805. When Austria ceded her Adriatic provinces after the Treaty of Pressburg, Bachelu's regiment moved into the region as part of the occupying force, the Army of Dalmatia. In the province for three years, he served for a while with Lauriston's division where he distinguished himself against the Russians at Castelnovo on 30 September 1806. Then from January 1808 he operated under Montrichard, who led the division on Lauriston's move to Venice as governor, spending most of the following year engaged against Bosnian rebels.

AUSTRIAN CAMPAIGN OF 1809

The renewal of war against Austria in April 1809 saw his regiment serve with Clausel's 2nd Division of the Army of Dalmatia. As part of Marmont's advance guard during the march from Kurn he fought at Pribudic on 16 May. Here the 11th Line distinguished themselves when they overran Stoichewich's column of 8,000 men as it tried to leave the field after receiving a merciless bombardment from Marmont's artillery. In the vigorous pursuit that followed he fought again at Gospich on 20 May and on 3 June occupied Laibach. Promoted *général de brigade* on 9 June 1809 he retained command of the 11th Line, which was given brigade status. He took part in Marmont's march to Vienna and fought at Wagram on 6 July where his men spearheaded the assault by XI Corps that broke the Austrian line between Wagram and Aderklaa and won Marmont his Marshal's Baton.

After the disbandment of XI Corps in January 1810 a period of garrison duty followed in Illyria when he joined Carra St Cyr's 1st Division of the Army of Illyria based at Laibach before returning to Montrichard's division in October. He became a *baron de l'Empire* on 29 August 1810 and remained in Illyria till 6 February 1811 when he took a posting to

Danzig as deputy commander of the garrison.

RUSSIAN CAMPAIGN OF 1812

He remained in Danzig during the early part of the Russian Campaign till appointed commander of the 3rd Brigade of Grandjean's 7th Division with Macdonald's X Corps on 1 September 1812. His command, made up of the 13th Bavarian and 1st Westphalian Line Regiments at first played a quiet role in the campaign where based at Dunaburg on the Dwina it formed a link between Macdonald's corps and the Grande Armée. In late December with Macdonald in full retreat and the Prussians apparently about to defect, he being one of few reliable generals available took charge of the rearguard. He received great praise for the way he handled the command, particularly when he inflicted a sharp reverse against Wittgenstein near Tilsit on 27 December. His action for a short time steadied the wavering Prussians and gained valuable time for Macdonald. He halted the Russians at Labiau on 3 January, and then again at Rosenburg and Stublau on 12 January before reaching the safety of Danzig on 18 January 1813.

THE SIEGE OF DANZIG 1813-1814

In the period before the Russians invested Danzig his old engineering skills came to the fore as he made herculean efforts to strengthen its defences. On 3 March he repulsed Platov's Cossacks after they had occupied the suburbs. Two days later at the head of General Destrées's Neopolitans he debouched from the city and drove back 8,000 Russians of the Army of Moldavia that had occupied Sottland and Ohra, inflicted over 1,500 casualties and took a further 500 prisoner. On 24 March he led a raid to resupply the city and for a short time seized Matschakau and Borgfeld so enabling the garrison to gather much needed provisions.

His most successful exploit during the siege was on 27 April when Rapp ordered another sortie against the Isle of Nehrung. At the head of 1,200 infantry and 300 cavalry, and the support of a small flotilla of gunboats he routed the 3,500 Russians defending the position and returned to Danzig with further quantities of food and livestock. After such exploits it was with little surprise that on 26 June 1813 his promotion to *général de division* went through. For the remainder of the siege he continued to play a crucial role helping to shore up the defences till the final surrender of the garrison on 2 January 1814 when he was taken prisoner.

THE RESTORATION AND NAPOLEON'S RETURN

After his return to France the Bourbons made him a *chevalier de Saint Louis* on 14 August 1814. He remained unattached till 18 March 1815 when they called him to join the Duc de Berry's forces gathering south of Paris to block Napoleon's approach. Discretion the better part of valor he didn't appear, and only rallied to Napoleon after his proclamation as Emperor in Paris. There was talk of him receiving a command under Rapp with V Corps in Eastern France, but nothing came of it. At one stage it appeared he would not receive one at all, since he was a known republican sympathizer rather than an enthusiastic Bonapartist. In the end a command came up on 14 April after the old veteran Dufour, who headed the 5th Division of Reille's II Corps died suddenly at Lille after a short illness. He took over his command on 30 April and based at Valenciennes spent the next six weeks feverishly preparing his division for the coming campaign.

THE WATERLOO CAMPAIGN

With the opening of hostilities on 15 June 1815 his troops formed Reille's advance guard that crossed the border at Thuin and pushed along the banks of the Sambre. At Marchienne he fought a sharp action against Steinmetz's brigade, which delayed his crossing of the river till noon. Once a passage was cleared he followed the cavalry as they pushed cross country to the Charleroi-Brussels road, and by the evening he made camp at Mellet five miles south of Quatre Bras.

The next day, due to Ney's hesitancy and the late receipt of marching orders his division only appeared before Quatre Bras around 2.00 P.M.. Deployed to the east of the Brussels Road, between the road and Delhutte Wood Ney ordered him to attack the Allies' exposed left flank. After the French guns opened with a brief heavy cannonade he drove the Dutch from Piraumont. Then pushing up the slopes towards the Namur road his men came under a withering close range fire from Picton's division that had just arrived on the field and deployed on the crest. Exhausted, his men halted, wavered and then broke as Picton flung his troops at them with the bayonet. Order was not restored till his men when reached Piraumont, then with a defensive line established he held the French left till Wellington's counter attacks petered out at dusk. In the battle Bachelu had suffered over 1,500 casualties and his division was in no position to join the pursuit of Wellington's forces the next day.

At Waterloo on 18 June his troops deployed to the left of the Charleroi Road were at first held in reserve. Poor leadership by Reille allowed Jérôme Bonaparte to become embroiled in a struggle for Hougoumont, which ended with Bachelu being dragged into two fruitless assaults on the chateau. As he approached Hougoumont in column from the east after being deflected away from the Allied center, his troops fell back in confusion when caught in an enfilading fire from Cleeve's battery of the King's German Legion. Around 3.30 P.M. as he led a second assault through the Hougoumont orchard a shell splinter struck him in the head. Carried from the field and later evacuated to Paris he never rejoined his troops and on 1 August 1815 was suspended.

LATER LIFE AND CAREER

Retribution was swift for failing to join the Duc de Berry's forces the previous March and suspected of treason the Bourbons imprisoned Bachelu on 15 October 1815, However no charges materialized and he was soon released. In order to exact revenge they then exiled him from Paris and later France, only allowing him to return in December 1817. Recalled after Saint Cyr's Army Reforms of 30 December 1818 he served on the General Staff. Still an unrepentant republican he was retired on 1 December 1824 following the crackdown against Bonapartist and republican sympathizers within the army after the accession of Charles X. Undeterred he entered politics and became a leader with the liberal opposition in his native Jura.

After the overthrow of Charles X the new government recalled him and on 4 August 1830 he became commander of the 19th Military Division at Lyons. His return started with much promise and on 13 October 1830 he was made a Commander of the Legion of Honor. Unfortunately he soon ran foul of his superiors for actively courting the republicans in such a politically explosive and anti-monarchist city as Lyons. In January 1831 he put his name to a pamphlet criticizing the government for retaining ex-Restoration officials and for failing to punish Charles's former ministers. He also demanded electoral reform and criticized the July Monarchy's weak foreign policy. He had also compromised himself further the previous October with his election to the Chamber of Deputies as a Republican representing the Jura. Pressure was placed on him to relinquish his command, which he did under protest in March 1831.

Active in politics full time and totally out of step with current thinking in the army, he never received another command. He lost his seat in July 1831 but managed to regain it the following year. In July 1834 he failed to gain re-election and was out of the Chamber till March 1838 when he became the representative for Chalons-sur-Saône. He gave up politics after he lost in the election of July 1842. As irony had it he retired from the Army on 30 January 1848, days before the July Monarchy finally fell to a resurgence of republicanism. His health failing he turned down offers of recall and lived quietly in Paris till his death on 16 June 1849.

CAREER ASSESSMENT

Bachelu was a highly respected, loyal, hard fighting professional. He represented the younger ambitious breed of general that would have gone far had the Empire survived or had he not persisted with lost causes and adjusted to the changed circumstances after Napoleon's fall. His uncompromising hard line republican stance, particularly during the more enlightened period of Louis Philippe, destroyed what could have been a very promising military career in later life.

FRENCH INFANTRY ASSAULT HOUGOMONT

BACKGROUND AND EARLY CAREER

Soult's deputy at Waterloo and for many years Berthier's understudy before Napoleon's fall in 1814, Bailly came from a minor noble family. His early upbringing was in the balmy colonial atmosphere of the Isle de Bourbon (Réunion) in the Indian Ocean where he was born in the capital Saint Denis on 27 January 1776.

On his return to France he received a commission as a *sous lieutenant* with the 74th Line on 24 February 1793. His first active service was with the Army of the Moselle and then the Army of the North. He fought at Saint Wendel and in the forests of the Argonne before taking part in the defence of Maubeuge. On 6 September 1793 he was suspended and imprisoned for no other reason than being a *ci-devant* noble. After his release he received a posting to the Army of the Pyrenees Orientales where irony had it, he became an aide to the notorious General Turreau. Anxious to cast aside his noble past he found little problem reconciling himself to following Turreau on his appointment as commander of the Army of the West on 22 November 1793. As a result he became a party to some of the more murkier episodes of the vicious war in the Vendée as Turreau's notorious *colonnes infernale* ranged far and wide through the region putting royalist sympathizers to the sword. He fought in a major action was at Noirmoutier when the Vendeans were driven back from the gates of Nantes on 3 January 1794.

In the rough and tumble of revolutionary politics Turreau was soon removed as army commander on 20 May 1794 and posted to the outpost of Belle-Isle-en-Mer and Bailly the loyal aide remained with him despite advice to the contrary. Then with Turreau's arrest the following September he returned to the mainland and joined the general staff of Hoche's Army of the West. The move turned out to be a good one, as promotion to *lieutenant* came on 20 January 1995 followed by that of *capitaine* on 5 October 1794. It was under the army's chief of staff Robert that he gained his first real experience as a staff officer and when Robert moved to the Army of the Sambre-Meuse in July 1796 Bailly followed.

Turreau's rehabilitation and later recall as a division commander with the Army of Mayence led to Bailly rejoining him as his aide on 17 September 1797. He remained with Turreau for another three years, seeing service with the armies of Switzerland and the Danube in 1799, and finally the Army of Italy in 1800. In the Marengo Campaign he was with Turreau's column that formed the left wing of the Army of Italy that captured Susa on 22 May. Then when Kaim the Austrian commander feared that Turreau's column was the advanceguard of the Army of the Reserve he turned on it with a large force at Avigliano on 27 May, forcing Turreau to fall back to the passes. He fought again at San Ambrosio on 5-6 June when Turreau failed a second time tried to break through to join the Army of the Reserve on the plains of Lombardy.

BERTHIER'S GENERAL STAFF 1800-1814

After Marengo, Bailly's long association with Berthier began when he was attached to the general staff of the Army of the Reserve on 27 June 1800. Two years followed in Switzerland as chief of staff of the Valais and Simplon Military Divisions during which time he gained promotion to *chef d'escadron* on 27 April 1802. In 1803 he joined Berthier at the Ministry of War in Paris as one of his aides where he built up a good working relationship with his chief. On Berthier's appointment as chief of staff of the Grande Armée in August 1805, Bailly joined his staff for the Austrian Campaign and served with distinction at Memmingen, Ulm, Hollabrünn and Austerlitz.

Promoted *adjudant commandant* on 1 March 1806 he found himself attached to Augereau's VII Corps as deputy to chief of staff Pannetier. In the Prussian and Polish campaigns he saw action at Jena on 14 October and then in Poland fought at Golymin on 26 December, at Hof on 7 February 1807 and Eylau the next day. The disbandment of VII Corps after its destruction at Eylau saw his transfer back to Berthier's staff where he served at Heilsberg on 10 June and at Friedland on 14 June 1807. Known as an efficient organizer, during the peace conference at Tilsit he held the post of the town's military governor.

On 21 February 1808 he became Murat's chief of staff in Spain. In the weeks that followed he acted as an intermediary during the negotiations and intrigue that finally led to the forced abdication of Charles IV at Bayonne on 10 May 1808. Promoted *général de brigade* on 22 May 1808 he later received his first independent command on 15 June when he took charge of the French garrison at Vitoria. As the rebellion in Spain spread he moved to Bilbao on 18 September to head a small force occupying the port. Then when the Army of Galicia 30,000 strong appeared two days later he hurriedly evacuated the city and scuttled back to the safety of Vitoria.

He rejoined Berthier on 15 November 1808 and accompanied the Grande Armée during Napoleon's

BAILLY DE MONTHION - FRANÇOIS GÉDÉON (1776 - 1850)

13

march on Madrid. On 28 January 1809 his services in Spain were recognized when he was made *baron de l'Empire*. When Napoleon and Berthier left Spain at the end of January he joined the party under Berthier that took charge of the Army of Germany. Having built up a reputation as a sound staff officer, it was at Berthier's insistence that he became his chief of staff when he took command of the army on 21 March 1809.

Berthier's brief spell as an army commander was not a success. When Napoleon took command on 17 April Berthier resumed his old post as *Major Général* with Bailly as Chief of Headquarters Staff, effectively his deputy in a restructured General Staff. It was a post to which he was ideally suited and held virtually without interruption till Napoleon's first abdication. The power and influence he wielded was considerable, when compared to his lack of seniority against that of more august commanders. As a result the enmity most field commanders held towards Berthier's staff officers was not spared on him and won him few friends.

In the years that followed it was through men such as Bailly that the Grande Armée's general staff operated to such a high degree of efficiency and immortalized the name of Berthier. He was responsible for running the headquarters itself. The preparation and despatch of orders, intelligence and personnel records all passed through his hands. His function was very much that of a headquarters commandant, controlling its movements and housekeeping, and also the movements of minor units not attached to any specific corps. His staff of *adjudant commandants* in turn handled such affairs as line of communication matters, prisoner of war escorts and guards. In the absence of Berthier he had to deputize for him before Napoleon, a role after Berthier's that was one of the least envied in the army.

He served at Eckmühl, Aspern-Essling and Wagram and shared in the honors and awards distributed after the campaign, becoming a *comte de l'Empire* on 15 August 1809 with a pension of 10,000 francs a year drawn on Westphalia. In addition, grateful Allies awarded him with the Grand Cross of the Order of Hesse and he became a Commander of the Military Order of Wurttemberg.

In 1810 based at Bayonne he served as an Inspector General responsible for drafts sent to Spain. In March 1811 he took charge of the General Staff set up in the city in anticipation of Napoleon leading a new army into Spain. On 5 February 1812 he rejoined Berthier in Berlin as his deputy to help prepare for the Russian Campaign. He served at Smolensk on 17 August, Borodino on 8 September, Maloyaroslavets on 23 October and was present during the crossing of the Beresina on 28 November. He was promoted *général de division* on 4 December 1812, the day Napoleon

left the army and returned to Paris. Two days later Murat made him the Grande Armée's chief of staff after Berthier's departure due to ill health. Then when Murat abandoned the army in January, he served in that capacity under Eugene Beauharnais during the difficult weeks of the retreat through Poland and Prussia to the Elbe.

On Napoleon's return to the Grande Armée at the end of April he continued as chief of staff till 12 May 1813 when he resumed his duties as Berthier's deputy on his return. He gave a faultless performance at Lutzen on 2 May and again at Bautzen on 21-22 May. In the Autumn Campaign he acted for another period as the Grande Armée's chief of staff when Berthier was ill from 24 August till 30 October heading the General Staff at the battles of Dresden, Leipzig and Hanau.

During the campaign in France he once again resumed his duties under Berthier, till once more he became Napoleon's chief of staff when Berthier was wounded at Brienne on 28 January 1814. He remained with Napoleon throughout the campaign and was at Fontainebleau when the Emperor abdicated.

NAPOLEON'S RETURN AND THE WATERLOO CAMPAIGN

He acknowledged the Bourbons and was made a *chevalier de Saint Louis* on 1 July 1814, but with the reduction in size of the army remained unattached on half pay. Not surprisingly he rallied to Napoleon on his return and received the appointment as chief of staff with the Army of the North on 8 April 1815. He worked hard to establish an efficient staff as feverish preparations were made to ready the army. Bad news arrived for him on 18 May 1815 when Soult superseded him on his appointment as the army's *Major Général*.

Bailly continued as Soult's deputy and Chief of the Headquarters Staff, but it was an arrangement that did not work. Soult was the wrong man for the job, who should have commanded one of the army wings. The army's staff for years had run like a well tuned machine used to working a certain way under Berthier and Bailly. A new chief of staff, in particular Soult whose loyalty was very suspect, had an unsettling effect. Bailly felt he qualified by experience to take over the mantle left by Berthier. Resentful at being overlooked, when from December 1812 till the First Abdication he had effectively acted as Napoleon's Chief of Staff due to Berthier's ill health. As a result he simply failed to show the enthusiasm and efficiency that was the hallmark of the General Staff under Berthier. His attitude throughout the campaign could be summarized as one of cold indifference when acting on Soult's orders. The campaign was punctuated by errors and omissions committed by the General Staff.

Omissions such as the failure to despatch orders in triplicate or even duplicate caused things to go wrong from the start. Vandamme's corps never received its marching orders on time on 15 June because the single staff officer sent fell from his horse. At Ligny VI Corps arrived very late in the day, largely because orders were late. Also on the same day had the staff work been up to standard d'Erlon would have been kept better informed of the army's movements, and certainly have avoided missing both battles on 16 June. Professional pride on Bailly's part did not want Soult to succeed, and as a result many failings of the Waterloo Campaign are due to the two simply not working in harmony. Wounded at Waterloo, Bailly was hastily patched up and remained with the army during the retreat to Paris. Then fearing he would be on the Bourbon hit list he left his post on 26 June and was then suspended on 1 August 1815.

Later Life and Career Assessment

Saint Cyr's Army Reforms led to his reinstatement and recall to the General Staff on 27 May 1818. He remained active till 10 December 1826 when with many other officers he was quietly relieved by a reactionary government that had become paranoid about the influence former Bonapartist generals held within the army.

He remained unattached till after the July Revolution when on 16 October 1830 he headed a commission to examine the workings of the General Staff. One of his achievements was to see the successful integration of the Army's Geographical Section into the General Staff. In 1831 he headed another commission to review entry requirements for officers who wished to serve as staff officers. In the years that followed he had much spare time and took an active interest in the Saint Cyr Military Academy where he lectured and helped set exams for future staff officers.

On 3 October 1837 he became a Peer of France. Appointed an Inspector General on 30 June 1838 he spent six years in the Arrondissements till placed in the Reserve Section on 28 January 1844. His career was marked with the award of the Grand Cross of the Legion of Honor on 19 April 1843. He was retired on 12 April 1848 when the new republic after the fall of Louis Philippe pensioned off hundreds of officers no longer required. He lived quietly in Paris till his death on 7 September 1850.

Bailly was possibly one of the most underestimated men to serve with the Army of the North in the Waterloo Campaign. As Berthier's deputy for many years he was the natural successor as Napoleon's Chief of Staff. Unfortunately he was faced with two problems; firstly protocol required a marshal to hold the post. That could have been overlooked but even as a *général de division* he still lacked seniority. Secondly and more serious, Napoleon didn't particularly like him. Not so easily bullied as Berthier, Napoleon found him stiff and formal and would have preferred not to employ him at all. Bailly however was loyal and very capable, that and Napoleon's awareness of some of Soult's failings made his presence even more vital.

The appointment of Soult in his place was a mistake not because he was incompetent in such a role, but rather that Bailly was better. As illustrated earlier, it affected Bailly's performance and that of the general staff under him, which in turn had grave consequences for the army. Soult's presence elsewhere would have had a major impact on the outcome of the campaign. His presence leading the left wing of the army instead of Ney would have been better since he was more familiar with Wellington as an opponent. Had Ney formed the right wing there is little doubt he would have marched to the sound of the guns on 18 June, being more malleable, had he wavered Vandamme and Gérard certainly would have convinced him. That leaves Grouchy, who at the head of the Cavalry Reserve certainly would not have led it to its destruction the way Ney did at Waterloo. In the end a heavy price was paid for not recognizing the ability of the most talented staff officer available at the time in the army to act as Chief of Staff for the Army of the North.

EARLY LIFE AND CAREER

The son of a baker, the Imperial Guard general Pierre Barrois was born in the village of Ligny-en-Barrois (Meuse) eastern France on 30 October 1774. His first active service was with the Eclaireurs de la Meuse, a volunteer light infantry formation he joined on 12 August 1793. A natural leader, on 12 September his fellow chasseurs elected him their *lieutenant*. Ambitious, he led the Eclaireurs at Wattignies on 16 October 1793 after ousting his less competent senior officers. His meteoric rise came to halt with the second *Amalgame* when the unit was broken up and merged into the 9th Légère on 21 March 1794. Serving on the regimental staff as part of Marceau's division of the Army of the North, he fought at Fleurus on 26 June 1794, the Roer on 2 October and was present at the capture of Coblenz on 14 October 1794. Promoted *adjudant major* on 14 April 1796 he was at the second blockade of Mayence where he gained further promotion to *capitaine adjudant major* on 22 September 1796. He took part in Moreau's successful crossing of the Rhine at Neuwied on 18 April 1797. In 1798 he moved to Brest and spent much of the year in preparation for the often delayed expedition to Ireland. When that was abandoned, he spent time as part of the garrison keeping an eye on the simmering unrest in the Vendée.

In April 1800 the 9th Légère transferred to the Army of the Reserve. At Marengo on 14 June, he was nearby when Desaix fell at the head of the 9th Légère as they charged home against Zach's grenadiers when the battle reached its climax. His steadiness noted during the battle, promotion to *chef de bataillon* followed on 26 October 1800. He remained in Italy for a further three years gaining promotion to *adjudant commandant* on 27 August 1803. A move came to the Camp of Montreuil when he was made *colonel* of the 96th Line on 5 October 1803. During this time on the Channel Coast he had the dubious distinction of being among the group of colonels that formed the military court, which tried and ordered the execution of the Duc d'Enghien after his abduction from Baden on 15 March 1804.

THE GRANDE ARMÉE 1805-1807

It was during the Austrian Campaign of 1805 that Barrois began to build up a reputation as a fine field commander. His regiment with Dupont's 1st Division of Ney's VI Corps at Haslach on 11 October, fought in one of the finest divisional actions of the Napoleonic Wars. Archduke Ferdinand with 25,000 men had left Ulm in order to break out of a threatened encirclement. Moving along the left bank of the Danube he hoped to fall on Napoleon's communications when he came across Dupont's division of 6,000 men on its own at Haslach. Dupont had earlier spotted the movement, and showing great generalship judged that if he retreated it would reveal his weakness. Instead he showed great

boldness and stood his ground in the hope that he might convince the enemy he was the advance guard of the French army. The bluff caused the enemy to move with caution, and helped the division to retire from the dangerous situation in which Murat had originally placed it.

BARROIS - PIERRE (1774 - 1860)

In the struggle that followed Dupont owed his reputation to the valor displayed by his regimental colonels, Barrois of the 96th Line, Darricau of the 32nd and Meunier of the 9th Légère who all later led divisions in Spain. On the left the 32nd held Haslach, surrounded by a small wood. Barrois backed by a wood formed the center while to his right the 9th Légère formed up in line. There was no chance they could withstand the firepower of the Austrians, so at a given signal the two regiments, the 96th and the 9th formed column, advanced and met the enemy with the bayonet. The charge broke the first Austrian line, threw it into complete disorder and they took some 1,500 prisoners. Angered by the reverse the Austrians advanced again and were hurled back with similar losses. Ferdinand then directed their efforts against the French wings by mounting assaults on Haslach and Jungingen till those assaults petered out with dusk. Not wishing to push his luck Dupont took advantage of darkness and retired to Albeck and then Langenau.

The contest, and Barrois's part in it, had a major impact on the campaign. Had Dupont not stopped the Austrians, a large part of their army would have made their way into Bohemia and one of Napoleon's finest operations, the Capitulation of Ulm, would never have occurred. The Austrians by encountering Dupont's force, which should never have been there on its own in the first place, could not believe it was unsupported, and so concluded Ulm was everywhere encircled. As the noose further tightened around Ulm, he fought at Albeck on 15 October when Dupont came across Werneck's corps of 20,000 men trying to steal away from the city. Whilst the 32nd Line and 9th Légère in close column threw themselves at the Austrian infantry, Barrois with the 96th in square received their cavalry. When the day ended Werneck had suffered some 1,000 casualties and a further 2,000 taken prisoner with several guns. Demoralized his corps fled north but harried by Murat's cavalry in the days that followed few made it to Bohemia.

Durrenstein on 11 November 1805 was Barrois's next major action, when he fought his way through the Russians and saved Marshal Mortier with Gazan's division isolated on the Danube's left bank. Exhausted after the battle, Barrois's regiment having been one of the most heavily engaged in the campaign, it retired to the southern bank with the rest of Dupont's division for a well earned rest. He was at Vienna where his troops were part of the city's garrison when Napoleon fought at Austerlitz on 2 December 1805.

During the Prussian Campaign of 1806 he continued to serve under Dupont when the division operated with Bernadotte's I Corps. His troops angered at missing the twin battles of Jena Auerstadt fought with renewed determination at Halle on 17 October, and Nossentin on 1 November 1806 during the pursuit of the vanquished Prussians. In Poland he won praise at Mohrungen on 25 January when his regiment drove the Russians from the heights north of the town. The affair resulted in his promotion to *général de brigade* on 14 February 1807. Bernadotte however, anxious not to break up a well knit command, insisted he remain with the 96th Line under his orders. On 26 February he fought at Braunsberg when Dupont repelled Lestocq's Prussians after they surprised Bernadotte's widely spread corps in winter quarters. He also distinguished himself at Friedland on 14 June 1807 when during Dupont's assault from the Sortlack Wood he drove through the Russian Guard and seized the bridge over the Mühlen Floss. The Russian front split, he surged down the Konigsberg road into Friedland causing panic in their rear as their line of retreat was cut. After the Treaty of Tilsit he spent a period of garrison duty in Silesia before being ordered on 7 September 1808 to join I Corps at Bayonne for Napoleon's invasion of Spain.

SPAIN 1808-1812

In Spain he served under Ruffin, who earlier in the year had taken over the division from Dupont. He fought at Espinosa on 10-11 November 1808 when Victor defeated Blake's Army of Galicia. At Somosierra on 30 November 1808 he led a diversionary attack on the pass, which distracted the Spanish infantry sufficiently to enable Napoleon's Polish light cavalry to effect a breakthrough and gain the glory. Before Madrid his troops were the first to overcome the city's outer defences, which led to its surrender on 4 December 1808.

His brigade then formed part of the Madrid garrison after the bulk of the army set off to deal with Moore's army in the north-west. In January, when Venegas thought it safe to advance on Madrid, his brigade was with the scratch force Victor pulled together to meet the enemy at Ucles on 13 January 1809. Ruffin's division with Barrois's brigade had

orders to envelop the Spanish right, but lost its way over the roads resulting in Villatte's having to bear the brunt of the fighting. Then as the Spanish fell back after a resolute defence by Villatte, Barrois fortuitously emerged in their rear causing panic, which resulted in over 5,000 enemy falling into his hands. Such pieces of good fortune coupled with a solid all round performance undoubtedly contributed to his elevation to the imperial nobility as a *baron de l'Empire* on 9 February 1809. He joined Victor's advance into Estremadura and fought at Medellin near the Portuguese border when Victor overwhelmed Cuesta's Army of Estremadura on the south bank of the Guadiana on 29 March 1809.

His good run in Spain ended at the Battle of Talavera on the night of 27 July. Here he took part in Ruffin's assault against Wellington's position in the Cerro de Medellin that nearly succeeded in turning the English left before being driven back. The next day Hill's division was ready for Ruffin's renewed assault and met the French in a classic column versus line action. In a vicious fire fight that lasted barely forty minutes Barrois's brigade suffered heavily as Ruffin lost over 1,200 men out of 4,900 engaged, before a bayonet charge sent the division reeling down the slopes.

In the three years that followed Barrois spent an uncomfortable time in Andalusia. For long periods his troops occupied the lines before Cadiz. Present at Barossa on 5 March 1811 he took over the division after Ruffin was mortally wounded during the battle. On 27 June 1811 his rank as *général de division* was confirmed. He campaigned briefly against Ballasteros and eventually forced the Spaniard to seek the safety of the guns of Gibraltar before he occupied Algeciras on 4 December 1811.

He moved on to Tarifa and took the part in the unsuccessful investment of the fortress that lasted from 20 December 1811 till 4 January 1812. In July 1812 he joined d'Erlon* whose forces were being hard pressed by Hill in Estremadura. Whilst his intervention helped to check the Allied advance, all was lost when the strategic situation in Spain changed after Marmont's defeat at Salamanca and Soult had to abandon southern Spain.

IMPERIAL GUARD 1813-1814

Recalled to France at the end of December 1812 to help rebuild the Imperial Guard after the disasters in Russia, Barrois carried out Napoleon's orders with amazing zeal and dedication. Existing cadres were expanded by depot personnel, while all line regiments in Spain and the rest of the Empire were scoured with each providing a fixed number of the best men available from their ranks. Out of virtually nothing he rebuilt the 1st Guard Infantry Division and received great acclaim when Napoleon reviewed the formation outside Paris on 15 February 1813. After crossing the Rhine, on 10 April he handed the division over to Dumoustier, who had arrived from Spain with elements of the Guard Voltigeurs, while he in turn took over the Guard's 2nd Infantry Division.

Present at Lutzen on 2 May he was held in reserve while Dumoustier's men broke the Prussian center between Rahna and Klein Gorschen. At Bautzen on 21 May his successful storming of the Kreckwitz Heights was decisive in forcing the Allies to quit the field. During the period of the Armistice he spent much time rebuilding and training his regiments, while the division increased to 14 battalions numbering over 9,000 men with 24 guns at the opening of hostilities. He fought at Dresden on 26 August after his troops completed one of the epic marches of the Napoleonic campaigns. Covering over ninety miles in three days, they were immediately thrown into the battle and drove the Allies from the Pirna Gate and recaptured the Great Garden.

In the weeks after Dresden his division was weakened by many battalions being detached till by

GUERILLA HUNTING IN SPAIN

the time of Leipzig he was down to ten battalions with some 5,470 officers and men. He fought with the Guard south of the city at Wachau on 16 October. Then as the situation deteriorated with the Allies tightening their grip around Leipzig he withdrew to Lindenau where he helped cover the Army's withdrawal during the 18-19 October. In the retreat to the Rhine his troops fought with the rearguard. The attrition was horrific amongst his troops, more of it due to illness and desertion than enemy action. Then as the weather deteriorated with the onset of winter the shaky morale of his men collapsed further. By 15 November 1813 when he crossed the Rhine only 2,600 men remained out of the 9,000 that started the campaign six weeks earlier.

Back in France another reorganization of the Guard took place and he moved to the 4th Young Guard Division on 13 December 1813. Posted to Brussels he reinforced Maison's I Corps that barred Bernadotte's invasion of the Low Countries. He spent the winter and spring months covering the Belgian fortresses where he achieved a fair measure of success, but his formation was sorely missed by Napoleon during the campaign in France. On 21 February 1814 he was made *comte de l'Empire*. One success he had was with Roguet* when at Courtrai they defeated Thielemann's Corps on 31 March 1814.

THE RESTORATION AND NAPOLEON'S RETURN

Not prepared to serve under the Bourbons he resigned from the army in May 1814 and retired to his home in the country. On news of Napoleon's return he immediately rallied to him and on 31 March 1815 accepted command of the 2nd Infantry Division based at Lille. It was an ideal posting since he knew the country well from the campaigns of the previous year. However before he could take up his post there occurred a change of plan on 2 April, which resulted in him being given orders to lead the 2nd Young Guard Division forming at Compiègne. He found the task particularly difficult since as soon as he formed a regiment it was invariably detached to quell the unrest in the Vendée. By June he had only four battalions of the 1st and 3rd Tirailleurs available so they were combined with Duhesme's 1st Division to form a single Young Guard division with him as Duhesme's deputy.

THE WATERLOO CAMPAIGN

At Ligny on 16 June he led the Tirailleurs when called to support Vandamme's faltering assaults against the Prussian right, and took Le Hameau in the late afternoon. At Waterloo on 18 June he was held in reserve before Rossomme during the early stages of the battle. Then as Bulow's IV Corps pushed Lobau back on Plancenoit and threatened the army's rear, he joined the desperate struggle for the village. With Duhesme his epic defence kept the jaws of the Allied vice open and enabled a large part of the army to get away after all was lost. Shot in the shoulder, and with Duhesme mortally wounded, he held together the remnants of the Young Guard as they struggled from the field. The intensity of the fighting was borne out by the returns submitted on 24 June, which recorded only 600 officers and men compared to 4,027 at the start of the campaign. With his men throughout the retreat, his wound made it impossible to continue and once he reached Paris on 28 June he handed over command.

LATER LIFE AND CAREER ASSESSMENT

On 1 August 1815 he was placed on the non-active list. Partial reconciliation with the Bourbons came on 1 September 1819 when he was made a chevalier de Saint Louis. Though fit and able he remained unattached throughout this period and retired from the army on 1 January 1825.

The accession of Louis Philippe resulted in his recall on 4 August 1830 when he took the post of commander of the 3rd Military Division at Metz. He later became Inspector General of Infantry for the 5th, 6th and 18th Military Divisions on 3 April 1831. Well in with the military establishment he secured command of the 1st Infantry Division of the Army of the North when Gérard invaded Belgium on 8 August 1831.

After the brief campaign numerous appointments followed with the Infantry Inspectorate and for his services he received the Grand Cross of the Legion of Honor on 30 April 1836. He also served as president of the Infantry and Cavalry Committee, which acted as an important advisory and policy making body for the Ministry for War. On 31 January 1840 in his sixty fifth year he moved to the Reserve Section of the General Staff.

With the fall of Louis Philippe he joined the hundreds of officers forced into retirement by the radical elements in the Second Republic who wished to cleanse the army of Orleanist influence. In his remaining years he was active and took great interest in the welfare of Imperial Guard veterans. He died at Villiers-sur-Orge (Seine-et-Oise) on 13 September 1860.

A fine leader, Barrois was another example of the younger breed of energetic commander to emerge in the latter period of the Napoleonic Wars. At first an unrepentant Bonapartist he managed to adjust to the new order in the years after Napoleon's fall. A true professional he generally managed to steer clear of politics and continued to have a long and distinguished career serving France well, and lived to see the Second Empire reach its zenith under Napoleon's nephew.

**BERTHEZÈNE - PIERRE, COMTE
(1775 - 1847)**

The son of a laborer, Berthezène was born in the village of Vendarques near Montpellier (Herault) southern France on 24 March 1775. He enlisted with the 5th battalion of the Herault Volunteers on 15 September 1793. His promotion through the ranks was swift and within a week he had reached *sergent major*. He served with the Army of the Pyrenees Orientales and fought at Peyestortes on 17 September 1793. In November he moved to Toulon and was present at the siege of the port where Napoleon as an unknown artillery captain first came to the fore. On 19 June 1794 he was commissioned *sous lieutenant* with the 7th Demi-brigade of Garnier's division serving in the Army of Italy. There was still an element of radicalism within the army with men electing their officers, and they made him their *lieutenant* on 18 November 1795. He saw little action in the opening campaign in Italy, as Garnier's division spent much of its time positioned on a quiet sector on the army's left north of Nice. The merger of the 7th into the 11th Demi-brigade in August 1796 saw him posted to the regimental staff where he gained a reputation as a conscientious staff officer, but again he had few opportunities to excel. When Grenier arrived to head a new division with the Army of Italy in March 1799 Berthezène managed to secure a place on the division staff. Active in the campaign that followed, he fought at Pastrengo on 26 March, Magnano on 5 April, Cassano on 28 April and at San Guiliano on 23 June 1799 where he was promoted *capitaine* on the battlefield. On 4 November he was at Fossano and Servigliano when Melas drove the French back to the passes over the Apennines.

In the Marengo Campaign he served as an aide to General Compans, who headed a brigade with Suchet's left wing of the Army of Italy. When Melas split the army in two and drove Suchet back to the Var, Berthezène distinguished himself at Saint Jacques on 19 April 1800. In the engagement he received a severe head wound that put him out of action for the rest of the campaign. His return to duty on 20 July 1800 saw his promotion to *chef de bataillon* with the 72nd Line. With the renewal of hostilities he took part in at Brune's crossing of the Mincio where he was shot in the leg during the struggle for Pozzolo on 25 December 1800.

After two years in garrison in Italy he then moved to the Camp of Saint Omer to prepare for the invasion of England. When the Grande Armée broke camp in August 1805 for the march to the Danube he remained with the 72nd Line on the Channel Coast as part of the Army of the North. As a result he missed the campaigns in Austria and Prussia, which was partially made up by his promotion to *major* on 6 July 1806 and transfer to the 65th Line.

After the horrific losses at Eylau, Soult aware of Berthezène's reputation as a fine regimental officer secured his transfer to Poland on 20 February 1807 as *colonel* of the veteran 10th Légère with Saint Hilaire's division. He fought at Heilsberg on 10 June and was present at the surrender of Konigsberg a week later. At the end of the campaign a period of garrison duty followed on the Oder and then he spent the Winter of 1808-09 at Rugen in Swedish Pomerania. During this time he also became a *baron de l'Empire* on 20 July 1808.

THE DANUBE CAMPAIGN OF 1809

Back with Saint Hilaire's division with Davout's III Corps, Berthezène's exploits with the 10th Légère were to bring his career to the fore during the opening days of the campaign. As III Corps fell back along the Danube from Ratisbon, Berthezène was near the head of the column that took part in the confused fighting in what became known as the Battle of Thann on 19 April 1809. Vukassovich's division forming the advance guard of Hohenzollern's III Corps barred Davout's march by occupying a wooded ridge between Tengen and Hausen. The 3rd Line at the head of the column deployed and hurled themselves at the enemy. It enabled the rest of the division strung out along the road to form up into attack columns. While the 57th Line made for the enemy's center to help the battered 3rd Line, Berthezène's 10th Légère on Saint Hilaire's right moved through the trees. They fell on the Austrian left, overturned a battery and forced the enemy to retire to the crest of a second ridge. An enthusiastic charge by his men was driven back and caused a lull in the fighting till Friant's division moving forward on Saint Hilaire's left supported by cannon forced the Austrians back.

Saint Hilaire in particular praised Berthezène and his other colonels for the fine way they deployed and attacked the Austrians. The enemy had they been more aggressive, could easily have rolled up his long column and defeated the division in detail. The cost was high; the division lost some 1,700 men including 500 from the 10th Légère. Most important of all, they secured a junction with Lefebvre's VII Corps and removed the immediate threat of III Corps being cut off.

At Eckmühl on 22 April he covered himself with glory as the main element of the Austrian army turned on Davout's forces. The marshal, aware Napoleon moving from Landshut was coming to his support, in his weak position felt attack was the best form of defence. He ordered Berthezène to spearhead the attack on Laichling and then to cut through to the Ratisbon-Eckmühl road. After III Corps launched a series of diversionary attacks the 10th Légère stormed into the village. Feeble resistance put up by the Bellegarde and Reuss-Greitz regiments resulted in the surrender of nearly 1,500 demoralized defenders. Behind the village he encountered severe resistance on the edge of the woods as the enemy formed up behind felled trees. His elite companies leading the charge weråe virtually wiped out. The rest of the regiment swept over the obstacles and plunged into the wood later to emerge in the open ground that stretched before the Ratisbon highway. By then the 10th Légère was a spent force, having lost another 600 men including Berthezène, who took a musket ball in the leg. Faced by enemy infantry, cavalry and artillery from all directions their position appeared hopeless, till Davout's chief of staff Compans arrived with two fresh regiments, followed by Friant's division, which ground the enemy down.

Many French units fought well that day. Davout received the title Duc d'Eckmühl. The official battle report singled out only one regiment for special mention - Berthezène's 10th Légère. It had in the words of the army bulletin, "covered itself with glory". The ferocity of the initial attack completely undermined the Austrian defence before Eckmühl and enabled other formations to later exploit this weakness.

Recovering from his wounds he missed the Battle of Aspern-Essling where the gallant Saint Hilaire died of wounds after the battle. Grandjean took over the division and on his return Berthezène fought for him at Wagram where in the savage fighting before Baumersdorf on 5 July he was once more struck down by musket balls, this time in the arm and leg. For his part in the campaign Napoleon rewarded him making him a Commander of the Legion of Honor.

THE IMPERIAL GUARD 1811-1813

Promoted *général de brigade* on 6 August 1811, Berthezène stayed with the 10th Légère till he replaced Radzout as commander of the Walcheren Isles garrison on 9 October 1811. A move then followed to Brussels on 6 December 1811 where he joined the Imperial Guard to help recruit and train the new Young Guard regiments. He was made commander of the 1st Brigade of Delaborde's Young Guard Division on 8 February 1812, which comprised the 5th Voltigeur and Tirailleur Regiments he helped form.

During the Russian Campaign his troops took part in the storming of Smolensk on 17 August 1812. At Borodino on 7 September he was only lightly engaged, his troops being held in reserve. When Moscow went up in flames he tried gallantly to maintain some semblance of order as the fire engulfed the city. His troops were the last to leave Moscow after Noury* blew up the Kremlin. At Smolensk in mid November he was in action defending the southern approaches to the city as the army struggled on. During the crossing of the Beresina on 28 November attached to Oudinot's command on the right bank he helped keep the line of retreat open. His men suffered terribly during the final days of the retreat. When he reached Konigsberg on 25 December his brigade had ceased to exist; only 40 men remained.

THE GERMAN CAMPAIGN OF 1813

In January 1813 he moved to Erfurt on the Elbe where he spent two months rebuilding the shattered Guard regiments. On 10 April 1813 he took charge of a brigade with Dumoustier's 1st Guard Division. He fought at Lutzen on 2 May 1813 distinguishing himself in the counter attack by the Young Guard divisions that broke the Allied center and secured victory. At Bautzen on 20-21 May he was in the fore during the assaults on the Kreschwitz heights.

He left the Guard after his promotion to *général de division* on 4 August 1813 when he took command of the 44th Division with Saint Cyr's XIV Corps. He won great credit for the way he handled his troops during the hard fighting as Saint Cyr fell back on Dresden. In the defence of the city from 25-27 August he held the area around the Great Garden before being pushed back to the Pirna Gate by Kleist's Prussians. After the French victory he spent September watching Austrian movements from the Erzgebirge till forced back to Dresden after Napoleon's defeat at Leipzig. Cut off with the rest of I Corps and XIV Corps he was a strong advocate of the garrison fighting its way back to the Rhine. When Saint Cyr surrendered on 11 November 1813 much to his disgust Berthezène became a prisoner with another notable Waterloo personality - Mouton comte Lobau*.

THE RESTORATION AND NAPOLEON'S RETURN

He remained a prisoner in Bohemia till the Restoration and returned to France in June 1814. He acknowledged the Bourbons but held no post under them. They graciously made him a *chevalier de Saint Louis* on 29 July 1814 before placing him on half pay on 1 September 1814. Under such straitened circumstances he naturally rallied to Napoleon after his return to Paris and accepted an assignment to head a commission set up to examine the records and find postings for former half-pay Imperial officers.

THE WATERLOO CAMPAIGN

On 3 June 1815 he replaced Duhesme as commander of the 11th Infantry Division with Vandamme's III Corps of the Army of the North. At Rocroi on 11 June, he received orders with the rest of III Corps to concentrate at Beaumont. He reached his destination on the night of 14 June and bivouacked three miles north of the town near the Belgian border.

Before dawn of 15 June he set out to cover the last few miles to the Belgian border. Almost immediately his division became involved in the huge traffic jam that delayed the French advance on Charleroi. Vandamme inexplicably absent from his headquarters failed to receive and pass on his marching orders to the formations that lay across Berthezène's path.

After involvement in a sharp action at Gilly on the evening of the 15th he took post the next day on the French left opposite the twin villages of Saint Armand. During the Battle of Ligny he joined the assault on the two villages after Lefol's division was driven back and by 4.00 P.M. had secured a tenuous foothold. When the Prussian brigades of Tippelskirsh, Kraft and Brause converged on his position from three sides he was overwhelmed and fell back in disorder. The French left was in danger of collapse but for the timely intervention of Duhesme with the Young Guard, which halted the triumphant Prussians as they debouched from the villages .

After a severe mauling at Ligny his troops took time to reorganize and join the pursuit. Only lightly engaged at Wavre on 18 June, he supported Habert's division after the Prussians brought it to a bloody halt it tried to storm the bridge over the Dyle in the town. The next day he moved upstream and crossed the river at Bierges in support of Gérard's corps, which had forced a crossing upriver the previous day at Limale. He moved up the Brussels road and reached La Bavette when news of Napoleon's defeat reached him. He reacted quickly, his force being the first to extricate itself and reach the safety of Namur. Here he barred the road before the town and led a skilful rearguard action against the Prussian cavalry as they tried to cut off the rest of Grouchy's force. During the evening of the 20th June he passed through Namur and handed the rearguard fighting over to Teste.

He fell back to Paris with the rest of the army, and repulsed a Prussian attempt to breach the northern defences of the city on 30 June. He was again in action near Chalons on 2 July when his 12th Line engaged advance elements of Blucher's cavalry as they tried to approach Paris from the south. After the Armistice he led his troops south to the Loire to be disbanded. From 7 August till 16 September he headed III Corps after Vandamme's removal.

The army disbanded, he retired to his home at Vendarques. He remained there till 9 December 1815 when he received news of his exile from France. Whilst a former general in the Guard, insufficient evidence was produced to implicate him as an irreconcilable Bonapartist and he was allowed to return from Belgium on 16 April 1816, but remained suspended without pay.

LATER LIFE AND CAREER

Reconciliation with Bourbons came with Saint Cyr's Army Reforms when on 30 December 1818 he was recalled and took a post as an Inspector General of Infantry. Eleven years followed with the Infantry Inspectorate, where he headed the 8th and 9th Military Divisions from 1819 till 1821 before spending eight years with the 10th and 19th Divisions. Throughout this period he gained great respect for his hard work and absolute professionalism, steering well clear of politics, a fault which bedeviled so many former Imperial generals.

On 21 February 1830 he received command of the 1st Division of the Army of Africa for the expedition to Algeria. He led the landing at Sidi Ferruch near Algiers, on 14 June 1830. After some fighting in the dunes he established a base camp at Staoueli. Here under Bourmont on 19 June he put to fight the Dey's forces with a similar decisiveness to that meted out to the Mamelukes over thirty years before by Napoleon. Pushing on to Algiers from 29 June till 4 July he took charge of the final investment and bombardment of the port.

In the early days he campaigned in Algeria with a fair amount of success and proved to be a stolid and reliable subordinate under Bourmont. Above reproach he was wise enough to distance himself from Bourmont on news of the July Revolution and refused to lend him support when approached to return to France with the Army of Africa and overthrow Louis Philippe. As a result he survived Bourmont's dismissal and the purge amongst Bourbon officers in the army. When Clausel took over, he again proved to be reliable deputy. Unfortunately Clausel was an erratic commander, more interested in an orgy of land-grabbing and property speculation that soon resulted in his dismissal. Berthezène replaced him on 21 February 1831.

In Algeria he was left with the thankless task of reversing his predecessor's policies of adventurism, neglect and mismanagement with an army that suffered from drastic cut-backs. His forces fell to around 16,000 men, with morale low amongst both military and civilian personnel. His command was further undermined by high handed subordinates such as Boyer who conducted a personal reign of terror in Oran. New expeditions he led to regain Medeah and Bone were both failures. The situation was neither helped by continued guerilla attacks, fever and the disease common to all soldiers waging an unpopular war, "homesickness". Inadequately supported at home he could see no clear way forward. The army suffered as a result and on 26 December 1831 he in turn was replaced by Savary, duc de Rovigo, Napoleon's former Chief of Police.

On his return to France he was quietly moved aside by a government that wanted no part with a man tainted by what appeared to be a costly and embarrassing colonial failure. His accomplishments were recognized on 11 October 1832 when he was made a Peer of France. He was called to the First Section of the General Staff on 23 March 1840 and given the honorary title of Commander in Chief before the Enemy. It was a title rarely used and was seen as an Orleanist attempt to honor former Imperial generals, who happened not to be quite in their dotage yet did not merit a Marshal's Baton. It secured

him full pay enabling him to continue on the Army General Staff till his death at Vendarques on 9 October 1847.

ASSESSMENT OF CAREER

Berthezène was certainly a fine soldier dedicated to his profession. He was no Bonapartist, his loyalty was to France and the interests of the army. He was one of the foremost regimental commanders and as a general his reputation remained untarnished during the years of Napoleon's decline. His experiences in Algeria were unfortunate and were due more to a government that lacked the will to continue an unpopular war than to his failings as a leader. His career can be best summarized by the tribute inscribed over the entrance to his chateau at Vendarques, "A man worthy of any position he might occupy, and merited the esteem and affection in which he was universally held".

THE ALGERIANS ARE REPULSED AT STAOUELI

BERTRAND - HENRI GATIEN
(1773 - 1847)

EARLY CAREER

Henri Bertrand the outstanding engineer general and loyal Imperial Aide was born at Châteauroux (Indre), Central France on 28 March 1773. He enlisted with the Paris National Guard and took part in the storming of the Tuileries on 10 August 1792. He then spent a year at the Mézières School for Engineers near Paris where he gained a commission as an engineer *sous lieutenant* on 17 September 1793.

His first active service was with the Army of the Pyrenees during the winter of 1794-95. He gained promotion to *lieutenant* on 13 November 1794 and then *capitaine* on 21 March 1795 when he moved to the Army of the North. A posting followed to Constantinople on 15 March 1796 as a military advisor on the ambassador's staff to help the Turks improve the defences of the Bosphorus. The assignment ended in January 1797 when he joined the Army of Italy, which brought him into contact with General Bonaparte. Apart from being a conscientious engineer, it was his knowledge of the Turks that helped secure him a place on the general staff with the Army of the Orient gathering at Toulon in April 1798.

Landing at Aboukir on 1 July 1798 he led the engineer detachment of Bon's division. He fought at the Battle of the Pyramids on 21 July 1798. Promoted acting *chef de bataillon* on 7 August, his rank was later confirmed by the Directory on 15 February 1799. He served in Syria and was present at the Siege of Acre from 17 March to 20 May 1799. Back in Egypt under Lannes, who led the division after the death of Bon, he fought the Turks at Aboukir on 25 July 1799. Needed by Menou, he was detached and directed the siege of the Aboukir fortress till a head wound forced him to give up his command on 28 July 1799.

Promoted *chef de brigade* and commander of the engineers at Alexandria on 4 August 1799 he spent much of the next year strengthening the defences of the city and improving the harbor. He sat on the military tribunal convened in Cairo to try the persons involved in Kléber's assassination on 14 June 1800. He was promoted *général de brigade* on 6 September 1800. On 21 March 1801 after the English landed at Aboukir, Menou appointed him Director of Fortifications at Alexandria, a post he held till the final capitulation of the city on 31 August 1801. His skilful maintenance of the defences went a long way towards prolonging the siege as the English dared not risk an attempt to take the port by storm.

Egypt evacuated he returned to France in December 1801. On 3 September 1803 he moved to the camp at Saint Omer as commander of the engineers. On 14 June 1804 he was made an Inspector General of Engineers and appointed an Imperial Aide. He spent much time at Wimereux supervising the construction of the small ancillary harbor, close to the invasion force's main encampment at Boulogne. In 1805 as the political situation deteriorated with Austria he carried out discreet reconnaissances across the Rhine travelling into Wurttemberg and Bavaria, then deep into Austria as far as the Tyrol. Dressed as a civilian he made detailed notes and surveys of routes the Grande Armée later used when war broke out.

IMPERIAL AIDE AND THE GRANDE ARMÉE 1805-1808

He accompanied Napoleon throughout the Austrian campaign and gained a certain notoriety for his part played in the seizure of the bridges over the Danube outside Vienna on 12 November 1805. The group of officers led by Murat brazenly crossed the series of bridges till they reached the main span at the northern end. Here at the last barricade an Austrian officer barred their way with a lighted torch ready to blow the bridge. Bertrand strode up to the confused officer declaring that an armistice was in effect and he wished to see Count Auersberg, the local commander. He then set off on his own with a bemused subaltern and repeated the story to a sceptical Auersberg. Meanwhile under the muzzles of the Austrian guns waiting for the return of Bertrand, Murat and Lannes kept up a lively banter with the Austrians on the bridge as if they were long lost friends. Then at a given signal on Bertrand's return, a column of grenadiers that had crept up unseen burst from their cover. The Austrians hesitated, but realized too late they had been tricked. The grenadiers overran the position after a brief scuffle and within minutes established a foot-hold on the northern bank, which enabled to pour across. At Austerlitz on 2 December 1805 he was back with Napoleon carrying out his duties.

During the Prussian campaign of 1806 he continued as an Imperial Aide. At Jena on 14 October 1806 he distinguished himself at the head of two cavalry regiments that rescued Ney˙ after a force of Prussian cavalry trapped him due to his own impetuosity. Later attached to Lannes's V Corps he led a force that took the citadel at Spandau on 25 October. He then spent time in Silesia where he helped Vandamme˙ to reduce Glogau. From 2-13 December he stayed as commander of the fortress till replaced by Verrières. In Poland a call came for him to help Lefebvre reduce the defences before

Danzig. After the successful conclusion of the siege he was promoted *général de division* on 30 May 1807. He rejoined Napoleon and served at Friedland on 14 June 1807.

He shared in the lavish distribution of honors and awards after the Treaty of Tilsit. On 30 June 1807 he received a pension of 36,000 francs drawn on the Duchy of Warsaw, another of 25,000 francs from Westphalia and a further 20,000 from Hanover in March 1808. The establishment of the imperial nobility saw him become *comte de l'Empire* on 21 September 1808. He served as an imperial aide during the Spanish Campaign being present at Somosierra on 30 November and the capture of Madrid on 4 December 1808.

AUSTRIAN CAMPAIGN 1809

War against Austria imminent he returned to France with Napoleon and on 14 March 1809 took command of the engineers with the Army of Germany. He rebuilt the bridge over the Salza at Burghausen as the French forces stormed down the Danube Valley. He directed the siege operations before Vienna on 11 May 1809. His most important contribution to the campaign was the construction of the pontoon bridges over the Danube before Aspern-Essling on 21-22 May 1809. In six days he completed the work, a notable feat as he spanned four sections of the river covering over 1,000 yards of water swollen by the spring thaw.

During the battle he desperately tried to keep the pontoons intact, but the Austrians on several occasions managed to break them by floating large obstacles downstream. Unable to receive adequate reinforcements Napoleon was forced to withdraw to the Isle of Lobau. In the weeks that followed Bertrand directed the construction of another greatly strengthened, properly protected pontoon bridge. His communications secure, when Napoleon broke out from Lobau. The Austrians were unable to halt his rapid deployment on the north bank and were soundly defeated during the Battle of Wagram on 5-6 July 1809.

After the campaign Bertrand was showered with awards, including the Grand Eagle of the Legion of Honor on 14 August, the Baden Grand Cross of the Order of Fidelity, and a further pension of 10,000 francs per annum drawn on Milan from 15 August 1809. On 25 March 1810 he replaced Marmont as governor of the Illyrian Provinces. Based at Trieste for three years he administered the provinces efficiently and with his wife Fanny brought a certain elegance and style to the previously austere surroundings occupied by his predecessor.

GERMANY 1813

With Eugene Beauharnais serving in Prussia and Poland Bertrand became commander of the Corps of Observation of Italy on 4 January 1813. After the

campaign in Russia, with 40,000 men it was the largest formation available north of the Pyrenees to halt the Allied advances. On 12 March he set out for southern Germany to reinforce the Grande Armée. He passed through the Tyrol via the Brenner Pass and Innsbruck, and reached Bamberg on the Westphalian-Bavarian border on 24 April 1813. The force was then redistributed amongst the army with Bertrand taking command of a reformed IV Corps with Morand's 12th (French), Peyri's 15th (Italian) and Franquemont's 38th (Wurttemberg) Divisions. He handed over his 13th and 14th (French) Divisions led by Pacthod and Lourencz to form XII Corps under Oudinot.

As the Grande Armée advanced into Saxony his corps formed its right wing. He soon showed his failings as a commander at Lutzen, when the Allies fell on Ney's III Corps as it neared Leipzig on 2 May 1813. Marching to the sound of the guns Bertrand was in an ideal position to hit the enemy in the flank, but at the vital moment his inexperience showed when he halted to await orders from Napoleon. For two hours he stood still while clearly before him less than four miles away a desperate struggle unfurled. He eventually joined the battle around 5.00 P.M., engaged Yorck's Corps as it withdrew from the field and in the struggle lost around 500 men. A great opportunity was missed. Had he ploughed earlier into the flank of the overextended Allies, a triumph rivalling that of Jena or Austerlitz was there for the asking.

While the Allies fell back on Dresden he again didn't impress. On the right wing he had the shortest distance to cover, yet allowed Miloradovich with a small rearguard to dog his advance and enable the Allies to fall back without much difficulty. The march discipline of his troops was particularly poor and losses due to stragglers and desertion were high. In addition he suffered in costly skirmishes where the enemy were able to take advantage with their superior cavalry.

On 19 May his advance guard comprising Peyri's 15th Division bivouacked at Konigswartha near Bautzen. Once camp was made, Peyri allowed his hungry command to fan out across country in search of food, which was normal practice provided they were covered by friendly cavalry. There was no cavalry with IV Corps and Peyri did not take the precaution to scout a nearby wood before allowing his division to break rank and forage. A strong force of Russian cavalry, acting as the advance screen of the Allied army, which was digging in at Bautzen, saw the foraging infantry, called for and patiently awaited reinforcements, then struck. Within minutes Peyri's Division lost 2,862 men, including 1,289 taken prisoner. The reverse threw Bertrand's plans into confusion and he only arrived on the battlefield late in the afternoon the next day.

Once at Bautzen on 20-21 May he again did not

shine when during the night of the 20th he failed to cross the Spree and pin Blucher to his position, while Ney approaching from Prelitz was to cut him off. Blucher saw the danger, retired in time and another possible hammer blow to the Allied cause turned out again to be a very ordinary victory for Napoleon.

During the period of the Armistice IV Corps joined Oudinot's Army of Berlin. On 23 August 1813 Bertrand's 22,000 men forming the right wing of Oudinot's advance on Berlin he came up against the Prussians at Blankenfelde near Gross Beeren. The Prussians held a good defensive position and in the unsuccessful attacks he lost 2,000 men, while Oudinot suffered defeat four miles away at Gross Beeren. Had both leaders adopted a common plan, shown more initiative and combined their forces both defeats could have been averted.

At Dennewitz on 6 September 1813 with the Army of Berlin now under Ney's command, Bertrand's corps fought with great skill and determination, but again met disaster. Coming up against Tauenzien, who was responsible for his undoing at Gross Beeren he exacted revenge and drove him from the knolls north of Dennewitz. The battle then fell apart as Reynier's VII Saxon Corps first gave way on his left. Oudinot smarting from his replacement by Ney, was slow to move XII Corps up in support of the Saxons. At the same time, Bertrand outnumbered and unable to hold back Bulow's corps on his own, was driven back in confusion. Oudinot's corps after a tardy march then came to his assistance, but arrived late on the field and was also swept away in the ensuing rout. As a result the French offensive against Berlin collapsed and Bertrand withdrew with the rest of Ney's army behind the Elbe.

Ordered to hold the line of the Elbe at Wartenberg and prevent Bulow's Prussians severing Ney's communications with Napoleon's main army centered on Dresden he again failed. On 3 October Bulow slipped across the river and established a bridgehead. Bertrand posted in a strong defensive position away from the river protected by a swamp and dyke, hesitated. The Prussians able to deploy in sufficient strength, drove him back from a seemingly unassailable river-line position. The result had a serious effect on the outcome of the campaign. Ney, his communications threatened, had to withdraw from the Elbe, while the rest of Bernadotte's Army of the North was able to cross the river unopposed at Rosslau and Barbey. Blucher with the Army of Silesia followed, while Bulow also poured through the gap. The front had collapsed and the way lay open to Leipzig and the Grande Armée's communications still centered on Dresden.

Bertrand fell back to Leipzig. With IV Corps down to under 10,000 men he performed well during the battle for the city and went a long way towards redeeming his rather tarnished reputation. On 15 October he passed through the city and took post at Lindenau. The next day he stopped the Austrians dead in their tracks as they tried to sever Napoleon's communications by entering the city from Lindenau. Then on 18 October as Napoleon ordered a withdrawal from Leipzig to avoid encirclement. Bertrand burst out from his defensive position. IV Corps completely overwhelmed Gyulai, drove the Austrians back from their positions behind the Luppe and Elster rivers and secured a line of retreat. As the army passed he held the gap open and formed the army's rearguard. He fought a tough action at Kosen on 21 October when Gyulai having re-gathered his composure, fell on his exhausted troops. The action cost Bertrand dearly. He lost another 2,000 men but won a brief respite for the Grande Armée as it struggled on towards the Rhine.

On 31 October he fought again at Hanau when Napoleon brushed aside a poorly contrived attempt by the Bavarians under Wrede to cut his retreat. In the action itself he occupied the town for several hours as the struggling army passed. He was with the rearguard when it reached Mainz on 7 November and crossed into France. IV Corps had suffered terribly during the Autumn Campaign. From its original strength of over 22,500 officers and men on 15 August 1813, only 2,983 infantry and a further 400 gunners from the original formations in his corps crossed the Rhine.

THE CAMPAIGN IN FRANCE AND EXILE ON ELBA

On 18 November Napoleon finally recognized Bertrand's limitations as a field commander, replaced him with Morand and recalled him to Paris as Grand Marshal of the Palace in place of Caulaincourt who became Foreign Minister. He was with Napoleon throughout the campaign in France. He served at Brienne on 29 January, La Rothière on 1 February, Champaubert on 10 February, Montmirail on 11 February, Craonne on 7 March, Laon on 10 March and Arcis-sur-Aube on 21-22 March 1814. He became very close to Napoleon during this period and was very supportive during the crisis that led to his abdication on 6 April 1814.

Totally dedicated to the Emperor he decided not to abandon him in times of adversity and accompanied him into exile on Elba. He landed with his wife and young family on the island on 4 May 1814. Over the next nine months he continued to carry the grandiose title of Grand Master of the Palace. He spent his time running Napoleon's household in Pontefaccio, at all times ensuring that proper protocol and a degree of normality was maintained.

THE HUNDRED DAYS

At the center of Napoleon's plans to slip back to France, Bertrand landed with him at Golfe Juan near Antibes on 1 March 1815. After Napoleon's march to Paris and arrival at the Tuileries he resumed his post as Grand Master of the Palace on 21 March. On 2 June 1815 he became a Peer of France.

Napoleon realized he could not risk Bertrand with a corps command for what was expected to be a brief campaign, so he remained at his post as master of the Emperor's household. Very close to Napoleon his views were often sought and given but there are no instances where they had any influence on events. He served at Ligny on 16 Ligny and at Waterloo two days later. He returned to Paris with Napoleon after the defeat and was with him during the crisis, which resulted in Napoleon's second abdication on 22 June 1815.

SAINT HELENA 1815-1821

Whilst the fate of Napoleon was determined by the politicians in Paris, Bertrand decided to remain with him and accompanied him to the dead Josephine's home at Malmaison. When Blucher's forces nearby threatened to take Napoleon prisoner he moved with him to Rochefort where he secured the frigate Saale to carry him to exile in the United States. The English naval blockade headed by the 74 gun Bellerophon soon stopped that plan. Instead Bertrand on Napoleon's behalf opened negotiations with Captain Maitland of the vessel for him to seek refuge in England. When Napoleon finally handed himself over to Maitland on 15 July Bertrand and his young family sailed with him for England. George IV not prepared to accept Napoleon, Bertrand after strong opposition from his wife decided they would follow him to Saint Helena. Travelling on the Northumberland they arrived off the island on 15 October 1815.

In charge of Napoleon's household his stay on Saint Helena was not easy. He had to act as the principal go between in the continual conflicts between Napoleon and the governor Sir Hudson Lowe. In addition, the idle days caused friction and jealousy amongst the small band of retainers that Napoleon brought with him. They vied with each other for his attention and favour in the vain hope that one day the exile would end. The personal demands Napoleon placed on Bertrand coupled with rumors that he was having an affair with his wife taxed his morale to the limit. He doggedly kept to his task always maintaining a degree of respect and dignity to which the Emperor was accustomed. He was at his bedside when Napoleon died on 5 May 1821.

LATER LIFE AND CAREER ASSESSMENT

Bertrand returned to France on 21 October 1821. His property, rank and privileges were restored. It was a rare display of generosity on the part of the Bourbons, who had previously condemned him to death in absentia in May 1816. He took no part in public life and lived quietly at his home at Châteauroux till after the fall of the Bourbons when he took up the prestigious post as head of the Ecole Polytechnique on 26 November 1830. Interested in politics he gained election as a left of center deputy for Châteauroux on 5 July 1831. His emergence into public life left him little time for himself or his family so he retired at his request from the army on 1 September 1832. He failed in his bid for re-election in 1834 and retired from politics altogether, being able to devote himself full time to the Ecole Polytechnique.

The revival of Bonapartism in the late 1830's saw Bertrand thrust to the center stage briefly for the last time in 1840. A campaign to have Napoleon's remains returned to France resulted in Louis Philippe asking Bertrand to accompany his son, the Prince de Joinville to Saint Helena to retrieve them. After his return he presided over the great ceremony that ended with Napoleon's internment in the Invalides on 15 December 1840.

He died at Chateauroux on 31 January 1844. The government of the day in recognition of his loyalty and service to Napoleon later had his remains reinterred beside Napoleon's tomb at the Invalides on 15 May 1847.

Bertrand was a brilliant general of engineers, who after putting away his sword, became in Dalmatia one of Napoleon's finest imperial governors. His devotion to the Emperor was unquestionable. Napoleon liked him and in turn always tried to treat him consideration in spite of his limitations. As a commander of large formations in the field he soon he was clearly out of his depth. It was hardly surprising that he soon came to grief in Germany in 1813 having never commanded as much as a brigade in the field before. To give him a corps command was a mistake. Though Napoleon never openly admitted it, he must have felt partly responsible for placing such a burden on a person so unsuitable. It goes a long way to explain why after his failures, Napoleon never consigned him to oblivion as he did to so many others for less.

To succeed in 1813 Napoleon needed experienced leaders of the highest order and not a man whose reputation for military prowess was founded on building bridges and governing remote outposts of the Empire. By Leipzig Bertrand showed that he had learnt from his mistakes and Napoleon owed him a great debt for keeping his communications open. This education was achieved however at a terrible cost and had the situation not been irretrievable he might have risen to greatness. A fussy man, a stickler for detail, a master of protocol and the ceremonial occasions he was the ideal choice for the post of Grand Master of the Palace. Whilst a conscientious, honest and brave man, as a field commander history must judge him as very mediocre.

BONAPARTE - JÉRÔME, PRINCE (1784 - 1860)

EARLY LIFE AND NAVAL CAREER

Napoleon's youngest brother, Jérôme was born in Ajaccio (Corsica) on 9 November 1784. Thoroughly spoiled as a child, he basked in the fame of his elder brothers. He was to become the most splendid and expensively idiotic of all the Bonapartes. He possessed brains, but was wayward, capricious, undisciplined and irresponsible. The tallest of the Bonapartes he was handsome and well dressed, but he was not a dandy, more the supreme exhibitionist. Napoleon as the self appointed head of the family had decided that Jérôme was the family member to embark on a naval career, which was much against his mother's wishes, who had a natural Corsican suspicion of the sea.

On 29 November 1800 he joined as a midshipman the Indivisible, a 74 gunner with Ganteaume's Mediterranean fleet at Toulon, about to sail for Egypt with reinforcements. Napoleon left strict instructions with the admiral to show no favours towards his wayward young brother. Ganteaume sensibly did nothing of the sort, as he was an excellent judge of men, particularly of the Bonapartes, and knew which ladder went up and which down. He instead catered to Jérôme's every need, lent him money and more or less allowed him to do what he liked.

Jérôme's time with Ganteaume lasted eight months. After many false starts the squadron broke the English blockade on 4 May, and made a landfall at Derna off the Libyan coast on 8 June 1801 within a day's sail of Alexandria. The local Bey mindful of the English threat from Egypt would not allow the French to land and Ganteaume, despite Jérôme urging to the contrary weighed anchor. He was aware of the complications involved in landing 5,000 troops, and then with news that a Royal Navy squadron was just over the horizon it made the venture too risky. On the way home Jérôme saw his first action when Ganteaume captured the English 74 Swiftsure off Crete on 21 June 1801. To cover himself, when they reached Toulon on 22 July 1801, the wily admiral allowed Jérôme to pose as the captor and receive much of credit for a mission that in effect had failed miserably.

Jérôme by now considered himself a naval veteran and an expert on maritime affairs, which soon bored both Ganteaume and Napoleon. Ganteaume recommended his transfer and promotion to midshipman first class, which he got when Jérôme joined the Foudroyant, flagship of Latouche-Treville's fleet fitting out for the expedition to Santo Domingo. The fleet left with great ceremony on 14 December 1801. Napoleon travelled to Rochefort to make sure his brother was with it and Jérôme left with his sister Pauline, who had recently married the expedition's commander and Napoleon's close colleague Leclerc.

Arriving off Santo Domingo at the end of January he took part in the bombardment of Port-au-Prince on 4 February 1802. Latouche, anxious to see the back of him, promoted him ensign and placed on the small sloop Epervier to return to France with despatches announcing the capture of Cap François, news that was supposed to herald the imminent fall of the whole island. He reached Brest on 11 April and his mission complete sat around the port basking in the glory of his brother's popularity and running up debts. Soon called to account by Napoleon, he was given a severe dressing down and sent back to the West Indies with the Epervier on 28 August 1802. He arrived off Martinique on 28 October, learnt the news of his promotion to lieutenant and appointment as commander of the vessel.

The resumption of hostilities with England gave him opportunities to distinguish himself, but after cruising around off Tobago for a few months doing little he received another severe rebuke from his brother in June 1803. Disgusted at what he considered was unfair treatment he quit the navy, abandoned his vessel at Martinique and sailed for the United States, landing at Norfolk, Virginia on 20 July 1803.

MARRIAGE AND DISGRACE 1803-1805

From Norfolk he went to Washington, where Pichon the French Consul General lent him money. It was a move that nearly spelt the diplomat's ruin when the First Consul heard of it. Jérôme soon became a celebrity amongst local society and before long was again in debt. The charms of a certain Elizabeth Patterson (1785-1879), a Baltimore belle of Irish descent, who came from one of the wealthiest families in the country, caught him. The matter was simple, Jérôme wanted her and she him, provided he first marry her. It was her passport out of Baltimore. Her father William Patterson was against it, but Elizabeth, a wilful woman had her way and on Christmas Eve 1803 they married by special license. Napoleon was outraged by the news, cut off any financial support, while the rest of his family thought it rather amusing. Before long William Patterson soon tired of carrying the financial burden and in October 1804 he hired a brig to ship them to Europe. It hit a mud bank, and apart from losing all their possessions the young couple nearly drowned. In March 1805 they left a second time for Europe with creditors on their heels and the threat that Napoleon would not allow "Miss Patterson" to land in France. Instead they headed for Lisbon arriving at the port on 8 April 1805. General Andoche Junot the French ambassador was on the quayside to meet him, with orders not to allow him to proceed any further with his wife. Jérôme left Elizabeth in Junot's care and travelled to Milan where he hoped to plead his case to Napoleon. The Emperor had other ideas and persuaded his malleable brother to have the marriage annulled by Imperial Decree on 6 May 1805.

Abandoned, Elizabeth found her way to London and in July 1805 she gave birth to a son Jérôme (1805-1870), with whom she returned to Maryland in October. It was the start of the American arm of the Bonaparte family. In 1905 their grandson Charles Joseph Bonaparte (1851-1921) was appointed the United States Naval Secretary in the Theodore Roosevelt's administration. The American side of the male line expired with the death of Charles's nephew, Jérôme Napoleon Charles Bonaparte in 1945.

NAVAL CAREER RESUMED 1805-1806

Napoleon ordered Jérôme to return to sea and on 18 May 1805 he took leave to join the frigate La Pomone. Promotion to *capitaine* followed on 5 June, with him also commander of the squadron of five vessels at Genoa. With Nelson away scouring the West Indies in search of Villeneuve's fleet, Jérôme was able to sail for Algiers on 7 July to rescue some two hundred and fifty French and Genoese held prisoner by the Bey. The threat of force secured their release and on his return the French press made Jérôme a hero.

On 1 November 1805 he took command of the Vétéran, a 74 gunner with Willaumez's squadron at Brest that sailed on one of the most useless voyages of the Napoleonic Wars. For fifteen months they zigzagged across the Atlantic, down to the Cape of Good Hope, up to Brazil, north to the West Indies, then up across the Atlantic and down again. The total bag for this frenzied activity was seventeen merchantmen. In the Bahamas on 29 July 1806, Jérôme tired of the useless cruise, left the squadron without orders and headed for France via Newfoundland and the Azores. In luck, he came across an English convoy on 18 August, captured eleven merchantmen and made port safely on 28 August 1806. Instead of being posted as a deserter, he was again proclaimed a hero in Paris and promoted *contre admiral* (rear admiral) on 6 September 1806.

Further recognition came on 24 September 1806 when Napoleon awarded him the Grand Cross of the Legion of Honor and made him an Imperial Prince. His naval career also at this time came to an end when *contre admiral* Prince Jérôme joined his brother for the campaign against Prussia. Given command of the 2nd Bavarian Division he was fortunate to have a competent general of the calibre of Wrede as his deputy. After the Bavarians stormed the fortresses of Plassembourg and Kulmbach near Bayreuth on 8 October he left his command and joined Napoleon at Jena on 14 October 1806. Satisfied Jérôme had won his spurs as division commander, Napoleon placed him in charge of the newly formed Wurttemberg and Bavarian IX Corps with the task of conquering Silesia. With the rough, hardfighting Vandamme* as his deputy he for once listened and acquitted himself reasonably well. Glogau fell on 3 December, Breslau on 8 January 1807, Brieg on 17 January and Schweidnitz on 8 February 1807. For this he was promoted *général de division* on 14 March 1807 and was still in Silesia when the campaign ended in Poland in June.

KING OF WESTPHALIA

Present at the signing of the Treaty of Tilsit on 7 July 1807, he found the terms of the treaty included the establishment of a Kingdom of Westphalia with him nominated as its king. He immediately handed over command to Vandamme and returned to Paris to enter into a crash course on kingship. He was also browbeaten into a second marriage and on 22 August 1807 took as his wife Catherine of Wurttemberg (1783-1835). An intelligent buxom young woman, with blond hair and blue eyes, her outstanding characteristic was her loyalty, unfortunately for her to Jérôme, with whom she fell fatuously in love. She had received the nod of approval from Napoleon since her father two years before had supported France against Prussia, and for his troubles had his electorate elevated to a kingdom. Three children came from the marriage. Prince Napoleon Louis (1914-), their great grandson, is the present Bonaparte pretender.

The young couple arrived at their capital Cassel on 7 December 1807 and were proclaimed king and queen the same day. Jérôme immediately set about removing the hard core of French generals who had established a regency on his behalf. In the years to follow his government was reasonably efficient. Amongst his ministers he had a good blend of locals combined with Frenchmen. His greatest fault was his extravagance, as the Cassel court tried to out rival its Imperial counterpart in Paris. His womanizing also became legendary, but was treated more with wry amusement than concern by his subjects. The military burden and enforcement of the Berlin Decrees were more the cause for potential unrest than Jérôme's profligacy or mishandling of Westphalia's economy. Apart from a brief crisis in 1809 his throne was never under any real internal threat. His subjects recognized a genuine attempt at liberal reform by him compared to the autocratic ways of the former princes.

In military matters, Jérôme set about building up the Westphalian army to the 25,000 men as required by his constitution. In addition he had to finance a French garrison of 12,500 men at Magdeburg. By April 1809 he had fulfilled his commitment and had 16,000 men in Germany and another 9,500 in Spain. Between January 1808 and October 1813 some 70,000 men were drafted into his army whilst a further 30,000 volunteered. The cost was great. In 1809 he lost 2,000 men in the German Campaign, and in the years that followed some 7,000 more from the original draft sent to Spain. In 1812 22,000 men joined the march to Moscow and only 1,500 returned while a further 2,000 were isolated in the Baltic fortresses. In 1813 he raised another 27,000 men for the campaign in Germany, though as events later proved they were mostly of questionable value.

His generals such as Ochs and von Hammerstein were among the finest non-Frenchmen in Napoleon's service. As the wars intensified he had to support increasing numbers of French troops in his country, rising to 30,000 in 1813. His Westphalian troops till 1813 were generally loyal and the desertion rate was no worse than that of France. Jérôme showed a genuine concern for the welfare of his troops and took his soldiering seriously. The people grew restive, due to the pressure of taxes, levies, conscription and news of heavy casualties. They recognized Napoleon as the lesser of two evils and somehow stoically accepted Westphalia's contribution to the war effort, though they paid a price far in excess to other countries of her size or population.

THE REVOLT OF 1809

Jérôme's first real test as a monarch came in April 1809, when the exiled Duke of Brunswick and the Elector of Hesse Cassel with Austrian backing tried to raise the populace. They were also joined by Pan-German nationalists from Prussia, who with the covert support of Scharnhorst, Blucher and Gneisenau, secretly encouraged von Schill and his followers to start a war of liberation.

Trouble first broke out within Westphalia on 22 April 1809 when an unstable fanatic, baron von Dornberg tried to persuade the Royal Guard at Cassel to defect. Meanwhile at Homberg, Sigismund Martin, a discontented local merchant raised a half armed mob of peasants, ex-soldiers and foresters and marched on the capital. Jérôme showing uncharacteristic decisiveness, had the previous day sent out two screening forces, each of less than three hundred men, under his reliable French generals Reubell and d'Albignac to cover the capital. Left with only his Royal Guard of 1,500 men in Cassel he faced a sleepless night before he confronted his troops the next day and called upon his officers either to go freely and join the enemy or swear an oath of loyalty. To a man they swore loyalty. Meanwhile Reubell's three hundred men faced Dornberg's five thousand five miles from Cassel. The rebels taunted the troops to desert, but Reubell not easily intimidated immediately opened fire with his two cannon and the rebels scattered. From that day Jérôme never looked back with regard to internal threats and showed that he was not lacking in a crisis.

The next threat came from Schill, who marched from Berlin on 28 April, supposedly leading his troops on an exercise. At the Westphalian border he announced they were the

JÉRÔME AND CATHERINE

vanguard of a Prussian Army in a "War of Liberation". Ignoring orders to return from an embarrassed King of Prussia he crossed into Westphalia and seized Halle. Jérôme pulled together the forces of the X Corps that were in Westphalia and marched against the invaders expecting to made short work of Schill. Gratien's Dutch division moved from the west while d'Albignac reinforced came from the north. Schill broke out of what appeared certain encirclement, by routing a small Westphalian-French detachment at Todendorf and headed with his 6,000 men for the comparative safety of Stralsund in Swedish Pomerania. Jérôme followed and on 31 May 1809 Gratien's troops broke into the port and Schill died in the struggle.

CAMPAIGN IN SAXONY 1809

Hardly had Schill been dealt with, when Jérôme was called to help Saxony. The Duke of Brunswick with his "Black Legion" 2,000 strong, supported by some 7,500 Austrians and 500 men under the Elector of Hesse had invaded the country. The Saxons were unable to contain the enemy as the bulk of their troops were on the Danube with Napoleon. On 11 June Dresden fell and Leipzig followed on 19 June 1809. Jérôme with a force of 11,000 crossed the border and joined 2,000 Saxons under Thielemann. Brunswick seeing the superior Westphalian force realized the game was up, hurriedly left Leipzig and two days later on 24 June Jérôme entered the city amidst pealing church bells and public jubilation.

On 1 July Jérôme took Dresden and then headed for Hof to join Junot's Reserve Corps operating against Kienmayer's Austrian XI Corps, which was causing mischief in Franconia. Kienmayer a capable commander checked Junot's 10,000 men at Berneck and Gefraess on 8 July, then turned on Jérôme and drove him from Hof on 11 July. Shaken, Jérôme fell back through Plauen to Schliez, where Kienmayer on 13 July with a few well placed shots had the Westphalian troops streaming from the town in confusion. Another reverse took place at Neustadt on 15 July. Jérôme's troops, showing every sign of panic, were cowering in Erfurt, when two days later news arrived of the armistice between Napoleon and the Archduke Charles. The Brunswick contingent, excluded from the peace terms, found themselves abandoned by Kienmayer. They slipped away from Erfurt on 26 July and for two weeks ranged free with Jérôme's generals at their wits end trying to corner this elusive force of two thousand men. They reached Elfleth on the Weser on 6 August, embarked on English ships and sailed to safety. Humiliated, Jérôme in a fit of pique turned on the unfortunate Reubell and dismissed him from Westphalian service.

THE RUSSIAN CAMPAIGN 1812

Jérôme in spite of his poor showing in 1809, was convinced he possessed martial talents second only to his brother Napoleon. While at Paris in March 1812 he demanded and received command of the right wing of the Grande Armée for the invasion of Russia. The force he commanded numbered 77,100 men in four corps. There was the Polish V Corps of Prince Poniatowski with 34,600 men, the Saxon VII Corps under Reynier of 18,500 men, his own Westphalian VIII Corps of 16,700 men headed by Vandamme*, while Latour Maubourg led the IV Cavalry Corps comprising a mixture of 7,300 Polish, Westphalian and Saxon cavalry. He assumed command at Kalicz on 22 April and then moved to Praga outside Warsaw when his troops moved closer to the Russian border. Rumors circulated he was to receive the Polish crown after the campaign. Not averse to such stories he dallied in Warsaw, tending to ignore his duties when his headquarters moved further east to Pultusk.

Napoleon's main army crossed the Niemen on 24-25 June 1812 and reached Vilna on 28 June. Jérôme held back with the bulk of his forces north of the Narew. This was in anticipation of Bagration's Second Army making a thrust at Warsaw to cut Napoleon's communications, leaving Jérôme the task to fall on the Russian's flank. Bagration did not fall for the trap and retired east. Napoleon then ordered Jérôme to press forward at all speed in order to pin the Russians. He reached his first objective Grodno on 30 June where he remained for four days to rest and reorganize his troops. The weather then broke, torrential rain made the roads impassable, lowered the soldiers' morale and made operations very difficult. Unsympathetic, Napoleon delivered another harsh rebuke to Jérôme for the delays.

Napoleon then received news that Bagration had started to head north-east to join Barclay de Tolly's main army. To keep the two forces apart Davout's I Corps and Jérôme's wing marched on a parallel course to force Bagration south-east away from Barclay. Near Minsk, Davout mistook Bagration's rearguard for the advanceguard and reported the Russian army in the city when it was to the north. Belatedly discovering his error, he occupied the city and awaited orders, complaining that had Jérôme appeared sooner they could have trapped Bagration.

Jérôme meanwhile was in hot pursuit of Bagration, who had veered south. His Polish cavalry on 10 July flushed the Russian rearguard from Mir and found Bagration's main body at Bobruisk. He sent a courier to Davout on 14 July requesting help. The courier returned with a message from Davout informing Jérôme that he was no longer commander of the right wing and that he Jérôme was his subordinate. Not prepared to serve under Davout, Jérôme sent a courier to Napoleon of his decision to resign. He waited two days for a reply, but none came so he left the army. The right wing leaderless, for a couple of days was thrown into confusion. Bagration, once more nearly brought to battle, had little trouble slipping away.

Jérôme appeared back at Cassel on 11 August, whilst his Westphalians went on to their destruction. They fought bravely, distinguishing themselves at Borodino in action with Ney against the Russian center. While Napoleon was in Moscow they kept open the army's main supply route to Smolensk. During the retreat they fell in with Poniatowski's Poles and the Imperial Guard. At the end of the campaign when Ochs mustered the remnant of VIII corps at Kustrin in January 1813 only 760 men remained.

Jérôme meanwhile cooled down during his journey from Russia and back in Cassel devoted himself to raising more men and money to support the war. Another 1,500 reinforcements joined the Grande Armée before it reached Borodino and by the year end he had sent another 3,500, bringing the total contribution to 24,400 men.

THE 1813 CAMPAIGN IN GERMANY

In January 1813 Napoleon ordered Jérôme to raise a new army of 20,000 men and provision Magdeburg for 15,000 French troops. Jérôme had anticipated the need and readily complied. At the end of February Magdeburg was ready and by May 22,800 Westphalians had joined the Grande Armée in Germany, increasing to 26,000 by the Autumn. Jérôme through intermediaries pleaded with his brother for a command, but Napoleon would have none of it after his behavior in Russia. As it was, Jérôme was more use raising men, horses and supplies. The Westphalians never fought as a single body, the divisions were broken up and scattered throughout the army, which did not help morale.

In April 1813 Jérôme steadied mounting alarm in Cassel when he repulsed a large force of Cossacks from Winzingerode's army that crossed the Elbe and seized Halle. Napoleon's successes at Lutzen and Bautzen in May followed by the Armistice gave him breathing space to prepare the defence of his kingdom.

In Dresden he pleaded with Napoleon to allow him to lead a Westphalian Corps using the argument they would fight better together. Nothing came of it and Jérôme left an embittered man. There was talk of offers made to Jérôme that he could keep his kingdom if he abandoned Napoleon. They proved groundless, the Allies in particular Prussia would never have accepted Westphalia's existence. It left Jérôme with little option but to stand by his brother's side. On 27 September 9,000 Cossacks arrived at the gates of Cassel forcing Jérôme to abandon the city. He returned on 16 October backed by French reinforcements from Mayence led by Allix. After the French defeat at Leipzig the situation became untenable and he left Cassel for the last time on 26 October 1813. His army reduced to a shell faded away. Some troops isolated in garrisons on the Elbe and Oder fought on, invariably kept in line with French bayonets. It was a shabby end to what had become a shabby kingdom.

EXILE AND NAPOLEON'S RETURN

He did not take part in the campaign in France but joined his wife and children at Compiègne. Then as the situation worsened he moved to Paris. On 29 March 1814 with the city about to fall he accompanied Empress Marie Louise to Blois. Both his and his brother Joseph's behavior as the Allied armies approached Paris bordered near panic. Had they kept their heads in the crisis and listened to Napoleon, Paris may not have fallen. Fearing for his family's safety after Napoleon's abdication he fled to Berne in Switzerland. Then after the dust had apparently settled, he moved to Graz in Styria and finally to Trieste on the Adriatic coast in August 1814.

In continuous contact with Napoleon on Elba, when news reached him of his brother's return to France he decided to leave as he was in danger of becoming an Austrian hostage. Catherine remained while on 25 March 1815 he left Trieste in secret on a Neopolitan vessel, only hours before Metternich's agents arrived with orders for his arrest. A storm forced him to disembark in Istria, which nearly ended in his capture before the vessel cast off. He landed at Ancona on 28 March and spent a few days with Murat, who was leading his army north against the Austrians to liberate Northern Italy. Not too impressed with his brother-in-law's madcap scheme to wage war against Austria while Napoleon was trying to pacify the Allies he moved on. A brief stay in Florence and Naples followed before he caught a French frigate and landed at Golfe Juan on 22 May 1815.

THE WATERLOO CAMPAIGN

Warmly welcomed by Napoleon, Jérôme was made a Peer of France on 2 June 1815. Napoleon not sure what to do with him posted him to the Army of the North. A quick reshuffle of commands followed and on 10 June he took charge of 6th Infantry Division of Reille's II Corps, replacing the experienced Rottembourg who moved to the Army of the Rhine. To help Jérôme or rather to keep him in check, Napoleon gave him Guilleminot, one of the army's foremost experts in infantry tactics, to act as his deputy. The quality of his division was excellent, numbering over 8,000 veterans and the largest in the army.

At Quatre Bras on 16 June he only reached the battlefield around 3.00 P.M., his division being the third in line on the march after Bachelu and Foy. Ney at once launched him against the Dutch positions holding Pierrepont Farm and Bossu Wood to the left of the Charleroi Road, after Foy's initial assaults had failed. His infantry drove the Dutch from the farm and forced them to retire into the wood. One brigade supported Foy's advance between the wood and the main road while the other pushed on to clear the wood. His troops drove the Dutch from Bossu woods, apart from one battalion, which tenaciously held onto its north east corner. He also beat back the Brunswick Corps to the edge of Quatre Bras. By this stage around 6.00 P.M., Wellington was pouring reinforcements into the battle and after a desperate charge by Kellermann's cuirassiers to break the British line the battle turned. Maitland's and Byng's brigades of the Guards Division advancing from the Nivelle road entered the wood and gradually forced back his division. When the battle ended around 8.00 P.M. he was back at his start point before Pierrepont. Jérôme lost that day over 1,000 men and received a nick in his left side from a musket ball.

Napoleon was pleased by Jérôme's performance at Quatre Bras and allowed him to lead the French assault on Hougoumont. The plan was to draw Wellington's attention to this secondary part of his line and induce him to draw forces away from his center where Napoleon intended to launch his main assault. The battle opened at 11.30 A.M. when Jérôme's cannon opened fire and the skirmishers of Baudin's brigade were thrown out in front of his columns. The copse and the field before the chateau were soon taken before the garrison halted the French attack. Guilleminot mindful of Napoleon's plan, urged Jérôme not to push the attack too far. Jérôme did not listen and sent in his second brigade, that of Soye against the chateau. This attack reached the northern end of the courtyard, where lieutenant Legros a man of great strength, broke open the gate with an axe and burst in with a handful of men. After a fierce struggle, Legros and his men killed, the gate was shut and secured.

Jérôme had totally misread Napoleon's plan and throughout the day

continued to batter away at Hougoumont, which was never at any time held by more than 2,000 men. Bachelu's division was committed, and finally even one of Foy's brigades was swallowed up in the pitiless fray, bringing to over 15,000 the number of French engaged. A sustained howitzer bombardment from the outset could easily have set Hougoumont alight and made it impossible to defend. Jérôme however persisted against the advice of his generals with the costly assaults. Wounded in the head during one attack, he surprised many that he was not wanting in courage. Unfortunately it was all for nothing, since during the day he showed neither sense, maturity or generalship.

After the battle he rallied the remnants of his division at Avesnes and led them to Laon where on 22 June he rejoined the army under Soult. The severity of the fighting was reflected in his regimental returns, as only 2,300 men remained.

LEGROS AT THE GATES OF HOUGOMONT

EXILE 1815-1847

On 23 June he left his troops to join Napoleon in Paris, only to find on his arrival his brother had abdicated. He thought of fleeing to the United States, having the temerity to think the Patterson family may help him. On Fouché's advice he quit Paris on 26 June and took refuge in Goeppingen Castle one of his wife's homes in Wurttemberg. It was the start of thirty two years of dissolute wandering.

Kept a virtual prisoner by his father-in-law, Jérôme was later joined by Catherine. She rejected all threats of being disinherited if she remained with him and it was only on her father's death in July 1816 that they were able to leave after Austrian help. They established a home in Trieste where much of the remaining fortune was lost in poor business ventures, including one to navigate the Danube. Jérôme also took an interest in a madcap scheme to have himself crowned king of Greece but little came of it.

Napoleon's death brought an easing of restrictions against the exiled Bonapartes and in 1823 Jérôme moved to Rome where most of the family had settled. He lived there till 1832 when the activities of his nephew Louis Napoleon who was involved in virtually every plot to overthrow an Italian ruling family resulted in his deportation to Florence. He tried to write his memoirs, but when Catherine died during a cholera epidemic that struck Florence in 1835 he moved to Switzerland with his young family. He asked permission to return to France, but the coup attempts by nephew Louis in 1836 and 1840 soon put paid to any thoughts of allowing members of the Bonaparte family to return. He returned to Florence and in 1840 he married a rich local widow Giustina Bartolini-Baldelli (1811-1903), who had once been one of Catherine's ladies-in-waiting.

THE FINAL YEARS - RETURN TO FRANCE

He was allowed to return to France in September 1847, after many petitions disassociating himself from the activities of his nephew. He lived modestly in Paris having spent his way through most of his third wife's fortune. Jérôme soon changed his tune after the fall of Louis Philippe. Louis Napoleon's return to France and the sudden resurgence in Bonapartism that resulted in his election as President of France on 19 December 1848 saw Jérôme vying for favours in the most nauseating manner. In need of money, his rank of *général de division* was restored on 11 October 1848. It secured him a salary of 12,000 francs a year. He became governor of the Invalides on 23 December 1848, which carried an annual pension of 45,000 francs and free accommodation. The most notorious appointment he received was on 1 January 1850 when he became a Marshal of France. He first demanded it from Louis, then when refused screamed at the President attacking his masculinity. To get rid of his troublesome uncle Louis gave in. The job was for life and carried a stipend of a another 30,000 francs a year. President of the Senate followed on 28 January 1852, but he took no interest in it as it meant much hard work for little reward.

When the childless Louis become Emperor Napoleon III on 2 December 1852, Jérôme became the First Prince of the Blood. That in itself was not alarming, since it was unlikely he would outlive his nephew. The prospect of Jérôme's equally nauseating son Napoleon Joseph Charles "Plon Plon" (1822-1891) becoming emperor cast grave doubts over the future of the Bonaparte line. The succession issue resolved itself in 1856 with the birth of the Prince Imperial. From then on Jérôme's influence diminished and he receded from public life. He had a stroke in 1858, which left him partially paralyzed. He died of another at his country home at Villegenis (Seine-et-Oise) on 24 June 1860, gambling with an aide when supposed to be in bed after a bout of bronchial pneumonia. Louis Napoleon graciously give him a state funeral, and interred him in the Invalides.

ASSESSMENT - SPOILT AND ROTTEN TO THE CORE

History's damning verdict of Jérôme Bonaparte as the spoilt younger brother of Napoleon who possessed little morale fibre is generally correct. As a military commander critics can be a little kinder, but not much. He certainly helped ruin the Emperor's chances at Waterloo by persisting with costly assaults against Hougoumont. However had a French victory resulted he would have been a national hero. Napoleon would have liked nothing better than his young brother to fix the Allied right whilst he delivered the hammer blow against their center. He was a capable commander, who did not lack courage in a crisis and genuinely cared for his men. He hero-worshipped his brother, his mannerisms were similar and he tried to emulate him in every way. This led him often to taking excessive chances against good advice, which invariably did not work simply because he did not possess the genius or ability of his brother. To conclude as a commander he possessed average ability, was brave, conscientious, tenacious and impetuous, but lacked strategic or tactical common sense. As for his moral character the comments at the beginning of this chapter remain unchanged. He was rotten to the core.

EARLY LIFE AND CAREER

Of noble birth Bourmont was born in Freigne (Maine-et-Loire) western France on 2 September 1773. He enlisted as an *ensign* with the Gardes Français on 12 October 1788. When the regiment defected to the National Assembly he was discharged on 31 August 1789. The position of nobles rapidly deteriorating, he emigrated with his father to Turin in 1791. He joined the emigré Army of the Princes and with the outbreak of war he campaigned for two years against the French forces in Savoy and Piedmont. A close confidant of the Prince de Condé on 16 February 1795 he became chief of staff with Scepeaux's royalist army operating in the Vendée. He carried out many missions for the Bourbons and made a significant contribution to prolonging the unrest in the region. In recognition, the Bourbons made him a *chevalier de Saint Louis* on 13 May 1796. The position in the Vendée untenable, he left France and settled in Switzerland where he became a colonel with the Comte d'Artois's forces. He returned to France in secret in August 1797 and was involved in drumming up Royalist support before the Fructidor coup d'etat of 7 September 1797. Fleeing to London after its failure he helped to organize another revolt in the Vendée in 1799. Victorious at Saumur and then seizing Le Mans on 15 October 1799 he achieved some success as commander of the rebels. Finally run to ground by Brune he concluded a peace with the First Consul on 4 February 1800 and was allowed to remain in France.

A die hard Royalist, his capacity for intrigue never waned. After the explosion of George Cadoudal's "infernale machine" on 24 December 1800 he was arrested on 17 January 1801 on suspicion of complicity in the plot against Bonaparte's life. Incarcerated in the Temple, on 11 July 1801 he was transferred to the citadel at Besançon. On 3 August 1804 he escaped and fled to Spain.

RETURN TO FRENCH SERVICE

He lived in Spain for several years before moving to Lisbon. When the French under Junot occupied the city in December 1807, as an emigré on the run his future looked bleak. Bourmont had earlier befriended Junot while he was French ambassador to Portugal and convinced him he was a reconciled noble and wrongly accused. To make his case more plausible he asked to serve in the French army so Junot posted him to Loison's division as its chief of staff. A period then followed as Junot's chief ordnance officer. After the Convention of Cintra and the repatriation of the army to France the authorities immediately arrested him on his arrival at Nantes on 30 October 1808. Only after persistent attempts by Junot, was his release secured on 3 February 1809.

Distrusted, only Murat's Neapolitan army would accept him so out of desperation he joined them on

20 April 1810 with the rank of *adjudant commandant*. In January 1812 the Neapolitan army forming part of the Corps of Observation of Italy he renewed his association with Junot, who was deputy commander to Eugene Beauharnais. In the months that followed he played an important role helping to organize the Italian troops for the Russian Campaign. He developed a good rapport with Junot and on 28 July 1812 se-

BOURMONT - LOUIS AUGUSTE VICTOR DE GHAISNES, COMTE (1773-1846)

cured a transfer to his staff with VIII Corps of the Grande Armée in Russia. He fought at Borodino on 7 September 1812 and during the retreat was seriously wounded on 10 November during the crossing of the Vop.

On 3 April 1813 he returned to duty and served on the staff of Macdonald's XI Corps. Wounded at Lutzen on 2 May 1813 he took no further part in the Spring Campaign. Promoted *général de brigade* on 28 September 1813 he served on the staff of XI Corps during the Autumn Campaign, distinguishing himself at Leipzig and in the retreat to the Rhine.

During the 1814 Campaign in France he served under Gérard* in the Paris Reserve Division till transferred in February to Victor's II Corps. At Nogent-sur-Seine on 11 February 1814 he received praise for the way he led the rearguard during the stubborn defence of the river crossing. The action resulted in his promotion to *général de division* on 13 February 1814. It disrupted Schwarzenberg's advance on Paris and was largely responsible for enabling Napoleon to concentrate his forces for the successful counterblow at Montereau on 18 February 1814. Wounded in the knee during the battle, he took no further part in the campaign.

THE RESTORATION AND NAPOLEON'S RETURN

After Napoleon's abdication, Bourmont was not out of a job for long and on 20 May 1814 was appointed commander of the 1st Sub-division of the 6th Military Division based at Besançon. On Napoleon's return from Elba he joined Ney's force sent to intercept him as its deputy commander. He soon became depressed as news of defections spread, and despite the demeanor of his troops turning sour, he doggedly kept to his task. Horrified by Ney's decision at Auxerre on 14 March 1815 to address the troops and rally to Napoleon, he left and returned to Paris. Once Napoleon was safely installed in the

Tuileries he came out of hiding and in the interests of France decided to offer his services.

Bourmont was totally distrusted by Napoleon's close followers, in particular Davout who strongly advised against giving him any command. The Emperor ignored it and on 4 June 1815 gave him command of the 14th Infantry Division with Gérard's IV Corps at Metz. Bourmont's loyalty became even more questionable when he later refused to sign the Act Additionel, which acknowledged Napoleon's supreme authority in time of war.

BOURMONT'S BETRAYAL AND ITS CONSEQUENCES

His division was the first to complete the long march from Eastern France and arrived on the Belgian border during the evening of 14 June. The rest of Gérard's corps had become strung out during the march from Metz, which caused the advance on Charleroi early the next day to be delayed. Forming the advanceguard his division soon ran into trouble when it bumped into Vandamme's corps, which had not yet broken camp. Amidst this confusion Bourmont with five members of his staff calmly rode across the border, replaced the Tricolor they were wearing with the White Cockade and announced their defection to the bemused Prussians barring their way.

With him he took a copy of Napoleon's operational orders and presented them to Gneisenau when he reached Blucher's headquarters. Outraged that any officer as a matter of honor could even consider desertion whilst also suspecting a trap, Blucher refused to see Bourmont and dismissed him with the curt rebuff, "A cur is always a cur."

Bourmont's desertion had little impact on the Allied plans. Wellington response was typical when questioned a few days before about the prospect of French defections. He commented, "We will pick up a marshal or general or two, but not worth a damn!". They picked up Marmont and Victor in the early days and at the last moment, potentially the most dangerous Bourmont since he knew Napoleon's operational plan.

Whilst the Allies hardly reacted at all, Bourmont's actions had more far reaching effects on Napoleon's plans. Gérard's advance on Charleroi was thrown into confusion with no one to lead the advanceguard. Further delays came as Gérard made out a lengthy report to Napoleon, appointed and briefed Hulot as the new commander and steadied his edgy troops. As the campaign unfurled they were to fight as well as any others in the army. Gérard was also acutely embarrassed as it was he who had recommended Bourmont's appointment in the first place.

Charleroi as an objective was abandoned. Gérard received new orders to cross the Sambre at Chatelet, five miles down stream. IV Corps cut across country and reached the river in the late afternoon and as a result was unable to help Vandamme's III Corps at Gilly. Had they crossed the river earlier, Gérard would have been in a magnificent position to turn the Prussian left flank and allow Vandamme to reach Ligny that night. It would have saved several hours and enabled the whole of the Army of the North to be across the Sambre by the night of 15 June.

As it was, Gérard's corps halted astride the Sambre, instead of bivouacking on the left bank. Bourmont's division under Hulot did cross the river till the next day. As a result IV Corps was late in concentrating at Ligny the next day and the opening of the battle was delayed. This delay contributed partly to the late hour at which the battle ended, and to the consequent impossibility in darkness to recognize the true direction of the Prussian retreat. Apart from the French not knowing how much of Napoleon's operational plan the Allies knew, Bourmont's defection caused Gérard's late arrival on the Sambre which had unfortunate and far reaching results.

BOURBON GENERAL

Louis XVIII accepted Bourmont with open arms at Ghent. Here he spent a few nervous days contemplating his future as he heard the distant rumble of cannon at Ligny and Waterloo. After news of the French defeat Louis XVIII on 21 June made him *commandant extraordinaire* of the 16th Military Division at Lille, the region closest to Ghent where the Bourbons could re-establish authority. Then not to miss the awards and positions to be handed to the faithful, he joined Louis's entourage that followed in the wake of the Allied baggage train during the march to Paris. Reward came with command of the 2nd Infantry Division of the Royal Guard on 3 August 1815.

During Ney's trial by the House of Peers he outraged most when he attended as a willing prosecution witness on 4 December 1815. His testimony went a long way to sealing Ney's fate when he insisted the defection at Auxerre was a premeditated act. Ney conducted a gallant defence and Bourmont was shaky under cross examination, but his testimony held up. Key witnesses such as Delaborde had died and the Peers needed a scapegoat.

Further recognition of his services came when he was made a Commander of the Order of Saint Louis on 24 August 1817. In 1823 when the situation started to deteriorate in Spain and French troops massed on the border he received command on 16 February of a division with Bordesoulle's Reserve Corps of the Army of the Pyrenees. He served with the expedition that marched into Spain and restored

Ferdinand VII's authority. In June he took over the 2me Colonne Mobile detached to put down liberal opposition in Andalusia. After a few minor skirmishes he restored order and on 8 October 1823 became commander of the forces at Cadiz. For a successful campaign, a grateful Louis XVIII then made him a Peer of France. When the main army pulled out of Spain, Bourmont remained as commander of the French corps that remained from 6 November 1823 till 17 April 1824.

On his return to France he resumed his duties with the Royal Guard. In addition he was appointed to the prestigious post of Gentleman to King's Chamber. On 13 May 1825 he received the Grand Cross of the Legion of Honor. On 17 February 1828 he became an Ultra member of the Council of War to counter the liberal element in that body that advocated far reaching reforms within the Army.

MINISTER FOR WAR AND THE INVASION OF ALGERIA 1830

The ultimate insult to the Army came on 8 August 1829 when he became Minister of War in Polignac's government. Tradition required a military man of repute who could command the respect of politicians and generals alike. A traitor, a member of an unpopular government and supporter of an equally unpopular monarchy his time in office was too short to do any lasting damage to the Army. Still, morale amongst officers and men alike plummeted and there was an alarming increase in desertion and republican activity. To deflect public opinion a successful war was needed, and a simmering dispute with the Bey of Algiers was magnified into a suitable cassus belli.

To get away from the hurly burly of politics, Bourmont secured himself command of the expedition on 11 April 1830. He worked feverishly

THE FRENCH ENTER ALGIERS

37

to prepare the expedition and at the end of May his work complete he sailed from Toulon with 37,000 troops carried on 350 merchant vessels, escorted by 103 warships. He landed west of Algiers at Sidi Ferruch on 14 June 1830. After a storm, which threatened to scatter the fleet and some skirmishing in the dunes he established a fortified base before commencing the advance on Algiers. He met the Dey's forces at Staoueli on 19 June. The battle itself was the largest fought by France in the period between Waterloo and the Crimean War, with over 30,000 men assembled by the Dey from Algiers, Medeah, Oran and Constantine against Bourmont's force. In common with other colonial battles of the nineteenth century success depended on repelling the first furious assault by the enemy. The Dey's forces were firmly held and then quickly repulsed. Rockets, and the new system Valée field and mountain guns did their work. The infantry then charged with great elan and took the Dey's camp for the loss of fewer than 500 men, one tenth of the enemy's figure.

He pushed on to Algiers and between 29 June and 4 July invested Fort l'Empereur, which dominated the city. The Turkish commander rather than surrender the fort after a furious bombardment ordered its evacuation and blew the magazine, flattening the work completely. News of the successes in Algeria reached Paris on 9 July, and the campaign effectively over the embattled government hoped to deflect attention from it towards Africa and duly awarded Bourmont with a marshal's baton on 14 July 1830.

When news of the Bourbons' fall reached Algeria in early August Bourmont failed to use his head. Outraged he made plans to intervene with the African Army, by landing at Toulon, then march to Lyons and finally Paris to topple the government. Unable to secure the co-operation of Admiral Duperre and his generals, on 2 September he quit his command and hurriedly departed to the Balaeric Islands when they threatened to arrest him. He remained a few weeks there before trying to rally support for Charles X in Barcelona, finally joining the exiled monarch later in the year in Scotland.

THE WANDERING EXILE

A die hard Ultra, he spurned all offers of reconciliation with Louis Philippe. In the years that followed he cut a rather pathetic figure as a follower of lost reactionary causes. He came under the spell of the Duchesse de Berry and helped plan a rising in La Vendée, always a royalist stronghold since 1789, supported by a military rebellion. Convinced the army was eager to avenge itself after its humiliation in the streets of Paris during the July Revolution he tried to rally the support of disaffected officers who had resigned in 1830. Pamphlets were circulated declaring the new government would slash taxes,

while the army would be increased to crack down on law and order. Louis Philippe lost patience with his wayward marshal and struck him from the list of marshals on 10 April 1832.

After a short period in Italy trying to gather further support Bourmont and the Duchesse de Berry landed near Marseilles on 28 April 1832. Support they expected in the south was nonexistent with neither the people or the garrisons rising. Undeterred they travelled secretly across France to La Vendée where an insurrection broke out in June. It was a lamentable failure because their supporters were no longer the fanatics of 1793, but passive royalists. Many had become land proprietors, who wanted to cultivate their estates in peace rather than ambush government troops in hedgerows. After a few skirmishes culminating at the siege of La Penissiere, which was turned into something akin to a second Saragossa by Bourbon propagandists, where sixty rebels held 500 soldiers at bay for several days, the rising fizzled out. Bourmont more astute than his duchesse had long since abandoned the cause and returned to England. Meanwhile De Berry built up a romantic Joan d'Arc image by keeping one step ahead of the police who were at their wits end trying to track her down in Western France.

Bourmont's next attempt at following lost causes was in Portugal where he led the forces of Don Miguel during the civil war till their defeat at Santarem on 16 May 1834. He lived a time in Rome and then moved to Germany where he lived till 1840 when under an amnesty he was able to return to France. Taking no part in public affairs he lived his remaining years at his chateau near Freigne, where he died on 27 October 1846.

AN ASSESSMENT - A MUCH MALIGNED AND UNDERESTIMATED LEADER

Bourmont was considered by Bonapartists and Allies alike a traitor, a reckless opportunist and a man who possessed few if any moral scruples. Propagandists for Napoleon have always played down Bourmont's defection explaining it away as an act of impulse rather than accept the possibility the Emperor had been duped by a rather brilliant, cool and calculating Napoleonic version of a double agent. In his later career Bourmont was destined to fail, due to the unpopularity of the Bourbons and the undeniable fact that he was a traitor in an army where a strong Bonapartist faction wanted revenge. Against expectations he proved a very capable army commander. He organized and led an expedition to Algeria, a logistical feat unparalleled since Napoleon's expedition to Egypt thirty-two years before. As a division commander in Napoleon's time he was no fool, and certainly in his post Napoleonic career he showed an ability to organize and command large forces above the ability of the average Napoleonic army commander.

CAMBRONNE - PIERRE JACQUES, VICOMTE (1770-1842)

BACKGROUND AND EARLY CAREER

Pierre Cambronne the brash outspoken Guard general, whose particular fighting qualities were only matched by his salty language, was a Breton born at Nantes on 26 December 1770. His father a merchant hoped he would follow him into a business career. He however grew up to be handsome tall and wavy haired, with fierce eyes in the Duke de Guise style that immediately commanded respect when he first enlisted as a volunteer with the 1st Battalion of the Loire Inférieure on 26 September 1791. From there he passed to the unit's grenadier company on 7 November 1791.

Soon after the outbreak of war in April 1792 he moved to the 1st Battalion of the Mayenne-et-Loire Volunteers serving in Dumouriez's Army of the North. He received his baptism of fire at Jemappes on 6 November and took part in the conquest of Belgium. He returned to Brittany the next year and served against the Vendeans, distinguishing himself at Louay on 20 June 1793 when he recovered a caisson that fell into rebel hands. The episode resulted in his promotion to *sergent-major* on 1 July 1793, followed by another to *lieutenant* on 10 September 1793. Cambronne's reputation for outspokenness arose in an incident on 6 October 1794 when he swore at and threatened a senior officer from another regiment, accusing him of cowardice. He was to be court-martialled, but the matter was dropped when it came clear he was right and would only damage regimental morale convicting a popular junior officer.

He served in the Vendée till 1796 with the Armies of the Coast of Cherbourg, the West and the Ocean Coasts. He fought at Quiberon on 21 July 1795 when Hoche fell on an emigré force of 3,500 that landed on the coast. In the weeks that followed he was active in pursuing Chouan bands, and in spite of being an ardent republican he tried his utmost to treat the populace with consideration, gaining a reputation for being hard but fair. In October 1796 he moved to the 46th Demi-brigade and joined Hoche's ill-fated expedition to Ireland that was forced back to Brest after being scattered by storms and English frigates in December.

In 1797 Cambronne moved to the Army of the Rhine-Moselle taking part in Jourdan's successful offensive over the Rhine in April. In March 1798 his regiment moved to the Army of England under General Bonaparte on the Channel Coast. In 1799 he joined the Army of the Danube and then Switzerland where he distinguished himself in the Second Battle of Zurich on 26 September 1799. In the battle he led his company in a furious bayonet charge against two enemy guns that were causing mayhem to a battalion of grenadiers approaching the city walls. The first discharge failed to stop his men before they were amongst the guns putting the crew to the bayonet. He then turned the pieces on the enemy and in the resulting confusion the 46th Demi-brigade took five standards. His part was a minor event in a major French victory, and though hailed by his men, it went surprisingly unnoticed and did not result in promotion.

He served with Moreau's Army of the Rhine the following year and showed exceptional bravery at Neuberg on 27 June 1800 when Montrichard's division was caught on its own and had to fight a stiff rearguard action. At the head of the regiment's 3rd grenadier company he was in the thick of the fighting. It prompted his colleagues proclaiming him the "Second Grenadier of France", after a fellow officer, lieutenant Latour d'Auvergne who led the second company, died in the action and was immortalized by the First Consul and given the principal honor. He then took part in the campaign that led to Moreau's triumph at Hohenlinden on 2 December 1800.

THE GRANDE ARMÉE 1807-1808

In the years that followed Cambronne served with the Dunkirk garrison from 1801-1803, then as part of the Army of the Coasts from 1804-1805. Recognition for past service came with promotion to *capitaine* and the award of the Legion of Honor on 14 June 1804. Promoted *chef de bataillon* on 29 August 1805 as the Grande Armée broke camp, he moved to the 88th Line. His regiment was with Vedel's brigade of Suchet's division in Lannes's V Corps. He took part in the encirclement at Ulm and was present at Austerlitz on 2 December 1805.

The following year in the Prussian Campaign he was in action with the 88th Line at Saalfeld on 10 October. Four days later he joined the struggle for Vierzehnheiligen as Suchet's division led V Corps' advance against the Prussians at Jena. In Poland he fought at Pultusk on 24 December and in recognition of his service in the campaign on 10 January 1807 was made an Officer of the Legion of Honor. The remainder of the campaign in Poland he spent with the French right wing on the River Bug watching Russian moves across the river that threatened Warsaw and French communications.

After Tilsit a period of garrison duty followed in Prussia till mid 1808 when V Corps commenced its long march to Spain. He served for a short period at the siege of Saragossa from December 1808 till January 1809 before Suchet's division was withdrawn to keep communications open with Madrid.

THE IMPERIAL GUARD

Napoleon's signing of a decree on 16 January 1809 to form the Guard Tirailleur Grenadier and Tirailleur Chasseur regiments had a profound effect on Cambronne's career. One of the few officers chosen from hundreds of candidates, he returned to Paris to help train some 3,200 Guard conscripts gathered from line and light regiments. Under the direction of Roguet* the new regiments were assigned excellent officers and NCO's from their parent regiments, and were after the Old Guard to become the finest infantry in the army. The fact Cambronne was chosen to lead 1st Battalion of the Tirailleur Chasseurs, showed the high regard his superiors had for him as a regimental officer. He was personally congratulated by Napoleon for his men's fine bearing when the Emperor reviewed the Tirailleurs on 19 March. On 2 April his men started their march to the Danube with Roguet's Tirailleur Brigade of Curial's Young Guard Division.

He distinguished himself at Aspern-Essling on 22 May 1809 when Roguet led the tirailleurs into Essling to rescue Boudet's beleaguered division. Enduring a tremendous bombardment he lost over a quarter of his strength but helped the Young Guard on their debut establish a reputation equal to their senior regiments. Present at Wagram on 5-6 July 1809 the Young Guard spent most of the two days with Napoleon's reserve at Raasdorf. Then as the battle reached its climax he was with Roguet when the Young Guard supported Macdonald's drive on Sussenbrünn that broke the Austrian center to ensure victory.

RETURN TO SPAIN 1810-1813

Cambronne returned to Spain in March 1810 as part of Roguet's Young Guard division based at Burgos. There his unit over the next two years was drawn into the vicious guerilla war that raged throughout Castile and Leon. Apart from his regiment remaining a tough well knit unit the only good that came from these operations was his elevation to *baron de l'Empire* on 4 June 1810. This was followed by his appointment as colonel-major of the 3rd Guard Voltigeur Regiment on its arrival at Burgos on 6 August 1811 serving under Dumoustier. He remained in Spain during the Russian Campaign and it was not till February 1813 that his regiment was recalled to help form cadres in Saxony after the disasters in Russia.

THE SAXON CAMPAIGN 1813

While on the march Cambronne received news of him being made a Commander of the Legion of Honor on 6 April 1813. In Saxony his troops were allocated to a new Guard division under Barrois*. Present at Lutzen on 2 May he saw little action as most of the fighting by the Guard fell to Dumoustier's division. A reorganization of the Guard infantry followed and Cambronne found himself back with Dumoustier for the Battle of Bautzen on 21-22 May where his troops again were not heavily involved.

His finest moment of the campaign was at Dresden on 25 August. The Allies occupied the suburb of Pirna and expected to drive the rest of Saint Cyr's XIV from the city with little trouble. Unknown to them, Napoleon had arrived in the city with elements of the Guard including Cambronne's regiment. As they were about to break in the Freiburg Gate it suddenly opened, and on the Emperor's orders out marched a column of infantry with Cambronne leading the first brigade. The enemy recoiled in terror at the ferocity of the counter-attack and their guns were overrun. Alarm spread along the Allied lines as the French issued from all Dresden's gates with similar results. Driven back, then pursued by French cavalry, 5,000 enemy were disabled and another 3,000 captured for the loss of 5,000 French.

Cambronne's bravery led to his transfer on 14 September 1813 to the 1st Old Guard Division under Friant* as commander of the 2nd Foot Chasseur Regiment. He fought at Leipzig, distinguishing himself on 18 October in the savage fighting for Probthayda as the Allied armies closed in on the suburbs. In the retreat to the Rhine he fought at Hanau on 30 October where at the head of three grenadier companies he drove back an attack by four Bavarian battalions. Promoted *général de brigade* on 20 November 1813 he was made commander of 1st Foot Chasseurs of the Old Guard. On 21 December 1813 he took over the division's 2nd Brigade from Michel* comprising the two Old Guard chasseur regiments.

FRANCE 1814

Campaigning in Champagne in the new year he took a shell splinter in the thigh at Bar-sur-Aube on 27 February 1814. It however did not stop him serving at Craonne on 6 March where he was hit in the left arm, then another shot severely bruised his left side. At the defence of Paris on 30 March he was again bruised by a spent shot that hit him behind the knee. He retired to Fontainebleau and on 2 April resumed his post under Friant. On 12 April after Napoleon's abdication the former emperor sent him on a mission to rescue the Empress Marie Louise at Orléans and bring her to Fontainebleau. Marching overnight through terrible weather with only a battalion of grenadiers he evaded enemy patrols and reached the city the next day only to find the Empress gone. It was one mission he failed, though through no fault of his own, as Metternich's envoy in the person of Prince Esterhazy had got there first and whisked her away to rejoin her father, the Austrian Emperor Francis at Rambouillet. Cambronne did however recover 2,500,000 francs of bullion, which the Empress had deliberately left behind to help her husband while in exile.

ELBA AND NAPOLEON'S RETURN

Napoleon chose Cambronne to head the 700 Guard Grenadiers that accompanied him to Elba, arriving on the island on 26 May 1814. Napoleon immediately made him the island's military commander with responsibility for maintaining law and order. He carried out his duties vigorously, it not being an infrequent occurrence to see his grenadiers bundling back to the mainland on the next available ship persons suspected as Allied spies. Totally devoted to Napoleon he longed to return to France. He found life on the island frustrating after having for so many years been most at home on the battlefield, and he shared the hope of the Guard for anything that would take them back into action. Napoleon took him into his confidence, and Cambronne helped in the secret preparations for his return.

When they slipped the Allied blockade and landed at Golfe Juan near Antibes on 1 March it was Cambronne who led the small force of 1,200 men ashore. He then led the vanguard as it struck inland from Cannes via Grenoble to Lyons. Napoleon's orders to him were strict but simple, "Cambronne you will go ahead, always ahead but remember, I forbid you to shed one drop of French blood in the recovery of my crown." At the start of the march the vanguard consisted of only fifty grenadiers, but it was enough to frighten a whole series of mayors into opening up their towns and supplying money and provisions. Then as the ranks swelled and the march became a huge triumphal progress, on 15 March 1815 he handed the command over to Girard* who had defected outside Lyons. For the rest of the march he continued to perform a masterly job in keeping the ever increasing army closed up. It had to live off the land and there was always the chance that there would suddenly be a battle.

Once in Paris Napoleon offered Cambronne promotion to *général de division* but he refused. The master of false modesty, he felt such a gesture would show undue favoritism and offend others, whilst also taking him away from his troops. Instead he was decorated a Grand Officer of the Legion of Honor on 1 April, followed by his appointment on 13 April 1815 as *colonel-major* of the 1st Foot Chasseurs of the Guard. On 2 June he received the title *Comte* and was made a Peer of France before departing for the frontier a week later.

THE WATERLOO CAMPAIGN

On 16 June Cambronne fought at Ligny leading a column of the Old Guard, made up of his 1st Chasseurs and Petit's 1st Grenadiers, that stormed into the town at the end of the day. The events at Waterloo two days later were to be the climax to his career. As the first echelons of the Old Guard attacked the ridge of Mont Saint Jean at the end of the day his 1st Chasseurs formed the second echelon ready to perform the coup de grace once Wellington's line was broken. When the assault failed and the army streamed past his men in panic, he with the remaining Guard battalions formed squares around which Napoleon hoped the army would rally. Fighting its way from the field, his regiment was soon reduced to the gory shambles of a single battalion. Forced to arrange his troops in a triangle two ranks deep he continued to fire in retreat as other units fled the field. Hemmed in near La Belle Alliance with his command reduced to some 150 men he was called on to surrender. It was here the celebrated response "Merde!" allegedly made by him during a brief lull went down into immortality. Wellington's men were unmoved by such posturing and blew away the square. Cambronne fell struck by a ball above the eyebrow. Taken prisoner, the wound not too serious, the English nursed him back to health in Belgium.

CAMBRONNE AT HANAU

"MERDE!" - WAS IT CAMBRONNE?

Did it happen? While Cambronne recovered from his wounds the controversy over his response at Waterloo gained momentum. Some alleged there was no such cry, it being the figment of an editor from the *Journal Général* who after Waterloo was looking for a good story. It was later immortalized by the young Victor Hugo in his writings. More likely words to the effect, "The Guard dies but never surrenders," as inscribed on Cambronne's monument are likely to be nearer the truth. Yet considering the absolute frustration and despair Cambronne must have faced when everything around him was falling apart, few men facing such a situation would restrain their language. There so few survivors from the square, no one of repute ever came forward to confirm or deny the claim. Cambronne not one to lack modesty, at first never denied the more salty version of events, whilst he certainly laid claim to those on his monument.

There was another story that gained ground, no doubt put out by Ultras to discredit him, that it was the dead Michel who led the Guard columns against Mont Saint Jean and having found his way to Cambronne's square uttered both defiant replies. Cambronne later had good reasons to let the matter rest. Considered by many as little more than a rough desperado, he needed to clean up his image. He had remarried into a wealthy English land-owning family in 1819 and wanted to be accepted as "a gentleman." Such salty language was not acceptable in polite society nor in any drawing rooms or salon in Paris.

LATER LIFE AND CAREER

Convinced his defiant response at Waterloo had made him a national hero and that no one would dare touch him, Cambronne rather impudently wrote to Louis XVIII asking permission to return to France. Not surprisingly he never received a reply and was among those proscribed by the Ordonnances of 24 July. Being suspended and struck off from the army list on 11 October 1815 did little to deter him and he appeared in Paris on 17 December 1815. Arrested and imprisoned in the L'Abbaye, he was brought before a Council of War on 26 April 1816. Helped by the brilliant lawyer Berryer, he put up a valiant defence. It undoubtedly saved his life and resulted in his acquittal with no loss of rank or privileges. Many thought he was doomed like Ney, and by returning to France to stand trial he had received one wound too many to the head, causing his behaviour to become a little irrational. To a certain extent that was confirmed when the army placed him on indefinite sick-leave and he returned to his home at Nantes. The tragic lesson to be drawn from

TO THE LEFT, THE FRENCH VIEW OF CAMBRONNE AT WATERLOO (SCOTT)

TO THE RIGHT, THE VIEW FROM THE OTHER SIDE: CAMBRONNE CAPTURED BY HUGH HALKETT AT THE HEAD OF HANOVERIAN LANDWEHR BATTALION OSNABRÜCK (KNOTEL)

Cambronne's trial is that the same may have happened to Ney had he gone through with a court-martial, instead of insisting he be tried by the Chamber of Peers.

In Nantes Cambronne was a local hero, his exploits at Waterloo had become a legend and in November 1817 he led a delegation of citizens that greeted the Duc d'Angoulême during a Royal Visit to the city. With no sign of an active command, he resigned from the Army in disgust on 1 July 1818 and was placed on half-pay. The Bourbons soon won him over by flattery when they made him a *chevalier de Saint Louis* on 18 August 1819. This was followed by a recall in April 1820 when he became sub-commander in the 16th Military Division based at Lille.

In a few years he moved from being rabid Bonapartist to committed Monarchist, which was evidenced further when he received the title *vicomte* on 17 August 1822. At the same time he became increasingly disturbed by the Ultra influence within the Army and it was the execution of the Rochelle Conspirators in September 1822 that was the final straw. On 15 January 1823 he retired at his own request and returned to Nantes where for the next twenty years he devoted much time to the welfare of former Guard veterans. He died there on 29 January 1842.

HIS CHARACTER AND ABILITIES

Brash and arrogant, brave to the extent of foolhardiness, and tough as nails, Cambronne possessed the attributes common amongst generals in the Guard. A charismatic and inspirational regimental commander, his abilities at a higher level were never really put to the test. Initiative was not a requirement in the Guard, rather bravery and brute force. Had he not uttered the immortal words at Waterloo there is little doubt he would have quietly faded away into the pages of history like many of his contemporaries in the Guard. The incident placed him on a pedestal he did not rightly deserve, and to an extent diminished the exploits of his fellow generals in their last campaign. The *sang froid* he displayed before a "whiff of grape-shot" sent him and his men to oblivion was a brave gesture, but it was wasteful and foolhardy. Until he was struck down at Waterloo his loyalty towards Napoleon was unquestioned. That he insisted in entering Bourbon service is difficult to understand, while even more surprising is that he remained a beneficiary in Napoleon's will. Perhaps the Emperor made allowances for, as suggested earlier, Cambronne becoming a little eccentric, having received one wound too many to the head.

CHASTEL - LOUIS PIERRE AIMÉ, BARON (1774 - 1826)

BACKGROUND AND EARLY CAREER

The heavy cavalry general Chastel came from Upper Savoy where he was born on 29 April 1774 in the village of Veigy. He fled to France in early 1792 during the unrest that broke out in the region. He enlisted at Grenoble on 13 August 1792 as a *lieutenant* with the cavalry of the Allobroges Legion, a force made up of exiled Savoyards formed to liberate their homeland from Piedmontese rule. In 1793 he served with the Army of the Alps during Kellermann's invasion of Savoy. The following year he passed to the Army of Italy and served at the Siege of Toulon, leaving when the port fell on 19 December 1793.

Chastel joined the 15th Dragoons in February 1794 serving with the Army of the Pyrenees Orientales. His regiment passed to the Army of Italy in July 1796 with the peace with Spain and the build up of General Bonaparte's forces in northern Italy. He took part in the crossing of the Tagliamento on 16 March 1797. A severe sabre blow to the head put him out of action for the remainder of the campaign. He was promoted *capitaine* on 9 July 1797.

Chastel joined the expedition to Egypt in May 1798, his squadron one of the two that went with the 15th Dragoons. Present at the Battle of the Pyramids on 21 July 1798, he later joined Desaix's punitive expedition up the Nile in December. He fought at Samhud on 22 January when Davout routed Murad Bey's Mameluke cavalry. On 11 February 1799 at Redecieh near Thebes his command was caught alone and nearly wiped out by Osman Bey, but was saved by Lasalle and the 22nd Chasseurs. An amateur Egyptologist while on patrol in the area Chastel came across the famous Zodiac on the ceiling of the temple at Dendera and is credited with its discovery. Returning to France he was present at the review of the Army of Egypt by Napoleon on 25 January 1802. He was promoted *chef d'escadron* on 4 February 1802.

IMPERIAL GUARD HORSE GRENADIERS 1805-1812

A period followed at the camp at Compiègne where on 29 October 1803 Chastel received promotion to *major* with the 24th Dragoons. When the 24th Dragoons moved south to join the Army of Italy in June 1805 he joined the Horse Grenadiers of the Imperial Guard. He distinguished himself at the head of a squadron at Austerlitz on 2 December 1805 resulting in confirmation of his rank within the Guard. He received the appointment on 18 December 1805 as the regiment's deputy commander under Lepic. He remained as deputy commander in the German and Polish campaigns. He was present with the Horse Grenadiers at Eylau on 8 February when they joined the charge by the Cavalry Reserve that saved the day after the rout of Augereau's VII Corps. Promoted colonel on 16 February 1807 he remained under Lepic despite the temptation to accept command of a line regiment. Chastel's recognition for being a brave and loyal member of the Guard came on 17 March 1808 when he was made a *baron de l'Empire*.

He served with the Horse Grenadiers during Napoleon's campaign in Spain, distinguishing himself before Burgos on 10 November 1808. The Guard cavalry recalled from Spain; he quit Valladolid on 24 March 1809 with the rear detachments. Joining the long march to the Danube he eventually reached Paris on 30 April. After a brief rest Chastel and the Grenadiers set out for the Danube and reached Vienna on 4 July. He received great praise for the fine condition of his men and horses after completing the 1,500 mile march from Spain.

His troops crossed the river and were present at Wagram the next day with Walther's division. Positioned on the French left with the rest of the Guard cavalry they were about to charge Liechtenstein's cavalry covering the Austrian withdrawal from Süssenbrunn when Bessières at their head was hit. Both division commanders Walther and Nansouty, dared not charge without orders, which never arrived. Napoleon later claimed the lucky shot that disabled Bessières cost him 20,000 prisoners, which the cavalry could have taken. It was an unfortunate incident and a rare occasion where the Guard's cavalry leaders failed to exercise good judgement.

Chastel spent a peaceful year in France after the Austrian Campaign, ending on his promotion to *général de brigade* and *Colonel major* of the Horse Grenadiers on 6 August 1811. He left for Spain the following day and replaced Lepic at the head of the Guard cavalry detachments at Valladolid serving under Dorsenne. A difficult few months followed trying to hunt down irregulars in the region. In February 1812 he was recalled to France with the last detachments of horsemen for the coming campaign against Russia.

THE RUSSIAN CAMPAIGN

Chastel left the Guard on his return to France. The Guard cavalry was swamped with general officers limiting the prospect of promotion. Promoted *général de division* on 26 April 1812, he replaced Kellermann* who had fallen ill as commander of the 3rd Light Cavalry Division. Chastel was on the Niemen two months later with his division as part of Grouchy's III Cavalry Corps for the start of the Russian campaign. His fine division comprised two brigades of Chasseurs à Cheval and a mixed brigade of Bavarian and Saxon Chevaulégers, numbering 3,343 men and six horse guns. He was detached from Grouchy in July and his unit spent time under Davout in pursuit of Bagration's Army. He fought a sharp action at Saltanovka near Minsk on 21 July and was responsible for the seizure of the bridge over the Dnieper at Orsha on 14 August. He fought at Borodino on 7 September engaging the enemy cavalry on the French left and covered Eugène's divisions in their assaults on the Raevsky Redoubt. After the assault he joined the great cavalry struggle as the Russians tried to repossess the Redoubt. He fought at Malojaroslavets on 24 October during the retreat from Moscow and at Krasnoe on 18 November. His division having completely disintegrated due to the harsh weather, he joined Grouchy's *Bataillon Sacre* as the remnants of the Grande Armée struggled on to the Niemen.

THE GERMAN CAMPAIGNS OF 1813

As the army fell back through Poland and Prussia Chastel gradually rebuilt his division gathering and reorganizing depot squadrons. On 13 February 1813 having managed to gather a force of some 931 men he took to the field serving with Latour Maubourg's I Cavalry Corps. At Lutzen on 2 May 1813 his troops were only lightly engaged as Napoleon was anxious to preserve his cavalry. During the days after the battle his division received further drafts from France increasing his unit's strength to 1,547 officers and men when he was detached to join Lauriston's V Corps. On 19 May engaged in the pursuit of the Allies he fought at Eichberg near Bautzen when Lauriston soundly defeated Yorck's Prussian corps. Chastel was held back once more at Bautzen on 21-22 May and grew overanxious to engage the enemy. He did not provide an adequate cavalry screen for the advance guard of V Corps and was largely responsible for the destruction of Maison's 18th Division at Hainau on 26 May. In the action over 3,000 Prussian cavalry led by Dolffs caught Maison's 16th Infantry Division foraging in the open. Chastel belatedly reacting to Maison's plight was also roughly handled and put to flight by the triumphant Prussian horse. Only after Lagrande's division formed square and Maison's survivors reach the safety of the nearby village of Milchersdorf did the Prussians draw off. In the action eleven guns and over 1,200 men were unnecessarily lost.

During the Armistice Chastel rejoined Latour Maubourg's I Cavalry Corps spending time building up the division, which including men en route numbered over 4,500 troops by the end of the peace. He served with Macdonald's Army of the Bober at Goldberg on 23 August where Blucher's Army of Silesia was driven back over the Katzbach. Then the threat from that quarter apparently removed, he rejoined Napoleon at Dresden on the 27 August with the rest of Latour Maubourg's force. He spent a brief period detached to Marmont's VI Corps in September before returning to Latour Maubourg for the Battle of Leipzig. He was heavily engaged around Wachau on 16 October 1813 when Murat with the Guard Cavalry and Latour Maubourg's Corps pierced the Allied center. In the battle they overturned Pahlen's Russian light cavalry and then sabred their way through the Russian grenadiers and foot guards of Grand Duke Constantine's Corps. Chastel's men were with the rearguard covering the Grande Armée's retreat as it fell back to the Rhine. Attrition during the campaign, more due to the weather than any other factor, was most severe amongst his men. On 15 November 1813 after he crossed the river only 337 officers and men remained. It was the second consecutive year in which he had lost a division, again due more to the weather than enemy action.

FRANCE 1814 AND NAPOLEON'S RETURN

During the 1814 Campaign in France the remnants of Chastel's division served with Marmont's VI Corps until 28 March 1814, when he was summoned to Paris on to head the cavalry of a provisional corps formed under Compans to defend the capital. He fought at Belleville on 30 March 1814 and helped slow the Allies as they broke into the city.

After the Restoration the Bourbons showed little sympathy towards Guard generals, and on 1 September 1814 Chastel was removed from the active list and placed on half pay. Unemployed it was easy for him to rally to Napoleon on the general's return taking command of 2nd Cavalry Division, II Army Corps of the Army of the North on 31 March 1815. Then with a final reorganization of the cavalry completed in early June he transferred to Exelmans's II Cavalry Corps taking command of the dragoons making up the 10th Cavalry Division.

THE WATERLOO CAMPAIGN AND ITS AFTERMATH

On 14 June Chastel's division concentrated at Walcourt. The following day he joined the march on Charleroi and encountered delays at Marcinelle as the Prussian rearguard offered unusually strong resistance. No supporting infantry available his troops had to wait till the Guard Engineers cleared the bridge over the Sambre. Chastel crossed the river around 2.00 P.M. and pushed on towards Fleurus until halted at Gilly where he encountered a strong Prussian force occupying the woods beyond the village. While troops gathered to deal with the Prussians, bickering broke out between Vandamme and Grouchy. No one had cared to inform Vandamme that Grouchy was his commander and took exception to a "commander of cavalry" directing his infantry assault on a wooded position. The attack was delayed for some time and Chastel lost a good opportunity to enfilade the Prussian left.

Napoleon's arrival soon resolved the dispute and as part of a general advance Chastel's dragoons moved against the Prussian left soon forcing Ziethen to withdraw from the woods.

At Ligny on 16 June Chastel's division with the rest of Exelmans's corps was positioned on the French right with Grouchy. He played an important role immobilizing the Prussian left by harassing the enemy with artillery fire and executing skilful cavalry feints. When the Prussian center finally broke in the evening his troops completed the rout by overturning Lottum's brigade as it retired from the field. The next day he headed east towards Namur in the hope of harrying the Prussian retreat, but was unsuccessful. He turned north and found Prussian columns heading north through Gembloux on the Wavre road. Chastel then showed his shortcomings as a commander when he failed to summon reinforcements and instead shadowed the Prussians at a respectful distance. Had he conducted a vigorous pursuit and delayed the enemy while reinforcements were on their way, his action could have brought Blucher to battle a day earlier than the Prussian general wished. Such an event would have had a decisive impact on the outcome of the campaign.

Chastel was present at Wavre on 18 June when Grouchy tried to force the Dyle. Held in reserve to exploit the situation once Grouchy breached the enemy line he did little all day. On 19 June he supported the renewed assaults across the river and was advancing up the Brussels road when news of Waterloo reached him. The situation critical, Chastel fell back with Exelmans to Namur and occupied the town and the bridges over the Sambre, securing Grouchy's line of retreat. Covering the army as it fell back on Paris he took part in Exelmans's attacks that mauled the Blucher's cavalry at Roquencourt and Velizy on 1 July.

LATER LIFE AND CAREER ASSESSMENT

The disbandment of the Army of the North saw Chastel placed on the non-active list on 1 August 1815. He lived in Switzerland till his return to France in December 1818. Considered a Bonapartist, the Bourbons never employed him and he remained unattached till his retirement on 1 January 1825. He moved to Geneva where he lived quietly till his death on 18 October 1826.

Chastel's greatest successes were in command of the Horse Grenadiers. He was unlucky in charge of larger formations, losing a division in Russia and another in Germany the following year. Both losses were due more to the appalling weather conditions rather than to his failings as a leader. He had always been considered an excellent soldier who combined great gallantry, with experience and coolness. At a tactical level maneuvering in the field, engaging the enemy with perfect precision or carrying out the Emperor's orders to the letter, Chastel was the master.

Chastel like many other generals who left the Guard seemed to lose confidence. They found it difficult to adjust to independent commands at divisional level. When exercising initiative at a strategic level, his record after Bautzen and Ligny certainly indicated he could have shown better judgement. Brave he undoubtedly was, had he remained within the close confines of the Imperial Guard where bravery was more a prerequisite than initiative, he may have had a more distinguished career.

Background and Early Career

The second of four brothers, Pierre David better known as Édouard Colbert was born in Paris on 18 October 1774. He came from a distinguished military family, where twenty-seven family members had served France as general officers. His father Louis Henry François, who died in 1790 held the rank of *maréchal de camp*. Édouard's youngest brother was the cavalry general Auguste Colbert (1777-1809), reputedly the most handsome and dashing man in the army. He died leading Ney's cavalry at Cacabelos during Moore's retreat to Corunna when taking one risk too many a well aimed rifle shot hit him between the eyes. Édouard's other younger brother Alphonse (1776-1843) headed a lancer brigade with Subervie's division during the Waterloo Campaign, and rose to *lieutenant général* during the reign of Louis Philippe.

It was Édouard's elder brother Amboise, a *lieutenant* in the 2nd Dragoons, who emigrated in 1792 with several regimental officers that caused the family the biggest problem. It destroyed the fortune of the Colbert family, as their properties were confiscated and left them in a perilous personal and financial position. Édouard's mother with his younger brothers sought refuge with relatives near Bordeaux, while he to convince people he was a good revolutionary, enlisted with the city's National Guard.

The danger the family faced due to their background persisted. In the end, on 13 October 1793 all three brothers joined the Eighth Battalion of the Seine (William Tell) in the hope they could conceal their origins by wearing a blue uniform. The formation was a motley assemblage of idealists, radicals from the revolutionary clubs, recently released criminals and a few cultured and educated men like the Colbert brothers, who enlisted to conceal or reject their origins. It was not the answer to Colbert's search for anonymity. The William Tell was full of zealous patriots ready to denounce them, which made Édouard determined to serve elsewhere. The first opportunity came in June 1794 when the unit furnished a detachment of men for the cavalry with Édouard among the first to volunteer as he was a very able horseman. He was rejected on the grounds that such a request by a *ci-devant* noble to enter the cavalry would only enable him to desert more easily. Consequently Édouard remained with the infantry serving in the Army of the Upper Rhine till April 1795 when he finally secured a transfer to the 11th Hussars.

A brief period followed with the Army of the Pyrenees. He was promoted to *sous lieutenant* on 28 September 1795. Peace with Spain saw his regiment move to the Army of the West. In La Vendée he showed little enthusiasm hunting down Chouan bands and recalcitrant priests. His cousins the Colbert-Maulevriers were active Chouans and known agents of the Princes. Not surprisingly he was once more denounced as a royalist and accused of giving information to the enemy. Convinced a simple explanation to the army's commander Hoche would suffice to clear his name and dissociate him from any plot, he received a rude shock. Hoche would not listen, ordered his suspension on 25 January 1796 and threatened to have him shot if he did not immediately leave the army.

COLBERT CHABANAIS - PIERRE DAVID, COMTE DE (1774-1853)

Having to renounce his hard won rank and flee was a bitter blow for Colbert. He spent the next two years in Paris always short of money and looking for business opportunities. In May 1798 he joined the expedition to Egypt as a civilian war supplies contractor. He joined the civilians that accompanied Desaix's expedition up the Nile that left Cairo in August 1798. A manpower shortage during its progress soon enabled him to re-enlist as an acting *capitaine* with the 3rd Dragoons. On 9 December 1799 he became aide de camp to the army's new Chief of Staff General Damas. He served under Damas till Menou arrested his chief of staff for insubordination during the army's final death throws in Egypt. Colbert had the fortune to return to France on 14 May 1801 when Menou decided to expel Damas and Reynier, expecting the frigate they were on to fall into English hands.

The incident had little effect on his career and on 15 April 1802 he became *adjudant major* to the Mamelukes of the Consular Guard. He found handling these men of differing cultures particularly difficult and was much relieved when he joined Junot based at Arras on 6 November 1803 as an aide.

The Grande Armée 1805-1807

The Austrian campaign saw him move to Berthier's headquarters staff as an aide de camp on 21 September 1805. Present at Austerlitz on 2 December 1805 he received a musket ball in the chest during the battle. His return to duty on 1 March 1806 saw his promotion to *chef d'escadron* with the 15th Chasseurs à Cheval attached to the Army of Italy.

He was recalled to the Grande Armée on 30 December 1806 when promoted *colonel* of the 7th Hussars with Lasalle's light cavalry division. Present at Eylau on 8 February 1807, he was with the rest of Lasalle's force kept inactive on the French left. In the new campaign, he fought at Heilsberg on 10 June 1807 when the French cavalry were mauled by the Russians. He redeemed his reputation four days later at Friedland when with Grouchy* on the French left his hussars did great damage during the Russian retreat. Over eagerness on his part, nearly led to his death near Konigsberg on 17 June when a detachment of Prussian Uhlans speared him three times in one of the last actions of the campaign.

THE ARMY OF GERMANY 1808-1810

After a period of garrison duty in France Colbert's regiment joined the Army of Germany in October 1808. He served with Pajol's* brigade till promotion to *général de brigade* on 9 March 1809 when he moved to Oudinot's II Corps as commander of the corps cavalry. His new command comprising the 9th Hussars, 7th and 20th Chasseurs à Cheval, played a very active part in the Danube Campaign. At the head of Oudinot's advanceguard he brushed aside a small Austrian force under Schaibler at Pfaffenhofen on 19 April 1809, that helped reopen communications with Davout's III Corps moving from Ratisbon. After Eckmühl he operated with Bessières's Cavalry Reserve and joined the pursuit of the Austrian army's left wing during the advance to Vienna. The advance guard's lead formation he crossed the Enns on 6 May and mauled the Austrians at Amstetten. At Neumarkt the next day he secured the bridge over the Ybbs. By the night of 8 May he reached Sieghardtkirchen only twenty miles from Vienna before 10,000 Austrians barring his approach caused his eager troops to halt. As II Corps gradually invested Vienna he moved south of the city to Wiener-Neustadt to cover the approaches from that direction.

In June he joined Grouchy's cavalry with the Army of Italy and on 14 June 1809 fought at Raab. After the battle he spent time clearing the southern bank of the Danube of isolated Austrian detachments. At the beginning of July he returned to Oudinot. He fought at Wagram on 6 July 1809 where he was laid low by a musket ball that struck his head. He shared in the awards distributed after the campaign, and for his role in the march on Vienna became a *baron de l'Empire* on 28 May 1809.

He remained with II Corps till August 1810 when posted to Holland as cavalry commander of the Observation Corps, force hastily put together after the abdication of Napoleon's brother Louis. The threat of a possible English intervention in Holland soon passed and in December 1810 he rejoined the Army of Germany.

THE IMPERIAL GUARD 1811-1814

One of the most flamboyant colonels in the cavalry, on 6 March 1811 Colbert took command of the 2nd Lancers of the Imperial Guard, formed from remnants of the Dutch Royal Guard Cavalry. From this date, his career was tied to that of this famous regiment till its final disbandment after Waterloo. Based at Versailles he whipped his new formation into shape. He was the prime motivator in introducing the famous red uniform and the Polish czapka. The unit then moved to Tours in the 22nd Military Division where it operated hunting down deserters and draft dodgers. On 21 September 1811 the Dutch Lancers received their eagle from Bessières at Versailles and then formed the escort for Napoleon's and Marie Louise's tour of Holland in the Autumn. Unfortunately still rather raw they did not form a good impression with the Emperor, who preferred the company of his Chasseurs à Cheval.

On 10 February 1812 Colbert left Paris for the long march to the Niemen. At the start of the Russian campaign he was with the Guard cavalry as it advanced on Vilna. On 6 July he was detached to Davout's corps to help trap Bagration's army falling back on Minsk. The operation failed, but he managed to seize a large convoy at Orsha, which enabled his horses and men to reach Smolensk in reasonable condition. Present at both the battles of Smolensk and Borodino, like the rest of the Guard cavalry remained well to the rear and well out of range. After the occupation of Moscow and the fire that engulfed the city, Colbert moved south to the Kaluga road to find suitable forage and quarters.

During the retreat from Moscow his cavalry covered the movements of Morand's* and Roguet's* guard divisions. At Gorodnia on 25 October he saved Napoleon when a detachment of Cossacks surprised his escort. Whilst he did save the Emperor, Colbert gained little credit from the affair since his pickets should have picked up the Cossacks in the area in the first place. His lancers suffered terribly during the retreat. It was with credit that he reached Konigsberg with 19 officers and 200 men compared to 57 officers and 1,095 men at the campaign's start.

Immediately ordered to Frankfurt-on-Main he set about rebuilding his regiment from scratch. To it he added a large contingent recruited from the Paris Municipal Horseguard. By the time the Guard Cavalry concentrated at Gotha in Saxony on 12 April his strength had reached 700 men. He was present at Lutzen on 2 May 1813, but with the rest of the Guard Cavalry remained in the rear and failed to exploit the Allied retreat. After Bautzen on 22 May he was more active during the Allied pursuit. Overzealousness landed him in trouble at Reichenbach on 24 May when a large body of Russian cuirassiers caught his men. It was the timely intervention by Latour Maubourg's Cuirassiers that saved his troops from being completely overwhelmed.

With the build up of Guard Cavalry during the Armistice his regiment joined Ornano's 1st Guard Cavalry Division. He took command of the division's 1st Brigade with his Dutch lancers brigaded with those from Berg, both regiments fielding strengths of 1,576 and 1,069 officers and men respectively. Present at Dresden on 26 August 1813 his men were held in reserve during the first day of the battle. The next day with the torrential rain, he came into his own. Small arms fire was virtually impossible, and apart from the bayonet infantry were unable to defend themselves against cavalry. Napoleon prepared for this eventuality, unleashed his cavalry. Colbert joined the great struggle before the city, where for the first time in the campaign French cavalry gained tactical superiority over the enemy. They then turned on the vulnerable enemy infantry causing the numerically superior Army of Bohemia to conduct a hurried retreat.

In October he was attached to V Cavalry Corps under L'Héritier* helping to organize a provisional brigade of dragoons. He was back with the Guard at Wachau on 16 October 1813 when under Murat the Guard Cavalry destroyed Eugen of Wurttemberg's II Russian Corps. After the battle Colbert's cavalry performed well during the retreat from Leipzig. Morale held and it was with credit that he crossed the Rhine with his regiment largely intact, over 500 of his Red Lancers remaining. On 25 November 1813 he was promoted to *général de division*. He spent December 1813 on the left bank of the Rhine covering the crossing points against incursions by marauding Cossacks.

In January 1814 Colbert returned to Paris and resumed command of his lancers, which had been strengthened to around 870 men. Dispatched to Brienne he distinguished himself at La Rothière on 1 March. With Guyot at the head of the Guard heavy cavalry and Colbert leading the Lancers they occupied the open ground between La Rothière and Petit Mesnil behind five batteries of Guard Horse Artillery. They repulsed an attack by Lanskoi's two hussar divisions, but broke when four regiments of Sacken's dragoons joined the melee. Duhesme's division in danger of being cut off was forced to evacuate La Rothière. Napoleon considered the battle lost and to gain time as the army retired ordered Colbert to attack Olssufiev's infantry as it debouched from the village. The Russians were driven back through La Rothière, but Colbert was unable to recover the twenty four guns lost by the Guard Horse Artillery earlier in the day. Napoleon meanwhile with the rest of the army retired to Troyes undisturbed, while Colbert's horse kept them at a respectful distance.

Colbert after the Battle of Champaubert on 10 February joined the pursuit of Olssufiev's Corps and in the evening drove the Cossack rearguard from the village of Montmirail. His men exhausted remained in reserve during the battle fought around the village the next day. He fought at Chateau Thierry on 14 February when Yorck and Sacken isolated found their corps bundled back over the Marne by Napoleon. The success was marred by Colbert just failing to secure the bridge over the river before it was destroyed, preventing a French pursuit.

On 15 February he became commander of the 1st Guard Cavalry Division, made up of remnants of Guard chasseurs and mamelukes, Polish and Dutch lancers and the scouts numbering barely 1,000 men after two weeks of hard campaigning. His next major action was at Craonne on 7 March. With Exelmans* in support they mounted the steep paths to the plateau to the rear of the Russian infantry. They routed the Cossack and hussar regiments barring their way causing Blucher to start a hurried retreat after he had successfully halted the initial French infantry assaults. He took part in the recapture of Rheims on 14 March. Then his cavalry leading the way during Napoleon's southern sweep that fell on Schwarzenberg's exposed flank he seized the crossing over the Marne at Epernay on 16 March. He fought at Arcis sur Aube on 20 March and at the head of Napoleon's eastward advance was wounded at Saint Dizier on 26 March 1814. Ornano took over his command, while he was with the army at Fontainebleau during the Abdication Crisis.

THE RESTORATION AND NAPOLEON'S RETURN

Napoleon as a sign of his esteem awarded Colbert a pension of 50,000 francs, but the bequest was annulled by the new government. On 8 April with the other generals of the Guard he signed his allegiance to the Bourbons. He fought long and hard, and won the battle against having the 2nd Lancers disbanded using the argument that traditionally the army always had foreign regiments. He reorganized the Lancers, discharged the few remaining Dutchmen and replaced them with veterans from line regiments and other Guard units. Whilst the regiment had a foreign identity, it was entirely French. Based at Orleans, the regiment was merged into the Royal Guard on 24 August 1814 and renamed the Royal Corps of Light Horse Lancers of France with Colbert as colonel. On 1 April 1815 after Napoleon's return the regiment resumed its old title and Colbert was reconfirmed as colonel. The squadron of Polish Lancers from Elba was merged into the regiment, increasing its strength from four to five squadrons, numbering in all 964 men.

THE WATERLOO CAMPAIGN

Marching with the rest of the Guard he crossed the Sambre at Charleroi around 3.30 P.M. on 15 June 1815. He then headed with the rest of the Guard light cavalry under Lefebvre Desnoettes to support Ney's advance on Quatre Bras. Whilst Ney halted at Gosselies he probed ahead up the Brussels Road with Lefebvre Desnoettes. At Frasnes they received their first cannon blast from a force of Nassauers that barred the way. Ordered to push on to reconnoiter Quatre Bras, he skirted the village with the squadron of Polish lancers and arrived within musket shot of the deserted crossroads. Unsupported and with enemy troops approaching in the distance he withdrew to Frasnes, which Lefebvre Desnoettes had since occupied. By the time Ney arrived to decide on the next step, 4,000 Dutch troops with eight cannon had set up position a mile in front of Quatre Bras. It was a force too strong for unsupported cavalry to brush aside yet Ney ordered a few probing charges against the Nassauer infantry. During the skirmish the lancers suffered some thirty casualties including Colbert, who took a musket ball in the left arm leaving him with his arm in a sling for the remainder of the campaign.

Due to strict orders from Napoleon only to commit the Guard cavalry in an emergency Colbert stood idly by watching the proceedings as the battle for Quatre Bras raged the next day. At Waterloo on 18 June 1815 Colbert was again placed in reserve behind Milhaud's cuirassiers to the right of the Brussels Road. His lancers followed Milhaud's cuirassiers in the charges against the Allied squares that started around 4.00 P.M. when Ney mistook a movement on the crest of the ridge as a sign of an Allied retreat. Leading the waves of cavalry that flooded the plateau, he urged them on to break the Allied squares. Hammering at their sides at close range with sabres, lances and pistols his men were unable to make any impression. Driven back by fresh Allied cavalry, they regrouped in the valley and attacked again. Climbing the slopes under heavy fire they crashed against the enemy squares, which again held. Four times they charged and failed. The courage and fury shown by the Red Lancers was one of the most moving events in the annals of the Guard. With Colbert at their head it is vividly displayed in the fresco by the painters Robiquet, Malespina and Desavarreux in the Musée du Panorama de Waterloo.

Exhausted his men were unable to support the final attack of the Guard or help protect the army as it fled from the field. For this Colbert and Lefebvre Desnoettes must bear some blame, since they joined the cavalry charges when apparently they had received no orders to do so. Despite being wounded again, Colbert rallied the Guard cavalry at Avesnes and took overall command. A parade status of the Lancers taken on 23 June showed 30 officers and 525 men remaining compared to 51 officers and 913 men at the campaign's start.

LATER LIFE AND CAREER ASSESSMENT

Retribution was swift. Hardly had the lancers disbanded after they retired behind the Loire, before he was arrested on 25 July 1815. While in prison his suspension came on 1 August 1815. No charges were brought and the authorities released him in June 1816. After Saint Cyr's Army Reforms of 30 December 1818 he was reinstated but his loyalty suspect he remained unattached. On 17 May 1826 he became

a Cavalry Inspector General. A move on 31 December 1826 to the Camp of Lunéville as commander of the 2nd Cavalry Division. In 1828 he became Inspector General in the 1st, 4th and 13th Military Divisions. On 29 October 1828 he was made a Grand Officer of the Legion of Honor.

A thorough professional he did not allow politics to hinder his career and after the July Revolution resumed his post as an Inspector General with the 1st Military Division in Paris. On 2 February 1831 he moved to the 2nd at Châlons-sur-Marne and then the next year to the south of France with responsibility for the 9th, 10th, 11th, 15th and 20th Divisions. He became a Peer of France on 11 October 1832. In 1833 he returned to Paris in charge of the 1st Military Division. A close associate of Louis Philippe's son, the Duc de Nemours, on 12 June 1834 he took an honorary post as his aide de camp. He was one of the many wounded on 28 July 1835 by Freshi's Infernal Machine, the device intended for the King which killed Marshal Mortier.

He accompanied the Duc de Nemours to Algeria from October to December 1836 taking part in the disastrous expedition against the mountain stronghold of Constantine that revealed Clauzel's failings as a commander. For his service in Algeria he received the Grand Cross of the Legion of Honor on 30 May 1837. He retired to the Reserve Section of the General Staff on 31 January 1840. A loyal supporter of Louis Philippe he became very cynical of the new breed of republicans that emerged after his fall. It resulted in his forced retirement from the army on 31 May 1848. One of the few veterans who held the rank of général de division during the Napoleonic Wars and was fit and able when the Second Empire came into being, he readily accepted readmission to the General Staff on 1 January 1853. It was little more than paying a compliment to a brave old soldier. He died in Paris on 28 December 1853.

Colbert's rise to general rank was unusual since he was an aristocrat. That in his early days dogged his career as he was continually under suspicion. Bold but arrogant, aggressive but not reckless, he was a superb leader and motivator. He turned the disillusioned and disbanded Dutch Royal Guard Cavalry into one of the finest cavalry regiments of the Imperial Army. His loyalty to the Guard and his wish to remain with his regiment acted as a break on his career, preventing him using his abundant talents more freely at the head of a brigade or division. When he did operate on his own as in France in 1814 he showed daring and initiative. His early days taught him the art of survival and gave him an ability to adapt in the period under the Bourbons. Throughout his career he remained above politics and proved a sound leader in a period where loyalties were tested to the full and the army had to adapt to changing circumstances.

Background and Early Career

The cavalry leader Jean Corbineau was born in the small hamlet of Marchienne near Douai (Nord) Northern France on 1 August 1776. He was the younger brother of *général de brigade* Claude Corbineau, who as an Imperial Aide died at Eylau on 8 February 1807. His father was a well known horse breeder in the Nord, who had close ties with army contractors. As both brothers had become excellent horsemen with a good knowledge of horse flesh, it was not surprising with their father's help they secured commissions in the cavalry. Jean joined the 17th Cavalry Regiment (later the 25th Dragoons) on 13 October 1792, while Claude the year before had entered the 3rd Dragoons.

Corbineau saw his first active service with the Army of the North. Preferring the dash of the light cavalry to that of the medium or the heavy arm he moved to the 5th Hussars on 1 March 1793. On 1 July he was promoted *lieutenant*. He gained a certain notoriety on 24 April 1794 when after being shot in the shoulder by a Hessian major during a skirmish near Cambrai he proceeded to disarm the culprit and take him prisoner. He again distinguished himself during Moreau's defence of Bentheim on 13 March 1795.

During 1796-97 he was with the Army of the Sambre-Meuse and took part in the deep advance into Germany and the subsequent the retreat to the Rhine. The following year he was across the Rhine in the renewed offensive before the Treaty of Leoben brought an end to hostilities. The resumption of war with Austria in 1799 saw Corbineau with the Army of the Danube before passing to the Army of Switzerland in 1799. In 1800 he moved to Moreau's Army of the Rhine, when leading a detachment into Saint Bloise on 29 April he was shot in the thigh. Out of action for several months his exploits next came to light during the pursuit of the Austrians after Hohenlinden when with the cavalry of Richepanse's division he fought at Lautreck on 19 December 1800.

On 8 December 1801 he gained promotion to *lieutenant adjudant major* joining the 5th Chasseurs à Cheval led by his brother. Periods of garrison duty followed at Mayence and Coblenz where on 23 October 1802 he received promotion to *capitaine*. During June 1803 he took part in the French occupation of Hanover and remained with the occupation force for two years. Promoted *chef d'escadron* on 2 February 1804, he was seconded to the Hanoverian Legion, which resulted in him missing the Austerlitz Campaign.

The Grande Armée 1806-1808

His return to French service saw his promotion to *major* with the 10th Hussars on 16 May 1806. He served with Trelliard's cavalry brigade of Lannes's V Corps during the Prussian Campaign fighting at Saalfeld on 10 October and Jena on 14 October 1806.

In Poland he was in action at Pultusk on 26 December. Promoted *colonel* on 7 January 1807 he took command of the 20th Dragoons of Klein's 1st Dragoon Division. He fought at Hof 6 February and then charged with the Cavalry Reserve at Eylau on 8th February 1807 when with Murat at their head they saved the day after the rout of Augereau's VII Corps. His position unaffected on Klein's replacement in May

Corbineau - Jean Baptiste Juvénal, comte (1776-1848)

1807 by Latour Maubourg, he led his regiment at Heilsberg on 10 June and Friedland on 14 June 1807.

On 17 March 1808 he received the title *baron de l'Empire* and received an unusually large pension of 10,000 francs. The award was excessive compared to that given to others of similar rank. It no doubt was a way Napoleon tried to make up for the loss of his brother at Eylau, who was a genuine favorite of the Emperor.

Spain 1808-1811

A period of garrison duty followed in Germany till 8 September 1808 when his regiment rejoined Latour Maubourg's 1st Dragoon Division with the Grande Armée for the invasion of Spain. He was at the capture of Madrid on 5 December 1808 and fought at Ucles on 13 January 1809 when Victor defeated Venegas's Army of the Center. His string of successes continued at Medellin on 28 March when Latour Maubourg's dragoons smashed through the Army of Estremadura. Present at Talavera on 28 July 1809, the inhospitable terrain prevented the deployment of cavalry and left his regiment after the battle covering the French withdrawal.

In September 1809 he passed with the 20th Dragoons to the 3rd Dragoon Division led by Milhaud* operating with Sebastiani's IV Corps. Present at Ocana on 18 November 1809 he led the division's 2nd Brigade with distinction after his commanding officer Vial was mortally wounded. He joined the invasion of Andalusia in January 1810 where during the march on Seville he won a fine action at d'Alcala del Réal on 28 January. He remained with Milhaud till his promotion to *général de brigade* on 6 August 1811 and appointment as military governor of Granada.

RUSSIA 1812

In December 1811 he left Spain and joined Oudinot's II Corps on the Elbe as commander of the 6th Light Cavalry Brigade. Present in Russia he was at the crossing of the Drissa on 28 July and fought at Polotsk on 18-19 August 1812. He shot to fame for his role played in the crossing of the Beresina during the retreat from Moscow. Detached to Wrede's command, he was covering the approaches to Vilna when Oudinot ordered him to rejoin east of the Beresina. Reaching the crossing at Borisov, Corbineau found it barred by Tshitshagov's army. He moved upstream and found a ford at Studenka, crossed and rejoined Oudinot on 22 November.

Till then he was unaware of the crisis the army faced trapped with its back to the river. His arrival was a godsend to Napoleon, who was immediately told of the lightly defended ford's existence. As a result, Napoleon decided to use Studenka as the main crossing point for the Army. Whilst Oudinot made numerous demonstrations against other crossing points, during the early hours of 26 November Corbineau stormed across the ford and cleared the opposite bank. Eblé's engineers followed and working shoulder deep in the ice flows completed a pontoon enabling Oudinot's troops to cross and secure a bridgehead for the rest of the Army to follow.

Eblé and his bridging train won immortal fame for the sacrifices they made at the Beresina, but had Corbineau not found the ford the army to a man would have perished. Napoleon appreciated this and after his departure recalled Corbineau to Paris on 26 January 1813 and made him an Imperial Aide.

GERMANY 1813

Corbineau was with Napoleon throughout the Spring Campaign of 1813 in Saxony. He fought at Lutzen on 2 May 1813 leading one of the four Young Guard infantry columns that broke Wittgenstein's center at the end of the day. At Bautzen on 20-21 May he was again in the thick of the fighting. Promoted *général de division* on 23 May 1813, he renewed his association with Latour Maubourg when he took command of the 1st Light Cavalry Division of I Cavalry Corps. The renewal of hostilities in August saw him assigned to Vandamme's I Corps at Pirna as head of the corps cavalry. Engaged in the pursuit of the Allies after Dresden on 25-26 August he was with Vandamme* when I Corps was trapped at Kulm on 30 August. Wounded in the battle he led a bold charge that punched a hole through the Prussian line that managed to save a large number of troops. With Vandamme captured he led the remnants of I Corps until replaced by Lobau* on 3 September 1813. He resumed his duties with Napoleon for the rest of the campaign taking on the additional responsibility of commander of the Imperial Guard Gendarmes d'Elite.

FRANCE 1814

During the 1814 Campaign in France he won Napoleon's esteem after the battle of Brienne on 30 January where his quick thinking saved Napoleon and his staff from a band of marauding Cossacks that had overpowered his escort. Given command of the 2nd Guard Cavalry Division he recaptured Rheims on 5 March and was installed as governor. He led a valiant defence of city from 11-13 March when he held out against 15,000 Russians with a garrison of 3,000 men, comprising half-armed National Guardsmen and two battalions of the Young Guard. After Rheims he was made *comte de l'Empire* and rejoining Napoleon fought at Arcis-sur-Aube on 20-21 March. Wounded by a shell splinter to the head, whilst unfit he remained with Napoleon during the final desperate race to save Paris, the mutiny of the Marshals, his abdication and final farewell to the Guard at Fontainebleau.

THE RESTORATION AND NAPOLEON'S RETURN

He was reconciled with the Bourbons who made him a *Chevalier de Saint Louis* on 19 July 1814. The award meant little as it was followed on 1 September 1814 by him being placed on half pay. Out of work, not surprisingly he rallied to Napoleon on his return and resumed his duties as an Imperial Aide on 20 March 1815. In April he was sent to Lyons to deploy the National Guard against the Duc d'Angouleme's forces. He spent May in the Vendée at the head of the Gendarmes d'Elite helping Delaborde put down the unrest in the region.

THE WATERLOO CAMPAIGN AND AFTERMATH

He joined Napoleon with the Army of the North and served at Ligny on 16 June and Waterloo on 18 June 1815. Part of the great entourage of officers that surrounded the Emperor there is nothing to suggest that he did not carry out his duties as expected. He was at Napoleon's side at the end of the battle and with Bertrand urged him to leave the field. His devotion towards Napoleon in the final days was only matched by that of Bertrand. He was with Napoleon during the flight to Paris, then at his side throughout the Second Abdication Crisis, which ended on 22 June and then moved with him to Malmaison.

Napoleon advised him not to seek exile with him, so he rejoined the Guard before Paris on 2 July. After the conclusion of the Armistice on 4 July he retired with the army to Orleans. On 1 August 1815 he was placed on the non-active list, stripped of his titles and forbidden to wear his uniform or decorations. At the end of 1816 his titles were restored but he remained unemployed. He saw no future under

Charles X at his own request retired from the army on 1 December 1824.

LATER LIFE AND CAREER ASSESSMENT

Immediately after the July Revolution of 1830 he received notice of his recall and took the appointment as commander of the 16th Military Division at Lille on 1 August 1830. He became a Peer of France on 11 September 1835 and was then awarded the Grand Cross of the Legion of Honor on 5 May 1838. He remained loyal to Louis Philippe and was responsible for the arrest of Louis Napoleon on 6 August 1840 after his landing at Boulogne in a theatrical attempt to overthrow the monarchy.

During his latter years he maintained a keen interest in the welfare of Imperial Guard veterans. An example was on 15 August 1841 when he presided over the dedication of Boiseu's statue of Napoleon in Boulogne commissioned by former members of the Marines of the Guard. Seeing little future after the fall of Louis Philippe he readily retired from the Army on 30 May 1848. He survived long enough to see Louis Napoleon become President of France on 2 December 1848 and died in Paris on 17 December 1848.

Corbineau, whilst very capable and brave, had a marked degree of luck. He had the knack of being at the right place at the right time. His discovery of the ford over the Beresina was a classic case. This, followed a year later by his rescue of Napoleon at Brienne and his brief defense of Rheims, all went towards building a reputation as a commander, who could handle a difficult situation. On closer examination he never commanded a formation greater than a brigade for any period of time. He was an Imperial Aide throughout the period he held the rank of *général de division,* and in this role he carried with him the direct authority of the Emperor to ensure orders were carried out. This often left little room for exercising imagination or initiative, but was a solid foundation to build a reputation if lucky. Fortune was kind to him, he had the lucky breaks, which resulted in recognition from Napoleon that was greater than he rightly deserved.

CORBINEAU AT NAPOLEON'S FAREWELL TO THE GUARD IN THE COURTYARD AT FONTAINEBLEAU, 20 APRIL 1814.
FROM LEFT TO RIGHT, CORBINEAU, BARON FAIN, NAPOLEON, GENERAL PETIT.
(DETAIL FROM VERNET'S PAINTING)

DEJEAN - PIERRE FRANÇOIS
(1780 1845)

BACKGROUND AND EARLY CAREER

The imperial Aide and cavalry general Pierre Dejean was born at Amiens on 10 August 1780. He was the son of Jean François Aimé Dejean, the celebrated general of engineers who served as Napoleon's Minister for Administration of War from 1802 till 1810. Pierre's early career, like that of the younger Kellermann, whose father had also won fame in the early years of the Revolution, rode on the shoulders of his father's successes in the new egalitarian era.

His career started in March 1795 when his father the newly appointed Inspector General of Fortifications for the Army of the Sambre-Meuse made him his unofficial aide de camp to teach him the rudiments of military life. He then gained a commission as an infantry *sous lieutenant* on 5 August 1796 and exactly a year later was promoted *lieutenant*. Still with his father, who replaced Beurnonville as commander of the Army of the North on 16 September 1796 he saw service in Holland. The benefit of patronage that helped young Dejean came to an abrupt end on 12 September 1797 when his father was ousted as army commander after he refused to support the conspirators during the Fructidor coup d'etat. Unemployed and the son of a disgraced general, his future looked bleak. He however secured a posting with the 8th Demi-brigade and remained with them till 30 September 1798 when he passed to the regimental staff of the 28th Line.

Dejean's fortunes changed dramatically under the Consulate, when his father after a period of disgrace had his rank and privileges restored and became a Councillor of State. Pierre benefited immediately being called from his regiment in May 1800 to act as his aide. After the Marengo campaign he spent time in northern Italy with his father, who headed a commission set up to ensure compliance with the terms of the Convention of Alessandria. A period followed in Genoa where his father was responsible for restoring the Ligurian Republic. He was promoted *capitaine* on 2 January 1801. The appointment of his father as Minister of Administration for War resulted in the continuance as his aide from 12 March 1802 till 3 October 1803. Then anxious to obtain a cavalry command he secured a transfer to the 20th Dragoons with the Army of the Ocean Coasts at the Camp of Montreuil.

THE GRANDE ARMÉE 1805-1807

Within the close confines of regimental life in a peacetime army he found advancement difficult. His promotion to *chef d'escadron* on 22 September 1805 and a move to the 3rd Dragoons undoubtedly had the involvement of his father. During the Austerlitz Campaign his regiment was with the 2nd Brigade of Walther's 2nd Dragoon Division of Murat's Cavalry Reserve. He fought at Hollabrünn on 16 November and was present at Austerlitz on 2 December 1805. During the Prussian Campaign his regiment still with the 2nd Dragoon Division now led by Beker he fought at Jena on 14 October 1806, Prenzlow on 28 October and was present at the capture of Lubeck on 6 November 1806. In Poland he fought at Pultusk on 26 December 1806 and at Eylau on 8 February 1807 where he took part in Murat's great charge of the Cavalry Reserve that saved the day after Augereau's corps had been routed. He was promoted *colonel* of the 11th Dragoons on 13 February 1807 replacing their colonel who had died at Eylau. He adjusted easily to the new command as the 11th Dragoons were brigaded with his old regiment under Cavrois. He served with distinction at Friedland on 14 June 1807 where his horsemen operating under Grouchy's command were very effective on the French left and later in the pursuit of the Russians to the Niemen.

SPAIN, PORTUGAL AND RUSSIA 1808-1812

His regiment, still part Cavrois's brigade, was formed into a division under Kellermann* for the French invasion of Spain. For two years he was based at Valladolid with the Cavalry Reserve. He did not take part in any of the major campaigns of the period being more involved in secondary operations against guerillas, which gradually hardened his troops into a fine force. In September 1810 he joined Massena's invasion of Portugal, when his regiment over 700 men strong was attached to Montbrun's Cavalry Reserve. The attrition during the campaign was terrible, and by July 1811 he mustered barely 160 men, having not fought in a single major action. His only consolation was his promotion to *général de brigade* on 6 August 1811.

On 25 December 1811 he took command of the 3rd Brigade (11th Cuirassiers) of Valence's 5th Cuirassier Division of the Army of Germany. In Russia the division served with Nansouty's I Cavalry Corps. He fought at Mohilev on 25 July 1812 and at Borodino on 8 September where posted on the French right he supported the assaults of Poniatowski's VIII Corps. Surviving the retreat from Moscow he briefly led a provisional brigade of Polish lancers (1st and 3rd) in February 1813 before being recalled to Paris.

IMPERIAL AIDE DE CAMP 1813-1814

On 20 February 1813 he was made an Imperial Aide. There was little reason why he merited the appointment before others, but then his father calling on past favours obviously had some influence. He accompanied Napoleon throughout the Spring Campaign in Germany being present at the battles of Lutzen and Bautzen. On 6 August he took command of the 1st Brigade of the Guard Light Cavalry Division formed from the Garde d'Honneur, which he led at Dresden and Leipzig.

The Autumn Campaign in Germany lost, as an Imperial Aide he was sent to Hunningen on the Rhine in December to strengthen the defences of that key river crossing. In January 1814 he moved to Lorraine where he organized a successful levee en masse. He rejoined Napoleon's army in Champagne and fought at Montereau on 18 February 1814. On 23 March he gained promotion to *général de division*. Four days later Dejean was ordered by Napoleon to Paris to stiffen his brother Joseph's resolve and ensure the capital was held at all costs. When he reached Paris on 30 March Joseph had fled and he was unable to prevent its capitulation.

THE WATERLOO CAMPAIGN

The Bourbons retained his services after Napoleon's abdication, and even went as far as to confirm his rank as *lieutenant général* back dated to his original promotion by Napoleon. He was also made a *Chevalier de Saint Louis* on 17 September 1814, but it all meant little since he remained unattached with no command. As a result, on Napoleon's return he had little crisis of conscience rallying to him and resumed his post as an Imperial Aide on 21 March 1815.

Present at Ligny on 16 June 1815 he was party to the confusion that led to d'Erlon's Corps not appearing during the battle. When it was reported to Napoleon in the afternoon that a hostile column threatened the French left, the Emperor sent Dejean to identify the force. Noting Napoleon's agitation and the need to obtain a prompt response he rode only close enough to identify it as d'Erlon Corps and immediately returned to report his findings. Had he used more initiative and spent another ten minutes trying to locate d'Erlon in order to urge that the Emperor required him to move through Wagnelée to attack the Prussians, the whole outcome of the campaign could have been very different.

At Waterloo on 18 June 1815 as an Imperial Aide he spent the day ensuring the Imperial will was carried out. He also found himself party to one of the most despicable acts of the day, which revealed the callous side to Napoleon's nature. In the evening Napoleon aware the troops visible in the distance were Prussian ordered Dejean to pass a message on to Ney that the approaching troops were Grouchy's. He duly galloped off, found Ney who in turn sent his aides down the line urging d'Erlon's battered divisions forward for one last effort as help was to hand. The ruse backfired horribly as Prussian shells started to plough into their rear. Assured victory was theirs, whipped up to fever pitch, screaming "Vive l'Empereur" as they advanced, raked by British volleys, their brittle morale snapped. Their first thoughts were that Grouchy had defected, to cries of "traison" they turned and fled before the terrible truth dawned that the Prussians had arrived. It arguably had a more devastating effect on the morale of the weary infantry than did the repulse of the Old Guard, which happened at the same time.

Dejean left the field with Napoleon and on 20 June arrived with him at Laon. Here the Emperor sent him to Avesnes to rally the troops there and organize some form of defence to delay the Allied advance. He spent a week stiffening the morale of the border garrisons and preparing them to meet the Allied onslaught before he returned to Paris.

LATER LIFE AND CAREER ASSESSMENT

Suspended and banished from France as a result of the Ordinances of 24 July he lived for varying periods in Styria, Croatia and Dalmatia till allowed to return to France at the end of 1818. On 1 January 1819 he was reinstated but remained unattached. On the death of his father on 14 June 1824 he became a Peer of France and sat with the Liberals in the Chamber. The accession of Louis Philippe resulted in his recall to duty on 29 October 1830. He led a cavalry division during the invasion of Belgium in August 1831. In 1832 he became an Inspector General of Cavalry in the 16th Military Division at Lille but soon left to rejoin the Army of the North. He served at the siege of Antwerp, which ended with the surrender of the citadel on 23 December 1832.

After Belgium on 15 August 1833 he was appointed commander of the 2nd Cavalry Division based at Saint Omer. In July 1837 he moved to the Camp of Compiègne as cavalry commander. He became Inspector General of the 5th Arrondissement in May 1838 and in June 1840 moved to the 7th. He was appointed president of the Cavalry Committee of the Council of War in December 1840. In June 1841 he became Inspector General in the 1st Arrondissement followed by command of the 3rd Cavalry Division in April 1842. In June 1843 he took over the 3rd Arrondisement. He was awarded the Grand Cross of the Legion of Honor on 14 April 1844. He was still on the active list when he died in Paris on 17 March 1845.

Despite relying on nepotism during his early career Dejean went on to become a fine regimental and brigade commander. As an Imperial Aide he was devoted and brave. He never led a division on the battlefield during the Napoleonic Wars, which leaves the question as to his true capabilities in charge of large formations unanswered.

DELORT - JACQUES ANTOINE ADRIEN, BARON (1773-1846)

EARLY CAREER

Delort was the fine Peninsular veteran and dragoon general, whose cuirassiers were recklessly thrown against the Allied squares by Ney at Waterloo. He was born in the village of Arbois (Jura), Eastern France on 16 November 1773. He enlisted with the 4me bataillon du Jura on 14 August 1791 and quickly rose through the ranks. On 16 June 1792 he gained a commission as *sous lieutenant* with the Austrasie Regiment (the 8th Line), after his previous unit had disbanded earlier in the year. His first active service was with the Army of the Rhine and then Kellermann's Army of the Center. Promoted *lieutenant* on 18 September he fought at Valmy on 20 September 1792.

From June 1793 till January 1797 he served as a staff officer for varying periods with the Armies of the Côtes de la Rochelle, Alps, Pyrenees Orientales and the Interior. During this time he distinguished himself in several minor engagements and gained promotion to *capitaine* on 28 August 1793. Ill health later forced him to retire from the Army on 10 January 1797.

ITALY 1797-1803

He refused to accept a discharge and on 21 October 1797 returned to duty. Posted to the Army of Italy he joined the general staff of Serurier's division, which occupied Venice. Keen to serve in the cavalry he secured a transfer to the 22nd Cavalry Regiment on 8 January 1798 as regimental assistant chief of staff.

The renewal of war with Austria saw him distinguish himself on 26 March 1799 at Pastrengo on the shores of Lake Garda. Promoted acting *chef d'escadron* on the battlefield, the Directory later confirmed his rank on 23 April 1799. Present with Serurier's command throughout the campaign he was at the crossing of the Adige on 30 March and the French descent on Verona. When Serurier's force of 6,000 was surprised by over 15,000 Austrians outside the city and repulsed he fought with the rearguard distinguishing himself at Villafranca on 5 April. Then as Suvorov's Russians joined the offensive Delort's horsemen helped cover the retreat of the out-numbered French. On 27 April he avoided capture at Verderio where the remnants of Serurier's force were trapped with their backs to the Ticino and surrendered.

Still with the 22nd Cavalry he was promoted senior *chef d'escadron* on 21st January 1800. His regiment spent the Winter and Spring of 1799-1800 covering the Var riverline with the battered remnants of the Army of Italy. During the Marengo Campaign his troops continued to maintain a defensive role behind the river and took no active part in the events that unfurled further east in Northern Italy. After Marengo on 14 June 1800 and the signing of the Convention of Alessandria he remained in garrison in Northern Italy. The renewal of hostilities the following December saw his regiment join Brune's invasion of Venetia. He distinguished himself at Monzembano on 26 December 1800 and was at the recapture of Mantua the next day.

He remained in Italy after the peace from 1801 to 1803 spending varying periods of garrison duty at Lodi, Saluces and Turin. The disbandment of the 22nd Cavalry on 23 September 1803 resulted in promotion on 29 October 1803 to *lieutenant colonel* and a transfer to the 9th Dragoons with the Army of the Ocean Coasts at Brest.

THE AUSTERLITZ CAMPAIGN

The activation of the Grande Armée in August 1805 saw his regiment with Beaumont's 1st Dragoon Division of Murat's Cavalry Reserve. In the campaign against Austria the 9th Dragoons joined the great cavalry screen that plunged into the Black Forest to mask the Grande Armée's envelopment of Mack's advance into Bavaria. He played a key role in the first major action of the campaign at Wertingen on 8 October 1805. After they crossed to the Danube's southern bank the previous day his cavalry surprised an Austrian force of 6,000 under Auffenburg near the village. He took charge of the regiment after his colonel, Maupetit fell in the initial charge. Then by the skilful use of feints and counter feints pinned down the enemy infantry till the arrival of reinforcements ensured their surrender.

He retained command of the regiment for the rest of the campaign fighting at Austerlitz on 2 December 1805 where he distinguished himself during Murat's charge of the Cavalry Reserve that drove back Bagration's threat to the French left. In the battle he took two severe lance wounds at the hands of Cossacks, which laid him low for several weeks.

ITALY AND SPAIN 1806-1810

On 8 May 1806 he was promoted *colonel*. He returned to Italy as commander of the 24th Dragoons serving with Massena's Army of Naples. Here he helped to reduce the few remaining fortresses in Calabria. His regiment later rejoined the Army of Italy in April 1807 and he spent a comparatively peaceful time in garrison at Lodi. In October 1808 he moved to Perpignan near the Spanish border as French forces under Saint Cyr gathered to relieve Duhesme's forces cut off in Barcelona.

After crossing the border he fought at Cardedeu on 16 December, exploiting the Spanish rout after Saint Cyr's infantry punched a hole through Vives's center that opened the way to Barcelona. He was again in action after the relief of the city when Saint Cyr on 21 December 1808 at Molins del Rey cut Reding's army to pieces south of Barcelona as it tried to retake the city. In a vigorous pursuit that followed his dragoons continued the chase till they reached the gates of Tarragona.

Delort returned to Northern Catalonia and spent time covering the army's communications as they laid siege to Rosas and Gerona. On 25 February 1809 he routed a force of Spanish cavalry at Valls. In the action Reding was slain by his dragoons as he tried to rally his fleeing troops. Delort was shot in the right leg during the engagement and was put out of action for several weeks.

Delort spent the Winter of 1809 with Augereau's forces covering the third attempt to reduce Gerona. He helped seal the city's fate on 17 November when his horsemen joined a raid by Pino's Italian division on the town of Hostelrich. His troops put to the torch vast supplies of provisions gathered by the local Junta to replenish the starving garrison at Gerona.

Attached to Souham's command, he played a vital role overcoming a force led by O'Donnell that surprised Souham's division at Vich on 20 February 1810. Timing his charge to perfection as the enemy were committed against the French infantry, he broke up their assault, leaving over 800 killed and wounded, and taking another 1,000 prisoner. Again at Villafranca and Manresa near Tarragona on 30 March he saved two French columns under Schwartz that O'Donnell had badly mauled, escorting them to the safety of Barcelona. On 4 September 1810 he won another sharp action at Cervera. Finally on 30 October 1810 for his stalwart service in Catalonia he was made a *chevalier de l'Empire*.

THE ARMY OF ARAGON

He joined Suchet's Army of Aragon and served at the successful siege of Tortosa during December 1810. On 4 January 1811 further recognition came when he was made a *baron de l'Empire*. Present during Suchet's advance on Tarragona on 15 January 1811 at Valls he rescued the Italian brigades of Eugenio and Palombini after they had rashly taken on a Spanish force over twice their size led by Sarsfield. At the head of only two squadrons, being all he had available, he joined the action and overwhelmed 700 enemy horse. It forced Sarsfield's infantry to halt and enabled the demoralized Italians to seek the safety of the town. Seriously wounded in the action Delort returned to France for a period of leave to recover.

On 28 June 1811 back in the saddle he took part in the final assaults on Tarragona. His dragoons covered the seaward exit from the city and caused mayhem as the enemy tried to reach the safety of the British ships in the roadstead. Suchet was awarded his marshal's baton, and a reorganization of the Army of Aragon followed with Delort on 21 July 1811 receiving long overdue promotion to *général de brigade*. He took charge of Boussard's brigade comprising his old regiment the 24th Dragoons, the 13th Cuirassiers and the 4th Hussars. On 23 September he captured the bridge over the Mijares at Villareal enabling Suchet's army to pour across the river into the province of Valencia. He fought at Saguntum on 25 October when Suchet crushed Blake's attempt to relieve the fortress. In the battle Delort's cavalry posted on the French left drove the enemy horse from the field. Then rampaging behind the enemy lines he helped Harispe, who had met strong resistance from the Spanish center, breach their line. During the crossing of the Guadaluviar that led to the final investment of Valencia on 25 December 1811 his cavalry crossed upriver and took Torrente sealing the southern escape routes from the city. He then moved to Alcira on the Xucar on 29 December, securing the last major river crossing before Alicante. On his return to Valencia he joined the brief siege that lasted till 11 January 1812.

OPERATIONS IN VALENCIA AND ALICANTE

During the first half of 1812 Delort was engaged in several punitive sallies into Alicante, which yielded little due to the limited number of troops available. Suchet's plans were further curtailed by recurring troubles in Catalonia and troops being recalled to France for the invasion of Russia. In this difficult time Delort did however achieve his finest performance in the field at Castalla on 21 July 1812. O'Donnell in charge of three converging columns numbering over 11,000 infantry and cavalry tried to trap his small force of 1,200 men made up of two infantry battalions and a squadron of 13th Cuirassiers at Castalla. Warned of the enemy's approach Delort evacuated Castalla and took up a defensive position outside the town. His defensive position on a hillside was covered by a stream and a ravine crossed by a narrow bridge. Here he waited for reinforcements while O'Donnell in a very dilatory manner launched his assaults. The first Spanish column as it skirted the ravine was charged by 400 men of Delort's 24th Dragoons. A large olive grove had screened their arrival and approach. Driven back in confusion they crashed into the second column crossing the bridge. Delort's infantry then charged the third column of 6,000 infantry that formed O'Donnell's center, which soon turned into a mass of fugitives. In the action the Spanish lost over 3,000 men, including 2,300 prisoners. Delort then proceeded to lay waste the country between Alicante and the Xucar, occupying Aspe, Jumilla and Villena. Bad generalship aside on the part of O'Donnell, the action was still a masterpiece in deception and timing. Delort conducted a withdrawal to a strong defensive position, then surprised an overconfident, numerically superior enemy when they were at their most vulnerable.

With opposition in the region broken, Delort's troops spent a quiet winter along the Xucar. In April he took part in Suchet's offensive against the Anglo-Sicilian forces that landed at Alicante. He was responsible for the destruction of two Spanish infantry regiments under Elio at Yecla on 11 April 1813. When Suchet moved north with the bulk of his army to counter Murray's landing near Tarragona on 2 June 1813, Delort remained on the Xucar under Harispe. When his outposts detected a cumbersome Spanish advance across their front in support of the Allied landings to the north, Harispe launched a lightning counter stroke across the river. Delort in charge of the advance guard on 13 June caught the Spanish columns at Carcagente near Alcira. He routed three Spanish brigades that lost over 1,500 men. The Spanish threat to Suchet's rear eliminated, Delort won valuable time for Suchet to counter Murray's landing further up the coast.

News of the French defeat at Vitoria changed the whole strategy in eastern Spain and Suchet had no alternative but to abandon Valencia. Delort covered the army's withdrawal as it fell back to new positions on the Llobregat south of Barcelona. The Spanish recovering from their reverses on the Xucar allowed the French to complete an orderly evacuation of Valencia province. Once the position stabilized, Delort recrossed the Llobregat on 15 August and helped in the relief of the Tarragona garrison. He fought at Ordal on 13 September when Suchet brought Bentinck's slow advance on Barcelona to a rude halt. The English force surprised in camp was hurled back to Tarragona in confusion and remained no particular threat for several months.

FRANCE 1813-1814

In December 1813 with further calls on French forces in Spain after the Autumn disasters in Germany, Delort was recalled to France. On 9 January 1814 he took command of a mixed brigade of the Paris Reserve Division. He served in Champagne and fought at Montereau on 18 February 1814. He distinguished himself in a decisive charge led by Pajol* that drove Schwarzenberg's forces back across the Seine and Yonne rivers. Promoted *général de division* on 26 February 1814 he replaced the wounded Pajol as head of the 2nd Cavalry Division with Saint Germain's II Cavalry Corps. His last action of the campaign in France was on 26 March 1814 at Saint Dizier where with Trelliard they broke Winzingerode's cavalry.

THE RESTORATION AND NAPOLEON'S RETURN

As he was not an avid Bonapartist, the Bourbons made him a *Chevalier de Saint Louis* on 19 July 1814, but it did not stop them placing him on the non-active list on 1 September 1814. News of Napoleon's landing resulted in his recall on 6 March 1815, and with foreboding he joined Ney's forces that marched south to oppose Napoleon's advance on Paris. On 18 March when Ney defected at Auxerre his sense of duty prevented him rallying to Napoleon and he disappeared. He only made himself available when Louis XVIII was safely in Belgium and the new government formally proclaimed.

Napoleon's inner circle was against his employment, in particular Flahaut who had some influence over who received commands. However Davout on the recommendation of Suchet, who recognized Delort's talents, approved his recall on 23 April 1815. He received a fine command, the 14th Cavalry (Cuirassier) Division with Milhaud's IV Cavalry Corps. It was one of the best equipped and highly motivated cuirassier formations to take the field for many years.

THE WATERLOO CAMPAIGN

Present at Ligny on 16 June, his division was held in reserve during the early stages of the battle. In the gathering dusk he moved forward and supported the attack by the Old Guard that broke the Prussian center. Pursuing the Prussians he came up against several regiments of cavalry led by Blucher, who was trying to rally his troops. In the action around Le Loup Woods Blucher was unhorsed and trapped on his own under his horse. Delort's horsemen put the enemy to flight but failed to notice the plight of the Prussian leader. Had luck been with them and Blucher fallen into their hands the outcome of campaign would have been very different.

At Waterloo on 18 June 1815 his troops were first held in reserve in the rear to the left of the Charleroi Road. In the early afternoon two of his regiments, the 6th and 9th Cuirassiers supported Jacquinot's lancers against Ponsonby's Scots Greys and Dragoon Guards. Charging down the slopes from La Belle Alliance they swept the valley in pursuit of Wellington's heavy dragoons. They continued up the slopes of Mont St Jean to beyond La Haye Sainte before being halted by enemy fire.

Disaster came around 3.30 P.M.. when the sequence of events began that led to the destruction of Delort's division. Ney misread a slight change of alignment by Wellington as a sign he was about to order a retreat. In an excitable state of mind he ordered Delort's cuirassiers to charge and occupy the heights. Delort sensed a cavalry charge was unwise due to the muddy terrain. No infantry or artillery support was nearby, and convinced the enemy was not retreating, he objected. The affront maddened an already jittery Ney, who countered with an order for Milhaud to attack with the whole of his corps. In turn they were followed by the Guard Light Cavalry, who instinctively trotted forward without being ordered to do so. Thus from a small disagreement between Delort and Ney, over 5,000 cavalry were inexorably committed prematurely without adequate infantry or artillery support.

Preceded by artillery fire they started towards the slopes of Mont St Jean in an oblique movement that in turn made their flanks vulnerable to enemy artillery. The horses made heavy weather of the climb through the muddy, heavily trodden terrain and the column's advance was brought to a crashing halt with each salvo. Spurred on by trumpets sounding the charge, the cuirassiers rushed the cannon and captured the crewless batteries one after the other. When they reached the crest they were met a wall of bayonets formed by the Allied infantry in square. Lunging and plunging at the infantry whilst suffering from their fire they were hit by Wellington's cavalry before falling back to reform and repeat the charge.

After four charges, his men exhausted, they failed to effect a breakthrough. Delort shot in the leg and sabred in the arm pressed forward each time with his men. His troops needlessly thrown away were too disordered and exhausted to support the Guard in their final assaults. Delort managed to restore some order by the end of the day and withdrew his division from the field intact. During the retreat he received replacement squadrons and helped cover the Army's retreat as it regrouped and fell back by stages to Paris.

LATER LIFE AND CAREER ASSESSMENT

He remained with the Army during its disbandment. Suspended on 18 August 1815 he retired to his home at Arbois. On 1 April 1820 he was reinstated but remained unattached. He was retired with several hundred other officers as a result of the ordinance passed on 26 January 1825. It was the work of a government that had become increasingly paranoid about Bonapartist influence within the Army.

A supporter of the Orleanists after the July Revolution he was recalled on 6 August 1830 and became commander of the 8th Military Division at Toulon. On 3 July 1831 he moved to the 3rd Military Division at Lyons where he replaced Bachelu, whose republicanism had caused considerable unrest within the garrison. He had the unenviable task of restoring discipline and repairing the damage in morale, which he did and won the gratitude of an appreciative Minister for War Soult. Interested in politics he had previously gained election as a Centrist Deputy representing the Jura on 30 October 1830.

Many military appointments over the years followed. They ranged from commander of Military Divisions to Inspector General of the military schools at Saint Cyr and Saint Germain. He also sat on numerous commissions engaged in proposing reforms within the Army. Honorary posts he held included being aide-de-camp to Louis Philippe in 1832 and 1834. He was re-elected deputy for Poligny in June 1834. On 30 May 1837 he received the Grand Cross of the Legion of Honor, which followed with him becoming a Peer of France on 3 October 1837. He retired to the Reserve Section of the General Staff on 17 November 1841. He spent most of his remaining years at Arbois where he died on 28 March 1846.

Delort was neither a Bonapartist or a Royalist, his duty was firstly to his profession, the Army. He was one of few commanders whose reputation remained unblemished throughout his career. He was a fine example of a medium and heavy cavalry commander, whose daring and initiative became a legend in Spain. He had a good eye for terrain and knew how to use it to maximum effect. On the battlefield in charge of cavalry, he had a wonderful sense of timing for the decisive charge. As the years passed he built up a reputation as one of the foremost cavalry generals in the Army. His great misfortune was that he was largely side-lined to Italy and Spain after 1806. Had he campaigned with the Grande Armée after then, there is little doubt he would have risen to higher command far faster than he did.

**DESVAUX DE SAINT MAURICE –
JEAN JACQUES, COMTE
(1775-1815)**

EARLY CAREER

The artillery general Jean Desvaux was born in Paris on 26 June 1775. He attended the Chalons Artillery School where he gained a commission as a *sous lieutenant* on 10 March 1792. Posted to the 4th Foot Artillery based at Chalons he spent a period at the regimental depot. Promoted *2me lieutenant* on 1 September 1792 he then joined the Army of the Alps. He gained credit for the way he handled his men during the occupation of Savoy in March 1793, then at the difficult siege of Lyons in July. Undoubtedly talented promotion was swift, *lieutenant* on 1 December 1792, *adjutant major* on 31 July and then *capitaine* on 22 September 1793, barely months after his eighteenth birthday.

The following year he moved to Dugommier's Army of the Pyrenees where he played a key role during the bombardment of fort Saint Elmé before its fall on 23 May 1794. In January 1796 with Spain knocked out of the Coalition he became aide de camp to General Saint Rémy, who commanded the artillery in Paris. On 19 December 1797 he joined the 2nd Horse Artillery Regiment and the following month came into contact for the first time with General Bonaparte when his company joined the Army of England. The invasion plans for England were a diversion for the expedition to Egypt, and Desvaux transferred to the Army of Italy during the Winter of 1798-99.

After the outbreak of hostilities in Northern Italy, he took part in the heavy fighting with Serurier's division covering the retreat to the Adige on 30 March 1799. On 23 April 1799 he gained promotion to *chef d'escadron*. Present throughout the campaign he tasted the despair of defeat for the first time on 15 August 1799 at Novi when Suvorov's combined Russo-Austrian army crushed Joubert.

In March 1800 on Saint Rémy's appointment as the Army of the Reserve's artillery commander, Desvaux joined him as his aide at Dijon. When Saint Rémy fell ill and was replaced by Marmont on 15 April, Desvaux continued as the new commander's aide. He took part in the crossing of the Alps and fought at Marengo on 14 June 1800. In November 1800 he joined the 8th Horse Artillery Regiment, leading a company during Brune's offensive by the Army of Italy across the Mincio in December 1800.

He distinguished himself at Treviso on 26 December 1800. After the campaign he rejoined Marmont, who headed the army's artillery, remaining on his staff till January 1802 when he returned to his regiment. Promoted *major* on 1 May 1803 he moved to the 8th Foot Artillery Regiment. He remained with them till 29 October 1803 when he was promoted to colonel of the 6th Horse Artillery Regiment. He resumed his association with Marmont when he rejoined him as his senior aide in February 1804 at the Camp of Utrecht.

THE GRANDE ARMÉE 1805-1808

Desvaux served with Marmont during the Austrian Campaign, receiving a slight shrapnel wound in the arm before Ulm on 19 October 1805. Captured by the Austrians in a skirmish near Judenburg on 3 November he was taken to Graz. He remained prisoner till freed by Massena's Army of Italy when it occupied the city on 27 November 1805. Resuming his duties with Marmont, in August 1806 he became Artillery Director for the Army of Dalmatia. In March 1807 he returned to France as *colonel* of the 4th Foot Artillery Regiment till a posting with the Army of Italy followed in February 1808.

THE AUSTRIAN CAMPAIGN OF 1809

During the Austrian campaign he initially served as the artillery commander for Grenier's corps operating as the left wing of the Army of Italy. He fought at San Daniele on 11 May, Tarvis on 18 May and Saint Michael on 25 May 1809. At the Battle of Raab on 14 June his supporting fire was crucial in helping the divisions of Durutte and Severoli overcome the Austrian center around the strongpoint of Kis-Megyer. He fought at Wagram on 5-6 July 1809, where his cannon did great work supporting Macdonald's assault on Sussenbrunn that secured victory and won a marshal's baton.

The campaign concluded, Desvaux received promotion to *général de brigade* on 9 July 1809. A week later, on Marmont's recommendation he was appointed commander of the Imperial Guard Horse Artillery and given the task to form four companies. On 30 October 1810 he became a *baron de l'Empire*.

THE RUSSIAN CAMPAIGN OF 1812

He led the four batteries of the Guard Horse Artillery during the campaign in Russia. He fought at Smolensk on 17 August and Borodino on 7 September 1812, suffering heavy losses in both battles. During the retreat from Moscow he managed to keep his guns intact till they reached Krasnoe on 17 November. With most of his horses dead, he had to abandon many guns as they could not be manhandled further. During the crossing of the Beresina on 27-28 November his remaining guns put up a sterling performance, where with the Guard Foot Artillery they managed to keep the road to Vilna open.

GERMANY 1813

Surviving the retreat, he worked ceaselessly to cobble together into some form of order the few remaining artillery pieces with the army. His task was made all the more difficult by the perilous state of the army. It never remained in one place more than a few days as it fell back through Poland and Prussia to the left bank of the Elbe. Recalled to Mainz in February 1813, he set about rebuilding the Guard Artillery at temporary depots established there. By mid April 1813, six horse companies were able to take to the field. On 10 April he joined Dumoustier's 1st Guard Division in charge of five foot and two

horse companies making up 52 cannon. He fought at Lutzen on 2 May 1813, where his guns were part of the 80 piece battery that dominated the French center. His guns gave good support to the Young Guard assaults that broke the Allied lines between Starsiedel and Kaya. Heavily engaged at Bautzen on 21-22 May his softening up of the Prussian redoubts between Besankwitz and Kreckwitz was vital again in enabling the Young Guard successfully to drive back the Allies.

After the end of the Armistice he resumed his place at the head of the Guard Horse Artillery. He fought at Dresden on 26 August 1813 supporting Mortier's assault against Leubnitz. At Leipzig on 16 October 1813 he supported Murat's mass cavalry charges before Wachau. During the retreat to the Rhine he fought at Hanau on 30 October. His guns were part

HORSE ARTILLERY OF THE GUARD IN ACTION

61

of the great battery formed by Drouot* that cut its way through the woods and blew away Wrede's attempt to cut off the French retreat.

FRANCE 1814

On 6 November 1813 he was promoted *général de division*. He left the Army on 30 November and proceeded to Saint Etienne to raise further artillery companies for the campaign in France. On 5 January 1814 Augereau appointed him artillery commander for his forces gathering at Lyons. In the campaign Augereau found Desvaux a continual sore in his side. The young general made himself unpopular by continually urging more aggressive action against the Austrians in order to relieve pressure against Napoleon further north. Desvaux fought Hesse Homberg's forces at Saint Georges on 18 March 1814, and again at Limonset two days later, before weight of numbers drove the French from Lyons on 23 March 1814.

THE RESTORATION AND NAPOLEON'S RETURN

He accepted service under the Bourbons and received the title *Chevalier de Saint Louis* on 20 July 1814. However as with most other generals from the Guard his loyalty was suspect. Unable or unwilling to curb the unrest within the Guard Artillery caused by their reduced status within the Bourbon Army, he found himself placed on half pay on 1 September 1814. Not surprisingly he rallied to Napoleon on his return and on 11 April 1815 became Imperial Guard artillery commander. He spent the next two months feverishly reorganizing the foot and horse companies, which had largely been disbanded and the personnel scattered among the line regiments. Material and men were brought from all over France. Whole batteries were taken from the line. By 5 June 1815 he had amassed 104 guns comprising ten foot and four horse batteries and sent them on their way to the Belgian frontier.

THE WATERLOO CAMPAIGN

While overall commander of the Guard Artillery during the campaign, he only had direct control of the 32 guns of the Guard Artillery Reserve. These pieces comprised the four Old Guard companies of 12 pounders nicknamed the "Emperor's Beautiful Daughters". His remaining cannon were distributed amongst the various Guard infantry and cavalry divisions. At Ligny on 16 June his guns were deployed in a mass battery, and pounded away throughout the day against the Prussian positions centered on Ligny itself, before the Guard Infantry stormed the position in the late afternoon.

At Waterloo, he sided with Drouot in urging Napoleon not to open the battle till the ground dried sufficiently to enable the artillery and cavalry to be used to maximum effect. He positioned his guns on the slopes before La Belle Alliance and concentrated his fire against the Allied center. Wellington wisely kept the bulk of his men out of view to the rear of the slopes. Whilst Desvaux's fire was accurate against those targets visible, the damp terrain reduced its effectiveness.

Shortly after, the repulse of d'Erlon's first assaults by the Union Brigade and Picton's division disrupted the fire of Desvaux's guns. Napoleon joined him, and around 3.00 P.M. while at the Emperor's side redirecting of the Grand Battery, a ball tore him in two, killing him instantly. It was a grievous loss, for his deputy Lallemand was unable to direct the artillery with the same effect as he. In the confusion of the defeat his remains were lost and with thousands of others he was buried in one of the unmarked mass graves.

CAREER ASSESSMENT

A modest man devoid of all controversy, Desvaux was besides a brilliant logistics expert an administrator who possessed all the virtues required of leadership. The role of the Artillery, in particular that of the Guard had become increasingly crucial to Napoleon's style of warfare and he adapted with ease to its more exacting role. He rarely had the opportunity to spring to prominence like some of his contemporaries in the more glamorous cut and thrust roles reserved for arms like the cavalry. Whilst an elite unit, the Guard Artillery unlike its sister infantry formations was never held back in order to deliver the final knockout blow and win all the laurels. In battle after battle his guns were always to the fore bearing the brunt of the enemy's counterbattery fire before moving forward in support of the infantry assaults. It was a cruel irony that Desvaux died in such a horrendous way after having for so many years been exposed to such dangers without suffering any serious injury.

DOMON - JEAN SIMON, BARON (1774-1830)

EARLY CAREER

The cavalry general Domon was born at Maurepas (Somme) on 2 March 1774. He enlisted as a volunteer with the 4e bataillon de la Somme on 6 September 1791 and the same day his men elected him *sous lieutenant*. From 1792 until 1795 he served with the Army of the North. He fought at Courtrai on 10 October 1792, and was present at the siege of Lille. He then took part in Dumouriez's invasion of Belgium that culminated in the battles of Jemappes on 6 November 1792, Anderlecht on 10 November and the occupation of Brussels.

He benefited from the purges amongst the officers initiated by the Representatives after Dumouriez's defeat at Neerwinden on 18 March 1793 and subsequent defection. Promoted *lieutenant* on 12 May, then *capitaine* on 4 June 1793, his battalion soon after merged to form the 2nd Demi brigade as a result of the first *Amalgame*. Appointed aide de camp to General Compère in April 1794 he served under him in Moreau's division. He fought at Tourcoing on 18 May and was wounded during the assault on Nechin on 22 May. Present at the Siege of Nijmegen in November 1794 he then took part in the crossing of the Waal in December. The following year he fought at Appeldorn, Loo, Greveborn, Telegen, Oldenzael and Bentheim as the Army of the North established control over Holland.

He transferred with Compère to the Army of the Sambre-Meuse in March 1797 and served in Olivier's division. During the crossing of the Rhine at Neuwied on 18 April 1797 he received a citation for bravery. He spent 1798 with the Army of England. The following year with the focus of attention back to Germany, he moved with Compère to the Army of the Danube firstly with d'Hautpoul's division and then under Vandamme*. Wounded by a shell splinter in the leg at Liptingen on 24 March 1799 he returned to duty on 20 May 1799 on promotion to *chef de bataillon*.

His career took a change in direction when he moved to the cavalry on 31 May 1799. He took the rank of acting *chef d'escadron* with the 5th Hussars, the rank later confirmed on 27 December 1799. With Moreau's Army of the Rhine during the march into Bavaria he fought at Hohenlinden on 2 December 1800. With peace he spent a period of garrison duty at Metz before joining in May 1803 Mortier's invasion of Hanover. In December 1803 he transferred to the 3rd Hussars at the Camp of Montreuil to prepare for the invasion of England.

THE GRANDE ARMÉE 1805-1807

The mobilization of the Grande Armée against Austria in August 1805 saw his regiment with Tilly's light cavalry division of Ney's VI Corps. Shot through the neck at Elchingen on 14 October he was out of action for the remainder of the campaign. He returned to the 3rd Hussars as regimental *major* for the Prussian Campaign, serving under Auguste Colbert, who headed Ney's corps cavalry. He fought at Jena on 14 October 1806. In January 1807 while in Poland he moved to the 7th Hussars and serving in a brigade headed by Pajol* with Lasalle's light cavalry division. He fought at Eylau on 8 February and Friedland on 14 June 1807. From July 1807 a period followed as commander of a provisional regiment later formed into the 11th Hussars. At the end of the year he returned to the 7th Hussars joining them in garrison at Ruremonde in Holland.

THE DANUBE CAMPAIGN OF 1809

The opening of the campaign saw Domon and the 7th Hussars with Pajol's brigade attached to Montbrun's light cavalry division operating with Davout's III Corps. He helped cover Davout's columns as they fell back to Ratisbon from their positions on the Bohemian frontier. He fought at Schierling on 21 April and then at Eckmühl the next day. At Wagram on 5-6 July he was part of the great cavalry movement behind Marksgrafneusiedel that turned the Austrian left on the battle's second day. He was in action at Znaim on 10 July 1809. After the campaign he received his promotion to *colonel* on 10 August 1809 with command of the 8th Hussars. The withdrawal of the Army of Germany from Austria saw him return to Holland with the 8th Hussars. On 22 October 1810 he became a *baron de l'Empire*.

THE RUSSIAN CAMPAIGN AND SERVICE WITH MURAT

In the campaign, his regiment was brigaded under Piré* with Bruyere's 1st Light Cavalry Division of I Cavalry Corps. He fought at Ostrovno on 25-27 July where his bravery noticed by Murat on 27 August resulted in his promotion to *général de brigade*. He retained command of his regiment till Borodino on 7 September 1812 when he took command of the brigade after Piré was wounded.

His apparent dash and style still had the eye of Murat, who secured his release on 22 October 1812 to enter the Neapolitan Army with the rank of *lieutenant général*. He was with Murat during the retreat from Moscow and when Murat on 17 January 1813 abandoned the remnants of the Grande Armée in East Prussia, Domon returned with him to Naples. In March 1813 he took the appointment as cavalry commander of the Neapolitan Army and set about raising new drafts for the campaign in Germany.

Suitable material very limited, by the time he returned to Germany for the Autumn Campaign he had only managed to raise a single brigade of dubious quality. Shot in the thigh on 21 August during a skirmish near Lowenberg he took no further part in the campaign and returned to Naples with Murat at the end of October.

Disillusioned by the intrigues of Murat and Caroline Bonaparte to secure their thrones, on 21 January 1814 he resigned from the Neapolitan Army on Naples's defection to the Allies. He returned to France and tried to secure a posting with the Imperial Guard but his association with Murat caused him to be treated with suspicion. Turned down he held no command during the final weeks before Napoleon's fall.

THE RESTORATION AND NAPOLEON'S RETURN

On 19 August 1814 the Ministry for War recognized his Neapolitan rank of *lieutenant général* for service in the French army. The gesture meant little since he was another who found himself on the non-active list and placed on half pay. The Bourbons did recall him on 15 March 1815 to help speed up raising volunteers to oppose Napoleon's return. Despatched to Chalons-sur-Marne he took charge of the eight cavalry regiments at the depot. As events moved fast, Domon played for time, did little and said even less. By waiting to see how events would develop he took the right course. On 6 April he received command of the 6th Light Cavalry Division with III Corps of Observation responsible for covering the Belgian frontier between Mons and Luxembourg.

THE WATERLOO CAMPAIGN

The mobilization of the Army of the North saw his formation become the 3rd Cavalry Division attached to Vandamme's III Army Corps. By 14 June his men had massed on the Belgian border near Beaumont. The next morning he formed the Vandamme's advance guard for the march on Charleroi but things immediately began to go wrong when III Corps failed to receive their marching orders. Domon acting on his, as a result plunged into Belgium without any support. The Prussians delayed his advance by barring the road through the thick woods with fallen trees and trenches. His first contact with the enemy came at Ham-sur-Heure where he cut down a few isolated detachments of landwehr and regulars. Pushing on, he brushed aside a battalion of the 9th Prussian Infantry Regiment outside Couillet and then occupied Marcinelle. Without support Charleroi was too strongly defended. He was driven back trying to storm the dyke leading to the bridge across the Sambre. Only after mid-day did the advance resume when the Imperial Guard's sappers and marines

arrived and dislodged the enemy defending the crossing. Exhausted by his efforts, Domon waited at Charleroi till Vandamme's strung out columns caught up. He then crossed the river and pushed onto Gilly where he halted after fresh elements took over the pursuit.

At the Battle of Ligny the next day he took post on the extreme left of the French line covering Marbais. In this position he was to exploit the situation once d'Erlon's I Corps arrived to turn the Prussian flank. When d'Erlon didn't arrive, he could do little but watch and wait as Vandamme's infantry battered their way against the Prussian positions. Finally late in the day when the Young Guard joined the struggle and effected a breakthrough, he was able to use his cavalry to good effect. As the enemy fell back from the villages of Saint Armand he routed the cavalry brigades of von Söhr and von Thümen as they covered the Prussian retreat. The gathering dusk and lack of clear orders prevented him doing more damage and enabled Pirch to pull his battered II Corps from the field.

Detached from Vandamme on 17 June he moved up to Quatre Bras to fall on Wellington's flank, but the Duke had slipped away. Throughout the day his men harried the Allied retreat. At dusk as he reached La Belle Alliance, a heavy cannonade from the ridge of Mont Saint Jean brought his overconfident troops to a halt. The weight of the bombardment unmasked Wellington's positions and revealed to Napoleon his intent to make a stand before the Forest of Soignies.

At Waterloo Domon was attached to Lobau's VI Corps as part of Napoleon's reserve. At 1.30 P.M. when the approach of Bulow's corps was seen in the distance, Napoleon ordered his and Subervie's divisions to move across country to delay the Prussians. He deployed before the Bois de Paris and waited for the Prussians to debouch from the woods. With Lobau he committed a serious tactical error in not pushing on beyond the Bois de Paris. Had he moved forward and taken up a position behind the Lasne he would have seriously delayed the Prussian advance. The Prussians deploying across a river in the face of enemy cavalry would have found it a very time consuming and risky business. Faced with such a scenario, it is very possible they would not have intervened in the battle once they saw Wellington's positions in the distance apparently overwhelmed by French cavalry.

In spite of the error, with Subervie's cavalry they conducted a brilliant delaying action as the Prussians advanced from the woods. Using feints and counter feints his cavalry pinned down Prussian attempts to outflank Lobau's infantry. A decisive charge scattered Crown Prince William of Prussia's uhlans and hussars. Finally weight of numbers told and when Pirch's II Corps entered the fray they forced the French to fall back on Plançenoit as the Prussians began to envelop their right. Wounded during the battle he remained with his men during the retreat to Paris.

Later Career and an Assessment

After the Army retired behind the Loire he was made responsible for the disbandment of the cavalry regiments gathered at Tours. His task completed Macdonald then sent him to Montpellier on 5 August 1815 to do the same with the five regiments of cavalry attached to the Armée du Midi under Decaen. On 1 October 1815 he was placed on the non active list. Further humiliation came when he was exiled from Paris and ordered to remain at Peronne. The grounds for his exile being that whilst he held a command under the Bourbons at Chalons he failed to carry out his orders with sufficient vigor.

The Bourbons recalled him as an Inspector General of Cavalry on 21 April 1820. The French intervention in Spain to restore Ferdinand VII to his throne saw him on 12 February 1823 given command of the dragoon division with Molitor's II Corps of the Army of the Pyrenees. Active in Grenada he fought at Compillo and El Castillo on 27 and 28 July 1823. For his services to Spain he received the Order of Saint Ferdinand on 20 October 1823. His return to France saw him made a Commander of the Order

of Saint Louis on 2 November 1823. He resumed his duties with the Cavalry Inspectorate and spent 1824 as Inspector General in the 15th and 16th Military Divisions. In 1825 his duties with the Inspectorate covered the 9th, 19th and 21st Military Divisions till 1827 when he moved to the 2nd, 15th and 16th. On 29 October 1829 for his distinguished service to France and the Army, he became a Grand Officer of the Legion of Honor. He was serving as Inspector General for the 5th, 6th and 18th Military Divisions when he died in Paris on 5 July 1830.

At regimental level Domon was a very capable and experienced light cavalry commander. The Waterloo Campaign was his first chance in charge of a division and events showed he lacked initiative when not given orders that were clear or precise. His selection as a division commander during the Waterloo Campaign was surprising considering the neutral stance he took on Napoleon's return. It showed the dearth of good light cavalry leaders available at the time. As a soldier he was another who considered his first duty was to France and the Army, rather than Napoleon. It left him with a clear conscience to serve the Bourbons and hold a succession of senior commands in the years that followed up till his death.

The Prussians pour out of Paris Wood, while Domon's cavalry tries to stop them

DONZELOT - FRANÇOIS XAVIER, COMTE *(1764-1844)*

EARLY CAREER

Donzelot was the rather unfortunate and inept general whose division was overridden by the Household Brigade at Waterloo. A native of eastern France he was born at Mamirolle (Doubs) on 3 January 1764. He enlisted with the regiment de Royal-Marine (later the 60th Line) on 28 October 1783, which at the time was stationed in Corsica. He then spent periods in Alsace and Franche Comte before finding his way to Paris as an orderly at the Ministry for War in Paris in 1791.

The outbreak of war in April 1792 saw him posted to the Army of the North where he briefly served on General Dumouriez's staff. On 22 September 1792 he gained promotion to *sous lieutenant* and joined the 21st Cavalry Regiment with Champmorin's brigade of Miranda's division. With the army during the invasion of Belgium he was at the sieges of Lille in October and Antwerp the following month. With Champmorin during Dumouriez's ill fated invasion of Holland he was present at the French defeat at Neerwinden on 18 March 1793. The resulting purge amongst the officer corps by the Representatives turned to his advantage. He proved a good patriot and as a result gained promotion to *lieutenant* on 24 March followed by that of *adjutant général chef de battalion* on 15 May 1793.

He transferred to the Army of the Rhine on 1 December 1793 serving with Ferino's division that took part in the capture of Wissembourg and the relief of Landau. Returning to the Army of the North in March 1794 he gained promotion to *adjutant général chef de brigade* on 4 June 1794. Back with the Army of the Rhine in 1795 he served in Desaix's division and on 30 October 1795 was wounded during the defence of the bridge over the Rhine at Hunningen. On his return to duty he joined the 3rd Division of the Army of the Rhine Moselle led by Duhesme[*] and served as his chief of staff at Neresheim on 11 August 1796. In December 1796 after Duhesme's suspension he continued to serve under Vieux. The division was with Dufour's right wing of the Army of the Rhine Moselle, and Donzelot distinguished himself during the crossing of the Rhine at Kehl on 21 April 1797.

EGYPT 1798-1801

In January 1798 he joined the Army of England based at Brest. Whilst there, he renewed his acquaintance with Desaix, who was the Army's acting commander. When Desaix moved to Italy to ready the forces designated for the expedition to Egypt, Donzelot joined him as his chief of staff. He sailed from Civitia Vecchia on 26 May and took part in the capture of Malta on 10 June 1798. He landed with Desaix's advance guard at Marabout on 1 July, fought at Ramanieh on 12 July, Shubra Khit the next day and at the Battle of the Pyramids on 21 July. He campaigned with Desaix in Middle and Upper Egypt and took part in many actions trying to hunt down the elusive Murad Bey. The most notable engagements being Sediman on 8 October, El Faiyum on 8 November and Samhud near Gerga on 22 January 1799. He acted as deputy commander of Belliard's force that captured the Red Sea port of Kosseir on the 29 May 1799, effectively cutting off the Mamelukes' source of supply and reinforcements from Jiddah. On 23 June 1799 General Bonaparte promoted him acting *général de brigade* and commander of the region. Left with a small garrison he improved the defences and put the port in order. He successfully repulsed a Royal Navy attempt to land a force at Kosseir from 14-16 August 1799.

Recalled to Cairo in February 1800 he made plans for the evacuation of Egypt. Then when the Convention of El Arish turned sour and Kléber turned on the Grand Visier's army at Heliopolis on 20 March 1800 he led the 61st and 21st Demi-brigades of a division under Friant[*]. After the battle his troops put down the uprising in Cairo. He led the final assault on the citadel on 27 April 1800. Order re-established he became commander for Upper Egypt with responsibility for the regions around Asyut and Mynieh. Recalled to Cairo in March 1801 by Belliard, he brought with him all troops from outlying garrisons in order to face the Anglo-Turkish invasion. While at Cairo, the Consulate confirmed his promotion to *général de brigade* on 29 March 1801. It was news he did not receive till after his return to France. He served under Belliard in the brief campaign that followed as the Anglo Turkish forces advanced up the Nile after bottling up Menou in Alexandria. On 15 May under Belliard he fought the Turks outside Cairo but unable to put them to flight they withdrew into the city. Faced by overwhelming numbers of English troops arriving daily, much to Donzelot's disgust and in spite of the garrison being well provisioned, Belliard surrendered on 27 June 1801 after offering token resistance.

CHIEF OF STAFF TO AUGEREAU AND MASSENA 1803-1807

He returned to France in September 1801 and after a period of leave took a posting at the Ministry of War where he renewed his acquaintance with Berthier from Egypt days. In August 1803 he moved to the Camp of Bayonne as Augereau's chief of staff, then spent a brief period with II Corps of the Army of the Ocean Coasts at Brest. When the Grande Armée was set in motion towards the Rhine he rejoined Augereau as chief of staff of VII Corps on 22 September 1805. He campaigned in the Tyrol and was with Augereau when Jellacic's Austrians capitulated at Feldkirch on 15 November 1805.

The following year he served as Massena's chief of staff with the Army of Naples during the leisurely occupation of that country. Present at the siege of Gaeta he spent a tough time from 25 May till 18 1806 July when the reduction of the fortress proved unexpectedly difficult.

CORFU 1807-1814

On 6 September 1807 he took the post as deputy commander of the force headed by César Berthier to occupy the Ionian Islands in terms of the Treaty of Tilsit. After a brief period on Santa Maura he moved to Corfu on 9 October 1807. He soon found himself in conflict with Berthier, who was a poor comparison to his elder brother. He was a fussy, insufferable, pompous little man, whose sharp vexations were equally unacceptable to his men and the local population alike. Despite the difficulties Donzelot was promoted *général de division* on 7 December 1807 and became a *baron de l'Empire*. Soon after, Berthier was recalled and on 28 March 1808 Donzelot replaced him as governor of the islands.

In the years that followed he proved to be a man of talent and integrity, who took constant care of the island's economy and defences. At first he had eighteen months of idyllic peace in this Mediterranean back-water till the Autumn of 1809. Then General Stuart commanding the British army in Sicily sent a force under General Oswald to establish a bridgehead at the entrance to the Adriatic as the first step to squeeze the French out of the region. Without any naval support and unable to refurbish the outlying islands from Corfu, Donzelot adopted the prudent policy of not spreading his forces too thin by needlessly trying to hold all the islands. Zante soon capitulated on 30 September 1809 and a week later Cephalonia, both without a struggle. Ithaca and Cerigo fell in October before Oswald called off plans for assaults on the main islands of Santa Maura and Corfu till the new year. It gave Donzelot the winter to strengthen his defences. When Oswald landed with a force of 2,500 on Santa Maura on 22 March 1810, the 1,000 strong garrison held them back for two weeks before they withdrew to the fortress of Amaxichi. Only after a ten day bombardment did the 800 survivors surrender. After such stout resistance, Oswald abandoned hope of taking Corfu without having to mount a major expedition. Donzelot with his remaining 4,000 men made Corfu virtually impregnable and kept the state of defences and economy in good order. The Allies never tried to take the island and Corfu only surrendered to a British force under General Sir James Campbell after being ordered to do so by Louis XVIII in June 1814.

THE RESTORATION AND NAPOLEON'S RETURN

On his return to France he received the title *chevalier de Saint Louis* on 8 July and then that of Grand Officer of the Legion of Honor on 23 August 1814. Under the Bourbons he became commandant of the 2nd Sub-division of the 12th Military Division based at La Rochelle on 31 August. On Napoleon's return he accepted service under him and on 6 April 1815 received command of the 2nd Infantry Division with Drouet d'Erlon's I Corps of the Army of the North. The appointment was a surprising since he had never led a formation larger than a brigade in the field and that not since Egypt fourteen years before

.

THE WATERLOO CAMPAIGN

After Durutte's division, which formed the advance guard, his was the second division of d'Erlon's corps to cross the Sambre at Marchienne late in the afternoon of the 15 June. The next day with the rest of d'Erlon's corps he had the indignity of missing both battles of Quatre Bras and Ligny, due to the series of confused and conflicting orders d'Erlon received from Napoleon and Ney.

At Waterloo on 18 June he was posted to the east of the Charleroi Road between Quiot and Marcognet's divisions opposite the Allied strong point of La Haye Sainte. After a lengthy artillery bombardment he joined d'Erlon's assault on the Allied center around 1.30 P.M. His division advanced in a massed battalion column twenty four ranks deep past the eastern side of La Haye Sainte and drove back Bylandt's Dutch Belgian brigade placed before the crest of the Ohain Ridge. The formation so cumbersome, it was unable to deploy in line before it reached the crest. Hit by Picton's division lying in wait behind the row of hedges skirting the Ohain Road it was decimated by rolling volleys from a range of less than forty paces. Then hit in the flank by the Inniskilling Dragoons his column became a disorganized mass. Unable to fire back effectively or use their bayonets his men fell back down the slopes in confusion. It took till after 4.00 P.M. to restore order amongst his demoralized troops before they were able to renew the attack. Advancing up the slope they broke through Kruse's Nassauers but Pack's brigade of Picton's division again brought them to a crashing halt.

Around 6.00 P.M. Wellington's army came under the most severe pressure of the day. Under Napoleon's orders Ney headed the third attack of the day on La Haye Sainte. Donzelot's 13th Légère supported by cavalry and a few guns carried the position. They pushed on and occupied the crest of the ridge behind the farm. From there at two hundred yards range a horse battery poured a devastating fire into Wellington's line. The situation could not be exploited since no further reinforcements were available. Donzelot's men grimly held onto the position till the final assaults by the Old Guard failed to their left an hour later. Panic then spread along the whole French line and his division broke. He managed to rally one brigade in the valley to cover the retreat, but Ney foolishly led it into a hopeless counter attack and after that it ceased to exist as an organized formation.

He fled from the field and reached Loan on 24 June where he managed to rally the remnants of his division. Only 1,552 men remained compared to 5,317 and 8 guns at the start of the campaign. A terrible attrition rate when up till Waterloo he had not engaged the enemy at all. He fell back to Paris and then behind the Loire with the declaration of the Armistice. He dutifully remained at his post as the army disbanded and briefly served as its chief of staff after Guilleminot was replaced on 20 July. The army no more, his job complete, he was placed on the non active list on 1 August 1815.

DONZELOT'S INFANTRY ASSAULTING LA HAYE SAINTE
(DETAIL FROM A KNÖTEL PICTURE)

LATER CAREER AND ASSESSMENT

His reconciliation with the Bourbons was comparatively swift and on 18 August 1816 he became an Inspector General of Infantry. His past reputation as a sound and fair administrator led to his appointment as Governor of Martinique on 13 August 1817. He took up the post on the island on 8 January 1818 and in the years to follow restored it to its former wealth. He was made comte on 22 August 1819. Frustrated by increasingly reactionary directives from an out of touch government in Paris, he demanded his recall and on 23 August 1826 arrived back in France. He took no part in either political or military affairs and retired to his estate at Ville Evrard. Placed in the Reserve on 7 February 1831 he retired from the Army on 1 May 1832. He died at Ville Evrard on 11 June 1843.

Donzelot's fine reputation as a chief of staff and military governor was largely clouded by his poor showing at Waterloo. Though never an Alexandre Berthier, his career did in many respects show similarities to that of his mentor. A sound organizer and able administrator, he found it needed different skills with large formations in the field to put theory into practice. Berthier's problems on the Danube in 1809 before Napoleon's arrival immediately spring to mind, compared to Donzelot's interpretation and implementation of orders at Waterloo. Egypt was the last time he had shown his potential in charge of formations on the battlefield. Then it was at the head of little more than a thousand men. Under Augereau and Massena he had little serious fighting to do. After that seven years in the Mediterranean left him totally out of practice for the hard campaigning expected during the Waterloo Campaign. Through no real fault of his, it unfortunately marred what was up till this time an otherwise exemplary career. His choice as a division commander was a surprise. A possible reason why Davout forwarded it to Napoleon was due to time constraints and the need wherever possible for continuity in the key commands. Donzelot secured his command largely due to the 2nd Infantry Division comprising regiments based in his 12th Military Division. In the final analysis, the poor tactics he displayed at Waterloo made his appointment rank as one of the more pitiful on the French side, possibly only exceeded by that of Jérôme Bonaparte.

EARLY CAREER

Drouet's misfortune in failing to be present at either of the battles of Ligny or Quatre Bras has resulted in him being labelled by most as incompetent. Coupled with Grouchy's failure to appear at Waterloo, Drouet's action was probably one of the major factors that caused Napoleon to lose the campaign. Before this unfortunate sequence of events Drouet had a long and distinguished career as a general officer going back to the Wars of the Revolution. Born at Rheims (Marne) on 29 July 1765, his first service was in the old Royal army when he enlisted with the Beaujolais Regiment (the 74th Line) on 21 October 1782. After five years and no promotion he left in September 1787 after obtaining an honorable discharge due to ill health. The Revolution caused him to re-enlist and he next appeared on 7 August 1792 as a *caporal* in the Rheims Chasseurs with the Army of the North.

The purge amongst the officer corps after the defeat at Neerwinden resulted in his men on 1 April 1793 electing him their *capitaine*. With the Amalgame his unit merged with others on 20 April 1794 to form the 13th Demi-brigade.

On 2 May 1794 he became an aide de camp to General Lefebvre, who commanded the advance guard of the left wing of the Army of the Moselle. He fought at Fleurus on 26 June 1794 and with the emergence of the Army of the Sambre-Meuse continued to serve Lefebvre, who took command of the army's 2nd Division. He was at the sieges of Valenciennes and Conde in August 1794 and later fought at Aldenhoven on 2 October. Promoted *chef de bataillon* on 20 September 1795, then *adjudant général chef de brigade* on 17 February 1797 he remained with Lefebvre till transferred to the Army of England on 18 March 1798. When most of the army dispersed after the abandonment of the invasion plans he returned to Lefebvre on 5 March 1799 as his chief of staff serving with the advance guard of the Army of the Danube. He fought at Ostrach on 21 March 1799 and when Lefebvre fell wounded during the battle, continued to serve as division chief of staff after Souham took command.

SWITZERLAND AND GERMANY 1799-1800

Promoted *général de brigade* on 25 July 1799 he led a brigade with Mortier's 4th Division of the Army of the Danube. He distinguished himself at Zurich on 25-26 September 1799 when on the French left he led a diversionary attack on Wollishofen. It enabled Lorge's division to cross the Limatt lower down at Dietkorn and push on to the city helping to seal the fate of Korkasov's army. In the days that followed he joined the pursuit of Suvorov's army, which had entered Switzerland by the Saint Gotthard Pass and unaware of Korkasov's fate found itself cut off. Drouet caught Rosenberg's rearguard at Muotahal on 1 October and Mortier brimming with overconfidence ordered him to steamroller his way over the enemy. The Russians in a strong defensive position across a narrow valley, brought his assault to a crashing halt, then slipped away. It was the first of many set backs to his career.

The renewal of the campaign in the Spring saw Drouet with Legrand's 2nd Division of the left wing of the Army of Germany. He fought at Erbach on 16 May and was present at the blockade of Ulm on 5 June 1800. In November he passed to Richepanse's 2nd Division of the Center Corps of the Army of Germany. He distinguished himself under Richepanse during the assaults against the Austrian left that went a long way to winning the Battle of Hohenlinden on 3 December 1800. Actively engaged in the enemy's pursuit he fought at Herdorf on 15 December, at Strasswalchen the next day, and Lambach on 19 December before the Armistice of Steyer concluded on 25 December 1800 brought the campaign to a close.

In May 1803 he led the advance guard of Montrichard's division during Mortier's invasion of Hanover and on 2 June 1803 was present at the defeat of the Hanoverian forces near Borstel. He remained in Hanover for the next two years during which time on 27 August 1803 he gained promotion to *général de division*.

THE AUSTERLITZ CAMPAIGN

The mobilization of the Grande Armée on 29 August 1805 saw the French forces in Hanover become Bernadotte's I Corps. Drouet initially led its 1st Division before he handed it over to Rivaud on 17 September and assumed command of the 2nd. Marching through Franconia, he crossed the Danube at Ingolstadt on 10 October and with the rest of Bernadotte's corps entered Munich two days later. Engaged in the pursuit of Kutuzov's army he crossed the Inn on 26 October and after a brief struggle on 30 October took Salzburg. Kutuzov shook off his pursuers, and recrossed the Danube with Drouet and the advance guard following at a distance. Drouet crossed the river near Molk on 14 November but soon lost contact with the Russians. Pushing into Moravia he halted at Trebitsch to observe the movements of Archduke Ferdinand, who falling back from Prague intended to unite with Kutuzov

DROUET - JEAN BAPTISTE, COMTE D'ERLON (1765-1844)

at Olmütz. On 30 November all hope of keeping the Allies apart gone, Drouet marched on Brunn with the rest of I Corps to rejoin the Grande Armée.

Present at Austerlitz on 2 December 1805, Drouet's first objective was to support Saint Hilaire's assault against the Pratzen Heights. The plan went awry when the French cavalry covering Napoleon's weak center was driven back. He had to move forward to fill the gap. His division advanced by half battalion columns, allowing the enemy cavalry to pass through the intervals to attack the French cavalry rallying in the rear. He then greeted the Russians on their return with well-timed volleys of musketry causing heavy losses. The Russian Guard infantry in support, unable to open fire for fear of hitting their own horsemen, fell back with the fleeing cavalry. The situation ripe for a vigorous pursuit, he advanced and joined Soult's divisions on the heights. Then with the Allied line apparently torn in two, Bernadotte inexplicably ordered Drouet to halt before Krezenowitz, allowing the remnants of the Russian Guard to file away unmolested. It was a move that compromised Drouet in spite of the fine performance put up by his division that day.

PRUSSIA AND POLAND 1806-1807

During the Prussian Campaign he returned to the 1st Division of I Corps. He fought in the first major action of the campaign when he drove Tauenzien from the Saxon town of Schleiz on 9 October 1806. He missed the twin battles of Jena and Auerstadt on 14 October 1806 when Bernadotte failed to follow the simple maxim of "marching to the sound of the guns" to help Davout at Auerstadt. Had it been any commander other than Davout, Auerstadt could have been a major French disaster with the whole campaign turning into a very drawn out affair for the French. Drouet was in no way to blame for Bernadotte's behavior, but his division did take its time first marching to Dornberg and then Apolda while the sounds of cannon were booming in the distance. Had he marched with a bit more snap it would have been impossible for Bernadotte not to have intervened in one or the other of the battles. The simmering rivalries between Napoleon's followers from Italy and Egypt with the generals who served under Moreau were still there. The affair had echoes eight years later, when Drouet in command of another I Corps with unclear orders did nothing rather than do what he felt was right, and so missed another battle - Ligny.

To make amends in 1806 his division performed with determination for the remainder of the campaign. He destroyed a Prussian column under Treskow outside Halle on 17 October and after crossing the Elbe on 21 October joined the relentless pursuit of the broken Prussian army across Northern Prussia. The campaign ended when his troops maddened by earlier missed opportunities, humiliations and hardships sacked Lübeck out of hand on 6 November 1806.

He moved into East Prussia, spent a few days at Posen before force marching to Thorn on the Vistula to counter a Russian threat to Warsaw. On 25 January 1807 at Mohrungen he held off an attempt by Bennigsen to isolate the French left wing formed by Bernadotte till the rest of I Corps fell back to link up with Ney's VI Corps. When Lefebvre took command of the newly formed X Corps after the capture of Victor and his staff by partisans, Drouet became the formation's chief of staff on 26 January 1807. From 18 March he was at the siege of Danzig. On 7 May he went a long way to concluding the operation when he seized the neighboring island of Holm. It led to the opening of the negotiations that ended in the final capitulation of the fortress on 27 May 1807. Lefebvre's corps disbanded after the siege, Drouet became Lannes's chief of staff of the Reserve Corps on 29 May 1807. He fought at Heilsberg on 10 June and was seriously wounded when struck by a musket ball in the chest at Friedland on 14 June 1807.

THE AUSTRIAN CAMPAIGN OF 1809

Drouet's recovery was slow, and it was not till 18 January 1808 that he returned to duty when he took a quiet posting as commander of the 11th Military Division based at Bordeaux. He was well rewarded for his part played in the previous campaigns, receiving two pensions of 25,000 francs each drawn respectively on Hanover and Westphalia. On 28 January 1809 he took the title *comte d'Erlon*.

With trouble in Germany, when Lefebvre took command of the VII Corps of the Army of Germany in March 1809, he requested d'Erlon rejoin him as his chief of staff. D'Erlon accepted and after a long journey from South-west France he arrived at Salzburg on 5 May. The immediate crisis on the Danube had largely passed with the Austrian army split in two and driven from Bavaria by Napoleon in a lightning seven day campaign. It enabled Lefebvre with d'Erlon to turn their attention to the serious revolt in the Tyrol. In April little was done to quell the troubles since everything had been committed against the main Austrian army in the Danube valley. On 10 May the first offensive opened and d'Erlon accompanied the main Bavarian force that made its way up the Inn valley. All went well, and outnumbered Andreas Hofer's insurgents quit the low ground and allowed Innsbruck to fall on 19 May.

A new Austrian threat to Napoleon's communications on the Danube at Linz resulted in Lefebvre and d'Erlon's recall with two of the three Bavarian divisions in the Tyrol. It was the opportunity Hofer waited for. Wrede's 8,000 troops were widely spread and were unable to contain the 18,000 insurgents that descended on Innsbruck on 30 May. They beat a hurried retreat along the Inn. At

Linz during most of June, and then after Wagram back at Salzburg, d'Erlon's found his job not without its difficulties. The Bavarian military machine was creaky, and the command structure inefficient. The Bavarian generals, especially the young Crown Prince Ludwig, whilst they tolerated d'Erlon, in particular resented Lefebvre's overbearing manner.

Napoleon after concluding a six week armistice with Archduke Charles on 12 July turned to solving the problems in the Tyrol. On 27 July, under Lefebvre's direction two divisions led by Wrede and the Crown Prince moved up from Salzburg to occupy the region, while another force under Rouyer came from the south. Three days later they reached Innsbruck. An ominous sense of foreboding gripped the formations as they moved deeper into the mountains. Rouyer moving over the Brenner Pass was defeated at Brixen on 1 August and cut off. A relief column headed by Lefebvre and d'Erlon was beaten back, revealing to them for the first time the practical difficulties of fighting in the region. Wrede suffered defeat at Bergisel on 13 August and a week later Innsbruck fell again, with Lefebvre and d'Erlon humiliated back at Salzburg.

D'Erlon adroitly deflected the blame for the poor handling of operations away from himself. Pleas to Napoleon by King Max Joseph that Lefebvre was not the man to lead his Bavarian troops, resulted in the Duke of Danzig's departure in mid October after d'Erlon took command on 11 October 1809. Given a free hand d'Erlon launched a well planned offensive from the north and occupied Innsbruck for the third time on 28 October. Hofer chose to make another stand at Bergisel on 1 November where this time his 8,000 insurgents were overwhelmed by Wrede's artillery. Rather than put the region to the torch as had happened before, d'Erlon took a more conciliatory line and managed to persuade most of the insurgent bands to give up. A link up with French forces from the south at Brixen on 11 November sealed the fate of Hofer, who was captured by a detachment of the 72nd Line on 8 January 1810. Taken to Mantua and tried by a Council of War, he was shot on 20 February.

The region pacified, d'Erlon returned to France in June 1810. The campaign was d'Erlon's first and most successful as an independent commander. He proved equal to the task compared to the period of vacillation under Lefebvre. He drew lessons from the two previous failed offensives, adapting to the tactical problems of waging mountain warfare. He trained the troops in night operations, and showed the importance of outflanking the enemy rather than making costly frontal assaults in confined spaces. He used swarms of light troops as control of the high ground became essential to win any campaign. Successful, they were tactics that he somehow failed to refine and develop later in Spain. More approachable that Lefebvre he largely overcame the problem of strained Franco-Bavarian relations. He involved the Bavarian commanders in the planing and decision making process and as a result gradually gained their confidence and respect. Whilst not all bad will and suspicion disappeared, he at least received their co-operation.

MASSENA'S ARMY OF PORTUGAL 1810-1811

With Spain in 1810 the main focus of France's military effort, d'Erlon moved to Bordeaux to organize drafts being readied for dispatch to that theater. The new formation he organized became the IX Corps of the Army of Spain and on 30 August 1810 he became its commander. The corps was a rag tag assembly of some twenty newly raised 4th battalions belonging to regiments already in Spain. Eleven were from regiments serving with Soult's Army of Andalusia and the rest were from Massena's Army of Portugal. His orders were to do no more than conduct these battalions, which were little better than a mass of drafts, to join the regiments to which they belonged. They were divided into two provisional divisions under generals Claparède and Conroux. Thrust into Spain without any regular organization, destitute of battalion transport, with improvised and insufficient staff they made very slow progress, mainly due to difficulties of the commissariat. The leading battalions had only reached Salamanca by 10 November.

He left detachments behind to keep his communications open and eventually linked up with Massena's Army of Portugal outposts at Espinhal on 26 December 1810. His arrival with 8,000 men and a moderate train of ammunition was of little comfort to Massena, who had hoped for more to resume the offensive. After a few weeks Drouet's men like the rest of the Army were living on the edge of daily starvation. Ordered back to Spain by Massena on 11 March, he led a huge convoy of sick and wounded to the frontier.

Still unable to break away from Massena and deliver the rest of his corps to Soult he was caught up in the campaign to relieve Almeida and Ciudad Rodrigo. Present at the Battle of Fuentes de Onoro on 5 May 1811, his assault against Wellington's Center nearly carried the village of Fuentes itself to win the day for Massena. It led to the famous quip from Wellington, "If Boney had been there we should have been beat".

SOULT AND SOUTHERN SPAIN 1811-1812

After the battle he managed to break away from Massena and on 13 June joined Soult in Estremadura with 10,000 much needed reinforcements. With Soult and Marmont he took part in the relief of Badajoz. On 24 June 1811 he received command of V Corps from Soult. Marmont faced by the full weight of Wellington's Army convinced Soult to give him V Corps till the threat passed. As a result d'Erlon ended back with the Army of Portugal, eyeing Wellington's movements across the Guadiana. D'Erlon then spent a month ranging around the region victualling Badajoz before he settled down to watch the movements of Hill's corps after Wellington had moved north with the rest of his army.

For three months there was very little activity apart from keeping convoys on the move and the roads open to Badajoz. The quiet soon lulled d'Erlon and his commanders into a false sense of security and their forces became too widely spread. Hill broke the impasse, crossed the river and on 28 October 1811 surprised Girard* at Arroyo dos Molinos and destroyed his division. Girard was recalled to France in disgrace; the affair also resulted in the severing of communications between the Armies of Andalusia and Portugal.

It was a very difficult time for d'Erlon since a power struggle ensued between Soult and Marmont as who should have operational control over his corps. D'Erlon tried to distance himself from Soult and even managed to obtain King Joseph's help. Soult became so angered he threatened resignation. His performance in the field at this time also deteriorated; he was uncertain to whom he was responsible. Soult resolved the problem on 7 February 1812 when he abolished the corps structure within his army. This left d'Erlon with the reduced status as commander of the Army of the South's 5th Division. His pride was partly restored by the fact that as Soult's most senior general he was in effect his deputy, and retained responsibility for the 18,000 French troops in Estremadura.

He found it increasingly impossible to cover such a vast area and was continually out maneuvered by Hill. An example being on 11 April when Le Marchant's 5th Dragoon Guards caught his division at Villagarcia and gave it a severe mauling.

After Marmont's defeat at Salamanca on 22 July 1812 the strategic situation in Spain changed dramatically and Soult found himself forced to abandon Andalusia. D'Erlon meanwhile held on in Estremadura while the Army of the South abandoned the siege works at Cadiz and retired eastwards picking up its isolated garrisons. He then slipped away from Hill on 26 August and crossed the Sierra Morena for the last time. He caught up with Soult at Cordoba on 31 August and joined the rest of the army a week later at Granada, where it was decided to retire to Valencia.

After arrival at Valencia on 2 October he took part in King Joseph's offensive to recapture Madrid. Operating on its own his division invested Chinchilla on 9 October the last Spanish stronghold in Murcia. It took divine intervention to cow the garrison after a seven day bombardment for it only surrendered after lightning caused by a severe thunderstorm struck down the commander and fifteen men.

Moving on he then occupied Cuenca on 20 October 1812.

COMMANDER OF THE ARMY OF THE CENTER 1812-1813

Back in his capital King Joseph for a short time reasserted control over the army commanders. A reorganization of the armies in Spain followed and d'Erlon took command of the small Army of the Center, while Soult retained his old command. D'Erlon's force of 18,000 men consisted of the infantry divisions of Darmignac and Palombini, with Trelliard's dragoons forming the cavalry. This direct interference by Joseph and a further splitting of the command structure was a disaster. The army commanders simply failed to co-operate. Soult disliked d'Erlon, having previously cut him down to size by taking away his corps status in Estremadura, he now had to treat him with the respect afforded a fellow army commander. For Soult it was too much to swallow. From the outset he considered d'Erlon a mere "Palace Politician", and to add insult to injury continued to address him as "comte" rather than "général" when conferring on military matters. After Madrid's recapture on 2 November 1812, it was certainly due to their differences that Wellington in his most vulnerable strategic position of the entire Peninsular War was able to scurry back to Salamanca and the safety of the Portuguese border.

In the Spring of 1813 d'Erlon had the impossible task to hold onto Madrid at all costs. His forces held a two hundred mile front between the capital and Valladolid when Wellington struck in June. He managed to withdraw intact and concentrate at Vitoria on 20 June 1813, but Joseph being no leader allowed the generals to make their own dispositions. When Wellington attacked the next day great gaps appeared in the French defences. His troops fought well, particularly Darmignac in holding La Margarita and La Hemandad, till Leval's division on his left gave way. He regrouped at Zuazon, but his divisions again broke when Gazan failed to come to his support and the Allies enfiladed his left. Only after he crossed the Bidassoa into France did he restore order.

THE LIEUTENANCY OF THE CENTER 1813-1814

On 16 July 1813 Soult took command of the battered French forces that had fallen back behind the Pyrenees and the Bidassoa. He set about their reorganization, abolished the command structure he inherited, by scraping the four armies that had fought at Vitoria and forming a single Army of Spain. D'Erlon took the Lieutenancy of the Center, in effect a grandiose title for an army corps devised to placate the reduced circumstances of the former army commanders. He retained Darmignac's command, which became the army's 2nd Division and took

Abbe's 3rd and Maransin's 6th Division. Within a few days his forces took part in Soult's offensive against Pamplona. They stormed the pass at Maya on 25 July, then drove Hill from Lizaso on 30 July. After Reille failed to effect a breakthrough at nearby Sorauren, d'Erlon went on the defensive and covered the army's retreat as it pulled back across the Pyrenees.

On 31 August he covered the French left when Soult failed to force the Bidassoa in a second attempt to relieve San Sebastian. He withdrew to the Nivelle after the Allies successfully countered and crossed the river on 7 October. Behind the Nivelle he held the French left, but again was forced to retire when Clauzel failed to hold the center positions after the Allies crossed on 10 November 1813.

Positioned behind the Nive he was nearly the cause of Wellington's undoing during the four day battle that followed. On 9 December he allowed Hill's corps to occupy the east bank virtually unopposed, as he fell back on Saint Pierre d'Irube. Then on 13 December with the river in flood, the bridges broken and Wellington's army split he attacked Hill's isolated corps on the east bank. Abbe's division in the center and Darmignac's on the right drove Hill's forces back towards the river and certain destruction. Victory seemed assured when for some inexplicable reason Maransin's and Taupin's divisions, which were following in support stood by and watched as Hill's weight of numbers gradually regained the upper hand and drove d'Erlon back. Hill from the jaws of defeat snatched a victory, which in no way was due his prowess on the field, but to the lassitude of the French generals, and once more the actions of others appeared to compromise d'Erlon.

The renewal of the campaign in February 1814 saw d'Erlon's forces driven eastwards till Soult decided to make a stand at Orthez on 27 February. In charge of Soult's right he drove back the English assaults on Saint Boes but when the French center and left gave way he had to conduct a hurried retreat.

At the Battle of Toulouse on 10 April 1814 he again commanded the French left. Darricau, one of his generals, was deployed before the Royal Canal at Petit Granague farm and gave Picton's division a severe mauling as it tried to cross. Darmignac also drove back assaults by Freire's Spanish corps from the Great Redoubt on Monte Rave. When the center divisions of Taupin and Harispe further along the line to his right gave way, d'Erlon fell back behind the canal ready to renew the struggle the next day. Soult did not wish to remain bottled up in a pro Bourbon city, withdrew to Carcassone, and was there when news of Napoleon's abdication arrived ending hostilities.

THE RESTORATION AND NAPOLEON'S RETURN

D'Erlon acknowledged the legitimacy of the Bourbons and on 1 June 1814 was rewarded with the title *Chevalier de Saint Louis*. This was followed by his appointment on 22 June as commander of the 16th Military Division based at Lille. He soon became involved in controversy when he presided over the Council of War that acquitted Exelmans of spying on 25 January 1815. Criticism of his stand drove him into the arms of Bonapartist conspirators. When news arrived of Napoleon's landing at Golfe Juan he joined the conspiracy hatched by Lefebvre Desnoettes, Exelmans and the Lallemand brothers to rally the northern garrisons to Napoleon.

On 8 March he called out the Lille garrison and rather prematurely set it in motion towards Paris. He was thwarted the next day by the unexpected arrival of Marshal Mortier, whose exhortations were enough to cause the garrison to return sheepishly to their quarters. D'Erlon was locked up in the citadel. After his confused jailors released him on the 21 March, he immediately rallied the garrison to Napoleon again and seized Lille.

THE WATERLOO CAMPAIGN

An appreciative Napoleon on 6 April 1815 appointed him commander of I Corps of the Army of the North. He set about preparing for the coming campaign the one cavalry and four infantry divisions placed under him. His task was not easy. None of his division commanders Allix, Donzelot, Marcognet, Durutte and Jacquinot had served under him before. By 14 June his corps with that of Reille's II Corps had secretly concentrated at Solre-sur-Sambre on the Belgian border where they formed the Left Wing of the Army. The next day after Reille had crossed the border at Thuin he soon encountered delays. The advance elements of I Corps only crossed the Sambre at Marchienne in the late afternoon due to a stubborn defence of the village by Steinmetz's Prussians, who held out against Reille's advance guard.

The next morning he gathered in his strung out troops at Jumet. Then at noon he received orders to head for Frasnes to support Ney, who was concentrating at Quatre Bras. Further delays occurred as he could not move till II Corps had defiled. It was after 2.00 P.M. when his advance elements reached Gosselies. False reports about the presence of Anglo-Dutch troops on his left caused him to halt at the village whilst patrols were sent out to investigate. Satisfied there was no threat, he moved on.

It was after 4.00 P.M. as he approached Frasnes and there occurred the series of events that were disastrous for the campaign and marred his career. They were forever to raise questions as to his competence, subject him to ridicule and in the eyes of ardent Bonapartists consign him to eternal damnation. I Corps was astride the junction of the Roman road and the Charleroi-Brussels highway. His men were ready to deploy and be in action against Wellington within half an hour when one of Napoleon's aides General de la Bedoyere arrived with contradictory orders to move east towards

Ligny. The execution of this move though no fault of Drouet's was disastrous. The approach of I Corps behind the French left engaged at Ligny caused great alarm and confusion amongst Vandamme's III Corps, who thought a hostile Allied column was approaching their rear. This gave the Prussians a brief respite to regroup and drive Vandamme's men from the twin villages of Le Hameau.

Had d'Erlon shown more initiative, or the order de la Bedoyere carried been more precise, the problem could have been avoided. D'Erlon could have headed for Marbais, warned Vandamme of his approach, and deployed behind Blucher's right, not towards Ligny and Napoleon's left.

In spite of disrupting Vandamme he was still in a good position to make a decisive entry into the battle. Then at 5.30 P.M. as Durutte's division was deploying on the far left of the battlefield, he wheeled the remainder of his corps around after receiving a desperate appeal for help from Ney at Quatre Bras. By the time he neared the cross-roads it was dusk and Wellington had driven Ney's forces back to the outskirts of Frasnes. Thus d'Erlon had managed with his 20,000 men to miss both battles and, after Grouchy, was to bear the brunt of abuse as the man held most responsible for losing the campaign.

During the morning of 17 June Wellington fooled the French by slipping away up the Brussels Road. On hearing the news Napoleon again severely rebuked both Ney and Drouet for not pinning the Allies with fresh troops so he coming up the road from Ligny could have rolled up their left.

At Waterloo on the morning of 18 June his corps deployed to the right of the Charleroi Road. He cannonaded the Allied line for an hour and a half before I Corps began its advance around 1.00 P.M. The approach formation used was a clumsy one, especially considering the tactics known to be favoured by Wellington. Three of his four divisions, those of Donzelot, Quiot and Marcognet advanced with their battalions deployed one behind the other, each of the three divisions moving forward on a front of one battalion. As these unwieldy columns moved across the 1,300 yards that separated the armies and began to climb the slopes they were met with heavy fire. Allied gunners opened up on them with every cannon they could bring to bear. The two flank divisions were entrusted with reducing La Haye Sainte and Papelotte, which had to be captured before Wellington's line could be overcome. Durutte headed for Papelotte, which he seized, whilst Donzelot headed for the farm of La Haye Sainte.

The remainder of I Corps advanced against Wellington's left center. D'Erlon's left flank covered by Traver's cuirassier brigade of Delort's division pounced on a battalion of King's German Legion moving to La Haye Sainte and cut it to pieces. It was the solitary success. The assault on the farm was met

D'ERLON'S MEN ASSAULT LA HAYE SAINTE (A KNÖTEL PRINT)

by deadly musketry, as Donzelot's infantry unaccompanied by guns and not even supported by cavalry, failed to storm the position. Deflected past the farm the French struck Picton's division and Bylandt's Dutch Belgian brigade. Picton's troops lined a hedge along the Ohain Road and the fire they poured in caused the French to falter still a hundred yards short of their position.

Picton ordered a charge and as his men moved forward he met his death. The engagement became a savage musketry duel as the French slowly tried to deploy. A brigade of Donzelot's division opposed Kempt's brigade, Marcognet confronted Pack and Quiot tried to drive his way between the two. The French with their attention diverted towards the enemy infantry, failed to notice that Somerset's Household Brigade had driven back Traver's cuirassiers covering the assault and left the infantry exposed. Ponsonby's Union Brigade saw the opportunity, swooped down on Drouet's infantry and surprised them whilst engaged against Picton. The Royals and Inniskillings overthrew Quiot and the 105th Line lost its Eagle. Meanwhile the Scots Greys charged Marcognet and within three minutes his formation was reduced to a fleeing disorganized rabble, with the 45th Line also losing an Eagle.

D'Erlon rallied his divisions by 3.30 P.M. and again sent two columns forward to renew the assault on La Haye Sainte. Wellington beat them back just before Ney launched the mass cavalry attacks against the Allied Center. D'Erlon's attack was a misfortune for had he been able to support Ney with his infantry and artillery the outcome of this combined arms mass assault could have been very different.

After Ney's cavalry attacks had failed, d'Erlon launched the battered remnants of his divisions into another attack. La Haye Sainte fell to Donzelot around 6.30 P.M. From here the French artillery enfilading the Allied line enabled the remaining divisions to breast the ridge and gain the Ohain Road. By 7.30 P.M. they controlled the ridge, but were too exhausted to move forward any farther. The Allied line buckled, but managed to hold and remain unbroken.

When the attack of the Old Guard failed around 8.00 P.M., panic spread along d'Erlon's line like wildfire and his men save for Durutte's division were transformed into a disorganized rudderless horde. He managed to rally the remnants of I Corps at Laon on 22 June. His returns showed that of the 19,838 men who were present at the start of the campaign only 5,573 remained. Falling back he passed through Soissons on 25 June and then to the outskirts of Paris on 29 June. On 4 July he retired with the rest of the army behind the Loire.

EXILE AND LATTER CAREER

Under no illusions that he would not be arrested for his part in the Lille conspiracy, he fled the country.

After a brief stay in Munich he settled in Bayreuth where he ran a very profitable inn. He was proscribed by the Ordinance of 24 July 1815 and tried in absentia by a Council of War. Found guilty of treason for the part he played in raising the northern garrisons against the King, the court sentenced him to death on 10 August 1816.

Charles X granted him an amnesty at his coronation on 28 May 1825 and soon after he returned to France. The Bourbons never offered him a command and he remained unattached on half pay till his retirement from the army on 2 December 1827. His fortunes changed after Louis Philippe became King and on 7 February 1831 the General Staff recalled him to duty. On 19 November 1831 he became a Peer of France and on 30 June 1832 took the appointment as commander of the 12th Military Division at Nantes, where he effectively put down the Duchesse de Berry's attempts to raise the Vendée.

The appointment as governor general of Algeria followed on 27 July 1834. His stay in North Africa was not a success. He made little progress in pacifying the colony or eliminating the widespread corruption inherited from his predecessors. Recalled on 8 August 1835, on his return he resumed his command at Nantes where he remained till 27 October 1839 when he was placed in the Reserve Section of the General Staff. Well liked by Louis Philippe, he was another general who in his dotage was used to foster the "Napoleonic Legend" and on 9 April 1843 became a Marshal of France. He died in Paris on 25 January 1844.

A FINE SOLDIER, GROSSLY UNDERESTIMATED

Take away his shortcomings during the Waterloo Campaign, overall Drouet had a very distinguished career. From when he attained general rank during the Revolutionary Wars through to serving with Bernadotte's I Corps of the Grande Armée he was sound and reliable. As chief of staff with X Corps and then VII Corps he proved to be a sound administrator. His first and most successful independent command in the Tyrol could have so easily cost him his career yet he showed he could handle pressure. In Spain he won the respect and friendship of Joseph Bonaparte, but this turned to his disadvantage when faced by Soult's ambition and personal vendetta against Joseph. Criticism of his conduct during the Waterloo Campaign is justified; he made a bad error of judgement. In his memoirs he tried to explain his actions, and when faced by the volleys of criticism fired back he at least stoically accepted his misfortune with dignity, unlike Grouchy.

In the final analysis, he did deserve the high office of Marshal of France. Take away the two days of the Waterloo Campaign, he did his duty and his career was devoid of controversy. He was a credit to France and the Army.

DROUOT - ANTOINE, COMTE (1774 - 1847)

BACKGROUND AND EARLY CAREER

The son of a baker Drouot was born in Nancy, eastern France on 11 January 1774. He studied at the Ecole de Nancy before entering the Chalons Artillery School where on 1 January 1793 he gained his commission as a *sous lieutenant* after graduating at the top of his class.

Posted to the 1st Foot Artillery Regiment serving with the Army of the North, he gained promotion to *2me lieutenant* on 1 July 1793. Active on the French frontier and in Belgium he fought at Hondschoote on 8 September and Wattignies on 16 October 1793. Further promotion to *lieutenant* came on 22 February 1794. He was with Jourdan's second offensive across the Sambre and fought at Fleurus on 26 June 1794. After the battle his company joined the newly formed Army of the Sambre-Meuse. Promoted *capitaine 3me classe* on 25 February 1796, he later moved to the Army of the Rhine and on 20 April 1797 took part in the Rhine crossing at Neuwied.

In Italy during the Winter of 1798-99 the bellicose posturing of the King of Naples became a threat to French interests. A French build up in the region followed and Drouot moved south to join the Army of Naples gathering under Championnet. He took part in the invasion of the country and was present at the capture of Naples on 22-23 January 1799. The French action led to the combined forces of Austria and Russia turning on the outnumbered French Armies in Italy with disastrous results. He fought at Trebbia on 18-20 June 1799 when Suvarov defeated Macdonald and drove the French back across the Apennines.

During 1800-01 he moved to the Army of the Rhine and served on General Eblé's staff. It proved a valuable training ground for him and he always had the highest respect for this fine commander, a hero at the Beresina who died exhausted after the Russian Campaign. Promoted *capitaine 2me classe* on 23 October 1800, Drouot fought at Hohenlinden on 3 December 1800. Promoted *capitaine commandant* on 21 January 1802 for two years he led the 14th Company of the 1st Foot Artillery Regiment.

NAVAL SERVICE AND THE ARMAMENTS INDUSTRY 1804-07

Posted to the naval base at Toulon in July 1804 he joined General Lauriston who commanded the troops assigned to accompany the fleet. His role was principally to upgrade the gunnery within the Mediterranean Fleet. He was with Villeneuve's fleet on the Indomptable when it slipped the Nelson's blockade on 30 March 1805, joined the Spanish Fleet at Cadiz and sailed for the West Indies. When the fleet returned to European waters join the Atlantic squadrons under Ganteaume he was in action on 22 July against Calder off Cape Finistere. Promoted *chef de bataillon* on 20 September 1805, he never received the news while at Cadiz. His orders to quit the fleet and return to France to join the General Staff of the Grande Armée also went astray. As a result he was with the fleet when it met Nelson off Cape Trafalgar on 21 October 1805. His tutelage had some effect, for Indomptable put up a good fight against Bellisle, managed to survive the battle and make its way back to Cadiz. From there Drouot immediately returned to France, while the Indomptable in a sortie from Cadiz soon after sank with the loss of all hands. On his arrival in Paris Drouot found the Grande Armée was deep in Austria. Reassigned as an Inspector of Armaments to the armaments factory at Mauberge, he spent a frustrating year testing and examining weapons to ensure that their quality was satisfactory.

Drouot's reputation as an excellent organizer and trainer had an unfortunate side effect for a man who wanted to see action. Whilst further promotion came, to *major* on 19 January 1807, he found himself in charge of the 3rd Foot Artillery at Metz responsible for the depot companies. In September 1807 he was back in the armaments industry as an inspector for the musket factory at Charleville. His prospects for seeing action picked up in February 1808 when he took the post as director of the artillery park with the Army of Spain.

THE IMPERIAL GUARD ARTILLERY

A dramatic change in fortune came for Drouot on 12 April 1808 when an Imperial Decree to form the Foot Artillery of the Imperial Guard was signed. Larobisière, at the time the Grande Armée's artillery commander, was aware of Drouot's organizational and administrative skills, secured his release from Spain and gave him the task to raise and organize the fledgling unit. At La Fère Drouot raised and trained the first four foot companies of Guard artillery from the cream of applicants sent by the line regiments.

The unit saw its first active service during Napoleon's invasion of Spain in November 1808. Recalled in March 1809 to face the threat from Austria he led the Guard batteries on their long march to the Danube. Present at Wagram on 5-6 July his skilful positioning of the batteries under heavy fire was decisive in breaking the Austrian assaults between Breitenlee and Süssenbrun. Later in the battle he paved the way for Macdonald's epic attack that won him his marshal's baton. He received the one and only wound of his career when a shell splinter tore into his right foot that left him with a marked limp for the remainder of his life. He bore a charmed life, made even more remarkable since it was Napoleon's habit to send his artillery well forward. Drouot would always dismount and advance on foot with his men wearing his full regimentals, attracting fire. On 9 July 1809 promoted *colonel*, he became official commander of the Guard Foot Artillery.

Showered with further awards he received the title *baron de l'Empire* on 14 March 1810 and took a pension of 10,000 francs drawn on Rome. Due to the vast expansion within the Imperial Guard in 1811, his companies were increased to six and the number of cannon per company to eight. Napoleon took great interest in the unit, continually subjecting it to detailed inspections and reviews, and soon the formation acquired the nickname, the "Emperor's Beautiful Daughters".

THE RUSSIAN CAMPAIGN

In March 1812 he left La Fère for Poland to prepare for the Russian Campaign. In Russia he fought at Smolensk on 18 August. On 7 September at Borodino he earned great respect from the rest of the army for the way his men handled their cannon. Astride the French center they first endured a relentless bombardment from the Russian redoubts and then suffered as they moved forward in support of Ney's assaults on the Semyonovskaya Redoubt. After Ney carried the position he redeployed further south and turned his attention to the Flèches, which finally fell to Davout later in the day. In the retreat from Moscow his guns were used at Krasnoe on 17 November before most were abandoned after the draft horses succumbed to the mud, snow and ice.

THE GERMAN CAMPAIGN OF 1813

His return to Paris saw him promoted to *général de brigade* on 10 January 1813 and made an Imperial Aide. Given the task of rebuilding the Guard Foot Artillery, by a superhuman effort he had four companies trained, equipped and manned with 32 guns at La Fère ready to march to Germany at the end of March. In Saxony he led the Guard Artillery in the action at Weissenfels on 1 May. The next day at Lutzen after the Allied advance was checked he formed a massive eighty piece battery near Kaja. His guns softened the Allied center till 5.30 P.M. when at a given signal the Young Guard advanced. Formed in four dense columns with Drouot's guns beside them giving support fire, they moved forward like a great battering ram. There was no disputing the passage, marching firing marching again, the guns blew their way though the enemy's lines with the infantry following. It was artillery doing its deadly work at its classic best. Yard by yard the enemy were pushed back on the Elster and only lack of adequate cavalry prevented the French gaining a resounding victory.

At Bautzen on 21 May his guns wrought havoc amongst the Prussians around Kreschwitz before the Guard infantry drove them back. At Dresden on 27 August the arrival of his guns late in the battle was decisive in support of the Guard's final attack on Lubnitz. On 3 September 1813 he was promoted *général de division.*

At the battle for Leipzig on 16 October he formed a massive battery of 150 guns south of the city before the Galgenberg plateau between Wachau and Liebertwolkwitz. The fire completely disrupted the four columns sent forward by Schwarzenberg, particularly the advance on Wachau by Raevski's 10,000 Russian Guard Grenadiers. When the French counter attack came around 2.00 P.M. he moved the guns forward in support to complete a successful day. On 18 October he helped stem the Allied advance on Probstheyda and the next day managed to withdraw his guns intact from Leipzig.

At Hanau on 30 October Drouot put up another fine performance, when Wrede tried to cut the French retreat. The Bavarian crossed the Kinzig and deployed his troops in the open with their backs to the river. From there he planned to advance through the forests and fall on the French columns as they passed by. Napoleon saw the danger to his forces but also that Wrede was equally vulnerable. He sent Drouot through the forests to met the threat with his cannon.

So confident was Wrede, he encouraged the French to give battle by allowing them to deploy before him without interruption. On the other hand Drouot knew once the Bavarians attacked, it would be impossible to retreat through the forests on the narrow tracks available. He immediately opened a hot fire against the enemy as one by one his guns emerged from the woods, unlimbered and gradually extended his line. Smoke concealed an additional danger as French cavalry formed up behind the guns. Confident his time was ripe and angered by an increasingly effective French fire Wrede's cavalry charged. Overwhelmed by the charge, Drouot drew his sword, rallied his gunners and defied all attempts to capture or dismantle his pieces. His resistance gave Napoleon time to launch the counterattack. In a confused mass every mounted man available charged into the melee. Grenadiers à cheval, lancers, dragoons and chasseurs, cut their way through the enemy surrounding Drouot's fifteen guns and then galloped on to break Wrede's squares. The Bavarian army then became a stampeding mass as it tried to reach the safety of Hanau by the single bridge over the Kinzig.

Even before Hanau for his part in the campaign Napoleon had made Drouot *comte de l'Empire* on 24 October and awarded him a further pension of 26,000 francs drawn on Rome.

THE 1814 CAMPAIGN IN FRANCE

He was with Napoleon throughout the campaign in France. He fought at Brienne on 29 January. At La Rothière on 1 February his guns covered Napoleon's retreat after the Guard pulled out when faced by overwhelming odds. He was with Napoleon at Montmirail on 11 February. Craonne on 7 March was his tour de force, during the desperate struggle in France. At the time he only had four batteries under him posted before the mill at Craonne facing ninety-six enemy cannon placed wheel to wheel before Heurtebise. Swab in one hand and range finder in the other he went from gun to gun encouraging his young, inexperienced gunners who were mercilessly pounded before the enemy withdrew.

He fought again at Laon on 9-10 March where his cannon covered the army's retreat. At the capture of Rheims on 13 March he fought on the French right before Mont Saint Pierre. When all appeared lost at Arcis-sur-Aube on 21-22 March he greatly enhanced his reputation. The French rout on the second day was stemmed only by Napoleon's personal intervention and the steadiness of Drouot's guns, which raked the Allied infantry and brought their advance to a crashing halt. At sunset his weary men then again covered Napoleon's retreat. Present with Na-

DROUOT AT HANAU

poleon during the last dash west to save Paris, he was with him at Fontainebleau during the abdication crisis.

EXILE ON ELBA 1814-1815

A bachelor, totally devoted to the Napoleon, he renounced his French citizenship and insisted on accompanying him into exile on Elba as a private citizen. When Napoleon offered him the post of governor on Elba he only took it on the condition that he could resign soon as Cambronne arrived to replace him. "I have entirely renounced the great things of the world", he wrote to General Evian on 5 May 1814, "and am going to devote the days of my exile to study."

Within a few days of his arrival Napoleon persuaded him to act as governor of the island. The dull tedious life on Elba did little to change Drouot's attitude. While Bertrand and Cambronne considered their stay on Elba as only temporary, Drouot hoped to retire to a life of reading and reflection, continuing his studies in philosophy and theology. The old bachelor also fell in love and hoped to marry the young attractive Henriette Vantini, who had been teaching him Italian.

When Napoleon told him of his planned return to France only a few days before he sailed, he strongly opposed it. The prospect of a new military adventure that might end in further exile or death for them all worried him. Once the die was cast, his strict sense of duty made him throw aside all misgivings and ensure Napoleon's plans were successful. It showed in a statement made by him at his court-martialled a year later for his part in the escape when he said, "I was bound to Napoleon by oath, not France". On 1 March 1815 he landed with Napoleon at Golfe Juan near Antibes and accompanied him on the march to Paris. On 25 March he resumed his place as commander of the Guard Artillery till replaced by Desvaux de Saint Maurice on 11 April when he became *Aide-major Général* of the Guard. The post gave him responsibility for organizing all arms of the Guard. On 2 June he was made a Peer of France. By 10 June his preparations complete, the Guard secretly concentrated on the Belgian frontier. On 14 June he replaced Mortier, who had fallen ill as senior commander of the Guard in addition to his duties as its Chief of Staff.

THE WATERLOO CAMPAIGN

He was at Napoleon's side throughout the campaign. On the morning of Waterloo he strongly advised Napoleon to delay the start of the battle till the ground had hardened. Napoleon took the advice and the delay probably cost him the campaign. Later in the day he urged Napoleon to withdraw when it be-

came apparent that Grouchy had failed to intercept the Prussians, but this time he was ignored. He was with the Guard when they launched their final attack in the evening. He persuaded the Emperor to leave the field in the evening when the battle was lost.

He remained with the army as it fell back towards Paris. When news arrived of Napoleon's abdication, the provisional government on 23 June appointed him commander of the Imperial Guard. He rallied the Guard before Paris and opposed any form of capitulation since he felt the Allies widely dispersed could easily be defeated in detail. On the declaration of the Armistice on 4 July he had great difficulty keeping order amongst the Guard, who were spoiling for one last battle, but managed to lead them south to the Loire to be disbanded.

LATER LIFE

Proscribed by the Ordinances of 24 July 1815 he refused to flee and marched up to the gates of L'Abbaye prison in Paris and offered to stand trial. Brought before a Council of War and charged with treason he defended himself with great dignity and secured his acquittal on 6 April 1816. Recalled to the army on 19 February 1820 and offered back pay since his suspension in 1815 he refused any position or compensation. He retired from the army on 6 February 1825 after accepting a pension of 5,400 francs a year. He settled in Nancy where he led the quiet life of a country gentleman.

Soon after his accession Louis Philippe awarded him the Grand Cross of the Legion of Honor on 18 October 1830. He became a Peer of France on 19 November 1831. He took great interest in the welfare of former members of the Imperial Guard and headed an association made up of former officers that helped improve the pensions of old soldiers. Other interests he had included that of president of the Nancy Agricultural Society. Blind in his final years, he remained active till the end caring for his men. He died in Nancy on 24 March 1847.

A FINE SOLDIER - "THE SAGE OF THE ARMY"

Drouot was a brilliant artillery general and administrator, who typified the later generation of Bonapartist general, ambitious, energetic and loyal. His rise in Napoleon's final years was spectacular and there is little doubt had Waterloo been a French victory he would have received his Marshal's Baton. Marshal Macdonald described him as, "the most upright and honest man I have ever known, well-educated, brave, devout and simple in his manner". A mark of Napoleon's esteem was Drouot's inclusion in his last will.

DUHESME - PHILIBERT GUILLAUME, COMTE (1766-1815)

BACKGROUND AND EARLY CAREER

Duhesme was born at Bourgneuf-Val-d'Or (Saone-et-Loire) on 7 July 1766. The son of a laborer he had little formal education. Highly intelligent, he did possess a keen wit and sharp tongue that gave him an animal instinct for survival. A man with a great physical presence he had a natural flair for rabble rousing and an ability to exhort others to great feats of bravery. A wild radical hothead he eagerly embraced the virtues of liberty proclaimed by his class and the need for change, no matter how violent. With such noble attributes in July 1789 the local mob elected him head of the National Guard in his district. When the Saone-et-Loire Volunteers formed on 25 September 1791 he became *capitaine* with its 2nd Battalion.

The gathering war clouds saw him with the Army of the North where on 6 August 1792 he emerged at the head of a volunteer company of chasseurs that he raised himself. Two days later the unit merged with the Hainaut Volunteers. His leadership qualities plus a lively instinct for intrigue resulted in the dismissal of the senior officers in the regiment and his promotion on 26 October 1792 to *lieutenant colonel* and command of its 4th Battalion.

He took part in Dumouriez's invasion of Holland in February 1793 where his troops formed the garrison at Ruremonde. Then after Dumouriez's defeat at Neerwinden on 18 March 1793 he helped to cover the army's retreat by burning the bridge over the Rhine at Loo. The army felt a sense of betrayal after Dumouriez's defection and Duhesme's exploits during those dark days were not forgotten. He fought stubbornly as the enemy drove the army from Holland through Belgium to the French frontier. Wounded by two musket balls in the leg during fighting in the forests of Villeneuve on 3 July he was out of action for several weeks. Then at a time when the influence of the Representatives was at their peak, they appointed him *général de brigade* on 7 October 1793, in command of the 3rd Brigade of Fromentin's Division. He fought at Wattignies on 15-16 October when Jourdan's forces relieved Maubeuge.

GENERAL OF THE REPUBLIC

His rank confirmed on 12 April 1794, he led the Army of the North's right wing advance guard formed by Desjardin's division during Jourdan's first crossing of the Sambre on 10 May 1794. The troops performed well and gained their objectives but had to fall back after three days as the promised support never arrived.

He returned to his command under Fromentin on 19 May, and with that leader indisposed he led the division till Kléber's arrival on 3 June 1794. His division was one of two from the Army of the North that joined the Army of the Moselle for the renewed offensive across the Sambre. He crossed the river at the Abbey of Alnes on 13 June and drove the Austrians from Trazegnies where one of his regimental commanders Jean Baptiste Bernadotte of the 71st Demi-brigade particularly distinguished himself. However with the divisions of the Army of the Ardennes panicking under Marceau and forced to fall back over the river at Le Chateau, the divisions of the Army of the North on their own, were forced to retire.

Ten days later during the Battle of Fleurus on 26 June 1794 he again crossed the Sambre and occupied the same positions behind the Piéton. This time with the left wing now under Kléber they successfully drove the enemy from the field. When the various armies reorganized after Fleurus his formation became the 7th Division of the Army of the Sambre-Meuse. He followed the enemy as they withdrew from Belgium and then spent four months directing the Siege of Maestricht till its capitulation on 4th November 1794.

CONFLICT WITH MOREAU

His rank of *général de division* confirmed on 8 November 1794, he took command of the 5th Division of the Army of the Sambre-Meuse. In January 1795 he left his new command and headed a contingent of 12,000 men sent to La Vendée to help pacify the unrest in the region. He returned to the Rhine in December 1795 and took over the 10th Division of the Army of the Rhine-Moselle. He fought at Obermedlingen on 5 December 1795 and covered the retreat of the army to Mannheim. He then moved to the 6th Division in April 1796 when the morale of that formation needed stiffening. The next month he passed to the 7th Division where he served under Gouvion Saint-Cyr who commanded the right wing of the Army. With the advance into Germany he seized Wolfach and Schramberg on 14 July.

He operated on the extreme right of the Army and on 11 August reached Dillingen on the Danube. Then cut off by a superior force, he cut his way back to Medlingen but in turn left the remainder of the Army dangerously exposed at Neresheim where the Archduke Charles defeated Moreau the same day. The result was a headlong retreat to the Rhine. Moreau furious at his disastrous change in fortune accused Duhesme of cowardice and had him suspended. Saint Cyr intervened on his behalf and he retained his command till the Army was safely behind the Rhine.

Duhesme conducted an eloquent defence and all charges dropped, he was exonerated. He returned to the Army in April 1797 and received command of the 3rd Division with Desaix's Corps. He took part in the crossing of the Rhine at Diersheim on 20 April. Doubts as to his courage were soon dispelled during the battle. At the head of his division, leading from the front, he picked up the drum from a drummer killed at his side and urged his men forward using his sword as a drumstick.

ITALY 1798-99

On 28 February 1798 his rehabilitation complete, much to the chagrin of Moreau, his fellow generals chose him to present the banners captured by the Army of the Rhine to the Directory. In the summer of 1798 he moved with his division to central Italy to counter threats from the Kingdom of Naples. His formation was one of two divisions of the Army of Rome under Championnet, the other Macdonald led He was soon in action when the Neapolitans invaded the Papal States and Duhesme countered by advancing from Ancona on the Adriatic coast. He seized the fortress of Civitia del Trento on 3 December and then Pescara on the 24th. After cutting across the peninsula to Naples, he distinguished himself during the capture of the city on 22 January 1799.

With Bourbon rule ousted and the Pathenopeon Republic set up he took the appointment as governor of Apulia. His behavior during this period was corrupt and cruel to the extreme. He was very much responsible for bringing the wrath of the Directory down on Championnet's head when he roughed up several Representatives who were with the Army after they brought some of his misdemeanors to light. The Directory recalled him with Championnet and other senior commanders on the 25 February 1799. Arrested when he reached Milan with the others on 16 March, he was brought in chains to Grenoble where accused of having arbitrarily removed vast sums from the Naples Treasury he was imprisoned. Brought before a tribunal, he was conveniently cleared of all charges. It was a dangerous precedent to bring charges of pillaging against generals in a time of a "National Crisis", as few if any in the army were innocent. It was more a matter of degree.

THE MARENGO CAMPAIGN 1800

He joined the Army of the Alps on the 23 June 1799 and continued to serve under Championnet, who on his reinstatement had taken over command of the Army. In the Autumn he campaigned in the mountain passes of northern Italy. He seized Susa, fought the Austrians at Pignolo on 29th October and took Savigliano two days later. He fought again at Genola on 4 November. When he heard news of

Grenier's defeat at Fassano he retired over the Apennines via Col di Tenda and made his way back to France via Nice.

A period of well earned leave followed during the Winter of 1799-1800. Soon after his recall on 19 April 1800 he moved to Dijon as acting commander of the Army of the Reserve. When Berthier later took command on 9 May he was given a provisional corps made up of the divisions of Loison and Boudet. He crossed the Alps by the Saint Bernard Pass and descended into Lombardy. At Bufalore on 31 May he crossed the Ticino and created the diversion that enabled Napoleon to take Milan with such ease two days later. Ordered to cover the Army's left flank as it cut across Melas's communications he seized Lodi on 4th June and Cremona three days later. He fought a sharp action against Vussakovich at Castelleone on 10 June, which removed the Austrian threat from the east. Moving south-east he then took Piacenza on 12 June, which covered any threat to the Army of the Reserve's rear from Genoa, whilst Napoleon sixty miles away to the west defeated Melas two days later at Marengo.

After the Convention of Alessandria he remained in Italy for a period. Under Massena, who resumed his place at the head of the Army of Italy he led the Reserve Corps with the divisions of Miollis and Rochambeau under him.

In September 1800 he joined the Army of Gallo-Batavia as Augereau's deputy. With Augereau he covered the left of Moreau's Army of the Rhine as it pondered across Germany towards the Danube. He fought at Burg Eberach on 3 December 1800, then Bamberg and Forcheim before Klenau turned on him at Nuremberg on 14 December. After nearly being encircled his troops with the rest of Augereau's army were ignominiously bundled back towards the Dutch border. News of Moreau's victory at Hohenlinden and the Armistice of Speyer on 25 December 1800 conveniently saved his and Augereau's reputations.

PERIOD OF LOST OPPORTUNITIES 1801-1806

In September 1801 he took over the 19th Military Division based at Lyons. It was a posting that reflected he was very much out of favour. A poor showing in Germany coupled with his outspoken republicanism made him a marked man. What probably saved him was his deep hatred of Moreau which made it impossible to link him with any conspiracy from that direction. Still he was a sufficient thorn in Napoleon's side not to offer him an infantry command with the Grande Armée. He remained at Lyons till 20 September 1805 when he was given the 4th Division of Massena's Army of Italy. During the campaign in Italy he took the citadel at Verona on 18 October. He followed up this success taking part in the victories at San Michele and Caldiero on 29 and 30 October 1805. Engaged in the pursuit of the Austrians he was present at the crossing of the Tagliamento on 12 November. In December he moved south and occupied Istria, which then became part of the Kingdom of Italy the next month.

February 1806 saw him in central Italy. He commanded III Corps of Massena's newly formed Army of Naples, which quickly ousted the Bourbons from that kingdom. Various postings in the Italian peninsula followed. They included that of commander of the coastal region from the Neapolitan border to Piombino, commander at Civita Vecchia and then military governor of Ancona province till his recall to France in September 1807.

CATALONIA 1808-1810

He took up the appointment of commander of the Pyrenees-Orientales Division based at Perpignan on 28 January 1808. On 9 February 1808 he crossed the Spanish frontier into Catalonia under the dubious pretext that his troops were part of the invasion force heading for Portugal. As they paraded through Barcelona on 29 February his troops suddenly ousted the startled guards from the citadel and occupied the city. The vacillating Spanish government under Godoy swallowed this blatant violation of its territory. Only when the uprising, which started in Madrid on 2 May spread to Catalonia did Duhesme become alarmed. The Catalans issuing from Gerona cut Duhesme's communications with France. Advancing from Barcelona he scattered a force of Catalan irregulars at Matero on 10 June 1808. After putting the countryside to fire and sword, he reached Gerona on 20 June. He made two treacherous though unsuccessful assaults on the city whilst trying to negotiate its surrender. The country was in uproar around him, and fearing for his communications he scurried back to Barcelona.

As the "somatenes" tightened their grip around the city, he managed momentarily to relieve the pressure when he scattered a mass of irregulars at El Rey on the Llobregat on 30 June. He marched out of Barcelona a second time on 10 July when he received news that Reille was coming to his relief. He pushed aside the somatenes and linked up with Reille at Gerona on 24 July. Even with a properly equipped siege train he still failed to reduce the city's defences. On 16 August a relieving force under Caldagues hit his lines. Forced to abandon the siege he fell back to Barcelona leaving his train and artillery to the enemy.

Duhesme's conduct during this brief campaign was a disgrace equal to that of the worst Spanish generals. The Spaniards had come against him with a mere detachment of some 7,000 irregulars, which a covering force of 5,000 men could easily have stopped. Yet he and Reille with a combined force of some 13,000 allowed itself to be caught strung out around the walls of the city after adequate warning of the enemy's approach.

Napoleon was determined to replace him. He appointed Saint Cyr as the new commander for Catalonia and sent another relieving force of two divisions. Saint Cyr broke Vives's blockade and entered Barcelona on 17 December 1808. Duhesme was pushed aside and given the post of military governor of the city. His rule during this period was cruel and corrupt to the extreme. Reports arrived of his personal involvement in torture and murder in Barcelona's prisons. Drunkenness and robbery was rife amongst the troops. The unauthorized confiscation of private property, the sale of army supplies and the submission of false returns was common place. The final straw came when he carelessly strung out three battalions in villages twenty miles north of Barcelona on the Granollers Road. The purpose of the exercise was to protect Augereau on his arrival as the new commander for Catalonia. The force was cut to pieces by O'Donnell's Catalans when they stormed down from the hills during the night of 21-22 January 1810. An old enemy of Duhesme from the campaigning days in Germany, Augereau seized on this and other misdemeanors and dismissed him from his post on 10 February 1810. He foolishly attempted to get Napoleon to intercede on his behalf but on reaching Paris the Emperor declined an interview and banished him from the city.

FRANCE 1814

He spent the next three years at his home in Bourgneuf. Then with France in desperate straits he was recalled on 3 December 1813 to command the garrison at Kehl on the Rhine. Forced to abandon the town he took charge of the 3rd Division of Victor's II Corps. He fell back from the Rhine into Champagne where he was driven from Saint Die on 10 January 1814 and then Saint Dizier on the 27th. His division was again badly mauled at Brienne on 29 January when caught in the open by Russian cavalry. Only the gathering darkness saved his raw troops from total destruction. At La Rothière on 1 February he redeemed his reputation when his 4,000 men, battered by over sixty cannon held the town throughout the day against Blucher's main assaults. His defense enabled Napoleon to disengage successfully later in the day. At Montereau on 18 February, after the late arrival his troops they were hurled piecemeal by Victor into an unsuccessful assault on Villoron that was only saved by the arrival of artillery support.

His reconciliation with Napoleon appeared complete when Napoleon made him comte de l'Empire on 21 February 1814. His division joined Oudinot's command where he took part in the latter's foolhardy offensive across the Aube on 26 February. He seized the bridge over the river at Dolancourt and occupied Bar. When Schwarzenberg realized that he was not facing Napoleon in person he in turn went on the offensive. Duhesme formed the rearguard as

Oudinot fell back hastily through Troyes and down the Seine Valley to Nogent. In late March he took part in Napoleon's last daring thrust against the Allied communications in eastern France. At Saint Dizier when Napoleon learnt of the Allied advance on Paris he joined the march on the capital via Bar-sur-Aube and Troyes, before retiring with the main army to Fontainebleau.

THE RESTORATION AND NAPOLEON'S RETURN

After Napoleon's abdication the Bourbons retained his services and made him an Inspector General of Infantry on 1 June 1814. He also received the title chevalier de Saint Louis on 27 June 1814. He remained loyal to the Bourbons on Napoleon's return till the last moment, only abandoning the Duc de Berry's forces before Paris on 19 March 1815. Napoleon hesitated to give him any command as a result, but the fighting qualities he displayed the previous year overcame doubts as to his loyalty. Reconciliation came on 2 June 1815 when he became a Peer of France and the following day was offered command of the 11th Infantry Division with III Corps instead of Lemoine, who Vandamme considered too old for a vigorous campaign. A week later he left III Corps when appointed commander of the Young Guard.

Considering he was one of the most senior generals in the Army it was a wise move, since it was unlikely he would have readily accepted orders from anyone, let alone such a difficult character as Vandamme. Whilst throughout his career he had shown he was no master strategist, on the battlefield he held a reputation as a tenacious leader and a foremost expert in light infantry tactics. During the period of his disgrace he had written a widely acclaimed book on the subject, which became a standard work at military colleges during the Restoration and the Second Empire. His appointment to the Guard where he could be kept under a tight rein by Napoleon made good sense. Since there was no doubt as to his bravery, coolness under fire and ability to follow unhesitatingly the Emperor's orders to the letter the Guard was the ideal place for him.

THE WATERLOO CAMPAIGN

He remained in reserve at Ligny on 16 June, and his troops only joined the battle around 6.00 P.M. when the combined assaults by Pirch II and Tippelskirsch drove Vandamme's divisions from Le Hameau and Saint Armand. As the triumphant Prussians debouched before the villages they were hit by Duhesme's men, who charged with great elan. Checked, the Prussians were driven back into Wagnelée, removing the threat to the French Left.

At Waterloo his troops were again held in reserve during the early stages of the battle, where they remained in position to the right of the Brussels highway behind La Belle Alliance. When the Prussians pushed Lobau to the edge of Plancenoit around 5.30 P.M. and the army's line of retreat appeared threatened, Duhesme was ordered forward with the Young Guard to support Lobau. In a furious attack that lasted little over half an hour he cleared the village and its vicinity of the enemy.

The Prussians re-entered the fray and with Lobau he heroically kept open the closing jaws of the Allied vice by stubbornly denying Plancenoit to Bulow's IV Corps. His courage afforded Napoleon a chance to rally the Army after the repulse of the Old Guard. Then when all hope was gone, his heroic defense kept open for Napoleon and the remnants of the Army of the North an avenue of escape from the field. Struck in the head by a musket ball he insisted on remaining on the field, while an aide propped him up on his horse. The wound proved fatal. In the confusion of the final rout he was left at the Auberge du Roi d'Espagne in Genappe. During the night the inn became Blucher's headquarters and he was cared for by Blucher's staff till he died on 20 June 1815.

His remains lie in the graveyard of the church of Saint Martin in the adjacent village of Ways.

CAREER ASSESSMENT - A ROUGH ROTTEN OLD REPUBLICAN

Duhesme was the epitome of the elder generation of Republican general. A rough diamond, an incorrigible looter, brutal and crude to the extreme, he was out of place amongst the new order that arose on Napoleon's rise to power. He was typical of many of the older generals, who fell foul of Napoleon and were consigned to oblivion. He considered the Emperor an upstart and a threat to the Revolution.

He was however a born survivor and offered his services repeatedly to Napoleon not out of love or personal loyalty to him but because he saw little credible alternative for France. Given his chance in 1814 and 1815, his career reached a high point at Waterloo when he paid the ultimate price. The tenacity he displayed is shown in the returns of the Young Guard where only some 600 men survived the campaign compared to some 4,300 at its start. As a strategist he was limited. As a governor or administrator he was a disgrace. His greatest weakness was placing personal gain before military considerations. Yet when in a tight corner, under the direct orders of Napoleon as he was in the Guard, he proved he was one of the most stubborn and accomplished division commanders in the Army.

DURUTTE - PIERRE FRANÇOIS JOSEPH, BARON (1767 - 1827)

EARLY CAREER

Durutte was born on 13 July 1767 in the market town of Douai (Nord) near the Belgian border. He responded to the National Convention's call for volunteers and on 1 April 1792 joined the 3rd Battalion, Volunteers du Nord at the Famars military camp near Lille. With the Army of the North when Luckner invaded Belgium he was present at the capture of Menin on 12 June 1792, Courtrai a week later and then the hurried retreat to Lille when Beaulieu's Austrians appeared in force.

Promoted on 22 August 1792 *sous lieutenant* he joined the Army's general staff at Dumouriez's headquarters. With Dumouriez during the renewed invasion of Belgium, he distinguished himself at Jemappes on 6 November 1792, which resulted in promotion to lieutenant after the battle. He joined the ill prepared invasion of Holland in February 1793 and whilst wandering too close to the siegeworks at Klundert by the Maas on 13 February 1793 was shot in the leg. Out of action he was fortunate to miss the army's defeat at Neerwinden on 18 March 1793 and the resultant purge amongst the officer corps after Dumouriez defected. Both events improved his prospects and promotion to capitaine with the 19th Dragoons serving with Landrin's division came on 6 March 1793. Present at the siege of Wilhelmstadt, he later fought at Hondschoote on 8 September 1793 when the Duke of York was defeated and had to abandon his siege of Dunkirk. The great bravery he showed during the battle ended in Durutte receiving a double promotion first to *adjudant général chef de bataillon* on 15 September followed by *adjudant général chef de brigade* on 30 September 1793.

When Michaud took over the 1st Division of the Army of the North on 19 March 1794, Durutte became his chief of staff. He was with Michaud at the capture of Ypres on 18 June 1794 and the occupation of the Flemish coastal towns. On 21 March 1795, appointed the Army of the North's deputy chief of staff under Moreau, his career looked set but events elsewhere proved the fighting days of the Nord were largely over. Moreau moved on to take command of the Army of the Rhine-Moselle and when the less flamboyant Beurnonville took over Durutte's prospects looked grim. He remained with the Nord for a further three years till 25 August 1799. Then with the army in the final throws of being disbanded he moved to Brune's Army of Batavia.

He first served as chief of staff of the 3rd Division under Desjardin till Daendels took over. At first very critical of the discipline shown by Dutch troops within the division he was responsible for wholesale sacking of several officers. It had the right effect, and after a poor showing at Zyp on 9 September, Daendel's division excelled at Bergen on 19 September. Durutte's promotion to général de brigade followed on 26 September 1799.

THE ARMY OF THE RHINE 1799-1801

The following month he joined the General Staff of the Army of the Rhine and renewed his friendship with Moreau. Active in the Spring Campaign in Germany, in April 1800 he led a brigade with Richepanse's division. He fought under him at Engen on 3 May 1800 and then at Biberach six days later. On 4 June 1800 Decaen took over the division, Durutte joined the advance into Bavaria. At the crossing of the Danube at Dillingen on 18 June, he then fought at Hochstadt the next day. Decaen was later detached to take Munich and Durutte had the honor of leading the troops into the city on 28 June 1800.

In the Winter he played a key role under Decaen at Hohenlinden on 3 December 1800. Marching through the forests in appalling weather they supported Richepanse's flank attack on the enemy's left that secured victory. Afterwards active in the pursuit of the Austrians, his troops were the first to cross the Salzach at Leufen on 13 December 1800.

THE MOREAU AFFAIR 1804

With peace in Europe and the Consulate consolidated, it was a time for most generals to bask in their recently found fame. This they did, but underlying the feeling of harmony a deep resentment grew towards the First Consul, particularly amongst officers of the Army of the Rhine who felt they had not received the recognition they deserved. Moreau who supported Napoleon at the time of Brumaire soon began to constitute a rival to the First Consul. It also stemmed from a jealousy between Moreau's ambitious wife and Josephine, both Creoles who never saw eye to eye. Durutte, a close friend of Moreau and certainly no admirer of Napoleon, sympathized strongly with Moreau's view that the gains of the Revolution depended on the maintenance of the Republic.

At first this undercurrent did not affect Durutte's prospects. Generals from the Army of the Rhine were regarded as the most professional within the army and any qualms they had towards the Consulate were at first not taken too seriously. The disbandment of the Army of the Rhine after the Peace of Lunéville saw him employed in the 16th Military Division as commandant for the Department of Lys based at Lille from October 1801 till August 1803. Promoted *général de division* on 27

August 1803 he moved to Dunkirk as commander of the vast camp preparing for the invasion of England. It was a much sought after command and his career appeared set for great things.

Secret meetings with the royalist Pichegru in February 1804 resulted in Moreau's imprisonment. The event itself upset many in the army including Durutte. Then at his trial in June 1804 when it became clear Moreau was no royalist, followers like Durutte became openly critical of the ambitions of the First Consul. When the court passed its verdict many of Moreau's followers suffered persecution. Durutte's outspokenness resulted in him losing his command. Later satisfied his mutterings were simply those of a nostalgic republican and that he was not part of any plot, Napoleon decided to spoil his career in another way. He recalled him on 22 August 1804, and gave Durutte command of the 10th Military Division based at Toulouse. It was a dead end posting, the last one an ambitious general would want, but then there were few prospects open except for devotees of the Emperor. An appeal to the Minister for War for a more active command resulted in another sideways move when on 28 May 1805 he became governor of Elba. He was to languish on the island for another four years.

THE AUSTRIAN CAMPAIGN OF 1809

On 25 March 1809 with war against Austria imminent Durutte received orders to leave Elba and join Eugene Beauharnais's Army of Italy gathering behind the Isonzo. With his fine record against the Austrians in 1800, the army needed experienced generals of his calibre in Italy. By the time he arrived to lead the Reserve Division, Eugene had already received two severe reverses at Sacile and Caldiero. Grenier's corps soon felt the influence of Durutte's formation, renamed the 2nd Division of the Army of Italy, when it joined the rest of the army on 6 May. Two days later he successfully led Eugene's assault across the Piave. After a rapid advance his force blockaded Venice on 13 May and then moving on seized the fortress of Malborghetto on the Styrian border a week later. In Styria on 25 May he caught Jellacic's division in detail at Saint Michael as it fell back from the Tyrol. In the action that followed he eliminated 5,000 of original force of 7,000 Austrians.

He played a major role at Raab on 14 June 1809 when Eugene caught up with Archduke John's army. During the battle Serras's division became bogged down in their approach to the mound and the Kismeyer farmhouse. To break the deadlock Durutte led the advance against the Austrians on the plateau before Szabadhegy. After a tremendous struggle on the heights as Severoli's and Puthod's divisions joined in, the last of Archduke John's reserves were committed; then the enemy gave way. When he linked up with the Army of Germany his division became the 3rd division of the Army of Italy. Present at the Battle of Wagram on 5-6 July 1809, he supported Macdonald and Marmont's attacks on the second day that broke the Austrian center and won both men their marshal's batons.

Napoleon, appreciative of the part Durutte had played in the campaign, made him baron de l'Empire on 15 August 1809 and awarded him a pension of 4,000 francs drawn on Rome. Durutte then moved to Amsterdam as military governor on 10 August 1810 after the incorporation of Holland into metropolitan France. On 24 December 1810 he took up the additional responsibility of commander of the newly formed 31st Military Division based at Groningen.

THE RUSSIAN CAMPAIGN 1812

The coming campaign against Russia saw Durutte move to Berlin on 12 April 1812 on his appointment as military governor of the city. He also received the additional task to prepare the 4th Reserve Division forming in the city for Russia. The force comprised some of the most unpromising material any general could wish for. There were five penal regiments formed entirely from pardoned deserters and an understrength one from the Principality of Wurzburg. All units were chronically short of supplies and equipment. On 4 July appointed commander of the division he joined Augereau's IX Corps with his formation becoming the 32nd Division of the Grande Armée.

He remained in East Prussia during the early months of the campaign. Then in early October, detached from Augereau he moved to Warsaw to help Schwarzenberg's Austrians and Reynier's Saxon VII Corps, who were being pressed by Tchitshagov's army in the Ukraine. From Warsaw he entered the Ukraine and linked up with Reynier at Bialystok on 11 November. Soon in action at Volkovisk on 15-16 November, he helped the Saxons out of a difficult situation by offering a bold defence and delaying Sacken's corps of 25,000 men till Schwarzenberg's arrival with reinforcements.

GERMANY 1813

Hearing the fate of the Grande Armée, he withdrew westwards with VII Corps and reached Warsaw on 3 January 1813. Forced to evacuate the city on 29 January he covered Reynier's retreat as VII Corps abandoned Poland. He fought at Kalisch on 13 February when Miloradovich surprised Reynier and inflicted a sharp reverse on the demoralized Saxons. He reached Glogau on the Oder on 19 February where he remained a month, before being ordered to fall back to the Elbe. On 19 March he took over the defence of Dresden after Reynier fell ill. The command structure had broken down; the Saxons had broken away on their own to cover the crossings lower down the Elbe at Torgau. Unable to hold out with just the remnants of his division numbering barely 3,000 men, he evacuated the city on 26 March when the Russians appeared in force and fell back to Merseberg on the Saale.

When Napoleon launched his Spring Campaign in Saxony, he was unable to lend support at Lutzen on 2 May. His force faced by Bulow's Prussians on the opposite bank of the Elbe was tied down around Merseberg covering a possible crossing. He joined the pursuit of the Allies after the battle. Again he was unlucky when he didn't reach Bautzen on 21-22 May when Ney's command, which he now fell under, failed to comply with Napoleon's orders.

On 13 August 1813 he became a *comte de l'Empire*. The renewal of hostilities saw his division still with Reynier's Saxon Corps. He took part in Oudinot's advance on Berlin and won a sharp action at Wittstock on 22 August. The next day at Gross Beeren he met disaster when coming to the relief of von Söhr's 25th (Saxon) Division, His troops were swamped by the fleeing Saxons before they could deploy. At Dennewitz on 6 September in the second attempt to take Berlin his troops fought well and with Reynier's Saxons nearly won the battle on their own against superior odds. However Ney, in the thick of the action elsewhere, acting more like a cavalry lieutenant than an army commander, ignored Reynier's pleas for help. When Bernadotte arrived and turned on Reynier with the bulk of his army, his exhausted corps including Durutte's division disintegrated.

He fought heroically at Leipzig, in particular on 18 October 1813 when his division covered the area between Paunsdorf and Schönfel. Here he was placed in a desperate situation when von Söhr's Saxons on his right went over to the enemy en masse and left a great hole in the French line. He plugged the gap and later in the day retook Sellerhausen, which gives little credence to Napoleon's exhortations that the battle was lost due to Saxon treachery. The next day his division formed the rearguard that valiantly held onto the northern suburb of Halle. Only after Bulow's corps had driven back Marmont's divisions on his right and threatened his retreat did he feel compelled to retire. Fighting street by street his troops crossed the bridge over the Elster just before it was blown. He joined Bertrand's IV Corps and was with the rearguard of the Army as it fell back to the Rhine reaching Mayence on 1 November.

FRANCE 1814

In early November his division joined Marmont's VI Corps. He held Coblenz till 1 January 1814 when forced to withdraw after Blucher crossed the Rhine upriver. He fell back to Metz where on 13 January as military governor of the fortress he received orders to hold it at all costs. He withstood repeated assaults by the Russians and with his division and the small garrison tied down over 40,000 troops before successfully breaking out on 24 March. He tried to join Napoleon's Army advancing from the East, but when the Allies turned on Paris and forced Napoleon to follow, he was left on his own. After a long march evading Allied columns he reached

BEFORE THE BATTLE OF WATERLOO, NAPOLEON'S LAST INSPECTION

Fontainebleau on 6 April only to find Napoleon had abdicated.

THE RESTORATION AND NAPOLEON'S RETURN

The Bourbons, aware of his difficult times under Napoleon, treated him fairly and made him commander of the 3rd Military Division and governor of Metz on 23 May 1814. He later became a chevalier de Saint Louis and a Commander of the Legion d'Honneur on 23 August 1814. Well out of the way at Metz he was able to stand back and wait as the dramatic events unfurled during Napoleon's march on Paris. The position he took was understandable, for he was employed and had everything to lose if Napoleon failed. In the end he only offered his services once Napoleon was installed in the Tuileries. For this Napoleon's inner circle considered him suspect, but Davout as Minister for War was more realistic. He thought him sound and backed his appointment as commander of the 4th Infantry Division with Drouet d'Erlon's I Corps on 28 March 1815.

THE WATERLOO CAMPAIGN

In the campaign his division formed the head of d'Erlon's column with the Army's left wing as it crossed the border into Belgium. Delayed by Reille's Corps preceding him, his troops only crossed the Sambre at Thuin in the late afternoon of 15 June. Anxious the next day to make up lost time he reached Jumet on the Charleroi-Brussels Road by 11.00 a.m. but could move no further till Reille's II Corps had defiled. He reached Frasnes after 4.00 p.m. and was ready to join in the struggle for Quatre Bras when d'Erlon was ordered to Ligny. An hour later Durutte had reached the edge of the battlefield when d'Erlon turned back to Quatre Bras and he was left on his own with only three regiments of Jacquinot's cavalry in support. Unfortunately d'Erlon gave him the paralyzing orders to "be prudent", rather than report his arrival direct to Napoleon and the nearest troops. Instead, the appearance of his division to the rear of Vandamme's III Corps caused great alarm as they first thought it to be the arrival of Wellington with a hostile column. The assaults on the twin villages of Saint Armand on the Prussian right were completely disrupted as a result. Only after he received direct orders from Napoleon did he start to move his force. He crawled forward to Wagnelée and pushed the Prussian outposts from the village. Had he and Jacquinot been more decisive and ignored d'Erlon's orders they could have struck the hard pressed Prussians a telling blow.

Present at Waterloo on 18 June 1815 he took post on the extreme right of the French line opposite Papelotte. After a lengthy bombardment his infantry drove Saxe-Weimar's Dutch brigade from the position around 2.00 P.M. He was forced to retire after Picton's rout of Donzelot's and Marcognet's divisions left him dangerously exposed. He renewed the assaults and by 6.00 P.M. had regained Papelotte, worked his way forward up the slope and threatened to envelop Wellington's left. Then as Ziethen's Prussian corps approaching from Wavre deployed on his flank he was forced to halt and face the new threat. He grimly held onto his gains and still was in possession of Papelotte when the attack by the Old Guard failed. Whether it was the Prussians working their way to his rear between his and Jeanin's division so cutting off their retreat, or his men hearing of the fate of the Guard remains unclear, but his men broke. With Ney he tried to rally one of his brigades but Vandeleur's cavalry caught him in the valley. In the desperate struggle that followed his right hand was severed at the wrist by a single sabre blow. Defenceless, a dragoon then split his head open by a second blow that left a hideous wound that disfigured him for life. Unconscious his horse bolted and somehow brought him bloodstained to La Belle Alliance where a surgeon tended him.

FINAL YEARS AND CAREER ASSESSMENT

He survived the horrors of the retreat and was nursed back to health in Paris. Missing an arm and blind in one eye, he retired disabled on 18 October 1815. Dissatisfied with the state of affairs in France he chose to emigrate to Belgium and lived quietly in Ypres till his death on 18 April 1827.

Durutte was one of the most underestimated of Napoleon's commanders. His association with Moreau and his fervent republican sympathies cost him a rightful place amongst the leading commanders after a distinguished career during the Revolutionary Wars. When recalled the roles he played invariably appeared to be side shows in major campaigns, as for example in Italy in 1809 and with Augereau's rear echelons during the Russian Campaign. The result was fine performances that did not receive the recognition they deserved. He performed well under Reynier in 1813 but by this time he was with an army that was badly led and continually beaten. In 1814 operating on his own he proved a stubborn and elusive adversary. He was by far the most capable of d'Erlon's commanders.

Durutte's feelings towards Napoleon were similar to those of Duhesme. They were both from the old school of Republican Generals, who jealously guarded the gains of the Revolution. They considered him an upstart and his ambition a threat to the future of the Republic. He did recognize Napoleon's genius, and as with many others he accepted Napoleon's return in 1815 as there was no credible alternative. It took Napoleon a long time to realize the worth of a man, who for totally different motives would remain loyal to him, and made an exceptional fighting general.

EXELMANS - RÉMY JOSEPH ISADORE, COMTE (1775-1852)

EARLY CAREER

Exelmans was born on 13 November 1775 in Bar-le-Duc (Meuse). On 6 September 1791 he enlisted as a volunteer with the 3rd Battalion of the Meuse, which at the time was led by *lieutenant colonel* Oudinot, later duc de Rovigo and *maréchal de France*. His talents soon recognized, by January 1792 he had risen through the ranks to *sergeant major* with the regimental artillery company only two months after his seventeenth birthday.

He saw his first active service with the Army of the Moselle where he served for two years. He spent time in garrison at Thionville from November 1792, was then present at the defence of Bitche 10-11 November 1793 and fought at Kaiserlautern on 23 May 1794. In June 1794 after the victory at Fleurus his battalion became part of the 34th Demi-brigade with the Army of the Sambre-Meuse. He served with the Sambre-Meuse as it made its spectacular advances, notably, the triumphant march to Brussels, the seizure of Namur, and in the autumn the advance to the Rhine. In 1796 he was with the Army during the crossing of the Rhine and its advance to the Bohemian frontier. He was present at its check at Amberg on 16 August, the defeat at Wurzburg on 3 September and the desperate scramble back to the Rhine and safety. On 22 October 1796 he gained promotion to *sous lieutenant*.

With Austria knocked out of the Coalition in 1797 and the main war effort turned against England, he passed to the Army of England in January 1798. He served on the regimental staff of the 43rd Demi-brigade and on 19 June 1798 was promoted *lieutenant*. On 22 October 1798 he became aide de camp to general of engineers Eblé, the hero of the crossing of the Beresina in 1812. When Eblé later moved to the Army of Rome in December 1798 Exelmans followed him. He took part in the brief campaign against Naples, was present at the capture of Capua on 10 January 1799 and distinguished himself in the operations that led to the capture of Naples itself on 22-23 January.

Keen to serve in the cavalry, he secured a transfer to the regimental staff of the 16th Dragoons on 13 April 1799. Here he struck up a good rapport with General Broussier and in July 1799 returned with him to France as his aide serving at the cavalry depot at Valenciennes. He was with Broussier during the Marengo Campaign, serving in the cavalry brigade attached to Loison's Division. He was present at the storming of Fort Bard on 25 May 1800, fought at Pizzighettone on 5 June, was at the crossing of the Adda on 12 June and the capture of Cremona the next day. Promoted *capitaine* on 8 July 1800, he moved with Broussier to Murat's cavalry camp at Amiens. When hostilities resumed with Austria in December 1800 he returned to Italy with Broussier who headed a cavalry division of the Army of Italy. He received a commendation for bravery for the part he played during the crossing of the Adige on 25 December 1800.

AIDE DE CAMP TO MURAT

When Broussier became governor of Milan, Exelmans transferred to Murat's Staff on 21 March 1801. He served as one of Murat's aides when the Army of the South occupied Naples. He spent a pleasant period in southern Italy till June 1802, when on the army's dissolution he returned to France.

For the next four years his career revolved around that of Murat's. He was with him during his tenure as military governor of Paris and gained promotion to *chef d'escadron* on 3 October 1803. Murat being commander of the Cavalry Reserve during the Austerlitz Campaign led to Exelmans's attachment to its advance guard as it masked the Grande Armée's approach to the Danube. He was with the first squadrons that reached the river at Donauworth on 6 October 1805, an event that tore aside the curtain to reveal Napoleon's army astride the river between the Austrian army and Vienna.

On 8 October Exelmans's name shot to the fore when during the first serious encounter of the campaign at Wertingen he won a reputation for bravery that bordered on recklessness. In the early hours at the head of a squadron of dragoons he encountered several hundred Austrian infantry and cavalry guarding Hohenreichen. The village lay in the Army's path. Annoyed by the sporadic fire from the buildings, he dismounted his two hundred dragoons and led them muskets in hand into the village.

Fresh detachments of dragoons came up and pressed the Austrians harder, and drove them back to Wertingen nearby. After they passed through the town they came upon a huge Austrian square formed on the high ground beyond. It was Auffenberg's force, comprising nine battalions of infantry, four squadrons of cavalry with cannon, numbering some 7,000 men. Mack had sent them out from Ulm to reconnoiter, based on vague reports that had spread of French troops appearing on the Danube.

Murat arrived, surveyed the scene and sent

Exelmans forward with orders for the cavalry to charge. Maupetit commander of the 9th Dragoons hesitated, thinking the odds too great and that the young aide carrying Murat's message was mad. Undeterred Exelmans charged on his own, and Maupetit aghast with his effrontery followed his lead with the 9th Dragoons. Exelmans's horse was brought down thirty yards from the enemy squares by the first crashing volley. Remounting another he joined the melee. For a while the square held while the dragoons tried to sabre the Austrian grenadiers, who replied with musketry and bayonet thrusts, one of which killed Maupetit. Attracted by the noise of the fire Murat's main body of cavalry arrived plus Lannes with Oudinot's grenadiers. Alarmed at the size of the approaching forces the Austrian square tried to move to a nearby wood. Charged by cavalry in the front and threatened by infantry in the rear, the Austrians in a compact mass fell into disorder. The dragoons swept the field as they took several colors and cannon, with over 2,000 prisoners.

Lannes and Murat had seen Exelmans's bravery, and sent him to Napoleon with news of the first success and to present the captured standards. Napoleon assembled the entire General Staff and after giving a stirring speech decorated Exelmans with the *Legion d'Honneur*.

After the Army surrounded Ulm, Exelmans joined Murat's pursuit of Archduke Ferdinand's forces that had slipped the French noose and made for the upper Palatinate and Bohemia. He was at Neresheim on 18 October when Werneck was surrounded and surrendered with 8,000 infantry after Ferdinand with the cavalry abandoned him.

He fought at Lambach on 31 October when the Cavalry Reserve after crossing the Inn engaged the rearguard of Kutuzov's army for the first time. He was in action again at Amstetten on 5 November, when the Russians made a brief stand before retiring to Krems, where they crossed to the north bank of the Danube. With Murat's march along the Danube to Vienna he entered the city with him on 13 November 1805. At Austerlitz on 2 December 1805 he was at the fore during Murat's epic struggle on Napoleon's northern flank to stem the advance of the Russian cavalry.

Davout's III Corps: Prussia and Poland 1806-1807

On 27 December 1805 he left Murat's service when appointed *colonel* of the 1st Chasseurs à Cheval with Vialannes's cavalry brigade of Davout's III Corps. During the Prussian Campaign of 1806 with Vialannes's cavalry he screened III Corps's advance into Saxony. On 12 October, near Naumberg on the Saale, he encountered three Prussian squadrons escorting a large pontoon train. The pontoons, and forty baggage wagons fell into his hands after a short struggle and proved invaluable later in the campaign. On 13 October he crossed the river and by evening reached the Unstrutt at Freiburg. At the same time, Napoleon ordered Davout to march on Weimar and join him there as he expected a battle there the next day. Early in the morning Davout sent Exelmans ahead from Freiburg to reconnoiter the route.

With the 25th Line in support he quickly secured the head of the defile on the edge of the Hassenhausen Plateau. His patrols moved on and located some Prussians in Hassenhausen itself and took several prisoners. They then beat a hasty retreat when pursued by a large cavalry force led by Blucher. As the Prussians followed him and emerged through the mist, Gudin's battalions deployed in square caught them in a vicious fire. Then faced by the rest of Vialannes's brigade the enemy retired taking with them several battalions of infantry moving up to threaten Gudin's right.

The epic battle of Auerstadt for the French was mainly an infantry affair, simply because Davout's cavalry was so hopelessly outnumbered. During the struggle Davout held Exelmans back till near the end of the day when he played a destructive role in the pursuit. He drove hard against the Prussian left flank in the hope of pushing them towards Apolda into the arms of Napoleon. After ten miles, exhausted he broke off the pursuit and bivouacked at Buttstadt. During the pursuit of the Prussian Army in the days that followed he seized the bridge over the Elbe at Wittenberg on 19 October. Davout diverted him from the main advance on Berlin to secure the crossings into Poland at Frankfurt-on-Oder.

His regiment was the first to cross into Poland on 1 November 1806 and occupied Posen four days later amidst a tremendous welcome from the local populace. After resting for several days while the Army caught up, he then set out on 16 November as part of Murat's screen that preceded the Army's advance to the Vistula. He entered Warsaw on 28 November 1806. He crossed the Ukra on 22 December, fought at Czarnovo the next day and was again in action at Golymin on 26 December. At Eylau on 8 February 1807, he cleared the Cossacks before Serpallen and then supported III Corps's advance against the Russian left covering Davout's assault on Anklappen.

Return to Murat, Capture and Parole

On 14 May 1807 promoted *général de brigade*, he returned to Murat as his senior aide in time to be caught in the rout of Murat's cavalry by the Russians at Heilsberg on 10 June. Four days later on 14 June 1807 he did have the satisfaction of witnessing the Russian Army's destruction at Friedland while Murat was with the bulk of the Cavalry Reserve before Konigsberg.

On 17 March 1808 Napoleon awarded him the title *baron de l'Empire*. He accompanied Murat to Spain and was with him during the occupation of Madrid by the French in April. After the Dos Mayos uprising he joined Moncey's expedition against Valencia. On 16 June 1808 while leading a patrol near Cuenca, the Spanish took him prisoner. Imprisoned in Valencia, when the city's fall appeared imminent the Spanish placed him on an English vessel that shipped him to England. After a short period of imprisonment, he gave his parole and lived in Cheltenham, but after two years broke it and escaped to France in April 1811.

SERVICE WITH MURAT AND THE IMPERIAL GUARD

He rejoined Murat, now King of Naples as his Grand Master of the Palace. Life as a courtier not to his liking he returned to France and took an inferior posting as *adjudant commandant major* with the Chasseurs à Cheval of the Imperial Guard on 24 December 1811.

During the Russian Campaign he changed regiments on 9 July 1812 and took the title *major* with the Horse Grenadiers of the Imperial Guard. He served at Smolensk on 17 August and was present at Borodino. Attrition amongst senior officers after the battle a major problem, he gained promotion to *général de division* the day after the battle on 9 September 1812 and replaced the wounded Pajol* as head of the 2nd Light Cavalry Division.

During the retreat from Moscow he shot to prominence at Krasnoe on 15 November. The Army at the time showed signs of complete disintegration. Miloradovich with a column of 20,000 men barred the road west and started to shell a horde of Westphalian stragglers eager to reach Krasnoe. The Westphalians froze in panic, and they made no attempt to deploy as thirty squadrons of Russian cavalry prepared to bear down on them. Exelmans nearby at the time galloped into their midst, assumed command and urged them to defend themselves. The jaded troops responded to the orders of this strange general who had arrived in their midst and began to form them into columns. Exelmans at their head, they joined the Vistula Legion and the Old Guard in an attack on the Russian positions. Surprised by the ferocity of the assault, Miloradovich withdrew and contented himself with falling on stragglers.

THE GERMAN CAMPAIGNS OF 1813

Shot in the thigh near Vilna on 10 December, Exelmans returned to France to recover. On 15 February 1813 he assumed command of the 4th Light Cavalry Division forming at Mainz, where an acute shortage of horses delayed his preparations. Part of Sebastiani's II Cavalry Corps he only reached Saxony in late May, so missing the battles at Lutzen and Bautzen, where Napoleon's shortage of cavalry was critical.

During the Autumn Campaign his reckless bravado proved his undoing during operations against Blucher's Army of Silesia. On 14 August 1813 Blucher struck westwards in violation of the Armistice. When Napoleon turned on him, Blucher suffered a reverse at Goldberg and fell back on Jauer behind the Katzbach. Napoleon then faced by a new threat to Dresden, quickly reorganized the Army on 23 August. He gave Macdonald the new Army of the Bober to shadow Blucher, with orders not to pursue him in force beyond the river. Convinced by reports that the enemy was in full flight, Macdonald crossed the river. Poor reconnaissance by the cavalry failed to reveal that Blucher had tumbled to Napoleon's departure. As a result, the Prussian turned his army about with the aim to recross the river and engage the weakened French. The two armies bumbled into each other on 26 August. More by luck than design Blucher caught Macdonald's widely spread columns in detail as they crossed the Katzbach.

Exelmans crossed the Katzbach at the ford of Chemochowitz in the morning and climbed the steep slopes to the vast Janowitz plateau. At first he scented success, as several French infantry divisions had crossed and were posted prudently near the woods and coppices that covered the plateau. Though there was no sign of the enemy, prudence dictated a thorough reconnaissance of the countryside, but neither he nor Sebastiani ordered one. Roussel D'Urbal* pushed ahead through the driving rain and disappeared from view with the heavy cavalry division and corps artillery.

It was at this stage that Exelmans's dogmatic and overbearing manner, which failed to allow his regimental or brigade commanders to exercise any initiative proved his undoing. The corps guns had been taken without his permission, which angered him so he called the division to halt, and left it with strict orders not to move till he personally recovered the guns. His two brigades five hundred yards apart were drawn up in columns of regiments. In the circumstances this was a very vulnerable position, which worried his brigade commanders Maurin* and Wathiez, who dared not move.

From a nearby wood a large body of Prussian lancers charged the 23rd and 24th Chasseurs à Cheval regiments, who barely managed to deploy in time. After a furious melee the three regiments of Prussian horse fled. The 7th Chasseurs à Cheval on their own defeated another regiment. Marbot (the memoirist), colonel of the 23rd Chasseurs à Cheval, then fell on some Prussian infantry formed in square, but a wall of bayonets held his men back. To Marbot's assistance came the 6th Lancers, who outreaching the enemy's bayonets exacted a heavy toll. Their powder wet and unable to fire their muskets, the Prussians gave way and fled to a nearby wood.

Two miles away while this was all going on Exelmans neared Roussel's division. He saw to his horror his colleague's horse swamped by elements of the 20,000 Allied horsemen that flooded the field. He turned and fled white faced to warn his division of the new avalanche of Allied horse about to descend on them. Committed against one foe, his regiments had no time to prepare and face another. First the flight of Roussel's division unsettled them, and then when the Prussians appeared through the swirling rain they turned and fled.

Charpentier's 36th Infantry Division nearby, unable to form square or fire on the approaching horse as the rain had soaked their powder, broke. This unwieldy mass pursued by the triumphant Prussians then crashed into the remaining divisions of V and XI Corps on the plateau or debouching from the river. Macdonald lost 10,000 men killed and wounded and a further 15,000 taken prisoner. The defeat alongside those of Kulm and Gross Beeren had a major impact on the outcome of the campaign.

Macdonald took full responsibility for the disaster; only years later did he apportion some blame to his cavalry commanders on the publication of his memoirs. This went a long way to explain how Exelmans so soon after the reverse managed to become *comte de l'Empire* on 28 September 1813.

Exelmans fought at Leipzig where he took part in the confused cavalry struggle at Liebertwolkwitz on 14 October. In the three day battle he engaged the enemy before Wachau on 16 October. Here his cavalry's near breakthrough caused the Czar and his staff to scurry for safety.

THE CAMPAIGN IN FRANCE 1814

During the retreat to the Rhine his cavalry was with the army's advance guard and performed valuable service near Hanau as the Bavarians tried to bar their route. When the Allies crossed the lower Rhine in January he was with Macdonald's forces as they fell back through Belgium into Champagne. In February 1814 he was at the defence of Châlons-sur-Marne and Vitry-le-François as York's Prussians pushed Macdonald down the Marne towards La Ferte.

Summoned by Napoleon on 15 February, he assumed command of the 3rd Guard Cavalry Division after the unfair dismissal of Guyot[*], who had lost several guns to Pahlen's Cossacks. For the remainder of the campaign he was very active. At Vertus on 28 February he scattered Tettenborn's Cossacks. He seized the stone bridge over the Aisne at Berry-au-Bac on 6 March, which threatened Blucher's communications and enabled Napoleon to win the costly battle at Craonne the next day. Guyot

THE ATTACK OF THE AUSTRIAN HUSSARS ON THE DIVISIONS OF EXELMANS AND COLBERT AT ARCIS-SUR-AUBE IN 1814

reinstated, Exelmans moved to the 2nd Guard Cavalry Division and took part in the recapture of Rheims on 12-13 March.

After a brief rest he headed south, joined Napoleon's offensive against Schwarzenberg and seized the crossings over the Seine at Méry-sur-Seine on 18 March and at Plancy the next day. At Arcis-sur-Aube on 20 March the Russians severely mauled his cavalry after Sebastiani, inexplicably commander of the Guard cavalry, ordered his and Colbert's divisions to cross the river. Totally outnumbered, Colbert* who formed the first line was driven back and in turn he carried away Exelmans's dragoons. They fell back across the bridge into Arcis with Cossacks in pursuit. Only the personal intervention of Napoleon himself, who at one time had to seek refuge in a square, restored order that enabled the army to conduct an orderly retreat.

He was with Napoleon's Army as it moved east to fall on Schwarzenberg's communications and relieve the beleaguered garrisons in Lorraine. The Allies in turn broke their shackles of indecision, ignored threats to their rear and advanced directly on Paris. Exelmans joined the dash west to save the capital. After the fall of Paris the Army retired to Fontainebleau where he was when Napoleon abdicated on 6 April 1814.

BOURBON INTRIGUES AND NAPOLEON'S RETURN

He accepted service under the Bourbons and on 12 June 1814 became a Cavalry Inspector General with the 1st Military Division based in Paris. On 19 July he took the title *chevalier de Saint Louis*. He soon showed contempt for the monarchy when he refused to swear an oath of allegiance, which resulted in him being placed under police surveillance. His suspension and banishment from Paris soon followed on 10 December due to his correspondence with Murat in Naples. He ignored the order, was arrested ten days later and brought before a court martial presided over by Drouet d'Erlon* at Lille on 15 January 1815.

Here the Bourbon plans went wrong. The establishment hoped a trial outside Paris would not attract attention, but d'Erlon was a secret Bonapartist. The Bourbons demanded an example be made and brought a host of charges against him, including being in correspondence with the enemy (Murat) and espionage. He conducted a brilliant defence. The press took up the story, which included the plight of his pregnant wife. The wife of a war hero, whose husband the police snatched from her arms in the middle of the night, caught the public imagination. The Court acquitted him on 25 January 1815. The verdict rocked the military establishment. It questioned the loyalty of the Army and showed a need to purge it Bonapartist officers.

When news of Napoleon's landing reached Paris, Exelmans kept a low profile. He did not wish to draw attention to himself after a premature conspiracy hatched by Lefebvre Desnoettes*, d'Erlon and the Lallemand* brothers had failed. He bided his time, then on 19 March with Paris in confusion he struck and rallied bands of half-pay officers, who encouraged Royal troops in the city to desert. He seized the artillery depot under the command of the Duc de Berry and then established himself in the Tuileries with a battery of guns till Napoleon's arrival that evening. He was immediately despatched to pursue Louis XVIII to the Belgian border, with unofficial orders from Napoleon ensuring that Louis came to no harm or was in fact captured.

On 31 March 1815 he assumed command of the 1st Cavalry Division of II Observation Corps under Reille. He became a Peer of France on 2 June 1815. On 5 June he was given the II Cavalry Corps made up of the two fine Dragoon divisions under Strolz and Chastel.

THE WATERLOO CAMPAIGN

By 14 June he had concentrated his force on the Belgian border at Walcourt. The next day he crossed before dawn and headed for Charleroi to secure the bridges over the Sambre. He failed to achieve his objectives due to the confusion caused by Vandamme's infantry starting late and so was unable to support Chastel, who was left on his own to try to storm the bridges. He crossed the river in the afternoon, after the Marines of the Guard had cleared the way. Involved in pursuing the Prussians as they fell back on Fleurus, barricades across the road at Gilly supported by cannon fire brought him to a halt. In the action that followed his dragoons worked their way around the Prussian left flank and by evening had bivouacked astride the Fleurus Road between Lambusart and Wangenies.

During the Battle of Ligny the next day, he operated under Grouchy as part of the Cavalry Reserve. Posted on the French right south of the Ligny brook opposite Boignée, he maneuvered his command skilfully, holding in check the Prussian left, while Pajol watched their movements farther east. His two divisions in an open plain were under fire all the afternoon. With only twelve cannon to reply to over thirty guns, his finest work came around 5.00 P.M. when he managed to drive back five battalions led by von Hobe that tried to turn the French right. He caught the Prussians on the crest of a hill midway between Mont Potriaux and Tonginelle when his cannon opened up. Strolz's 5th Dragoons protecting the guns, concealed by undulations in the terrain deployed at the trot, then broke into a gallop and supported by the 13th Dragoons rode headlong into the enemy's column and overthrew it. They took five Prussian guns and halted any further threats to that sector.

Such work tied down 18,000 men of Thielemann's Prussian III Corps. His use of skilful feints, counter feints and the occasional charge first prevented these formations from threatening to turn the French line, and later when the battle reached its crisis from coming to the aid of Blucher's hard pressed center. In the evening his horsemen poured through Ligny and harried the Prussian retreat, but lack of clear orders and the gathering dusk prevented them from exploiting the situation to the full.

Directed to seek out the enemy the next morning, he located Thielemann's Corps at 9.00 A.M. at Gembloux retiring north along the Wavre Road. He never attempted to engage or harry the enemy, in spite of both Pajol's and Teste's divisions being nearby to lend support. Neither did he bother to notify Grouchy immediately of their presence. He completely missed the significance of such a move, that indicated a real danger of the Allied Armies effecting a link up before Brussels. Thielemann around 2.00 P.M. cleverly masked his force with cavalry, slipped away and Exelmans lost contact for twenty-four hours.

Had Exelmans engaged Thielemann with some vigor, the resulting commotion would have drawn Grouchy's forces to the action like iron filings to a magnet. If the Prussians were caught a day earlier, there is little likelihood they would have been able to intervene at Waterloo after extricating III Corps caught in a deadly struggle between Gembloux and Wavre.

On 18 June all that Exelmans did was engage a Prussian infantry detachment in the morning that was retiring on Wavre from Mont Saint Guibert. He was also present with Gérard and Grouchy when the opening cannonade at Waterloo was heard in the distance. In the altercation that followed, it is believed he offered to shot Grouchy rather than let the march on Wavre continue. Gérard calmed him down. In the afternoon he tried without much success to turn the Thielemann's left flank along the Dyle at Wavre. He later added to Grouchy's concern by pointing to the masses of Prussians moving west in the direction of Napoleon and Waterloo.

Grouchy held him in reserve the next morning when he renewed the assault on Wavre. When news of Waterloo reached Grouchy around 11.00 A.M. with further news that the escape route via Charleroi was about to fall into Prussian hands, Exelmans was ordered to secure the crossing over the Sambre at Namur. With amazing speed his cavalry reached Namur at 4.00 P.M., having covered the twenty-five miles to secure the crossing in little over four hours. He covered Grouchy's force as it fell back to Givet, then the rest of the Army as it fell back by stages from Laon to Paris. On 1 July he defeated Blucher's cavalry at Velizy and Roquencourt as they tried to approach Paris from the south.

EXILE AND LATER CAREER

After the Armistice he retired with the Army to the Loire. He stayed at Clermont till relieved of his command on 24 July 1815. Exelmans foolishly refused to keep a low profile, which resulted in his exile from France on 9 December after he publicly denounced the execution of Ney. He lived for varying periods in Brussels, Liege and the Duchy of Nassau till allowed to return in January 1819. Reinstated on 1 September 1819, the Army treated him badly and he remained unattached till 7 May 1828 when took the appointments as Cavalry Inspector General in the 9th, 10th, 12th and 21st Military Divisions.

Never reconciled to the Bourbons he actively conspired against them. On the outbreak of the July Revolution he secured the defection of the Paris garrison, and with Pajol on 3 August 1830 led them against the Royalist forces regrouping at Rambouillet. On 21 August 1830 he was awarded the Grand Cross of the Legion of Honor by Louis Philippe and then made a Peer of France for the second time on 19 November 1831. On 3 November 1840 he received a life appointment to the Army General Staff.

With the establishment of the Second Republic and the resurgence of Bonapartism, he found himself showered with awards and assumed the role of an Old Soldier and Elder Statesman, similar to that of Soult. He became a Grand Councillor of the Legion of Honor on 15 August 1849, Marshal of France on 10 March 1851 and was called to the Senate on 26 January 1852 where he sat as a Conservative. Active till the end, he died at Sèvres (Seine-et-Oise) on 22 July 1852 as a result of a fall from a horse.

AN ASSESSMENT - ARROGANT, RECKLESS AND OVERRATED

Exelmans possessed a reckless arrogance and luck that contributed to the notable successes of his career at Wertingen and Krasnoe. His failings as a division commander soon showed up in 1813 but went unheeded. For the first time in charge of a corps during the Waterloo Campaign his confidence deserted him and he was clearly out of his depth. His failure to notify Grouchy promptly of the direction of the Prussian retreat after Ligny was lamentable. The fact he was given a Marshal's Baton was as incomprehensible as the one awarded to Jérôme Bonaparte. Jérôme however was broke and needed the pension the award carried. Exelmans was on the other hand a wealthy influential man, who as a relic of the past could prove useful to Louis Napoleon. It confirmed the rank had been debased. Brave and loyal he undoubtedly was, but his overestimated talents proved costly in Napoleon's last campaign.

*FLAHAUT DE LA BILLARDERIE -
AUGUSTE CHARLES JOSEPH,
COMTE DE (1785 - 1870)*

BACKGROUND AND EARLY CAREER

Among Talleyrand's many liaisons was one with the wife of the comte de Flahaut, who presented him with a son Auguste, born on 21 April 1785 at the Louvre in Paris. Morals of the day dictated the comte show a brave face to his wife's indiscretions and he accepted the boy as his own. Talleyrand, like Napoleon, never loved or cared for anybody more than himself, and took little interest in the precocious child's welfare.

The Flahaut family fell on hard times during the Revolution. His mother and father were by now estranged. Auguste left France with her for England in September 1792. Comte Flahaut remained, was arrested and later guillotined at Arras in May 1793. Living in London and then Switzerland the comtesse Flahaut did her best for her son, and when her money ran out she resorted to writing novels.

In August 1797 with the easing of tensions in the period of peace that followed the Italian Campaign Flahaut's mother, with Talleyrand's help was allowed to return to France. Through Talleyrand who had become an influential politican, she joined the Barras-Talleyrand entourage in Paris society. It later helped Auguste secure a post in September 1799 with the Ministere de la Marine as a student hydrographical engineer.

He soon tired of a student life and on 24 March 1800 enlisted as a volunteer with the Hussars of the Consular Guard, a formation raised from reconciled emigrés and men of good families. Hardly had he completed his basic training when on 19 May he transferred to the 5th Dragoons with the Army of the Reserve. He crossed the Alps and faced his baptism of fire at Marengo on 14 June 1800 when Kellermann* made his legendary cavalry charge. Promoted brigadier on 18 March 1801, soon after he received a commission as *sous lieutenant* a week later.

AIDE DE CAMP TO MURAT

After the excitement and glory of Marengo a quiet posting with the Corps of Observation of the Gironde away from the hectic social scene of Paris followed in 1801. His mother anxious for his career soon intervened and through her contacts on 21 October 1802 helped him secure a place as one of Murat's aides, who was then Military Governor of Paris.

He remained with Murat for the next four years during which time he gained promotion to *lieutenant* on 16 November 1803. The social scene in Paris more to his liking than the hardships of military life he formed a close friendship with Josephine's daughter Hortense Beauharnais. Auguste had shared billets with her husband Louis Bonaparte during the Marengo campaign. That not enough, by September 1804 still not yet twenty, he was also bedding Caroline Bonaparte, Murat's wife. It was his leaving Paris for the Austerlitz Campaign which probably prevented him being caught and ruining his career. He carried his affairs with a brazeness only matched by his reckless bravado on the battlefield. He distinguished himself at Nuremburg on 21 October and then again at Enns on 3 November where shot in the arm he took no further part in the campaign.

Promoted *capitaine* on 10 February 1806 he served in the Prussian and Polish Campaigns. He fought at Jena on 14 October 1806 and was at the capture of Erfurt the next day. Engaged in the long pursuit across Prussia he was at Prenzlow on 28 October when Hohenlohe's force surrendered and at the capture of Lubeck on 7 November 1806. In Poland he distinguished himself at Golymin on 26 December.

His affairs began to occupy his mind, as mess-room gossip was an embarrassment and he wanted to distance himself before Murat found out. On his promotion to *chef d'escadron* on 15 January 1807 he requested a transfer and secured a posting to the 13th Chasseurs à Cheval with Lasalle's division. A particularly tough period followed compared to the heady pursuits across Prussia with the Murat entourage. The harsh conditions played havoc with both men and horses. Also that vital element called luck, that distinguished Lasalle from other fine cavalry generals, deserted him in Poland. At Eylau on 8 February 1807 Flahaut's regiment remained in reserve and did not join the charge by the Cavalry Reserve. At Heilsberg on 10 June he had another bad day when his regiment caught deploying, was overwhelmed by Russian cavalry with the rest of Murat's cavalry.

MORE AFFAIRS OF THE HEART

Back in France after the Treaty of Tilsit, affairs of the heart soon landed him in trouble. His affair with Caroline Murat continued. Murat suspected and the hot headed Gascon was after him, which made it expedient that he join Junot's expedition to Portugal in October 1807. He returned to France in March 1808 and soon after was appointed as an aide de camp to Berthier. He was with Berthier throughout the campaign in Spain, showing bravery particularly at Somosierra.

He left Spain in January 1809 when Berthier was recalled to head the Army of Germany. Throughout the Danube Campaign he was present at all the major battles including Eckmühl, Aspern-Essling and Wagram. Promoted *colonel* on 13 May 1809 he also received as an annual pension of 4,000 francs drawn on Rome.

A period of relative inactivity followed where his renewed liaison with Caroline, now Queen of Naples, continued under the nose of Murat. In the end it required him to leave Paris to avoid a scandal. Flahaut's old friend Eugene Beauharnais invited him to spend the summer of 1810 at the fashionable resort of Plombières. Hortense his sister, now the estranged ex-Queen of Holland was there. They renewed their affair and on 6 November 1810 Flahaut took up an appointment as her equerry. A son, the future Duc de Morny, one of the leading political figures of the Second Empire was born on 15 September 1811. To avoid a scandal the child was whisked away, given a false identity and brought up by Madame de Souza, Flahaut's mother. At first Napoleon was merely irritated by Caroline using one of Murat's young aides to make a cuckold of him. The irritation later turned to outrage when he learnt his brother Louis had received the same treatment from the same man. Pressure placed on him to move well away from Hortense, Flahaut resigned his position.

CAROLINE

THE RUSSIAN CAMPAIGN OF 1812

In May 1812 Flahaut joined Schwarzenberg's Austrian Corps as a liaison officer for the invasion of Russia. Once the campaign was under way he found his way to Warsaw where he joined Eugene de Beauharnais as his senior aide. He distinguished himself at Ostrovno on 28 July and was in the thick of the fighting at Borodino and Malojaroslavets. He was at Eugene's side throughout the retreat from Moscow and received promotion to *général de brigade* on 4 December 1812 for his conduct during the crossing of the Beresina.

NAPOLEON'S STAFF AND IMPERIAL AIDE DE CAMP

With Eugene as the remnants of the Grande Armée fell back through Poland and Prussia, he was at his side during Spring Campaign in Saxony. When Eugene departed for Italy in June 1813, he realised his ambition of many years when Napoleon called him to serve on his staff. His first task was to head a commission working with the Prussians and Russians to ensure compliance with the terms of the Armistice and resolve any disputes that arose. Despite differences in the past, Napoleon had a very high regard for Flahaut's charm, manners and efficiency. So much so, that he seriously considered him for the post of Grand Marshal of the Palace, in place of Duroc who had died in May. His liaisons past and present with members of the Imperial family in the end counted against him and tipped the scale in favour of Bertrand who took the post in November 1813.

In the Autumn Campaign he served at Dresden, Leipzig and Hanau. During the turmoil of the retreat from Germany he accepted the formal appointment as Imperial Aide de Camp and promotion to *général de division* on 24 October 1813. His return to France coincided with Napoleon making him a *comte de l'Empire* on 11 December 1813 and the award of a further pension of 18,000 francs a year drawn on Rome.

He served at Napoleon's side during the campaign in France carrying out many important assignments. One of his missions was to open negotiations with the Allies at Chaumont. His skill as a diplomat nearly secured peace, but for Napoleon's insistence that France retain her natural frontiers the talks collapsed on 23 February 1814. Another mission Napoleon gave him was to urge Marmont not to abandon Paris before the Allies entered the city on 31 March 1814.

THE RESTORATION AND NAPOLEON'S RETURN

After Napoleon's abdication he offered his services to the provisional government on 14 April 1814. He received the title *chevalier de Saint Louis* on 29 July 1814. Considered an ardent Bonapartist there was no place for him in the new army and he found himself placed on half-pay on 1 September 1814. He considered joining the military plot hatched by generals d'Erlon*, Lefebvre Desnoettes* and Exelmans* to overthrow the monarchy, but little came of it.

Known to be on Fouché's list of suspects, he kept a low profile when news of Napoleon's landing first reached Paris. He watched, waited and on 20 March like so many other waverers when events gathered momentum and Napoleon appeared before Paris on 20 March he rushed from the city to join him. It helped ensure that any hesitation he had previously shown was not misunderstood. Any misgivings Napoleon had were immediately put aside and he resumed his duties as an Imperial Aide. Flahaut's standing amongst the Austrians was high so Napoleon sent him to Vienna to negotiate the return of Empress Marie Louise. He turned back at Stuttgart when it became apparent the Allies were in no mood to talk.

Back in Paris he was seconded to the Ministry for War to examine the service records of officers Davout recommended to hold commands. Davout took exception to the move. To have a young upstart who had little experience of senior command second guessing his appointments was too much for the overworked marshal even if he was acting on Napoleon's orders. Flahaut equally could have handled the task with more tact but the assignment went to his head. In the end Davout had him literally thrown out of the Ministry for War and Flahaut had to complete his work at the Tuileries nearby. Napoleon in recognition of Flahaut's enthusiasm and loyalty on 2 June 1815 made him a Peer of France.

THE WATERLOO CAMPAIGN

As one of Napoleon's aides during the Waterloo Campaign, he was present at Quatre Bras on 16 June. He was guilty of a serious omission after the battle when he failed to appreciate the significance of Wellington's troops beginning to withdraw during the night. Had he immediately returned to Ligny to inform Napoleon, there is doubt that Wellington would have been able to slip away so easily with Napoleon bearing down on his flank from Ligny in the early hours of the morning.

Present at Waterloo he marched with the Guard during its final assault on Wellington's lines. Miraculously he survived the battle unscathed and fled the field with Napoleon. Once in Paris he made a stirring speech before the Chamber of Peers to accept Napoleon's son as Emperor if Napoleon abdicated. When all when all appeared lost, he successfully approached the provisional government to allow two frigates to be placed at Napoleon's disposal for him to travel to the United States. After Napoleon's departure for Rochefort, his work complete, he took his leave and joined Exelmans's cavalry nearby. He replaced Strolz as head of the 9th

Cavalry Division outside Paris on 29 June 1815. He fought at Rocquencourt on 1 July and was with the Army as it retired to the Loire.

MARRIAGE AND EXILE

Proscribed on 24 July 1815, mysteriously no charges were brought against him. Unofficially the old hand of Talleyrand was doing its work on his behalf. A final offer of marriage spurned by Hortense, on 28 July 1815 sick and disillusioned he left his post after being granted sick leave. On 1 September 1815 he was placed on the non active list. He wandered through Europe for several months spending periods at Aix-les-Bains, Geneva, Frankfurt and Amsterdam before seeking refuge in England.

As the former lover of two queens London society considered him a much sought after trophy. The notoriety he had acquired from these liaisons far exceeded his prowess shown on the battlefield. Moving amongst the Whig aristocracy he befriended Margret Elphinstone the daughter and heiress of Admiral Lord Keith, whose squadron had prevented Napoleon sailing from France. They were married on 19 June 1817. Flahaut resigned from the French Army and the couple settled in England. Surprisingly the marriage was particularly successful, and five daughters were born. They continued to live in England and Scotland for the next ten years bringing up their young family. By

HORTENSE

1827 both he and his wife wished to return to France and after discreet enquiries permission was given to settle in Paris.

DIPLOMATIC CAREER 1831-1848

The fall of the Bourbons meant his re-admission to the Army on 14 November 1830 as a member of the General Staff. On 5 May 1831 he became a Grand Officer of the Legion of Honour. Two days later he embarked on a diplomatic career as a special envoy to the King of Prussia. He had the twofold task of enlisting Prussia's support over the question of Belgian independence and also to plead for restraint in their handling of the Polish Question. He stayed in Berlin till September, was successful in the first mission but his pleas for the Polish cause fell on deaf ears. After his return he became a Peer of France on 19 November 1831. He joined the French invasion of Belgium where he served as aide de camp to the Duc d'Orléans.

In the years that followed his diplomatic career faltered. He hoped to become ambassador to London, but Talleyrand held the post. His father was now a wily octogenarian wielding immense influence from London and blocked his every move. In 1837 he became first equerry to the Duc d'Orleans. He was awarded the Grand Cross of the Legion of Honour on 5 May 1838. His past association with Napoleon for a short time threw him in the public eye. Considered part of the old inner circle he helped organise the great ceremony that celebrated the return of Napoleon's remains from Saint Helena in December 1840. From 1841 he served as ambassador to Vienna. His time at the Court of Metternich he enjoyed. Soon out of tune with the radical Republicans that took over after the fall of Louis Philippe in February 1848 he was recalled and retired from the Army on 8 June 1848.

THE SECOND EMPIRE 1852-1870

He went along with the revival of Bonapartism under Louis Napoleon as
a viable alternative to the successive unstable governments of the Second Republic. The creation of the Second Empire saw him become a senator on 31 December 1852. He became a Grand Chancellor of the Legion of Honour on 27 January 1864 and was admitted to the Reserve Section of the General Staff.

Whilst he received the honours and awards his role during the Second Empire was more that of elderly onlooker rather than a principal actor. He greeted the Second Empire without enthusiasm. He was never an intimate or much of an admirer of Napoleon's nephew, although he was on friendly terms with him. He did however accept the benefits, nomination as a Senator and Chancellor of the Legion of Honour.

He succeeded Persigny in 1854 as President of the Corps Législatif. In 1860 he realised his dream of the London embassy. But he was seventy-five and two years later he resigned after the death of his wife. He learnt what went on behind the scenes from his son De Morny who was numbered among the inner circle of politicans and advisers that helped Napoleon III formulate policy. Flahaut was one of the few in on the secret and on the spot when De Morny as Minister of the Interior carried through the fateful coup on 2 December 1851 that secured Louis Napoleon the Second Empire. De Morny's sudden death in 1864 surrounded by numerous financial scandals was a blow to him, and it did little good to the Empire's prestige or Flahaut's reputation.

Flahaut disapproved of the Italian Campaign in 1859 and the Mexican adventure, but did not indulge in public criticism. Except for his two years in the London he did little in any way to shape the fortunes of the new regime. He witnessed with concern the decline of the Second Empire and the drift to war against Prussia. He died in Paris on 2 September 1870 the day of Sedan. He was spared the sight of the Empire's humiliating demise but lived long enough however to know it was doomed.

ASSESSMENT - AN UNDERESTIMATED MAN

Flahaut was well mannered, impeccably groomed, handsome and brave, attributes essential for an Imperial Aide to possess. To assess his prowess as a military man is more is difficult. At the head of large formations his experience was limited, for he never headed anything larger than a cavalry regiment. Hard fighting commanders like Davout had little or no time for him, as to them he was all show and influence which meant little when the bullets started to fly.

The assignments he had under Napoleon showed he had the makings of an accomplished diplomat and this proved the case in later life. His amorous affairs resulted in him carrying the nickname le général de la boudoir, but contrary to the belief of fellow officers more likely injured his career than helped it. He inherited a fair portion of his father's brains and to his credit possessed a far warmer heart. Whilst Talleyrand was one of the cleverest men of the time, he was too sarcastic and avaricious to arouse the sympathy of anyone. Auguste on the other hand was the friend to all, a born charmer who began his series of conquests as soon as adolescence permitted. He was no playboy or feckless philanderer, a man with a distinguished military record as his could not afford to be. Women to him were a weakness rather than a vice.

A fervent believer in Napoleon to the end, he was the longest surviving of Napoleon's generals and outlived Waterloo by over fifty-five years. His grandson Lord Landsdowne was British Foreign Secretary from 1900-1906 and was largely responsible for concluding the Entente Cordiale with France in 1904.

FOULER - ALBERT EMMANUEL, COMTE DE RELINQUE (1769 - 1831)

BACKGROUND, EARLY LIFE AND CAREER

Son of the Royal Stable Master at Versailles, the cavalry general Albert Fouler was born in the village of Lillers near Béthune (Pas de Calais) on 9 February 1769. Following his father's footsteps, he developed a keen interest in horses and followed him into the Royal Stables as a stable boy. Good at his work he received the prestigious appointment as Royal Stable Boy to Louis XVI on 1 April 1786. The post later enabled him to gain a commission as a *sous lieutenant* with the Navarre Regiment (5th Line) on 12 September 1787, a rare occurrence for a person not of noble blood. His progress was steady, *lieutenant* on September 1791 and then *capitaine* on 1 May 1792.

His first active service was with Luckner's Army of the North in Belgium. In September 1792 he became an aide de camp to General Pully, where he saw service with Kellermann's Army of the Center as it pushed the Duke of Brunswick's Army back to the frontiers after Valmy. When Pully moved to the Army of the Moselle in December as commander of that army's cavalry, he followed. He took part in the capture of the Wavren heights on 15 December 1792. After Pully's arrest as a Royalist, and then his subsequent trial and disgrace Fouler moved to the Army of the Sambre-Meuse in March 1795. With the Sambre-Meuse for nearly four years he served for varying periods on the staffs of Drouet* and Mortier, who at the time both held the rank of division *adjutant général*. Shot in the foot at Schweinfurt on 24 July 1796 the wound caused him to be out of action for several months.

During 1798 he moved with Mortier to the Army of Mayence. His career took a change in direction when he joined the 19th Cavalry Regiment (later dissolved in 1803 to form the 1st Carabiniers) on 16 March 1799. Captured in a skirmish near Mayence on 16 May 1799, he spent several months as prisoner in Austria till exchanged in October 1799. Promoted chef d'escadron with the 21st Chasseurs à Cheval on 20 November 1799 he served for a time with the Army of Italy. He fought in the Spring Campaign when the Austrians trapped Massena in Genoa at the end of April 1800 but managed to retire west to the Var with Suchet's forces.

He moved to the Army of Gallo-Batavia when promoted chef de brigade of the 24th Cavalry Regiment on 26 October 1800. He served with Augereau's forces as they supported Moreau's march into Bavaria. On 20 November 1801 he became commander of the 11th Cavalry Regiment (11th Cuirassiers).

THE GRANDE ARMÉE 1805-1807

The establishment of the Empire saw him join the Imperial Household with the title Master of Horse to the Empress in May 1804. A reason for his selection was that Napoleon trying to establish a dynasty wished to copy many of the traditions of the old Court at Versailles. Having held an honorary post in the Royal Stables Fouler appeared a most appropriate choice for a similar post in the new Imperial household. The post being at the time not too onerous allowed him to continue his military career with his regiment at the time based at Compiègne.

During the Austrian Campaign of 1805 he led the 11th Cuirassiers with d'Hautpoul's 2nd Heavy Cavalry Division of the Cavalry Reserve. He distinguished himself during the Battle of Austerlitz on 2 December when d'Hautpoul's cuirassiers swept the Russian cavalry from the area behind Posorlitz and helped to break the deadlock with Bagration on the northern flank.

He fought at Jena on 14 October 1806 and was with d'Hautpoul during the pursuit of the broken Prussian army across Saxony and Prussia that ended on 6 November with the capture of Lubeck. Promoted *général de brigade* on 31 December 1806 he moved to Espagne's 3rd Heavy Cavalry Division as commander of its 2nd Brigade (7th and 8th Cuirassier regiments) when it arrived from Italy. He spent the winter months near Danzig covering the siege of the port. In the Summer his brigade rejoined the Cavalry Reserve and was present at Heilsberg when the Russian cavalry routed Murat whilst it was trying to deploy. Wounded by a lance in the action he was not present at Friedland four days later.

THE DANUBE CAMPAIGN

He spent the following year stationed in Germany and received the title *comte* de Relinque on 16 September 1808. In January 1809 when war with Austria appeared imminent he resumed his command with Espagne's cuirassier division in charge of the 7th and 8th Cuirassiers.

The greatest exploits of his career came at the Battle of Aspern-Essling between 21-22 May 1809. In the evening of 20 May with the rest of Espagne's division he crossed to the Danube's northern bank. His brigade with the rest of the division took post between the villages of Aspern and Essling while Lasalle's light cavalry fanned out ahead in search of the enemy's outposts. When the whole Austrian army suddenly appeared through the morning mist Bessières in command of the French cavalry deployed Espagne in four lines behind Lasalle, who totally outnumbered had hurriedly retired. With thirty two squadrons Bessières tried to stem the Austrian advance but could do little more than delay it. By 3.00 P.M. a column led by Hiller enveloped Aspern as further light cavalry under Marulaz tried to halt the progress of the Austrian infantry. Espagne and Lasalle were left to hold the center without infantry or artillery support. They furiously attacked the Austrians who threatened Essling but made little headway against the enemy infantry. The Austrians, confident that victory was within their grasp, received the charges with utmost steadiness holding their fire till the squadrons were barely ten paces away and then cut the French horsemen down.

Bessières in person led Espagne's division forward after Lannes had accused his cavalry of not charging with sufficient vigor. In all sixteen squadrons charged forward to relieve Essling. They broke the first line of enemy infantry, avoided a second line of squares but as they met the third their charges lost impetus as enemy cavalry fell on them. The task required too much and the assault failed. Espagne died, after being hit in the stomach by grapeshot and unhorsed. Fouler's brigade lost a third of its number that day. Then placed in charge of the division the next day he joined the bloody and confused cavalry combats that were repeated through a storm of musketry and grape. Napoleon asked for too much again when he called on Bessières for a final effort to cover the army's retirement to the Isle of Lobau. Bessières responded by sending Fouler forward with the remnants of six squadrons. The gesture was not in vain, and the bulk of the army managed to recross to Lobau. Fouler received several sabre wounds and after his horse was killed, a troop of hussars captured him while he was trying to make his way on foot to the river.

Released when hostilities ceased on 12 July Fouler returned to Napoleon's headquarters in Vienna on 22 July 1809. On 11 August 1809 he took command of the corps cavalry with Junot's newly created VIII Army Corps that had formed at Frankfurt. Ordered to Spain on 20 October 1809 he took charge of organizing the newly arrived cavalry drafts into the 2nd Cavalry Reserve Division of the Army of Spain. In February 1810 he was to pass to the 3rd Dragoon Division led by Milhaud* as commander of its 4th Brigade. Whilst passing through Paris Napoleon secured Fouler's recall to become one of his equerries and he never took up the post.

MASTER OF THE HORSE 1810-1814

On 17 April 1810 he became Master of the Horse, his main function being responsibility for the Imperial Stables, which included the procuring of horses for use by the Imperial household, their training, care and maintenance. The administration and running of the Imperial Stables that housed nearly a thousand horses caused his military career to take second place. The work kept him in France and for four years he took no part in any campaign.

In 1814 with France on its knees he persuaded Napoleon to allow him to join his staff as an aide de camp. He distinguished himself at Saint Dizier on 23 March 1814 when he joined a successful charge led by Lefebvre-Desnoettes* against Tettenborn's cossacks. The victory itself was comparatively unimportant, but raised the morale of the troops who had suffered great hardship and fatigue in the previous weeks. In the spirit of optimism that briefly prevailed at the Imperial Headquarters after the battle he was promoted général de division the next day.

THE RESTORATION AND NAPOLEON'S RETURN

The Bourbons immediately placed him on half pay after Napoleon's abdication. No Master of Horse from Napoleon's household was to look after the stables at Versailles and the Tuileries. A change of heart came on 6 July 1814 when they recalled him to serve as a chef d'escadron with the Royal Musketeers. It was a calculated insult for man who held the rank of général de division. To soften the blow he received the title chevalier de Saint Louis on 29 July 1814. Then later in a vain hope to secure his loyalty as Napoleon neared the gates of Paris Louis XVIII made him a Commander of the Legion of Honor on 19 March 1815.

Not surprisingly Fouler offered his services to Napoleon who accepted him. He returned to his role as Master of the Horse during the Hundred Days and in that capacity served in the Waterloo Campaign. He held no operational command. His role was very similar to that of other Imperial Aides with the exception that whilst they could be sent off at any moment to fulfil any task, he was always close at hand to provide Napoleon and his staff with fresh mounts when needed. His main claim to fame in the Campaign was that he had fresh horses available when Napoleon took to a carriage after fleeing from the battlefield. He retired from the Army on 9 September 1815.

He returned to his home at Lillers. He never took up an active command nor was he recalled. He lived quietly at his home till his death on 17 June 1831.

CAREER ASSESSMENT

As a general Fouler was destined more for the role of an administrator and ceremonial officer once he joined the Imperial Staff rather than that of an active fighting field commander. In both roles he proved to be sound. Whilst he never commanded a division in the field, Fouler certainly proved his worth as a brigade commander with the cuirassiers in 1809. He was reliable and showed good all round ability as a leader in the field. Later in an administrative capacity the horses and carriages of the Imperial Household were always at hand and in impeccable condition. His move to the Imperial Household offered him few opportunities to excel in the field and makes it difficult to assess objectively his abilities as a battlefield commander in the later days of the Empire.

FOY - MAXIMILIEN SEBASTIEN, COMTE (1775 - 1825)

EARLY CAREER

The indomitable infantry general and Peninsula veteran, Foy was born in Ham (Somme) on 3 February 1775. He entered La Fère as an officer cadet on 1 November 1790 and gained a commission as a sous lieutenant on 1 March 1792. To further his studies he attended the Chalons Artillery School. Promoted *2me lieutenant* on 1 September 1792 he joined a foot company of the 3rd Artillery Regiment serving with the Army of the North. Soon after in November 1792 he transferred to the 2nd Horse Artillery Regiment. During a period when the fortunes of the Army of the North fluctuated wildly and the Representatives cut a great swathe through faint hearted leaders promotion was swift, with Foy becoming a *lieutenant* on 6 March 1793, *capitaine* on 15 April 1793 and then *capitaine commandant* on 1 September 1793.

His career appeared assured, but dabbling in politics as a Girondin supporter he appeared amongst his colleagues as an outspoken critic of the Government for the way it handled the War. With the power of the Jacobins and their Representatives at its height it brought his rising star to an abrupt halt. Dragged before a Revolutionary Tribunal on 13 June 1794 on trumped up charges of illegally procuring and selling army rations he was found guilty, stripped of his rank and sentenced to an indefinite term of imprisonment. Tribunals had executed many for lesser offences, but his comparative youth and because Robespierre's government fell soon after undoubtedly helped to save his life.

Released on 1 August 1794, it was not till 25 March 1795 that he was reinstated and rejoined his regiment in charge of a foot battery serving in the Army of the Rhine-Moselle. In 1796 he took part in the offensive across the Rhine and fought with Abatucci's division at Offenburg, and then at Kambach on 13 August. When the Army fell back to the Rhine he distinguished himself during the defence of Hunningen from 26-30 November 1796. The following year he took part in Moreau's renewed offensive over the river at Diersheim on 20 April 1797. Across the river his guns provided vital support for Duhesme* as the latter's division strengthened its position on the right bank. Severely wounded in the battle he took no further part in the campaign.

Moreau recognizing Foy's bravery promoted him to *chef d'escadron* on 23 June 1797. In 1798 he joined the Army of Switzerland and served briefly in Schauenberg's division. He then passed to Oudinot's where on 7 March 1799 he fought at Feldkirch and was present at the capture of Schaffhausen. He was appointed acting *adjutant général chef de brigade* on 31 July 1799. He played a significant role during the Battle of Zurich on 25-26 September 1799 when in charge of the artillery attached to Lorge's division his fire halted the Russians as they tried to cross the Limatt.

THE ARMIES OF GERMANY AND ITALY 1800-1801

On 3 March 1800 his promotion confirmed, he became commander of the 5th Horse Artillery Regiment. The next year he continued to serve under Lorge this time with Lecourbe's Right Wing of the Army of Germany. He fought at Engen on 3 May, Moesskirch on 5 May and Biberach on 9 May 1800. Appointed Lorge's Chief of Staff on 24 May 1800 he played a key role when Lorge's and Lapopye's divisions under Moncey were detached to Italy and crossed the Saint Gotthard Pass to join Napoleon's Army of the Reserve.

The reinforcements reached Italy after Marengo but he remained with the division as chief of staff when Boudet took over on 5 July 1800. With the renewal of hostilities he fought at Monzambano during Brune's crossing of the Mincio on 26 December 1800. After the brief campaign he stayed in Italy till August 1801 when he returned to France and resumed command of his regiment.

THE MOREAU AFFAIR AND THE CAMP OF BOULOGNE

In the period of relative inactivity that followed during the final years of the Consulate his Girondin feelings once again came to the fore. He gained unwelcome attention when he refused to sign a petition backed by senior officers declaring their support for Napoleon becoming Consul for Life. An active supporter of General Moreau his career prospects looked bleak when the latter was brought to trial for treason in May 1804. Ironically many years later it worked in Foy's favour when Napoleon, reminded of the young officer's stance, tried to bully him in an interview and soon realized he was not easily intimidated by him.

Foy spent 1803 in the 16th Military Division organizing the coastal defences at Calais and Boulogne. Then given a new task, he took charge of the mobile coastal batteries he assembled to protect vessels moving along the coast from English warships. In 1804 he became Artillery Chief of Staff at the Camp of Utrecht and struck up a good rapport with Marmont, who as a fellow artillerist was sympathetic to his needs. When the Grande Armée became active and broke camp the formations at Utrecht became Marmont's II Corps with Foy continuing his role as the Corps Artillery Chief of Staff.

CONSTANTINOPLE AND LISBON 1807-1808

As II Corps played a secondary role in the campaign against Austria, so opportunities for further fame and fortune were limited. On the conclusion of peace he moved as artillery commander to the newly acquired province of Dalmatia. After a year there he joined Sebastiani's military mission to Constantinople in early 1807 where he helped the Ottomans improve their defences to the Dardanelles. When he left Turkey in September 1807 the Ottomans made him a Knight of the Turkish Crescent.

At the end of 1807 he travelled to Lisbon and joined Junot's Army of Portugal as Director of Fortifications. Present at Vimeiro on 21 August 1808 he commanded the Artillery Reserve and took a splinter wound in the leg during the battle. After the Convention of Cintra he was shipped back to France with the rest of the Army. Being one of the few to come out the campaign with any credit he received promotion to *général de brigade* on 8 November 1808.

THE SECOND INVASION OF PORTUGAL 1809

At this stage his career took a change in direction when he took an infantry command. He headed the 1st Brigade of Heudelet's 3rd Division of the old Army of Portugal, which under Junot became VIII Corps of the Army of Spain. Soon after its arrival in Spain the corps was disbanded on 2 January 1809 and Foy's brigade passed to Delaborde's division with Soult's II Corps. He took part in the pursuit of Moore's Army. At Corunna on 16 January 1809 he joined Delaborde's unsuccessful assault against Hope's position on the heights behind the Mero river at the eastern end of the British line.

During Soult's invasion of Northern Portugal in March 1809 his brigade formed the advance guard. He routed a scratch Portuguese force at Villaza on 10 March when it tried to surprise his column. Then pursuing the enemy he forced the passes above Chaves on 14 March and seized Carvaho d'Este. Pushing on, he gained a foothold on Monte Adaufe above Braga on 19 March. Ideally positioned when the rest of Soult's corps arrived the next day he led the assault against the Portuguese center. With La Houssaye's dragoons in support Foy's men broke the enemy and drove them back in confusion through the streets of the town.

On 27 March after reaching the outskirts of Oporto the enemy took him prisoner. He had acted rather rashly when he rode into the enemy lines and arrogantly demanded their surrender whilst negotiations were still in progress to hand over the city. Taken to the rear, a mob nearly tore him apart mistaking him for General Loison, who had been responsible for many atrocities in the region the previous year. Quick-witted he held both hands above his head to prove that he was not the notorious Le Maneta, who had lost an arm as his guards hustled him away. Soult freed him two days later when the French stormed the city.

DISASTER AT OPORTO MAY 1809

Like Soult, in the weeks that followed the period of relative inactivity lulled him into a false sense of security behind the Douro at Oporto. Whilst the French were too weak to resume their advance on Lisbon, he did not believe Wellington's Army would go on the offensive and attempt to cross the river. Unexpectedly at night on 12 May 1809 a small British force slipped across the river above the city and occupied the Seminary on the north bank. Foy rushed his troops to the crossing point and massed before the English position. The two battalions behind the seminary walls supported by devastating artillery fire from the opposite bank beat the assaults back. The result was catastrophic defeat for Soult. As more troops poured across the river further upstream they threatened to cut off his army. Soult abandoned the city and his army conducted a headlong retreat to the Spanish border. With him Foy's troops barely survived.

Back in Galicia, Soult in July 1809 sent Foy on a mission to King Joseph in Madrid outlining a plan to concentrate all the French armies against Wellington. After he gave a confident presentation Joseph and his commanders accepted it. However, personal rivalries during its implementation between the King and the marshals allowed it to fail and resulted in Joseph's defeat by Wellington at Talavera on 28 July 1809. When Foy rejoined his brigade, he took part in Soult's march south that nearly enveloped Wellington after his victory and forced the Duke to beat a hasty retreat to the Portuguese frontier.

From January 1810, based at Talavera with II Corps he had the task of covering Soult's communications during the invasion of Andalusia. Engaged in the operations against Romana's forces in the valley of the Guadiana he fought a successful action at Caceres on 14 March before being forced to withdraw due to an overwhelming concentration of the enemy in the region.

Awards long overdue due to his republican past began to come his way. The first on 15 August 1810 being an annual pension of 4,000 francs drawn on Rome. This was followed by him joining the Imperial nobility as *baron de l'Empire* on 9 September 1810.

MASSENA'S INVASION OF PORTUGAL 1810-1811

In September 1810 he joined Massena's ill fated invasion of Portugal, his brigade forming part of Heudelet's 2nd Division with II Corps now led by Reynier. He played a valiant role at Bussaco on 27 September. At the head of the 17th Légère he breached the Allied line to the left of the San Antonio Pass by driving back three Portuguese battalions and the 45th Foot. However, Leith quickly plugged the gap by deploying the 9th and 38th Regiments and raked Foy's exhausted column with a steady fire till it fell back. In the action he took a musket ball in the chest.

Having recovered sufficiently from his wound, Massena sent Foy on a mission to Paris to outline the Army of Portugal plight to Napoleon. Leaving Santarem on 29 October accompanied by a large escort of over 1,800 men that cut its way through to the Spanish border he reached Ciudad Rodrigo on 8 November. A fresh detachment of dragoons escorted as far as Valladolid. From there he rode straight through by post, braving guerilla bands and swollen rivers, through Burgos and Bayonne to Paris, which he reached on the night of 21 November. Napoleon the next day summoned him to the Tuileries, it was his first meeting with the Emperor. Napoleon read Massena's report and then put Foy through a sharp two hour cross examination of the situation in Spain and Portugal. Impressed by his forthright and intelligent explanations, Napoleon promoted him on the spot to *général de division*. His rank was later confirmed on 29 November 1810.

After a short period of leave Foy left Paris with a fresh set of orders on 22 December and rejoined Massena at Santarem on 5 February 1811. On 9 March 1811 the plight of the army worse Massena sent him again to Paris with another report and appeal for help. His cross examination by Napoleon, followed by news of Ney's abrupt dismissal by Massena, made it apparent that Massena had lost the confidence of the army and its commanders. He returned to Portugal with Napoleon's despatch dismissing Massena and appointing Marmont as head of the Army of Portugal. On 20 April 1811 he had the unfortunate task to present it to his chief. The resulting outburst from Massena was acrimonious. The accusations of disloyalty and treachery greatly disturbed him as he had always been a great admirer of Massena.

MARMONT'S ARMY OF PORTUGAL 1811-1812

After Massena's defeat at Fuentes de Onoro on 5 May 1811, Marmont took over the army and started to restore its flagging spirits. Wholesale changes

THE FRENCH IN THE MOUNTAINS OF PORTUGAL

were made and on 11 June 1811 Foy replaced Marchand as head of the 1st Division of the reorganized army. He spent several months at Toledo till January 1812 when he joined Montbrun's command that moved east to support Suchet's invasion of Valencia. He reached Chinchilla on 18 January where he heard the marshal had taken Valencia so he retraced his steps to Toledo. Marmont's army had meanwhile become widely dispersed and Wellington seeing the error of his ways struck. With Foy away and a mere screen left to cover Ciudad Rodrigo, Wellington surprised the fortress, which fell on 18 January 1812. Foy was then ordered to Talavera to maintain contact with Soult's Army of the South and the fortress of Badajoz, which became Wellington's next target. With 5,000 men he could only make feeble demonstrations against Wellington's communications in that direction as no further support came from Marmont.

In May 1812 Marmont gave Foy the task to re-establish contact between Soult and the Army of Portugal by securing the crossings over the Tagus at Almaraz. On reaching the river he found Wellington's troops had destroyed the bridge and nearby forts. Then with overwhelming Allied forces in the vicinity he had to beat a hasty retreat to Talavera.

A Mauling at Garcia Hernandez

In July he rejoined the main body of Marmont's army and fought at Salamanca on 28 July 1812. Positioned on the French right flank, he became involved in the fighting against Alten's Light Division around the church of Nuestra Senora de la Pena. Then in the afternoon when the French left and center collapsed he fought a staunch rearguard action covering the army's retreat as far as Alba de Tormes.

The next day his division suffered a cruel piece of luck and was badly mauled at Garcia Hernandez. His 76th Line had formed square to repel cavalry when a mount crashed to the ground and rolled into his troops. The animal thrashed about and crushed several men in its path and ploughing a hole through the ranks. The King's German Legion Dragoons saw their chance, converged on the gap and poured into the square completing its destruction. It was the only known instance in the Napoleonic Wars of a perfectly formed French square being broken by unsupported cavalry. Nearby, two battalions of the 6th Légère shaken by the sight of the havoc caused to their colleagues, rather that form square tried to reach the safety of a steep slope. The Germans bore down on them and cut up one battalion in column while the other managed to form square on top of a hill. Foy further ahead with the 69th and 39th Line had time to form square and drove off the exhausted enemy horse with a few well timed volleys.

The French Counter Offensive - Autumn 1812

After his battered division fell back behind the Douro, Foy now under Souham's leadership, made a remarkable recovery and undeterred he went on the offensive. He relieved the garrison of 800 men at Toro on 17 August. Then after demolishing that fortress, he headed for Astorga to relieve the defenders there. Finding it had fallen two days before on 18 August, he turned on Santoclides and cut up his rearguard before he led a relentless march on Zamora that relieved the beleaguered garrison of 1,200 men on 26 August. After destroying the fortifications he resisted the temptation to cut Wellington's communications by ransacking the Allied depots at Salamanca. Instead, he retraced his steps and rejoined Clausel with the Army of Portugal at Valladolid on 4 September. The Army's withdrawal eventually halted north of Burgos on 19 September and he took post at Briviesca 30 miles beyond the city.

He took part in Souham's successful offensive that relieved Burgos on 21 October and then advancing on Palencia stormed the town on 25 October before he seized the crossing over the Carrion that turned Wellington's line. The English in full flight, he headed the French Army's advance guard. He reached the Pisuergua at Simancas on 28 October, only to have the bridge over the river blown in his face. Undeterred the next day he reached the Douro at Tordesillas. In a bold move a volunteer detachment of the 6th Légère swam the river and secured the crossing, which forced Wellington to abandon Valladolid on 15 November 1812. He pushed on and a week later had occupied Salamanca before settling down to winter quarters at Zamora.

In January 1813 he moved to Avila in Old Castile, where on 20 February he led an unsuccessful raid with 1,500 men against the English garrison at Bejar. In March he headed north with orders to put down the insurrection in the Biscay provinces. He reached Bilbao on 21 April and set about the task with the support of the divisions under Sarrut and Palombini. He invested Castro Urdiales on 29 April and successfully stormed the town on 12 May.

Northern Spain 1813

Totally absorbed with his first real independent command, he failed to appreciate the wider strategic objectives of the campaign in northern Spain. His pursuit of the weak Spanish forces along the Biscay coast resulted in his failure to come to Joseph's aid at Vitoria after he admittedly received some rather ambiguous orders to do so at Bergara on 19 June. The end result was his 5,000 men were sorely missed, as were the divisions of Sarrut and Palombini. Later he came to the aid of Joseph's defeated armies as they fled north and with a few battalions he halted Longa's division at Mondragon on 22 June. He then picked up Maucune's division at Villafranca, and also the garrisons from Bilbao and Durango. With some 18,000 men he then made a stand at Tolosa on 25 June, which was enough to halt Graham's attempt to cut the French retreat.

His good work during the retreat was marred when he crossed the Bidassoa into France on 1 July. To him the French forces appeared in total disorder, and rather than attempt to maintain a fortified bridgehead, he rashly destroyed the bridge over the river at Urdax. The mistake proved very costly later when Soult tried to recross to relieve San Sebastian.

THE PYRENEES AND SOUTHERN FRANCE 1813-1814

He fought at Roncesvalles on 25 July where his division formed the head of a column under Soult in a push to throw the Allies off balance by relieving Pamplona. Beset by fog and unable to deploy in the narrow defiles, Cole's division drove his force back. At Sorauren on 27 July, posted on the French left he made a demonstration against Picton's division on the Huarte Heights, but with the appearance of cavalry on his flank and under heavy shellfire he retired. He joined Reille's second assault on Sorauren on 30 July, but again was driven back. Losing his bearings on the mountain tracks his division became separated from the main army, eventually emerged at Iragui, where still harried by Picton, it sullenly made its way back to France.

After the offensive failed he survived Soult's wholesale reorganization of the armies and sacking of commanders. Soult scrapped the four armies that fought at Vitoria and formed a single Army of Spain. Foy and his division fell under d'Erlon's command as part of the Lieutenancy of the Right. He fought on the French left at Urdax during Soult's attempt to force the Bidassoa on 31 August to relieve San Sebastian. He was then posted to Saint Jean Pied-du-Port on the extreme left of the French line to cover a possible southern flank attack through the pass at Roncesvalles. However when Hill moved his forces from the vicinity to join Wellington's assault across the Nivelle, he failed to move across in support of d'Erlon's forces on the upper reaches of the river. Instead on 10 November, in an effort to draw attention from his colleagues, he executed a sudden thrust towards Maya but was checked by some Spanish battalions left there by Hill, meanwhile d'Erlon's defences left unsupported crumbled behind him.

After retiring behind the Nive he was driven from his positions by Stewart's 2nd Division on 9 December, so allowing Hill's corps to establish itself on the east bank in force. On 10 December he supported Reille's attack on Barrouillet where he cut through Hay's 5th division and nearly succeeded in turning the Allied left. He lost sight of the overall objective, which was to cut off Wellington's forces before Bordeaux when he became embroiled in trying to seize the nearby chateau that formed a strongpoint rather than bypass it. The delay saved the Allies. Hope then arrived with his 1st Division and drove back Foy's exhausted men. On the night of 12 December Foy took part in the crossing to the east bank of the Nive and fought the next day at St Pierre d'Irube. He held out on the French left till d'Erlon's assaults against the Allied center collapsed and then withdrew to the entrenchments around Bayonne.

His next major engagement was at Orthez on 27 February 1814. He occupied the ridge to the north of the town and was badly wounded by a shrapnel splinter in the shoulder just as the enemy were closing. The sight of him being carried from the field unnerved his troops and they soon gave way. The gap his men created then in turn caused Harispe's division to his left to give way, followed by the rest of the French line.

THE RESTORATION AND NAPOLEON'S RETURN

He was still recovering from his wound at the time of Napoleon's abdication on 6 April 1814. Wellington visited him while he passed through Cahors on his way to Paris. They had a lengthy discussion over past battles and campaigns and Foy, greatly impressed by the kindness and concern the Duke showed, likened him to the great Marshal Turenne.

The Bourbons treated him fairly and in May 1814 he received the appointment as Inspector General with the 14th Military Division based at Caen. Awards were to follow, *Chevalier de Saint Louis* on 8 May 1814 and then a Grand Officer of the Legion of Honor on 29 July 1814. He moved to the 12th Military Division at Nantes as an Inspector General on 30 December 1814.

On hearing of Napoleon's return, Foy reacted cautiously, as the prospect of France plunging into a civil war weighed heavily on him. He only finally declared for him at Nantes on 23 March 1815 when he heard Louis XVIII had fled Paris and a new being government formed. On 23 April he received command of the 9th Infantry Division, which was later to form part of Reille's II Corps of the Army of the North. On 15 May he was elevated to the dignity of *comte de l'Empire*.

THE WATERLOO CAMPAIGN

At the beginning of June his preparations complete his division concentrated at Avesnes and then on 11 June left for Mauberge moving secretly by stages. Under Reille's command his division formed part of the left wing of the Army of the North's march into Belgium. He bivouacked near the frontier at Leers on the night of 14 June before crossing the next day. He saw little action at first, as Bachelu forming the advance guard bore the brunt of the days fighting, securing the crossing over the Sambre at Marchienne and pushing Steinmetz's brigade back beyond Gosselies.

At Quatre Bras on 16 June he was heavily engaged. His 1st Brigade advanced up the Brussels Road and drove three Dutch battalions from Gemioncourt farm. The 2nd Brigade soon ran into trouble and was only able to dislodge Prince Bernard's five Nassau battalions before Pierrepont farm and Bossu Wood when Prince Jérôme's division joined the action. In the afternoon he drove back a counter attack by the Brunswick Corps as it tried to push south between the Brussels Road and Bossu Wood. In the late afternoon he moved forward to support of Kellermann's cavalry charge on Quatre Bras. Ney's poor timing resulted in Foy's men being unable to keep up with the French horse and they were later carried away by the fleeing cuirassiers. By evening with Jérôme driven from Bossu Wood and Bachelu on his left falling back from Piraumont, he abandoned Gemioncourt and withdrew to the outskirts of Frasnes.

Having lost over 800 men during the day his troops needed time to reorganize with the result they took no part in the pursuit of Wellington the next day. He moved forward slowly and spent the night of the 17th at Genappe. Breaking camp early the next morning he reached the field before Mont Saint Jean around 9.00 A.M. and positioned his division between those of Prince Jérôme and Bachelu opposite Hougoumont.

When the battle began at 11.30 A.M he joined the assault on the Chateau. Supposed to outflank the position, he became embroiled in the affair when Jérôme appeared in difficulties. Sending in his troops from the east side of Hougoumont he managed to gain the orchard before converging fire from the east wall and on his flank repeatedly drove his men back. Around 1.00 P.M. he was shot in the shoulder and carried from the field. Fortunately the ball had only hit an epaulette and caused severe bruising. Able to return to the field he rallied his troops to support the attack by the Old Guard but they made little impact. At the end of the day he managed to lead a detachment of some 300 men from the field and marched them across country to safety. The intensity of the fighting was reflected in the returns taken at Laon on 25 June 1815 when only 1,501 men were on the division's roll, compared to 5,306 at the start of the campaign. He remained with the Army as it retired to the Loire and on its disbandment was placed on the non-active list on 1 August 1815.

POLITICAL CAREER - THE LIBERAL CONSCIENCE OF FRANCE

Reconciliation with the Bourbons came on 30 December 1818 when he became an Inspector General with the General Staff. In June 1819 he took over responsibility for the 3rd Arrondissement, which comprised the former 2nd and 16th Military Divisions at Mézières and Lille.

Interested in politics since a young man he gained election as a liberal Deputy for Aisne in September 1819. From his very first speech in the Chamber he made a dramatic impact. A great orator, he was forceful yet sincere. As a man he had a reputation for being honest and totally incorruptible. He also benefited from having a military career that was virtually untarnished. Through stirring speeches he recalled the glories of the past and the need for reconciliation that moved even the hardest of Bourbon hearts. Whilst the standing of the Liberals fell in the Chamber his own rose, and he was one of the few to gain re-election in February 1824.

He took up writing and wrote a celebrated account of the Peninsular War, which stopped at 1809 due to his untimely death in Paris as a result of a heart attack on 28 November 1825. His death reflected the true state of feeling in France at the time. Whilst the liberals were reduced to only a handful in the Chamber, due to Foy their support among the people was widespread. Over 100,000 people joined his funeral procession and carried his coffin head high to La Père Lachaise Cemetery. A public subscription was opened for his penniless young family. It raised over a million francs in a few weeks.

A MILITARY ASSESSMENT

A courageous man, Foy was a fine commander. Only becoming a *général de division* in 1810 when the fortunes of the French armies were on the wane, he came out of Spain with a virtually untarnished record considering it was the graveyard for many a reputation. As a tenacious rearguard commander he certainly showed his mettle after Salamanca and Vitoria. When he briefly held a small independent corps command after Salamanca he was brilliant as the French turned on Wellington at Burgos. In the Biscay provinces his failure to appreciate the strategic developments during Wellington's northern advance to Vitoria marred the successes of his first corps command.

During the Waterloo Campaign he had few opportunities to show initiative. The fact that he allowed himself to become unnecessarily embroiled in the assaults on Hougoumont as he had done eighteen months before at Barouillet showed a weakness for losing sight of the overall objective. He did not blench at the use of cold steel. At times his keenness come to grips with the enemy allowed critics to consider him too impulsive. On the other hand, he had the highest regard for the steady fire power of English infantry in defensive positions. He felt the only way to deal with them was to close quickly where French qualities in a melee situation were superior. A gallant soldier, he was undoubtedly one of Napoleon's ablest divisional generals in the Peninsula War. Had his life not been cut short when his stature as a politician had reached such heights, an interesting career would have awaited him in the more enlightened times of Louis Philippe.

FRIANT - LOUIS, COMTE (1758-1829)

BACKGROUND AND EARLY CAREER

The senior of the three infantry generals who commanded divisions of Davout's legendary III Corps, Louis Friant came from very humble origins. The son of a wax-polisher, he was born in the village of Morlancourt near the small town of Villiers-le-Vert (Somme) on 18 September 1758. He enlisted as a grenadier with the *Gardes Français* on 9 February 1781 and the following year on 1 July 1782 gained promotion to *caporal*. He became disillusioned with army life when after five years service his colonel turned down his promotion to *sergeant* and he obtained a discharge due to ill health on 7 February 1787.

He next emerged as a *caporal* with the Paris National Guard where the records show he enlisted on 4 September 1789. By 1792 he had risen to *adjutant major* with the National Guard in the radical Arsenal section of Paris. On 11 September he headed the detachment from the Arsenal that joined the Army of the Moselle massing on the frontier. During the march march his men elected him their *lieutenant colonel* on 23 September 1792. Detached again they joined the Army of the Center, where his unit renamed the 9th Paris Battalion had an active period. He fought at Arlon on 9 June 1793, at Abbaye d'Orval on 12 August and Kaiserlautern on 28-29 November. Shot in the left leg on 16 December 1793, he was out of action for several months before joining the 181st Demi-brigade de Bataille on 31 March 1794. In the Belgian Campaign of 1794 that culminated in the victory at Fleurus he fought at Arlon on 17 April, before Charleroi on 9 June and at Fleurus on 26th June. His demi-brigade joined Scherer's division of the Army of the Sambre-Meuse and settled down to laying siege to the Belgian fortress towns while Friant himself on 2 July became the division's chief of staff.

On 3 August 1794 his career began to take off when he received promotion as acting *général de brigade*, replacing Chevalier in Muller's division after the former's dismissal for drunkenness. Ten days later he in turn took command of the division after Muller failed to show the necessary revolutionary zeal expected. He led the division during the siege of Maestricht till 13 October 1794 when Chapsal considered a more experienced commander took over. He served at the siege of Luxembourg from 5 April 1795, and on the fall of the fortress on 8 June became its governor. The Committee of Public Safety confirmed his rank of *général de brigade* on 13 June 1795.

He rejoined the Army of the Sambre-Meuse on 21 March 1796 when he led a brigade with Poncet's 3rd Division. He was with the Army during the Rhine offensive and on 28 June 1796 took part in the crossing at Neuwied. Advancing up the east bank of the river he took part in the siege of Ehrenbreistein from 8 July till 15 September. In January 1797 he moved to Bernadotte's division that was sent to reinforce the Army of Italy. On its arrival in Northern Italy Bernadotte's command became the army's 3rd Division, with Friant as commander of the 5th Brigade (30th and 55th Line). He was present at the crossing of the Tagliamento on 16 March and the capture of Gradisca on 19 March 1797. After the Preliminaries of Leoben and the occupation of the Venetian States by French troops he remained with his brigade in Frioul. He took over the division from 9 August when Bernadotte departed for Paris and joined in the intrigues that led to the Coup d'etat of Fructidor on 4 September 1797.

MARRIAGE AND THE ARMY OF THE ORIENT

On 12 January 1798 he received orders to join the Army of England. Whilst in Paris he found time to marry Claire Leclerc, sister of General Victor Leclerc. Leclerc was the close friend of Napoleon and later married his sister Pauline. Friant's marriage gave him an entrée into the inner Bonaparte circle despite his past association with Bernadotte and the Army of the Rhine.

On 5 March 1798 he received secret orders to join the Army of the Orient. His arrival at Civita Vecchia near Rome on 14 April saw Friant join Desaix's command for the voyage to Egypt. On 23 June 1798 after the landing at Aboukir he took command of the 2nd brigade (61st and 88th Line) of Desaix's division. He fought the Mameluke Army in the first major encounter of the campaign at Shubra Khit on 13 July. At the Battle of the Pyramids on 21 July 1798 his brigade was with Desaix's great division square, which broke the Mameluke cavalry charges.

After the fall of Cairo he joined Desaix's legendary expedition to Upper Egypt that lasted more than a year. He headed south from Giza on 25 August 1798 at the head of his brigade, fought inconclusive actions against Murad Bey at Sediman on 8 October and then at Samhud on 23 December. Murad Bey reinforced by levies from across the Red Sea caught Desaix's force south of Girga on 22 January 1799. In the action that followed Friant with his brigade formed in a large square and routed the Arab cavalry as they tried to break through a wall of bayonets supported by rolling volleys of musket fire. Although the power of the Beys was broken in Upper Egypt as a result, the campaign continued unabated. The expedition was engaged in continuous marches, counter marches and minor clashes as far south as

Aswan and the Nile Cataracts as it tried to pacify the region. He won further actions at Samatah on 12 February and Aboumanah on 21 February.

His solid performances resulted in promotion to acting *général de division* on 4 September 1799 and governor of Upper Egypt after Desaix's recall to Cairo. In January 1800 he began the French withdrawal from Upper Egypt in anticipation of agreement being reached for the evacuation of Egypt. After the breakdown of the controversial Convention of El Arish his forces were fortuitously close at hand to join the forces at Cairo to defeat the Turks at Heliopolis near the city on 20 March 1800. When the uprising followed in Cairo he hurried to the city with the rest of the army. He distinguished himself during the seizure of Boulaq and the storming of the citadel on 15 April 1800.

A Consular Decree on 6 September confirmed Friant's rank of général de division and his appointment as governor of Alexandria. He reacted quickly on hearing an Anglo-Ottoman force was off the coast at Aboukir and offered stubborn resistance as they disembarked on 8 March 1801. Weight of numbers told and he withdrew inland where he delayed the English advance at Madieh on 13 March, winning valuable time for Menou to concentrate the rest of the army. At the Battle of Canopus on 22 March he fought on the French left under Reynier, taking part in Menou's ill advised night attack that failed to surprise Abercromby's forces. The British forewarned of the French presence drove back the renewed assaults during the day. Friant withdrew to Alexandria, where as governor under an increasingly eccentric and irrational Menou, he had a very difficult time during the siege that lasted till 31 August 1801.

At the end of 1801 he returned to France with the remnants of the Army of the Orient. For two years located in Paris he served as an Inspector General of Infantry. On 29 August 1803 he renewed his association with Davout, who had also served under Desaix in Egypt, when he received command of a division forming at the Camp of Bruges. Located near Dunkirk preparing for the invasion of England he was the senior member of the triumvirate of generals, who with Davout were to weld III Corps into such a formidable fighting machine. Gudin and Morand* were to join later.

THE AUSTERLITZ CAMPAIGN 1805

On the formation of the Grande Armée on 30 August 1805 his division became the 2nd Division of III Corps. His troops with the rest of Davout's corps played a relatively quiet role in the early stages of the campaign. III Corps completed its crossing of the Rhine on 27 September. Its line of march followed the Neckar, through Heidelburg to Neckarelz, which it reached on 29 September. Then moving south-east to Mosbach on 2 October, Oettingen on 6 October, it

reached the Danube at Neuburg on 8 October. Crossing the river it pushed south to Aichach and Dachau. At Dachau they took up a position astride the Ammer river in support of Bernadotte's I Corps, which had occupied Munich. They remained there whilst the trap closed around Mack's army encircled at Ulm. Well into the campaign his troops had still not met any enemy resistance when they resumed their advance on 26 October by crossing the Iser at Freising. The route he took was through Mühldorf, Burghausen, Steyer, and Gaming, to Lilienfeld. South of Lilienfeld he fought the first action of the campaign by III Corps when they caught and severely mauled Merfeldt's Austrian Corps on 8 November. On 15 November 1805 with Davout at his side, Friant's troops were the first to occupy Vienna. He remained in the capital for two weeks while a decisive battle was shaping up some seventy miles to the north near the small Moravian town of Austerlitz.

When it became apparent the Russians had concentrated their forces, Napoleon ordered Davout to march his divisions towards Brunn. Friant received the order at 8.00 p.m. on 29 November and within an hour and a half his entire division was on the move. Marching all night and the next day they arrived at Nikolsburg in the evening of the 30th, a distance of forty five miles. After a night's rest they continued to the Abbey of Raigern where they arrived at 7.00 p.m. on 1 December. At 5.00 a.m. Friant broke camp and started to march north to Turas to lend support to the French left as the main Russian attack was expected to fall on Kobelnitz. When it later became apparent that Kutusov's main assault was to the south of the Pratzen Heights, Davout ordered Friant south to hold a position between the villages of Tellnitz and Sokolnitz. Friant's 1st Brigade under Heudelet joined the action after the Russians had already driven back from Tellnitz one of Legrand's regiments of Soult's IV Corps. They caught the enemy by surprise and drove the Russians from the village in great disorder from the village. Suddenly another of Legrand's regiments, the 26th Line, which had also arrived at Tellnitz to lend support, opened fire on Friant's troops from the rear mistaking them for Russians. Thrown into confusion Heudelet's brigade was ejected from Tellnitz by the Russians after they received reinforcements. The Russian commander Doctorov leading the assault made a fatal error when he inexplicably halted his victorious troops to await the fall of Sokolnitz one and a half miles to the north before resuming the advance. Heudelet regrouped, managed to contain the Russians while Friant with the rest of the division moved on Sokolnitz with his remaining two brigades and Legrand's 36th Line in support. The fighting was fierce, and despite being outnumbered Friant's infantry drove the Russians from the village. They formed a new line and held on to the west bank of the Goldbach while victory was decided on the Pratzen Plateau.

The savagery of the fighting was reflected in the casualties Friant's division suffered, over 1,400 men, which represented over thirty percent of its total strength. Friant had four horses shot from beneath him during the day. In recognition of the part he had played Friant was awarded the Grand Eagle of the Legion of Honor on 27 December 1805.

HIS FINEST HOUR - AUERSTADT AND THE PRUSSIAN CAMPAIGN OF 1806

Friant as acting corps commander at the start of the campaign, as Davout was in France on leave with his family. On 26 September 1806 he ordered III Corps to concentrate at Bamberg, which it did without incident in the first week of October. The advance into Saxony was at first uneventful. Leaving Kronach on 9 October, III Corps marched north through Lobenstein and Schliez on the 10th, Auma on the 11th, and arrived at Naumberg on the 12th. On the 13th the divisions and took position on the east bank of the Saale astride the Erfurt-Leipzig road. At 3.00 A.M. on 14 October Davout received orders to march on Apolda where Napoleon expected him to fall on the left of the Prussian Army massed on the plains between Jena and Weimar. Gudin's division forming the advanceguard led the way followed by Friant then Morand. It unexpectedly crashed into a Prussian division led by Schmettau at Hassenhausen, which later proved to be advance elements of the main Prussian Army. He was soon fighting for his life as another two divisions came up and threatened to envelop him.

Friant's timely arrival enabled Gudin's hard pressed regiments to hold their ground. Davout launched Friant straight away against the Prince of Orange's division, which had threatened to engulf Gudin's right and he drove it back two miles to beyond Spielberg. The battle reached its climax when Morand arrived and beat off several Prussian assaults against the French left. Davout then seized the initiative and ordered Morand on the left and Friant on the right to move to the attack. Pivoting on Gudin in the center, the 1st and 2nd Divisions moved forward. The skilful deployment of their artillery on both wings brought the enemy under an interlocking cross-fire that swept the entire Prussian line. Friant swept back Kuhnheim's and von Arheim's divisions, which took refuge behind Gernstadt till later in the day when he again drove them back in disorder. By evening his exhausted troops were in possession of Ekartsberg whilst Morand had taken Auerstadt.

Unwittingly Friant had played a decisive part in the most glorious day in the annals of any single French corps. III Corps had engaged and beaten a well-trained force twice its number, inflicted over 15,000 casualties and taken 115 cannon. Exhausted the three divisions rested for two days. Casualties totalled 6,833 men, more than a quarter of their strength at the start of the campaign. On 17 October III Corps rejoined the pursuit of the Prussians. They passed through Leipzig on 18 October, Duben on the Mulde the next day and reached the Elbe at Wittenberg on the 20th. A poor attempt to fire the bridge was extinguished by Friant's men and the last barrier before Berlin was crossed virtually unopposed. On 25 October with Davout and his staff at their head, Friant's division led the victorious march into Berlin.

POLAND AND EAST PRUSSIA - WINTER 1806-1807

His the lead infantry division for the march into Poland, Friant broke camp on the morning of 30 October and reached Frankfurt-on-Oder the next day. He remained there till 6 November while the rest of the Grande Armée caught up. Moving on, he occupied Posen on the 9th, where after a week's halt while the army regrouped he set off for Warsaw on the 16th, which he reached on 1 December 1806. On 10 December he crossed the Vistula and established himself on the right bank. Engaged in the crossing of the Ukra on 23 December, he succeeded in storming the heights behind Czarnovo after Morand's division, having borne the brunt of the day's fighting, had failed to take the position. Struggling in the thaw that had turned the roads into rivers of mud he arrived late at Golymin on 26 December and was unable to offer any assistance in that indecisive battle.

At the Battle of Eylau on 8 February 1807 he played a key role in the events of the day. Davout's III Corps formed the right wing of the Army. Napoleon counted on him marching the fifteen miles to the battlefield to fall on the Russian flank, whilst Augereau and Soult's corps with Murat's cavalry in support engaged the enemy's right. The strategy was to outflank the Russians by taking Serpallen and drive them north against the sea. Breaking camp two hours before dawn III Corps hastened to join the main army. Friant led the way with his division with Morand close behind. By 9.00 A.M. Davout launched Friant against Serpallen and within the hour the Russians were streaming back to Klein Sausgarten. Morand's division at the same time began to form up on his left and quickly moved into line linking up with Saint Hilaire's division of Soult's IV Corps on Napoleon's right.

By 4.00 P.M. he had taken Kutschitten and started to roll up the Russian flank. In the gathering dusk his weary troops were hit in the flank by Lestocq's Prussian Corps, which suddenly emerged on the field. His men driven from Kutschitten after a stubborn defense, fell back to the woods on the outskirts of Klein Sausgarten. The Russians took heart, the whole weight of their army turned against the threat by III Corps to their left. For five hours

Davout's divisions grimly held onto their positions till the fighting gradually died away in the dark around 11.00 P.M..

Friant with the rest of III Corps had failed for the first time to achieve their objectives. Napoleon in the conditions had expected too much, Friant lost over 1,400 men in the battle, over a third of his strength. On the 10th Davout ordered Friant to follow the enemy. Then when it became apparent the Russians were not prepared to stand and renew the fight, he retired to the Vistula with the rest of the army to take up winter quarters.

In the brief Summer Campaign of 1807, which culminated in the annihilation of the Russian army at Friedland on 14 June 1807, Friant with the rest of III Corps missed the great battle. Napoleon had directed III Corps north to support Soult's IV Corps and Murat's cavalry in a separate march on Konigsberg. The Treaty of Tilsit saw his division remain in the newly created Duchy of Warsaw centered on Sochaczew till October 1808 when it moved to Southern Saxony as part of the Army of Germany. The creation of the Imperial Nobility resulted in him becoming a comte de l'Empire on 8 October 1808.

THE DANUBE CAMPAIGN 1809

During the winter months of 1808-1809 Friant took charge of III Corps whilst Davout was on leave in France. During this period, fortunately he kept the troops at a high state of readiness. Davout returned early at the end of March when war with Austria appeared imminent. Davout soon became aware from intelligence reports that the bulk of the Austrian army that had previously faced him had moved south-east from Bohemia and concentrated south of the Danube behind the Inn.

Berthier countered by ordering Davout with III Corps to start moving southwards on 6 April to face the new threat. Friant set out from Bamberg, passed through Nuremburg and was at Amberg on 10 April when he heard the campaign had started. On 16 April he was at Hemau protecting Davout's left rear after a series of confusing and conflicting orders from Berthier had found III Corps in the vicinity of Ratisbon. Bellegarde threatened III Corps by bearing down on its left from Bohemia, whilst the main Austrian army threatened was about to fall on its right having crossed the Inn and the Isser. Friant in turn had fallen back on Ratisbon and formed Davout's rearguard as III Corps retired along the south bank of the Danube to link with Lefebvre's VII Corps moving up from Ingolstadt.

On 19 April the ponderous Austrian advance caught up with Davout. Friant fought in the first major encounter of the campaign centered on Thann when the widely spread uncoordinated Austrian attacks tried to cut off III Corps and drive it into the Danube. On the French left with Montbrun's cavalry in support Friant held Hohenzollern's III Corps at bay and in the afternoon drove him back to the woods behind Hausen. At the same time the link up was effected with Lefebvre.

On 21 April whilst Napoleon split the Austrian Army in two at the Battle of Abensburg, Friant still held the French left with Saint Hilaire's division in support. As the day progressed he drove Rosenberg's IV Corps back beyond Schierling. The strategic situation turned dramatically the next day when the bulk of the retreating Austrian army turned on III Corps, who threatened their means of escape by way of Ratisbon. Friant and Montbrun faced converging attacks by two corps under Liechtenstein and Rosenberg as Archduke Charles fought to keep his route to the north bank of the Danube open. Montbrun's cavalry kept the enemy at bay with aggressive skirmishing, whilst Friant had to give ground and lost the gains of the previous day. Only the timely arrival of Napoleon from the direction of Landshut in the afternoon turned a near French disaster into a triumph.

After two days rest while Napoleon stormed Ratisbon, III Corps was given the task to pursue the enemy into Bohemia. By 28 April Friant had reached Nittenau when he received orders to halt and retrace his steps to Ratisbon. His division with the rest of III Corps followed the other French formations along the south bank of the Danube towards Vienna. Secure in the knowledge that Archduke Charles had retired deep into Bohemia they failed to notice that he started to retrace his steps. The new course the Austrian Army followed was to head east on a parallel course that was a direct threat to Napoleon's communications had they chosen to cross the Danube. Friant halted for several days at Saint Polten till 18 May when he was ordered to move on to Vienna, whose outskirts he reached at noon of the 20th. He was at Ebersdorf the next day but was unable to cross the Danube and join Napoleon's forces on the north bank as the river was in spate and the bridge cut. His division with the rest of III Corps sat idly on the south bank and waited to move while Napoleon outnumbered and for the first time outwitted on the battlefield fought for his life at Aspern-Essling.

WAGRAM 5-6 JULY 1809

During June Friant's and Morand's divisions remained at Ebersdorf whilst the rest of Davout's forces moved down river to cover the French army's right at Pressburg. Preparations complete for Napoleon's second crossing of the Danube, III Corps moved to its staging points on the Isle of Lobau during the night of the 4 July. Friant made the passage across the river the next day without incident and the corps established itself on the north bank. He then joined the advance across the open Marchfeld to occupy Markgrafsneusiedl on the French right flank.

On 6 July his division scaled the heights to the east of the village while Morand approached from the west. Around 11.00 A.M. a massive intervention by Charles at the head of his cavalry reserves nearly turned the French tide. The Austrians broke through the first lines of Davout's divisions and would have routed the entire corps had the second line of reserves not held their ground. The Austrian attack faltered, the French regrouped, re-established their line and supported by the cavalry divisions of Montbrun, Grouchy and Arrighi continued the advance. Threatened with envelopment the Austrians began to disengage down the length of their line and withdrew from the field. Friant struck down by a shell splinter in the shoulder at the end of the day was not with his troops as III Corps pursued Rosenberg's IV Corps to Brunn.

THE MOSCOW CAMPAIGN 1812

After the French withdrew from Austria in October 1809 Friant's division remained in Germany as part of the Corps of Observation of the Elbe. Europe at peace apart from the festering problem of Spain, Friant spent lengthy periods away from his troops till the months preceding the Russian Campaign. On 26 January 1812 the first sign of the pending campaign came when his division moved into Swedish Pomerania.

Great preparations were made for the Russian Campaign, including III Corps being renamed I Corps in recognition of its fighting qualities. The number of infantry divisions increased from three to five. Friant's division remained the second but he lost two of his regiments the 85th and 108th respectively to Dessaix's and Compans's newly formed divisions. His three remaining French line regiments each fielded five battalions and numbered 12,927 officers and men at the start of the campaign. His was the second division after Morand's to cross the Niemen at Kovno on 24 June 1812. Reaching Vilna on 28 June he was detached to Murat's command. Here under Lobau* he took part in the pursuit of Barclay's First Army of the West while Davout's depleted force moved south east in the direction of Minsk to join operations against Bagration's Second Army of the West.

On 7 August 1812 out of the blue he was appointed colonel in chief of the Foot Grenadiers of the Imperial Guard replacing Dorsenne who had died in France. The appointment was a great honor since he had never served in the Imperial Guard and showed the high regard Napoleon had for Davout's III Corps generals. He rejoined Davout before Smolensk and was slightly wounded during the storming of Roslavl on 17 August. He remained with his division and did not move to the Imperial Guard. He fought at Borodino on 7 September where his division led the French attack against the Russian center. With Davout's remaining two divisions, those of Dessaix and Compans, following he entered a desperate struggle for the flèches to the south of Semoyonovskaya. Around 11.00 A.M. preceded by a tremendous artillery bombardment and flanked by the heavy cavalry of I and IV Cavalry Corps he stormed the village itself. Latour Maubourg on his flank kept the Russian cavalry in check. The Russians exhausted, having committed their reserves, Friant was ideally positioned to achieve a breakthrough. Napoleon refused to commit the Guard. Dismayed and twice wounded in the battle he left the field.

Napoleon's decision possibly cost him the campaign, and Kutuzov was able to retire with his army. The retreat from Moscow is well known history. Friant with his division suffered the rigors of the retreat from Moscow. No sooner had he reached Poland when Napoleon recalled to France on 11 January 1813 to take up his post with the Imperial Guard. He carried with him not only a great reputation as a battlefield commander but also that of a great organizer and administrator.

THE IMPERIAL GUARD - SAXONY 1813

Friant immediately set about the prodigious task of rebuilding the broken Guard regiments. In May 1813 he set out for Saxony with the 4th Division of the Young Guard, but was too late for the major battles of the Spring Campaign. Another reorganization of the Guard during the Armistice resulted in Friant on 29 July 1813, as colonel of the Foot Grenadiers, taking his rightful place at the head of the Old Guard Division replacing Roguet*.

His troops forming the infantry element of Napoleon's, Friant was with him throughout the campaign. Present at the Battle of Dresden on 26 August the Old Guard formed the central reserve posted behind Marmont's VI Corps by the fourth redoubt where his men remained idle during the day. In mid October he retired on Leipzig with Napoleon and was present during the four day struggle for the city from 16-19 October 1813. On the 16th the Old Guard acted as a central reserve positioned at Probsthayda to help the forces barring the southern approaches to the city. On the 18th he took part in the savage fighting for Probsthayda itself as the Allied armies closed in. He lent support to Lauriston, who with Victor had repeatedly driven back the enemy's assaults and in the process they inflicted over 10,000 casualties.

As the French Army retreated to the Rhine, Napoleon marched with the Old Guard in the middle of the army. Friant was with this force ready to aid the vanguard in the event of resistance ahead or help the Young Guard in the rear being relentlessly pursued by Schwarzenberg and Blucher. He fought at Eisenach on 24 October and then on 27 October at Hunefeld when a battalion guarding the Imperial headquarters had to drive off a detachment of Prussian cavalry. At Hanau on 30 October his grenadiers held Neuhof and the bridge at Lamboy to keep the road to Frankfurt open. On reaching Mainz his division remained with Napoleon restoring order in the city and its environs, which were swarming with stragglers and deserters. By 7 November order restored, his troops moved into cantonments around Trier on the Saar for a rest.

THE 1814 CAMPAIGN IN FRANCE

Under Marshal Mortier he took command of the 1st Division of the Old Guard on 16 November 1813. On 14 December he retired with him into Belgium and spent Christmas at Namur. When news arrived that Schwarzenberg's Army had cut through Switzerland via Basle and poured into Eastern France, Napoleon on 2 January 1814 called Mortier with Friant to march south to Langres. For eighteen days they conducted a skilful delaying action from Belgium to Langres, then to Bar-sur-Aube with numerous minor victories including that at Rouvré on 24 January 1814.

Friant, back under the eye of Napoleon, fought at Champaubert on 10 February, Montmirail on 11 February and Vauchamps on 14 February 1814. His was the first division into action during the costly struggle outside Craonne on 7 March. His regiments ascended the plateau and deployed with great skill into the teeth of artillery fire from over ninety guns

of Woronzow's Corps. Then with Drouot's artillery in support they drove the enemy from their positions. He was in action again at Laon on 9 March covering the French withdrawal when the Russians fearing a trap called off their pursuit. Earlier, Napoleon had come up against the whole of Blucher's army, which at the time was twice the number of his. Then he took part in Napoleon's forty mile sweep across Blucher's front that led to the recapture of Rheims on 13 March. He fought in Napoleon's last battle of the campaign at Arcis-sur-Aube on 20-21 March, took part in the march east to sever the Allies' communications and then the mad scramble back to save Paris.

THE RESTORATION AND NAPOLEON'S RETURN

As was expected with a general of the Guard he was loyal to the end and was present at Fontainebleau during the Abdication Crisis. With most of the Guard generals who did not go into exile with Napoleon he signed his allegiance to the Bourbons on 10 April 1814. He was made acting commander of the Guard as it marched to the Loire from where it was to be dispersed. Under the Bourbons he took a very conciliatory stance. He went a long way to appeal to the old veterans to accept the release of their oath to Napoleon and accept the new order. His men were not impressed by such behavior and on occasions were near mutiny. On 18 July 1814 he was appointed

WATERLOO: NAPOLEON SENDS FRIANT AND THE OLD GUARD INTO THE FINAL ASSAULT

colonel of the Foot Grenadiers of France, the new name for the Foot Grenadiers of the Imperial Guard.

When he heard of Napoleon's return he sat by quietly to await events. One of Napoleon's after his installation in the Tuileries was to summon Friant on 21 March 1815 and re-appoint him to the Imperial Guard as commander of the Old Guard Grenadiers. The appointment was controversial, as Friant was the subject of much ridicule for the way he had collaborated with the Bourbons in order to retain a command. However, Napoleon chose to overlook this, as men of his calibre were rare. Some prompting from Davout helped, who also recognized his former colleague's excellent fighting qualities and sound administrative capabilities.

In the short time available Friant raised new third and fourth regiments of grenadiers from former veterans of the Guard Fusiliers and Tirailleurs. By the beginning of the campaign his division had seven battalions and two companies of artillery amounting to sixteen guns, totalling 4,317 officers and men.

THE WATERLOO CAMPAIGN

On 10 June 1815 his division secretly left Paris and four days later had concentrated on the Belgian frontier. While the battle raged at Ligny on 16 June Napoleon held the Guard in reserve most of the day before the mill at Brye. Then around 7.00 PM amidst the great thunderstorm that engulfed the battle, he led his 1st and 2nd Regiments of Grenadiers against Ligny itself and broke the wavering Prussian center to seal victory.

At Waterloo with the rest of the Guard he was held in reserve before Rossomme during the early part of the battle. As the day went by he lost the 1st Battalion of the 2nd Foot Grenadiers to Morand, who needed it to stiffen the Young Guard in Plancenoit. The two battalions of the 1st Foot Grenadiers were held back at Rossomme when the remains of his division, only four battalions strong, joined the Chasseurs in the Old Guard's final assault against Wellington's center. The attack came in several uncoordinated columns. Friant with Ney was at the head of the first, which had the 1st Battalion of the 3rd Foot Grenadiers at its head. They overwhelmed the English 30th Regiment and the 2nd Battalion/79th Foot of Halkett's Brigade, which was reeling under a bombardment from some of the Guard Artillery that had moved up in support. Friant was rendered hors de combat after being shot in his sword hand and returned to Napoleon to report all was going well.

The final repulse of the Guard is history. Friant took refuge with the remains of the Imperial Staff in a square formed by the 1st Foot Grenadiers. He remained with them as they retired from the field till the enemy broke off their pursuit. In the confusion that followed he gathered the remnants of his division, which when returns were completed at Laon numbered only 1,597 officers and men compared to 4,490 infantry at the campaign's start. He was with his men till the end, as the Army rallied before Paris, and then after the declaration of the Armistice as they withdrew behind the Loire.

RETIREMENT AND FINAL DAYS

Prepared to take the consequences for having served under Napoleon, charges were never brought against him since technically he had never defected. It was Napoleon who had summoned him to serve rather than he who had offered his services to Napoleon. Acutely embarrassed because that he had not been included in the wholesale dismissal of officers, he chose to resign from the Army at his request on 4 September 1815. In his fifty eighth year he had enough. He took no further part in military affairs or public life and died at the Chateau of Gailouet in the commune of Seraincourt (Seine-et-Oise) on 24 June 1829.

CAREER ASSESSMENT

Friant was one of the finest division commanders in the French Army. His long association with Davout was a major factor in both having such outstanding careers. Related by marriage, being brothers-in-law, their relationship appeared on the surface as always a formal one. A study of Davout's correspondence reveals how much strength he gained from having Friant at his side. His behavior after the First Abdication and his pandering to the Bourbons was sad for it was not expected from a general of the Guard. Unfortunately it was typical of many generals who had grown weary by war and were guided more by instincts of self preservation than anything else. It was a fitting acknowledgement of his talents, that Napoleon on his return from Elba had little hesitation in ignoring such shortcomings and gave him the most prestigious divisional command in the Army.

BACKGROUND AND EARLY CAREER

The son of the Royal Huntsman from the Royal estates at Damvilliers (Meuse), Gérard was born there on 4 April 1773. As he grew up he realized the future of the monarchy was precarious and rejecting prospect of a career on the Royal estates he instead enlisted as a volunteer with the 2nd Battalion of the Meuse Volunteers on 11 October 1791. His first active service came with Dumouriez's Army of the North where he fought in the forests of the Argonne. He was with the Army during the invasion of Belgium and was present at Dumouriez's victory over the Austrians at Jemappes on 6 November 1792. His progress through the ranks was rapid during the campaign in Belgium, *sergeant major* on 16 December followed by *sous lieutenant* on 21 December 1792.

The following year when the tide turned, he fought at Neerwinden on 18 March 1793 and was with the French forces as they fell back to the frontiers. He distinguished himself during Jourdan's counter offensive in the Autumn that culminated in the Battle of Wattignies on 16-17 October and received promotion to *lieutenant* on 30 December 1793. On 4 April 1794 his career took a significant turn when his battalion merged with the 71st Demi-brigade commanded by colonel Bernadotte. It was the start of a strong association between the two that was to have a significant influence on his career over the next fifteen years. He served under Bernadotte in Jourdan's invasion of Belgium and taking part in the crossing of the Sambre on 12 June, the capture of Charleroi on the 25th and the Battle of Fleurus the following day. A period followed serving with the newly formed Army of the Sambre-Meuse. Present at the crossing of the Roer in Holland on 2 October, he then served at the Siege of Maestricht.

CAREER UNDER BERNADOTTE

Bernadotte had watched the officer's progress and on his promotion to *général de division* on 28 April 1795 secured Gérard's appointment to his staff. An active period for Gérard followed as Bernadotte established his reputation as one of the foremost generals of the Republic. He was at the crossing of the Rhine at Neuweid on 15 September 1795 and the capture of Kreutznach on 1 December 1795. In 1796 with Bernadotte he campaigned deep in Germany. He fought at Limberg on 6 July and was at the capture of Wurzburg on 24 July. At Teining on 22 August he was with Bernadotte when the Archduke Charles nearly cut off the division. Then during the hasty retreat to the Rhine his devotion to his chief became very evident.

Gérard had his first encounter with Napoleon when Bernadotte's division crossed the Alps to re-inforce the Army of Italy at the end of the year. Temporarily attached to the 30th Demi-brigade he distinguished himself during the crossing of the Tagliamento on 16 March 1797 and then again at

Gradisca a week later when Bonaparte promoted him *capitaine* on the battlefield.

When Bernadotte became French Ambassador to Vienna in January 1798 Gérard accompanied him as one of his aides. Like Bernadotte he was not happy with the life of a diplomat. As relations deteriorated between France and Austria he showed with his chief a lack of restraint when a Viennese mob threatened the Embassy on 14 April. Outraged after the mob burnt a tricolor at the door, with sword in hand he with Bernadotte and several officers set about them with the flats of their swords. The incident caused howls of protest and when Bernadotte was ordered to quit the city the next day as ambassador he was much relieved. In the uneasy period of peace that followed he continued under Bernadotte for a period in the 5th Military Division before they moved to the Army of Mayence.

When the army became commander of the Army of the Lower Rhine and Bernadotte was named its commander on 5 February 1799, Gérard took up the appointment as his senior aide de camp. He served with Bernadotte at the blockade of Philippsbourg before political intrigue resulted in Bernadotte's dismissal by Minister for War Jourdan on 4 April 1799. After Massena took over the army, Gérard secured a place on his staff and distinguished himself in the first battle for Zurich from 4-6 June 1799.

In a dramatic change of fortune, Bernadotte became Minister for War, Gérard was recalled as his aide and on 13 July 1799 promoted *chef d'escadron*. Directory intrigues soon proved too much for Bernadotte and, coupled with reverses on all fronts, resulted in his dismissal two months later. Gérard continued to serve at the Ministry for War without fulfilling any particular role. When Bernadotte took a posting as commander of the Army of the West, Gérard joined him as his senior aide on 3 May 1800.

After Brumaire as a former rival to Napoleon, Bernadotte's long term prospects looked decidedly bleak. The difficult time that followed also affected Gérard who realized his career was very dependent on the fortunes of his chief. He saw with concern Bernadotte being ignored for any major command on the Rhine during the Hohenlinden Campaign. Two years later with Napoleon's power consolidated

GÉRARD - MAURICE, ETIENNE, COMTE (1773 - 1852)

as Consul for life, Bernadotte still with a dead-end command readily accepted the governorship of Louisiana. Gérard in turn out of a sense of loyalty agreed to accompany him to America. It all fell apart when Napoleon concluded the Louisiana Purchase followed by the breakdown of the Treaty of Amiens. With no American colony for Bernadotte to govern Gérard found himself without a post and placed on half-pay on 23 September 1803.

When Bernadotte was later appointed governor and commander of the French army of occupation in Hanover in May 1804 he did not forget his aide and Gérard returned to duty on his staff. Gérard gained promotion to *adjutant commandant* on 20 August 1805 and became Bernadotte's senior aide when I Corps was mobilized for the campaign against Austria. At the Battle of Austerlitz on 2 December 1805 he distinguished himself leading a column from Drouet's* division that stormed the Pratzen Heights. In the struggle he received a grape-shot wound in the leg that laid him low for several weeks.

PRUSSIA AND POLAND 1806-1807

In the Prussian Campaign he fought in the first action of at Schliez on 9 October. With the rest of Bernadotte's corps he missed the twin battle of Jena-Auerstadt on 14 October 1806. He fought at Halle on 17 October and was in command of a column of troops that routed a Prussian force at Strelitz on 31 October. He distinguished himself at Crewitz on 3 November and was at the fall of Lubeck three days later.

Gérard was acutely aware that Bernadotte's differences with Napoleon were compromising his career and requested a transfer. Bernadotte understood and soon after his promotion to *général de brigade* on 13 November 1806 Gérard joined Augereau's VII Corps in charge of a brigade with Desjardin's division. Luck again deserted him when his brigade was caught in the holocaust at Eylau on 8 February 1807. French and Russian artillery decimated his troops when Augereau's corps mistakenly wandered off its line of attack due to a sudden snow storm that engulfed the field. So damaged was Augereau's command that it was disbanded and on 15 March Gérard found himself back with Bernadotte's Corps in command of the 2nd Brigade of Villatte's 3rd Division.

When Bennigsen moved against I Corps at the end of Spring, Gérard fought Russians at Spanden on 5 June 1807. He was present at Friedland on 14 June, but the French victory was so complete his troops never fired a shot. After Tilsit when I Corps occupied Prussia and Bernadotte became governor of the Hanseatic towns of Hamburg, Bremen and Lubeck, Gérard on 23 August 1807 took the post as his Chief of Staff. After the Royal Navy bombarded Copenhagen and seized the Danish fleet he spent from March till May 1808 in charge of a mission sent to help the Danes strengthen the city's defences. To mark his efforts the Danes awarded him the Grand Cross of the Order of Denmark.

THE DANUBE CAMPAIGN 1809

Gérard joined Bernadotte for the campaign against Austria serving as IX (Saxon) Corps chief of staff with the Army of Germany. On 3 May 1809 he received the title *baron de l'Empire*. He fought at Durfort on 7 May and successfully stormed Raasdorf during the first day of the Battle of Wagram on 5 July. The next day he led a decisive charge by the Saxon cuirassier regiments against Hessen Homberg's columns that stemmed the rout of the Saxon infantry that panicked under Bernadotte. Alongside Bernadotte he later incurred Napoleon's wrath when his chief foolishly issued a bulletin praising the Saxons when in fact their behaviour nearly cost Napoleon the battle. Napoleon's patience with Bernadotte finally broke and he dismissed him, leaving Gérard the task of disbanding the Saxon Corps. On 30 October his task complete he moved to Tharreau's 1st Division of Oudinot's II Corps, replacing Conroux as head of the 1st Brigade.

SPAIN 1810-1811

Gérard again found himself unemployed on 19 July 1810 when with Europe at peace, apart from the war in Spain, there was a general reduction in the size of the army. A reprieve came when he was ordered to join d'Erlon's IX Corps with the Army of Spain. This rag-tag formation of some twenty odd replacement battalions set out from Bayonne in October 1810 to reinforce Massena's beleaguered Army of Portugal and Soult's forces in Andalusia. Serving with Claparède's division they made painfully slow progress through Spain. He reached Almeida on the Portuguese border on 15 November but remained there a further month till the rear elements of IX Corps caught up. On 14 December he moved to Conroux's division that set off down the Mondego Valley in search of Massena, while Claparède remained at Celorico to keep communications open with Almeida. He made contact with the Army of Portugal on 26 December 1810 and spent two months around Leiria before the army faced by starvation fell back to the Spanish border. The following Spring he joined the new offensive to relieve Almeida and fought at Fuentes d'Onoro on 4-5 May 1811. His brigade in particular suffered heavily in unsuccessful assaults on the town.

In June with Conroux and d'Erlon he left Massena's army and moved on with the remnants of IX Corps to join Soult's Army of Andalusia. He served under Conroux with Victor's I Corps before

Cadiz but soon fell seriously ill. Sent north to Valladolid to convalesce he obtained leave and returned to France on 15 August 1811. His recovery was slow with him going through a period of disillusionment when once more on 1 October 1811 he found himself unattached and placed on half pay.

The Russian Campaign of 1812

The increase in size of the army for the coming campaign against Russia meant his recall on 14 March 1812. Ordered to report to Davout's I Corps in Poland he joined Gudin's 3rd Division in commander of its newly formed 3rd Brigade. He revelled in the competitive atmosphere of Davout's command which gave his career a new lease of life compared to the negativeness of Bernadotte. On 17 August he distinguished himself in the struggle for Roslavl during the Battle of Smolensk. Then with Gudin at his side, his brigade was the first into action at the Battle of Valoutina on 19 August. Gudin with his legs taken away by literally one of the first cannon fired, handed command of the division to Gérard, not due to seniority, but rather that he happened to be on the spot. It worked to Gérard's advantage since Gudin's men were so maddened by the fate of their much loved and respected leader they drove all

before them. The success, attributed to Gérard's leadership, resulted in his formal appointment as commander of the division.

At Borodino on 7 September 1812 Gérard was detached to reinforce Eugene's IV Corps. His division was held in reserve as Broussier took Borodino with his first assault and crossed the Kalatsha without much difficulty. The storming of the Raevsky Redoubt behind the river proved more difficult and Morand* after gaining a foothold was driven back by furious Russian counterattacks. Gérard's troops still comparatively fresh then supported the mass cavalry charges by Caulaincourt that effected a breach around 11 A.M.. Occupying the position Gérard greatly enhanced his reputation as the Russians as he successfully drove back all attempts to dislodge him.

On 23 September 1812 he received promotion to *général de division*. In the retreat from Moscow Gérard's division formed Davout's rearguard as Napoleon fell back on Mozhaisk after failing to dislodge Kutuzov from Maloyaroslavets on 25 October. He successfully drove through Miloradovich's forces at Viazma on 3 November after they cut off Davout from Eugene. He suffered heavy casualties at Krasnoe on 17 November when the Russians again barred his route. After crossing the Beresina the remnants of his division joined Ney's rearguard and reached Kovno on the Niemen on 13 December. Here with Ney he fought in the last action of the campaign as Platov's cossacks tried to cut Ney off from the bridge over the river.

Ney and Gérard fire the last shots of the Russian Campaign

Poland and Saxony 1813

Eugene de Beauharnais, in command of the Army as it fell back through Poland and Prussia had the highest regard for Gérard and gave him command of a provisional division formed to provide the rearguard. With barely 1,600 men under him he delayed the Russian pursuit, in one notable action driving them from Bromberg on 21 January 1813. One of the few French commanders left showing real fight, Gérard replaced Lagrange on 6 March 1813 as head of the 31st Division of Macdonald's XI Corps, it being one of the few intact formations remaining.

In one of the opening actions of the Spring Campaign he distinguished himself in the capture of Merseburg on the Saale. He fought at Lutzen on 2 May when his troops excelled in taking the village of Kitzen. During the pursuit of the Allies he handed over his division to Ledru and took over the 35th still with XI Corps. He fought under Macdonald at Bautzen on 21 May and did well to establish a bridge across the Spree north of the town that enabled the corps to pour across and scale the heights. Engaged in the pursuit he received a leg wound in a skirmish before Laubon on 25 May.

Having proven himself in the Spring, Gérard received command of XI Corps on 23 August 1813 on Macdonald's appointment as commander of the Army of the Bober. He immediately came to grips with the Prussian division of Duke Charles of Mecklemberg and drove them from the heights of Goldberg, forcing the enemy to retire beyond the Katzbach. Confident that Blucher was on the run and taking advantage of an ambiguity in Napoleon's orders that directed him not to cross the river, Macdonald did just that on 26 August. Gérard was aware of the risks but had little alternative but to comply. The following day he found himself with his 27,000 men and Sebastiani's 3,000 cavalry confronted by over 55,000 Prussians. The pontoons behind him were washed away by a sudden storm and his infantry's muskets were wet and useless. Blucher's superiority in all arms was decisive as his forces were remorselessly driven back into the river. Shot in the thigh Gérard was one of the 14,000 casualties suffered, most from his corps.

On his return to duty he resumed command of the 35th Division while Macdonald was back at the head of XI Corps. At Leipzig on 16 October he took part in the fighting to the south of the city where his division positioned on the extreme left of the French line drove Ziethen's Prussians from Klein Posna. He withdrew with his division to a more defensible position during the night of 17 October and the next day received a bad head wound when he repulsed Bennigsen's assaults on Zweindorf. Evacuated from the city that night he was fortunate to survive the rigors of the retreat to the Rhine.

THE 1814 CAMPAIGN IN FRANCE

After a brief stay in Paris recovering from his wounds Gérard spent December 1813 as an envoy in the Departments of Meurthe and Vosges reporting on the preparedness of the eastern defences. On his return he took command of the Paris Reserve on 29 December, a formation comprising the city's two National Guard Divisions led by Ricard and Dufour. At their head on 25 January he joined Napoleon's army in Champagne and four days later fought at Brienne. Engaged again at La Rothière on 1 February he narrowly escaped after his troops were cut off in Dienville. The incident arose due to Napoleon's sudden withdrawal when he realized his army was vastly outnumbered, leaving Gérard to fight his way

out on his own. As Napoleon fell back on Troyes after the reverse Gérard fought a fine delaying action at Lesmont, holding the bridge till the army passed and then blowing it in the face of the enemy. He delayed the Austrians again at La Guillotiere, defending another river crossing before Troyes, enabling Napoleon to regroup. He then formed the rearguard as the army fell back to Nogent-sur-Seine.

When Napoleon turned on Schwarzenberg's Army of Bohemia after the Seine front was stabilized, Gérard with Grouchy destroyed Pahlen's command at Mormont on 17 February. He then turned on Wrede's advanceguard at Valjouan and drove it back through Nangis with heavy losses. At Montereau on 18 February he replaced Victor as head of II Corps after the latter's tardy arrival on the battlefield caused Napoleon to miss the opportunity of an early victory. In the battle Gérard gave one of his finest performances. He methodically massed all his available artillery and silenced the Austrian guns covering the twin bridges over the Seine and Yonne. With cavalry in support led by Pajol*, he then launched a devastating attack which carried the town and its key bridges. The battle's result was to send Schwarzenberg's forces scurrying eastwards and the Allies into fits of panic as thoughts loomed of Napoleon rampaging in their rear.

In Napoleon's following drive east Gérard passed through Troyes on 24 February and seized the bridge at Delancourt two days later. When Schwarzenberg regained his composure and turned on Oudinot's Corps at Bar-sur-Aube, Gérard took charge of the heavy rearguard fighting as Oudinot withdrew. The French fell back westwards towards Paris and he fought a stiff action at Vendeuvre on 1 March. Austrian cavalry gave him a rough handling on 3 March as he retired along the Barse. He defended Troyes the following day before retiring down the Seine to Nogent. Here he dug in and on 16 March drove the Allies back from the town. He rejoined Napoleon and was present at the last victory of the campaign when Winzingerode was caught at Saint Dizier on 26 March. He retired with the army to Fontainebleau and was there during the Abdication Crisis. No rabid Bonapartist, too many years with Bernadotte having left their mark, he decided in the interests of France to declare for the provisional government on 8 April 1814.

THE RESTORATION AND NAPOLEON'S RETURN

Under the Bourbons Gérard's first task was to replace Davout at the head of XIII Corps still resisting the Allies at Hamburg. He arrived before the city on 12 May and after exercising great tact and patience negotiated honorable terms with Davout for the city's surrender and the evacuation of the French troops. Out of respect to Davout and his men, he only took charge of XIII Corps as the troops entered France in June.

He received numerous awards, becoming on 1 June 1814 a *Chevalier de Saint Louis*, then receiving on 29 June the Grand Cross of the Legion of Honor. From his old friend Bernadotte, now Crown Prince of Sweden, came the Grand Cross of the Order of the Sword of Sweden. Despite becoming disillusioned he remained active under the Bourbons and on 25 February 1815 was made Infantry Inspector General for the 5th Military Division based at Strasbourg. He saw little good coming from the news of Napoleon's landing, and whilst sympathetic was not prepared to join any popular uprising against the king. He remained at Strasbourg to await the outcome of events and only when Napoleon arrived in Paris did he offer his services. On 31 March familiar with the troops and commanders stationed in Eastern France he was appointed commander of the IV Corps of Observation of the Moselle based at Thionville and of all troops in the 3rd and 4th Military Divisions. In April when it became apparent war was inevitable he moved to Metz to prepare IV Corps for the coming campaign. A diversion from his preparations came on 2 June 1815 when he became a Peer of France and attended the great ceremony on the Champ de Mars.

THE WATERLOO CAMPAIGN - BOURMONT'S DEFECTION

Ordered to concentrate his corps at Phillipeville on the Belgian border he started the long march from Metz on 6 June. His troops reached their destination undetected on the night of 14-15 June, their approach screened from Prussian outposts by the forests of the Ardennes. Forming the extreme right wing of Napoleon's advance he was late on the road to Charleroi the next morning as his men needed time to recover from their long march. Disaster struck when Bourmont at the head of the advanceguard division deserted to the enemy with his staff. Confusion arose among the columns and further delays occurred while Gérard completed a lengthy report explaining the misfortune to Napoleon. It was he who had personally backed Bourmont's appointment in spite of strong opposition. Hulot took charge of the lead division and reached Chatelet in the late afternoon after Soult redirected Gérard to cut across country to the Sambre.

Due to his late arrival on the Sambre, Gérard was unable to assist at Gilly. Had he crossed the river he would have been in a magnificent position to fall on the Prussian left flank enabling the whole army to be astride the Sambre by the night of 15 June. Crossing the next morning resulted in his late arrival with IV Corps at Ligny, so delaying the start of the battle. Had the Prussians suffered defeat earlier during daylight hours, in consequence cavalry in pursuit would have established the true direction of their retreat and punished them more. The delay though not any fault of his, had far reaching results on the final outcome of the campaign.

LIGNY AND WAVRE

At Ligny early in the day Gérard had a fortunate escape. As his troops formed up before the town he went forward to reconnoiter the enemy's positions accompanied by a few staff officers and hussars of the 6th Regiment. When he neared the Prussian lines a body of cavalry moved against him, causing his escort to retire at full speed. During the flight his horse fell in a ditch hidden from view by the high standing wheat. The escort seeing his plight turned back to defend him and in the desperate skirmish that followed Gérard's chief of staff Saint Remy was badly wounded. Gérard unable to remount in the hand to hand fighting, most certainly would have been killed or taken prisoner. It was only the arrival of a detachment of the 12th Chasseurs à Cheval led by Grouchy's son, attracted to the scene by the sound of firing that drove off the Prussian horsemen.

In the battle itself, after a heavy opening bombardment Gérard directed the assaults by Pécheux's and Vichery's divisions against Ligny around 3.00 P.M. They gradually wore down the Prussian center as Blucher had to commit his reserves and when Napoleon ordered the Old Guard forward around 7.00 P.M. they achieved the breakthrough. As part of Grouchy's right wing he spent the next day trying to establish the route the retreating Prussians had taken. Due to the conflicting and confusing reports he received from the cavalry during the day he had little idea of the enemy's line of retreat and by evening had only reached Gembloux.

The next morning he was with Grouchy at Walhain when they heard the opening cannonades from Waterloo. He urged Grouchy to march immediately towards the sound of the guns. Had Grouchy accepted this piece of unsolicited advice, Gérard's corps would have reached the edge of the battlefield by 7.00 P.M.. His mere approach moving against Bulow's IV Corps exposed left flank would have been enough to relieve the Prussian threat to Plancenoit and Napoleon's communications. Grouchy almost at the same time heard that Wavre four miles ahead was swarming with Prussians so he ordered Gérard forward. Nearing Wavre Gérard saw Prussians streaming westwards behind the Dyle. To disrupt their march he headed for the bridge across the river at Bierges. Leading the assault on the bridge a musket ball hit him in the chest. Carried from the field apparently mortally wounded he took no further part in the campaign.

His wound was not fatal, the ball had glanced off a decoration and caused severe bruising. He recovered sufficiently to rejoin the army before Paris. On 7 July he was nominated by Davout with Kellermann and Haxo to negotiate terms for the Army's submission to Louis XVIII. Their approaches rejected, the Army had no alternative but to submit unconditionally on 14 July 1815. Behind the Loire he presided over the Army's disbandment till his removal on 12 August

1815. Word leaked out that he was to be arrested so he hurriedly left France and settled in Belgium.

LATER POLITICAL AND MILITARY CAREER

A confirmed bachelor, peace brought a change of heart with him falling in love with the daughter of the old veteran General Valence. The couple married in 1816. He returned to Paris at the end of 1817 and with Saint Cyr's Army Reforms of 30 December 1818 was recalled but remained unattached throughout the period of Bourbon rule.

Gérard opted for politics and was elected a liberal deputy for the 1st Arrondissement in Paris on 28 January 1822. The rout of the Liberals in the elections of 1824 resulted in the loss of his seat. Further misfortune came that year when he lost an eye in a hunting accident. He gained re-election on 8 February 1828 as deputy for Bergerac in the Dordogne. He was amongst the deputies that led the revolt in the Chamber against the repressive Ordonnances of 27 July 1830 that hastened the downfall of the Bourbons. In a fiery speech in the Chamber he caused such a hew and cry against the government that Polignac resigned as first minister. In the provisional government set up on 1 August 1830 he replaced Bourmont, who was in Algeria as Minister for War. His was appointment ratified by Louis Philipe on 11 August 1830. On 16 August 1830 he became a Marshal of France. In the new elections on 21 October 1830 he was elected deputy for Clérmont (L'Oise). Ill health forced him to resign as Minister for War on 16 November 1830 though he continued to serve as a deputy.

On his recovery he resumed his military career. On 4 August 1831 he was made commander of the Army of the North. He led the army into Belgium on 8 August in support of Belgian claims for independence. Crown Prince William, the inept veteran of the Waterloo Campaign, decided it prudent to withdraw and settle the matter at the conference table. Gérard withdrew but when the Dutch refused to comply with the terms of the Treaty of London and hand over Antwerp to Belgium he invaded the country a second time on 15 November 1832. Within a fortnight he crossed the country and drove the Dutch from the trenches before Antwerp into the citadel. On 4 December he started a bombardment of the fortress that lasted till 24 December when Baron Chassé, another Waterloo veteran, capitulated. On his return to France he was made a Peer on 11 February 1833.

When Soult was out of office for a brief period from 18 July till 19 October 1834 he replaced him for second term as Minister for War, while also acting as President of the Council. He became a Grand Councillor of the Legion of Honor on 21 October 1835 after the death of Mortier. He held the post till 11 December 1838 when he took over that of Commandant Superieur of the National Guard on the death of Lobau. When Oudinot retired as Grand Councillor of the Legion of Honor on 21 October 1842 he held the post for a second time till 25 February 1848.

A close advisor to Louis Philippe, he remained loyal to him till the end. A strong advocate of law and order he looked on with dismay as the monarchy lost all credibility and control. Well into his seventies when Louis Philippe abdicated he resigned from all offices he held. Approached, he was not prepared to be associated in any way with the Second Republic, which in its first months had a very shaky existence. He retired from public life, only to emerge briefly in support of Louis Napoleon when he saw in a resurgence of Bonapartism a possible answer to the long term stability France needed. He accepted an appointment as a Senator on 26 January 1852, but died in Paris on 19 April 1852 a few months before the establishment of the Second Empire.

A CAREER ASSESSMENT

Of Napoleon's generals who became Marshals after his fall, Gérard possibly merited the rank before any others. His rise in the period from 1812 to 1815 was as meteoric as was Napoleon's decline and fall. His earlier career had been hampered by a long association with Bernadotte. Under Davout who was one of the most exacting of commanders his fortunes changed dramatically. Davout recognized talent and gave him Gudin's division in preference to his long serving brigade commanders. Gérard proved his worth to such an extent that within a year he was a corps commander.

At forty-one he was the youngest of the French corps commanders in the Waterloo Campaign. His record had proven that as a strategist he was the equal to others who led a corps in the campaign. In 1813 he saw the dangers of crossing the Katzbach but was ignored by Macdonald. The same happened in 1815 when Grouchy failed to heed his advise to march to the sound of the guns at Waterloo. An unfortunate failing was his impatience accompanied by a lack of tact. His brusque manner and the way he addressed his superiors often produced exactly the opposite to what he wanted to achieve. Had he suggested to Grouchy rather than demanded they march on Waterloo, his chief may have reacted differently and the whole outcome of the campaign may have been very different.

He was no avid Bonapartist. His loyalty was first to France and the Army, in that order. His changes in allegiance on Napoleon's return and again after the fall of Louis Philippe were justified as in the interests of France rather his own. As a result he kept his hands clean and rightly deserved the recognition he achieved in later life.

BACKGROUND AND EARLY CAREER

The son of a shopkeeper and a native of southern France, Girard was born in the village of Aups (Vars) north of Toulon on 21 February 1775. He enlisted on impulse on 23 September 1793 when at the nearby town of Barjols a recruiting battalion passing through the district called on him to volunteer. Having received some formal education, within a week he had risen to become battalion quartermaster and paymaster for the 3me Marathon Revolutionnaire.

In March 1794 his unit merged with the 46me Demi-Brigade de Bataille with Massena's division of the Army of Italy. He served on the regimental staff as an orderly in the expedition against Oneille and Saorgio in April 1794. His efficiency caught the eye of Massena's chief of staff *adjudant général* Monnier, who on 18 August 1794 secured Girard's transfer to the division staff. He served on Massena's staff throughout a period when the fortunes of the Army of Italy fluctuated wildly and Massena's reputation as one of the foremost generals in Italy grew. He fought at Melogno on 25 June and at Loano on 23-24 November 1795. He received a commission as *sous lieutenant* from Massena on 16 March 1796.

The arrival of General Bonaparte as the army's commander in April 1796 caused the tempo to pick up as a new lease of energy was injected into the troops and their commanders. In the campaign that knocked Piedmont out of the Coalition, Girard was present with Massena at the battles of Montenotte on 12 April, Dego 14 April, Cherasco on 25 April and the crossing of the Lodi on 10 May 1796. As the campaign moved on through the Duchy of Milan and into the Venetian Republic he fought at Castiglione on 5 August and then at Bassano on 8 September. Wounded at La Brenta on 8 November 1796 he was out of action for several weeks.

Promoted *lieutenant* on 23 May 1797 he became Monnier's aide, who now headed a brigade with Joubert's 5th Division of the Army. On 5 November 1797 General Bonaparte recommended his promotion to *capitaine* and approved his transfer to the 85th Line, one of formations making up Monnier's brigade. A particularly distinguished period followed when fine performances by the 85th Line brought his career to the fore. On 29 December 1798 Championnet, who now headed the Army, appointed Girard acting *chef de bataillon*. His rank was later confirmed by the Directory on 18 January 1799. During the 1799 Campaign in Italy he was trapped with Monnier's force of 3,000 at Ancona from 18 May till the garrison was forced to capitulate on 15 November 1799. He played a prominent part in the town's defence, particularly when he led an assault against Ascoli on 1 June. An honorable capitulation negotiated, he returned to France on parole with the recommendation of Monnier for his promotion to *adjudant général chef de brigade*. The promotion went through and his rank was confirmed by the Consulate on 28 March 1800.

ITALY 1800-1804

On 14 May 1800 Monnier now the commander of the 6th Division of the Army of the Reserve appointed Girard his chief of staff. Girard served with the Army during the crossing of the Alps. He distinguished himself at the crossing of the Ticino at Turbigio on 31 May when Monnier overcame the last obstacle before of Milan, before Napoleon entered the city on 2 June 1800. He fought on the French right flank at Marengo on 14 June and with Monnier's widely scattered units took part in the desperate struggle for Castel Ceriolo before the Austrians drove them back on San Guiliano.

After the campaign he remained in Italy under Monnier based at Bologna. The renewal of hostilities after the breakdown of the Convention of Alessandria saw Girard campaigning in the Italian peninsula against the Neapolitans. He took a leading role in the assault on Arezzo on 19 October that broke the Neapolitan grip over the Papal States north of Rome. Back with Brune's main army he was at the crossing of the Mincio on 25 December and then with Monnier on 17 January 1801 when the French captured Verona.

On the formation of Murat's Army of the South to knock the Neapolitans out of the Coalition he served with Monnier in the expedition against Ancona at the end of January 1801. Promoted *adjudant commandant* on 3 August 1801, his association with Monnier ended when the latter was dismissed on 13 September 1802 for his opposition to Napoleon becoming Consul for life. Girard, his career also in jeopardy survived the furore and remained in the Italian Republic till 20 June 1804 when he transferred to Paris as chief of staff to the 1st Military Division.

THE GRANDE ARMÉE 1805-1807

On 31 August 1805 Girard joined Murat's staff with the Cavalry Reserve of the Grande Armée. He distinguished himself in action at Nuremberg on 20 October and again at Austerlitz on 2 December 1805. At the end of the year he was serving as deputy chief of staff under Belliard with the Cavalry Reserve.

At Murat's side during the Prussian Campaign of 1806 he served at Jena on 14 October and during the campaign received promotion to *général de brigade*

GIRARD - JEAN BAPTISTE, BARON (1775 - 1815)

on 13 November 1806. He resumed his career with an infantry command in Poland when he led a brigade with Suchet's 3rd Division of Lannes's V Corps on 31 December 1806. With Suchet he covered the line of the Omulew and Narew rivers against a possible Russian advance on Warsaw to cut Napoleon's communications. On 12 May 1807 he repulsed a Russian attack led by Essen at the confluence of the two rivers and then a month later at Drenzeno on 12 June he beat back another attempt to cross the Omulew.

SPAIN - OPERATIONS IN ESTREMADURA AND ANDALUSIA

After the Treaty of Tilsit he spent a year based in Silesia before his brigade started their long march across Europe to join the forces concentrating at Bayonne for Napoleon's invasion of Spain. As he passed through Paris he was made a *baron de l'Empire* on 26 October 1808. Suchet's command, now the 1st Division of Mortier's V Corps, crossed the Spanish border in mid December 1808. Girard spent the early months of 1809 with his brigade at Calatayud covering communications between Madrid and Saragossa while the rest of V Corps was trying to fight their way into Saragossa. On 5 April he took temporary command of the division when Suchet replaced Junot as head of III Corps.

After the fall of Saragossa he moved with V Corps into Estremadura. He distinguished himself at Arzobispo on 8 August 1809 when he drove back an attempt by the Spanish under Cuesta to cross the Tagus to link up with Wellington's forces falling back into Portugal. Disappointment came when Darmagnac soon after was made commander of the division and he reverted to his post as head of its 1st Brigade. Mortier intervened on his behalf and the situation was rectified on 20 September when he formally became acting *général de division* and his command restored with Darmagnac sent back to Castile as military governor.

Engaged in the operations to catch Areizaga's Army of La Mancha he fought at Ocana on 18 November 1809. His division played a vital part in the battle when forming the reserve they halted the retreat by Leval's Germans and Poles after a furious Spanish countercharge. Mortier, forced to bring Girard forward to support his broken first line was hotly engaged in a costly frontal assault against the divisions of Lacy and Giron. The Spanish infantry was distracted by Girard's approach and failed to observe to their right friendly troops slacken their fire, waver and break. Girard's assault enabled one of the most brilliantly executed combined infantry and cavalry charges of the Napoleonic Wars to be delivered by Milhaud* rolling up the Spanish line from left to right, destroying five divisions. The Spanish lost 4,000 killed and wounded and 14,000 prisoners opposed to 1,900 French. Wounded in the battle Girard was out of action for a few months. During his time recuperating in France Girard's promotion to *général de division* was confirmed on 17 December 1809.

In early 1810 he returned to Spain spending time assisting Sebastiani in Murcia. He then moved to Southern Andalusia in May to help put down the guerilla bands in the region. He relieved the French garrison blockaded at Ronda after severe fighting at Albondonates on 1 May, and at Grazalema two days later. In June he was ordered to hunt down Lacy, who had landed a force of 3,000 men at Algeciras with a plan to raise the countryside and march on Ronda. As Girard closed from the north and Sebastiani from the east Lacy gave up his attempt to reduce Ronda and fell back to Marbella and Estepona before completing a hasty evacuation to Gibraltar on 12 July.

In August Girard moved from the Ronda region towards the passes of the Sierra Morena to counter an advance by La Romana's Army of Estremadura that threatened Seville. On 11 August 1810 at Villagarcia, with a combined force of 7,000 infantry and 1,200 cavalry, he drove back a Spanish force of 11,000 inflicting over 600 casualties for the loss of 200 men. In the pursuit he followed them as far as Zafra. The action itself was notable in that it removed any further threat from that quarter to Seville.

In January 1811 he took part in Soult's invasion of Estremadura that was supposed to lend support Massena in Portugal. On 11 January he appeared before the walls of Olivenza and after a twelve day bombardment the fortress surrendered. He then served at siege of Badajoz, where on 7 February he took part in a sharp struggle outside the town when the Spanish garrison tried to break out. Carlos de Espana sallied from Badajoz and hurled 5,500 men against Girard's entrenchments on the hill of San Miguel. In the struggle the first line was overrun before Girard regrouped and drove them back into the city for the loss of 650 men against 400 French. He fought at Gebora on 19 February when Mortier intercepted the Army of Estremadura, numbering some 12,000 under Mendizabal, trying to relieve Badajoz. Girard with his division led the French advance and destroyed a force nearly twice its size, inflicting 800 to 900 casualties and taking a further 4,000 prisoner.

ALBUERA - A FINE PERFORMANCE

After the fall of Badajoz on 9 March 1811 Soult left a garrison, which in turn was blockaded by Beresford at the beginning of May. In response Soult gathered an army to march to its relief. Beresford abandoned the siegeworks and took up a position at Albuera to face the new threat. In the battle that followed on 16 May 1811 Girard played a major part and fought his finest battle. In temporary command of V Corps, he delivered the main attack against Beresford's right, while Latour-Maubourg his corps commander led the army's cavalry. As diversionary attacks went in against the Allied left and center, Girard unleashed one of the most imaginative and skilful French attacks conceived of the entire Peninsular War. Out of sight, deep in olive groves and hidden by the heights between Nogales and Chicapierna he executed a circular sweep with V Corps that descended on the exposed right flank of Beresford's line.

Zayas's Spanish division from Ballasteros's army, which unknown to Soult had just joined Beresford, was the first to deploy in line to counter this flanking movement. For a while they fought surprisingly well, but Girard's columns gradually gained the upper hand as they neared the crest of the ridge. Colbourne's brigade of Stewart's division then came to the rescue and passing through the Spanish line halted the French attack with its fire. The smoke obscured visibility and they failed to notice the approach of French cavalry. In turn they were overrun by a brilliantly executed charge by Latour-Maubourg with the 2nd Hussars and Vistula Lancers, hovering nearby in anticipation of such a move. Cole's 4th Division quickly filled the breach, and there followed one of the bloodiest exchanges of volley fire of the entire Peninsular War between two evenly matched infantry formations. It lasted for forty minutes before Girard's two divisions began to fall back. Girard suffered 3,128 casualties out of 8,437 men engaged, while Colbourne lost 1413 out of 2,064 and Cole a further 2,110 in the savage exchange.

Girard received criticism for the column formation he used at Albuera. At the time his tactics were sound, compared to those adopted in the past by his more august predecessors against Spanish infantry. He had approached the Spanish line in an ordre mixte with two of his battalions in line three deep to counter the enemy fire. It was certainly one better than the formations used in the unsuccessful assaults by Victor at Talavera or Barrosa, or for that matter by Ney and Reynier at Bussaco. In these attacks the divisions were formed up by battalion in "column of division" formation, and whilst more mobile they restricted the formation's firepower even further. Girard's formation was a novel midway compromise that showed imagination and flair. He was unlucky in that he suddenly found himself facing steady troops who were able to employ a vastly superior firepower. One factor was V Corps had never fought English infantry before, the old tactics used for years and so successful against Continental armies had been more than enough against the Spanish.

DISASTER AND DISGRACE - ARROYO DOS MOLINOS

After Albuera the military situation in Estremadura developed into a stalemate as the arrival of Marmont's Army of Portugal forced Wellington to once more abandon the Siege of Badajoz. On 27 June 1811 Girard handed command of V Corps to d'Erlon*. His division was for a time under Marmont's orders with the task of covering Badajoz. In the Autumn he returned to d'Erlon and V Corps with the unenviable task of covering the whole front between the Tagus and the Guadiana rivers while Claparède covered the area from Badajoz to the Sierra Morena. It left him very exposed to incursions by Wellington, and with little prospect of help from d'Erlon who was at Zafra fifty miles south-east of Badajoz with the bulk of Claparède's forces.

Wellington saw Girard's weakness and sent Hill with 10,000 Allied troops to surprise him. Girard marching from Montanchez arrived at Arroyo dos Molinos on the night of 27 October. Hill had been marching in a parallel direction from Aldea del Cano and had halted at Alcuescar the same night. Girard's cavalry, which numbered over 1,000 horsemen, failed to detect this movement. Hill hearing of Girard's proximity only five miles away and being superior in all arms decided to march against him without delay. Breaking camp he set out at 2.30 a.m. through torrential wind and rain for Arroyo dos Molinos. Disguised by the elements Hill's force was within half a mile of Girard's camp before the alarm was raised. Unfortunately for Girard, Renard's brigade had already left, leaving him with only some 4,000 men made up of six weak battalions, two regiments of cavalry, and a half battery of guns.

Hill's 71st and 92nd Regiments first swept through the village and captured Girard's baggage, taking General Bron the cavalry commander prisoner, before they pressed on against Dombrowski's brigade. In the growing confusion Girard tried to break out along one road then another, but in vain. The leaderless French cavalry were defeated by the 9th Light Dragoons and the 2nd Hussars of the King's German Legion amidst a dense morning mist. Girard's infantry gave ground; 1,000 surrendered at the foot of the 1,000 foot-high cliffs, while the rest fled the 24 miles to Truxillo pursued by British cavalry. Girard himself scaled the cliffs then made his way to safety by foot. Girard lost over 1,600 men while Hill suffered barely a 100 casualties.

Strategically the reverse was a disaster, and Hill succeeded in cutting all communications between the Army of Portugal and Soult's Army of the South. Soult, intolerant of such a lapse, replaced Girard with Barrois* on 16 November and gave him the post of military governor of Seville. When news of the fiasco reached Napoleon he wanted blood and on 31 December 1811 immediately ordered Girard's recall to France. Girard departed from Andalusia in February 1812 and by the time of his return things had simmered down. The affair was looked at more objectively. The force Hill had thrown against him was overwhelming. An attack made in such appalling weather was unprecedented, it was not surprising the outposts could not raise the alarm. Instead of dismissal and disgrace Girard took a backwater posting under Claparède to train a new Polish division forming at Sedan. The task completed he then took over the formation upon the departure of Claparède on 5 April 1812.

THE RUSSIAN CAMPAIGN OF 1812

On 4 May his regiments became the 28th Infantry Division with orders to join Victor's IX Corps of the Grande Armée in Silesia. During the march across Germany he picked up two Saxon regiments, the Rechten and von Low, which formed his second brigade. On 22 July he crossed the Niemen at Tilsit and entered Russia where he carried out rear area duties between Vilna and Minsk whilst also covering the Army's communications. On 24 September he moved up to Smolensk where his three Polish regiments formed the city's garrison. In mid October he left Smolensk to support Saint Cyr's VI Corps, which was being hard pressed by Wittgenstein near Polotsk. He fought at Tchasniki on 29 October when Victor halted an attempt by Wittgenstein to cut the Grande Armée's communications with Poland.

His gallantry once more came to the fore on 26 November 1812 when he rejoined the remnants of the Grande Armée on the east bank of the Beresina. He fought a skilful rearguard action against Wittgenstein's forces covering the French retreat across the river. Two days later he managed to extricate the remnants of his division and crossed the river at Studenka after receiving a leg wound. The determination he showed on the Beresina went a long way towards Napoleon snatching a strategical victory against all odds from the jaws of defeat. During the final days of the retreat to the Niemen his troops formed the Army's rearguard. By the time he reached Vilna on 10 December his formation had completely disintegrated due to cold, hunger and marauding Cossacks.

POLAND AND SAXONY 1813

In January 1813 Girard took over a scratch division of 3,500 Poles and Saxons that tried to break the momentum of the Russian advance. Retiring by stages through Poland he reached Kustrin on the Oder where on 5 March the formation was disbanded and the units parcelled out to reinforce the fortresses on the river. Girard was recalled and took charge of the 10th

Infantry Division of Ney's III Corps forming at Wurzburg. He spent six weeks there building up and training the new formation before joining Napoleon's march into Saxony.

A repeat of the Arroyos dos Molinos affair nearly ruined Girard's career a second time when he failed to post adequate pickets, which allowed Prussian cavalry to blunder into his and Souham's divisions encamped near Lutzen on the morning of 2 May. Fortunately due to uncharacteristic caution on the part of Blucher, who did not pursue his advantage, Ney was able to rally the rest of the corps and stabilize his front. Girard rallied his green troops and in the afternoon retook the village of Rahna, but in the process was cut down by grape-shot and put out of action for the remainder of the Spring Campaign.

His return to duty saw him in command of the Division of Observation, a detached force centered on Magdeburg to cover the Elbe. When hostilities were renewed his 13,000 men joined Oudinot's northern march on Berlin. Then Oudinot's nerve failed him after Reynier's reverse at Gross Beeren on

THE FRENCH INFANTRY ATTACK AT LIGNY

23 August and his army fell back to Wittenberg. It left Girard advancing on his own from Magdeburg isolated as the Prussians closed in. Caught at Hagelberg on 26 August, he fought a desperate rearguard action. Wounded, he lost 3,000 men, several guns and was very fortunate to regain the safety of Magdeburg.

Girard held onto Magdeburg after the Prussians crossed the Elbe and took in isolated French formations as they fell back to the safety of the walls. At the end of October after the Battle of Leipzig the Magdeburg garrison increased this way to nearly 30,000 men before Bennigsen's Russians encircled the city. He stubbornly held on to the city till mid April 1814 only surrendering when confirmation of Napoleon's abdication reached him and he was guaranteed the safe repatriation of his troops to France.

THE RESTORATION AND NAPOLEON'S RETURN

The Bourbons made Girard a *Chevalier de Saint Louis* on 19 July 1814, but the hollowness of the gesture was shown the same day when he found himself unattached and placed on half pay. Not surprisingly he rallied to Napoleon on his return from Elba and joined him at Auxerre. Napoleon as a mark of his esteem gave him command of his advance guard as it neared Paris on 18 March 1815. Numerous assignments followed in the period before the campaign in Belgium. He was first given command of the 18th Infantry Division with the Army of the Alps. Then before setting out to take up the post, on 3 April he was ordered to remain in Paris to help Lobau to organize VI Army Corps. On 3 June he moved to the 7th Infantry Division of Reille's Corps replacing Lamarque, who was given the task of putting down the unrest in the Vendée. Girard was another who became a Peer of France on 2 June 1815 as part of Napoleon's plan to swamp the Upper Chamber with his followers.

THE WATERLOO CAMPAIGN

The 7th Division outside the Guard was one of the finest formations in the Army. In particular the 11th Light Infantry, formed from veterans of the Tirailleurs du Po and Tirailleurs du Corse, had one of the most distinguished records in the Army. At dawn on 15 June he crossed the frontier at Thuin. Following Bachelu's route de marche he only briefly came into action at the end of the day when shadowing the Prussians as they retired north-east from Gosselies he pushed them out of Wangenies. The next morning, separated from Reille's command and on the edge of the Ligny battlefield, he was placed under the

orders of Vandamme. Positioned in reserve on the extreme left of the French line his task was to support the attacks by Lefol and Berthezène against the hamlets of Le Hameau de Saint Armand and Saint Armand la Haye. After the Prussians repulsed the assaults several times Girard joined the battle around 4.00 P.M. and soon took Le Hameau endangering Blucher's right wing. Pirch's brigade entered the struggle and retook La Haye, which in the next two hours changed hands four times. Whilst leading one of the assaults Girard was struck in the chest by a musket ball and carried from the field. His troops held onto Le Hameau but lost La Haye when the unannounced approach of d'Erlon's Corps in the French rear caused alarm. Down to 2,500 men by the end of the day, both brigade commanders carried from the field, the division led by the only surviving regimental colonel, which bore testimony to the intensity of the struggle. Girard clearly dying, Napoleon visited him in the evening and conferred on him the title Duc de Ligny.

In no condition to carry on fighting, Girard's division remained at Ligny for two days as it needed time to recover and reorganize. It then moved to Quatre Bras on the night of 18th June to try to stem the Prussian advance but was carried away with the remnants of the Army fleeing from Waterloo. Girard nearby and not wishing to be taken prisoner was hastily evacuated. A tortuous coach journey to Paris followed that worsened his condition and he died in Paris of his wounds on 27 June 1815.

CAREER ASSESSMENT

A loyal, brave and tenacious commander, much has been made of Girard's failures at Albuera, Arroyo dos Molinos and Hagelberg. With independent commands he tended to be unlucky in the field, even accident prone, The closeness between victory and defeat was no better reflected than at Albuera, where had he succeeded a permanent corps command was the next step. The task facing Soult to attack a force nearly fifty percent stronger than his own, comprising a high proportion of British infantry in a defensive position was virtually impossible. Yet Girard nearly pulled it off for him. In Russia relegated to a rear area command he did sterling work during the retreat. At Ligny the punishment his division took was the heaviest suffered by any formation before Waterloo and bore testimony to the quality of his troops and his leadership. Napoleon made him a duke on the battlefield as he was dying, which was a fitting reward to a leader who despite all, stayed loyal to his Emperor to the end.

GROUCHY - EMMANUEL, MARQUIS DE (1766 - 1847)

BACKGROUND AND EARLY CAREER

Emmanuel marquis de Grouchy, commander of the Army of the North's right wing became the most controversial man of the Waterloo Campaign when he failed to appear at Waterloo on 18 June 1815. From a well connected noble family he was born in Paris on 23 October 1766. He entered the artillery at the age of 14 when on 31 March 1780 he enrolled at the Strasbourg Artillery School. After gaining a commission as a sous lieutenant on 24 August 1781 he joined the Besançon Artillery Regiment.

His career path changed direction when on 28 October 1784 promoted capitaine he joined the Royal Etranger cavalry regiment at Belfort. Family influence then led to his appointment on 25 December 1786 as a lieutenant with the elite *Compagnie Ecossaise* of the *Garde du Corps du Roi*. It was an important posting and carried the equivalent rank in line regiments of *lieutenant colonel*. The *Gardes du Corps* was the senior regiment of the Royal Guard cavalry and with the Swiss Guard formed the Royal bodyguard at the Louvre. Grouchy's company was the first and most privileged company of the four, with the special task of protecting the King's person. It was the most prestigious corps in the French Army to belong to.

The sheer arrogance and lack of competence exercised by his fellow officers soon made Grouchy unhappy with his position at the Louvre. He felt uneasy with the dispensation of patronage and influence in order to further peoples' careers. He too was part of the system and he did not like it. His elder sister Sophie had married Antoine de Condorcet, one of Louis XVI's junior ministers and a bright light during the early days of the Revolution. At the same time his father also held the elevated title of Chevalier, Page de la Grande Ecurie at the Louvre. As a result, with many of the younger generation from the aristocracy he held sincere views towards reform. It caused him to clash frequently with fellow officers, who did not share his views and they found reason on 27 January 1787 to discharge him from the regiment.

As the Revolution gained momentum he returned to a line command on 18 December 1791 as lieutenant colonel of the 12th Chasseurs à Cheval. The mass emigration of senior officers soon resulted in further promotion on 1 February 1792 to colonel of the 2nd Dragoons. The outbreak of war first saw him with the Army of the Center followed by a move to the Army of the North on 8 July 1792 as colonel of the 6th Hussars. He showed promise during the early days and on 7 September 1792 gained promotion to maréchal de camp. A brief period followed with the Army of the Alps before he moved to the Army of the Coast of Brest on 1 March 1793 first as commander at Rennes and then to Le Havre.

THE VENDÉE 1793-1796

When the revolt broke out in the Vendée at the end of May 1793 Grouchy fell back to Nantes to await reinforcements. He defended the city against a Vendean assaults by Charette from 31 August - 1 September 1793 and in the process took a musket ball in the arm. Soon back defending the city's wall it was ironic that as he was about to break out to join Canclaux's relieving force he received news on 7 October of his suspension as a noble. Outraged his troops surrounded his headquarters and swore to defend him, but Grouchy shirked from a possible confrontation and stoically accepted his fate. The city relieved he was allowed to retire to his home at Pontécoulant near Caen in Normandy.

He was soon back in action when a Royalist uprising threatened his district and after the local National Guard panicked the authorities asked for his help. He assembled the men, gave them a few fiery speeches then handed over command insisting he was only prepared to serve them as a fusilier. Asked his reasons, his reply was that as a noble he could not lead soldiers of the Revolution nothing could stop him shedding his blood for it. The local difficulties soon passed and he returned to civilian life.

The fall of Robespierre on 27 July 1794 and the more tolerant attitude towards ex-nobles resulted in a recall. Canclaux recalled to head the Army of the West, in turn asked for Grouchy's reinstatement to serve as his chief of staff, which was granted on 19 November 1794. The two operated well together and on 23 April 1795 Grouchy gained promotion to général de division. In July 1795 Grouchy became temporary commander of the army when Canclaux fell ill; sending men to help Hoche's Army of the Coast of Brest deal with the emigré landing at Quiberon. The massacre of emigrés that followed was a stain to the reputation of all commanders who served in the region, but there is no evidence to suggest Grouchy was in any way personally involved in the atrocities.

On 26 November 1795 he took command of the Army of the West from Canclaux. The prospect of an independent army command was short lived when on 1 January 1796 the three armies in western France were united under Hoche to form the Army of the Ocean Coast with Grouchy as its chief of staff.

The relationship with the mercurial Hoche was cool, and on 25 March 1796 he was happy to move to the largely inactive Army of the North as its chief of staff under Beurnonville. More interestingly he turned down a divisional cavalry command with the Army of Italy as he thought it not likely to be a very active theater. The irony came in the campaign's early days when Stengel the army's cavalry commander was killed. Since cavalry was his arm, Grouchy would very likely have succeeded to that post as he possessed the experience and seniority.

THE EXPEDITION TO IRELAND 1796

In the two years that followed Grouchy's career tended to lose its way as he showed flaws in character and a tendency to shirk away from responsibility. Hoche in a change of feeling towards him asked for his recall, and on 13 June Grouchy left Holland to take the post as commander of the Isle de Ré. Illness delayed his arrival till 2 September 1796 when he instead was given responsibility for the 12th Military Division at Nantes.

Grouchy's posting was fairly active, as plans were afoot for an invasion of Ireland and on 1 November Hoche appointed him his deputy for the expedition. On 15 December 1796 Hoche sailed from Brest with 13,000 men for the west coast of Ireland. Soon out of port English frigates discovered the French fleet and it dispersed. Storms followed and scattered it further, while Hoche on the *Fraternité* went missing. Grouchy on the *Immortalité* with fifteen vessels carrying some 6,000 troops, on 22 December anchored near Bear Island at the entrance to Bantry Bay. Instead of immediately landing he waited offshore for Hoche's arrival. Then two days later realizing he might lose a great opportunity he decided to land. As the wind was rising, Bouvet the fleet's commander refused to allow a landing as he considered it too dangerous. The wind rose further the following day, when suddenly without consulting Grouchy, Bouvet gave the signal to cut cables and head for the open sea. An indignant Grouchy tried to stop the admiral but failed, and the fleet made for Brest

How hard Grouchy tried to stop Bouvet is the subject of much conjecture. Unsympathetic critics concluded this was the first showing of the indecisiveness that so often dogged his career. He showed a timidity in assuming responsibility, and shrank from exercising independent command that was to dog him with such fatal consequences in 1815. Ireland was at his mercy, his plans included a proclamation to the people and a rapid march to Cork, which he expected to reach on 1 January. In light of what Humbert's small force of 1,500 men achieved the following year, a great opportunity was missed in spite of Hoche's non arrival. Bouvet did not survive the enquiry that followed and was dismissed. Grouchy was also censured for allowing the admiral to dictate to him the expedition's fate. His career suffered in that he remained at Nantes and did not as he hoped receive command of the forces in the West when Hoche moved on to head the Army of the Rhine.

THE ARMY OF ITALY 1798-1799

The west of France no longer a major center of operations, Grouchy on 29 September 1798 joined Joubert's Army of Mayence at Gissen as commander of the 2nd Division. Soon after on 21 October 1798 he moved with Joubert to the Army of Italy while Bernadotte took over his command. On 28 November he arrived at Turin as commander of the French forces occupying the citadel. In terms of the Treaty of Cherasco signed in May 1796 the French had retained the fortress dominating Turin and exercised considerable influence over the Kingdom of Piedmont. Grouchy was sent there to apply pressure on Charles Emmanuel with the intention of causing the monarchy's downfall. On 7 December 1798 Grouchy raised the stakes when he trained his cannon on the royal palace and with French troops advancing from Modena forced the King to abdicate. The whole affair was rather shoddy. Driving at gunpoint a monarch who had done little wrong out of his palace and his capital in such a humiliating fashion was outrageous. The populace became even more inflamed when Grouchy's men proceeded to loot the palace and later by proclamation the country was annexed into the Cisalpine Republic.

The move proved an Achilles Heel for the French as Grouchy spent much time dealing with the rebellion and unrest that resulted. The outbreak of war against Austria saw many French troops needed elsewhere tied down in Piedmont. After the first reverses in northern Italy Grouchy joined Moreau's army as it fell back through Turin towards the Apennines. For a short time he acted as Moreau's chief of staff before he was given command of a division. He played a leading role at the Battle of San Guiliano near Marengo on 20 June 1799. In one of the few French successes of the campaign his troops took the town and with Grenier's division in support drove the Austrians over the Bormida with losses of over 4,000 men.

Suvorov turned the tables at Novi on 15 August 1799 when he destroyed Joubert's army. Grouchy fought on the French left under Perignon defending Pasturana. In the savage hand to hand fighting for the village the Russians cut him down and took him prisoner. Wounded in fourteen places, the Grand Duke Constantine greatly impressed by the courage Grouchy displayed took care of him. It appeared he would be exchanged for the Austrian general Lusignan, but his worth diminished with the large number of senior French commanders held captive. He remained prisoner at Graz till after Marengo when he was exchanged in July 1800. In captivity Grouchy was allowed a certain amount of freedom, which he made good use of making detailed maps and notes of the area that proved most useful in 1805. Also while a prisoner he heard of developments in France, in particular Brumaire and the Consulate, which prompted him to write a letter of protest to Napoleon objecting to the whole affair.

MOREAU AND LOST OPPORTUNITIES 1800-1805

His release on 6 July 1800 saw him posted to the 2nd Army of the Reserve in Switzerland where he actually commanded the army for a few days in November 1800 when Macdonald fell ill. Grouchy also had a good rapport with Moreau, which on 12 November 1800 secured him a place with the Army of the Rhine as 1st Division commander of Richepanse's Center Corps. The corps included a division led by Ney*, which with Grouchy's together under Richepanse's direction virtually won the Battle of Hohenlinden by themselves on 3 December 1800. Grouchy's reputation, intact after the misfortunes of the previous year in Italy, now faced a new problem as it was not under Napoleon's eye that he won fame, but under that of his arch rival Moreau. To Napoleon Grouchy was from the "other camp", he had said as much in his letter to Napoleon objecting to the Consulate. His closeness to Moreau compromised him, especially when he openly championed his cause during his trial in 1804. During this time he was in the frame for a Marshal's Baton, but his behavior killed his chances. Another marshal from the Army of the Rhine would have helped break down the rivalry between them and the "Italians". Ney was the only man from the Rhine who received a baton, whilst others gained one for much less.

Grouchy returned to Italy in 1801 and was happy to serve under Murat, a fellow cavalryman. On 23 September 1801 he became Inspector General for all cavalry in Italy, a job he performed well till August 1803 when he left to join Augereau as his cavalry commander at the Camp of Bayonne. The idea of an invasion of Portugal shelved the following year saw him with Augereau at Brest preparing for the invasion of England. In March 1804 another disappointing move followed to the Camp of Utrecht in Holland where he found himself with another infantry command, the 2nd Division of Marmont's II Corps.

THE GRANDE ARMÉE 1805-1807

In the Austerlitz Campaign his first action was at Wertingen on 8 October 1805 where he lent support to the Cavalry Reserve who bore the brunt of the days fighting. At Günzberg the next day he helped secure the crossings over the Danube. His was one of the first formations to arrive before Ulm, and

drove an Austrian attack back into the city on 17 October. He was of great help to the young Marmont during the occupation of Styria where his knowledge of the country gained while a prisoner was invaluable. He served briefly with the Army of Dalmatia till 27 March 1806 when illness struck him down and forced to relinquish his command he returned to France.

It was in 1806-1807 that Grouchy really came to the fore as a cavalry leader. Napoleon was unhappy with the way the dragoons and the light cavalry had performed in the Austrian Campaign. The role played by the dismounted dragoons was particularly lamentable. The Imperial order was for specialization; no foot dragoons or mixing dragoons with light cavalry formations, or the transfer of officers from one body to the other. Grouchy returned from illness keen and ready to help Murat sort out the dragoons within the Cavalry Reserve. He did a good job and on 20 September was given command of the 2nd Dragoon Division for the campaign against Prussia.

The campaigns in Prussia and Poland marked a golden period for the French Cavalry and whilst not involved all the spectacular successes of that arm Grouchy still proved a very effective leader and greatly enhanced his reputation. In the pursuit of the broken Prussian Army after Jena he entered Berlin on 25 October and took part in the action at Zehdenick the next day when Murat caught Prince Hohenlohe's rearguard. He distinguished himself again at Prentzlow on 28 October when Hohenlohe capitulated, and was at the fall of Lübeck on 6 November. In the fore as Napoleon moved into Poland he took Thorn on 5 December. At Biezun on the banks of the Vistula on 23 December he made several successful charges in support of a mixed force under Bessières that ran into trouble.

While Eylau on 8 February 1807 was a disaster for the French infantry, it was a day that the cavalry covered themselves with glory. At the critical moment with the French center shattered and Napoleon narrowly escaping capture, to save the situation the Emperor ordered Murat and Grouchy to charge with the Cavalry Reserve. One of the greatest charges of history followed, with Grouchy and the dragoons earning their share of the glory. His division in the lead fell on the right flank of the enemy cavalry covering the Russian advance against the French center; completely routing them. The lines of Russian infantry exposed, for a moment held back d'Hautpoul's cuirassiers following with well directed volleys. Then with Grouchy's horsemen they effected a small breach and the line gave way like a burst dyke, first slowly then with increasing ferocity as the sides crumbled. From there they hacked, sliced and trampled their way through the disorganized enemy and reached the gunners who had earlier wrought havoc on Napoleon's infantry.

Grouchy his horse killed under him was left stunned and shaken. The prompt intervention of an aide with a spare horse enabled him to remount and continue. He rallied his men to lead a second charge, when Murat appeared and took the lead. Together they led the dragoons in a wide left wheel against the enemy cavalry that had reformed, and side by side with d'Hautpoul's cuirassiers drove them from the field. A charge like it had never been seen before, and for the loss of some 1,500 horsemen Murat's cavalry saved the battle. Grouchy's part was acknowledged in the 79th Official Bulletin: "General Grouchy, who was in command of the cavalry on the left flank, rendered important services." He also received the Order of Maximilian-Joseph of Bavaria from a grateful King Maximilian.

Grouchy's next notable feat was at Friedland on 14 June 1807. Detached with Lannes's Corps his dragoons, the 9th Hussars and two regiments of Saxon cavalry did valuable work in the early hours of the day. They delayed the Russians after they crossed the Alle and tried to push on beyond Sortlack Wood. Then as Uvarov's cavalry threatened Henrichsdorf on the French left he moved the four miles across the field to dispute the road to Konigsberg. As more French formations arrived on the field, as senior cavalry commander present he took charge of the French cavalry. With his dragoons and Nansouty's cuirassiers he drove Uvarov's horsemen from Henrichsdorf and stabilized the French left. During the day he exercised considerable skill with his cavalry at no time numbering more than 6,000 in covering Lannes's front between Henrichsdorf and Posthenen against twice that of the enemy. His use of feints and counter feints did much to delay the Russian deployment against Lannes's and Mortier's infantry, which left the enemy totally impotent and unable to exploit their initial advantage.

When Napoleon launched the main counter attack that led to the destruction of the Russians trapped in the elbow of the Alle at Friedland, Grouchy was with the French cavalry on the French left. A large number of Russians escaped along the river across his front and as a result he attracted some unfair criticism from Murat, who though not even at the battle expressed an opinion that Grouchy had not acted with sufficient vigor. The Duke of Berg had also ignored the simple fact that the French cavalry had been in action for over fifteen hours and were exhausted. This piece of carping soon forgotten Grouchy received the Grand Cross of Bavaria and the Grand Eagle of the Legion of Honor in the awards distributed after the campaign.

SPAIN, ITALY AND AUSTRIA 1808-1809

In February 1808 Grouchy moved to Spain as commander of the dragoon formations in Spain.

Based at Madrid when the mob came out onto the streets with the outbreak of the Dos Mayos Rebellion on 2 May 1808 Murat appointed him the city's military governor. It took him three hours, and several charges by his chasseurs and dragoons to restore order. As the uprising spread and the news of the French surrender at Baylen precipitated a withdrawal to the Ebro, Grouchy covered the evacuation of Madrid in late July.

Grouchy's period in Spain ended on the Ebro when in November 1808 he joined Prince Eugene in Italy. On 28 January 1809 he became a *comte de l'Empire*. In the campaign against Austria he headed the 1st Dragoon Division of the Army of Italy. He played a prominent part in the Battle of the Piave on 8 May when in the late evening Archduke John tried to re-establish the shattered Austrian line. Eugene brought up a battery of 24 guns, while at the same time Grouchy's dragoons charged the Austrian infantry. It proved decisive and led to the Austrian retreat, which continued till they cleared Italy altogether.

During the pursuit into Austria-Hungary, Grouchy led a mixed force made up of his dragoons, an infantry division under Pacthod and Sahuc's light cavalry. He took Graz on 30 May after a two day siege and then continued to harry the Austrian rearguard. He inflicted sharp reverses on them at Anger on 7 June, at Vasvaar on 10 June and at Papa on 12 June.

At the Battle of Raab on 14 June 1809 he led the French cavalry and played a key role in securing victory for Prince Eugene. In the battle's early stages Eugene made heavy weather as the divisions of Durutte' and Serras were repulsed trying to break Archduke John's center around the Kis-Megyer Heights. As the French and Italian infantry poured back into the valley pursued by triumphant Austrians, Grouchy's cavalry further to the right with Montbrun's division in support crossed the Pancza creek. They then defeated the entire Austrian left flank cavalry wing under Mescery. It was undoubtedly one of the finest performances of Grouchy's career. Mescery stationed behind the rivulet south of the Kis-Megyer Heights with 6,000 horse had orders to prevent the French cavalry crossing to the east bank. Undeterred by the numbers facing him, Grouchy moved forward to reconnoiter the enemy positions and discovered a small ford known to the Austrians who covered it with three small guns. Perturbed he brought forward two horse companies at the gallop that soon silenced the enemy guns before forcing the remainder of the enemy cavalry nearby to fall back. Under the cover of these twelve guns, Grouchy and Montbrun crossed to the east side of the rivulet to engage the 6,000 enemy cavalry, mostly of dubious quality, facing them. Within half an hour, the entire Austrian cavalry wing was put to flight and would have been destroyed but for a valiant rearguard action by the Erherzog Joseph and Ott Hussar regiments.

Prince Eugene's infantry was able to rally and renew their assaults. The Austrian line, now formed into a huge defensive L position which ran from the Pancza rivulet in front of Kis-Megyer then faced south along the Szabethegy heights, was in danger of being encircled by Grouchy's cavalry. Faced with little alternative as his route to the walls of Raab was threatened, John began to thin out his line and withdraw. Eugene's fresh attack supported by Pacthod's division slowly pushed back John's determined troops as they lost Kis-Meyer before falling back to Raab.

Grouchy fought at Wagram on 5-6 July 1809. The battle's first day he spent covering the French rear where his regiments watched Archduke John's movements as he approached the battlefield after moving up the Danube's northern bank from Pressburg. The Archduke posing no threat, Grouchy broke contact and crossing the Russbach brook joined Davout's forces on the French right wing. Montbrun's and Arrighi's horsemen in the morning had made heavy weather against a stubborn Austrian defence and it was only the fresh impetus of Grouchy's horsemen arriving that caused the enemy to break. With his cavalry pouring behind the Austrian positions, Davout was then able to roll up their line.

DISILLUSIONMENT AND RETIREMENT 1809-1810

On 31 July 1809 he received the significant post of Colonel-General of Chasseurs in place of Marmont. It was an appointment that would bring him into closer contact with Napoleon. Outside the marshalate he was the most senior cavalry general in the army. Any cavalry command he wanted was there for the asking in the next campaign that seemed likely as there was unfinished business in Spain. Success in Spain was but one step away from the marshal's baton he coveted. Surprisingly he shirked away from the responsibility, complaining ten years of campaigning had taken its toll and he needed a rest. The underlying reason was more likely he felt slighted at again being passed over when Macdonald and Marmont received their marshal's batons after Wagram. On 20 October 1809 he returned to France and unattached spent a long period on leave with his family. It saved him from a recall to Spain, and probably saved his reputation; the Peninsula had become a graveyard for reputations rather than a place to make one.

THE RUSSIAN CAMPAIGN AND ITS AFTERMATH

Grouchy returned to duty in April 1810 as commander of a light division in Italy where little was heard of him till January 1812. Recalled to the *Grande Armée* he replaced Latour Maubourg as the head of III Cavalry Corps gathering in Poland. The formations under him made up a formidable array,

comprising Chastel's 3rd Light Cavalry Division and La Houssaye's 6th Dragoon Division numbering in all some 6,800 men and 18 cannon. He started the Russian Campaign with Prince Eugene's left wing of the Grande Armée, then after the fall of Vilna was detached to Davout's command. Dessaix's infantry division of Davout's corps was attached to him and with this mixed force he formed the French right column that failed to encircle Bagration's army before Minsk. On 4 August he surprised the garrison at Orsha and seized the great magazines there before they could be set on fire. The forage in particular went a long way to saving much of the French cavalry, which were in appalling condition. After a brief rest he crossed the Dneiper and three days later he rejoined Murat near Krasnoe.

On 15 August he was in action to the east of the city when he was detached to seize the Liadny defile. Murat however deceived by a false report of a Russian movement towards Elnia diverted the bulk of Grouchy's cavalry in that direction, leaving him with only some 600 horsemen to block the defile. Neverovski's division forming the Russian rearguard moved on the defile where the 8th Chasseur à Cheval and the 6th Hussars were too weak to hold the position on their own. He charged Neverovski's column, which fought back with great gallantry and forced its way through and on to Smolensk, leaving behind it some 1,500 dead and wounded, 800 prisoners and seven guns.

At Borodino on 7 September Grouchy's cavalry formed a vital part of Napoleon's battle plan based on a straight forward frontal attack with diversionary operations on either wings. On the French left he was to support Eugene's corps as it seized the village of Borodino. After that as Eugene's infantry crossed the Kalatsha he was to support the attacks on the Great Redoubt, leaving a covering force to watch the northern bank of the river. The assaults at first went to plan as Morand[*] took Borodino in the early hours. Crossing the river and climbing the slopes Morand then gained a foot-hold in the Great Redoubt after a furious struggle. Grouchy was unable to lend support as his corps was driven back, first by shot and shell, then by two corps of cavalry led by Prince Eugene of Württemberg. His horse killed under him Grouchy was also wounded by case shot in the chest. He recovered in time to join the retreat from Moscow and was present at Maloyaroslavets on 24 October. He struggled valiantly to keep his cavalry going but it completely disintegrated. With no command, he then organized an escort to protect Napoleon and formed *le bataillon sacre* made up of unattached officers.

The campaign in Russia broke him and on 19 January 1813 Grouchy returned to France. When he heard of his reappointment as commander of the III Cavalry Corps forming at Metz on 15 February 1813 he pleaded with Napoleon for an infantry command. His reason he simply felt he was no longer robust enough to lead cavalry. Napoleon turned down the plea and Grouchy requested that he retire from the Army. It was accepted with effect from 1 April 1813, after Arrighi had taken over the corps the previous week.

By the Autumn a desperate situation had developed in the cavalry. Bessières was dead and Murat was about to head for Naples, leaving no one of their stature and experience to head the cavalry. On 7 November 1813 orders went out to recalled him, at first to join the Army of Italy. However with Murat's return to that theater it was soon realized that his talents would be wasted. He was of more use if he returned to the Grande Armée and on 15 December 1813 he took command of the remnants of cavalry behind the Rhine.

THE 1814 CAMPAIGN IN FRANCE

He fought at Brienne on 29 January 1814 when Napoleon surprised and defeated Blucher's overconfident Army of Silesia. Two days later on 1 February at La Rothière the position was reversed. After passing on reports from his outposts that the Allies had united, Grouchy failed to convince a now overconfident Napoleon that he was in danger. As the battle developed he posted his horsemen behind Victor's II Corps on the flat ground between Petit Mesnil and Chaumesnil. By evening the battle lost, he did valuable work with the divisions of Piré* and L'Héritier* to cover Victor's withdrawal. Then in the days that followed he shielded the rest of the army as it fell back to Troyes.

Allied indecision caused their advance on Paris to run out of steam. First Marmont outmaneuvered Wrede's Bavarians east of Arcis-sur-Aube. Then Grouchy effectively cut communications between the armies of Silesia and Bohemia when he stopped an attempt by Russian cavalry to cut the Troyes-Arcis road on 3 February. Blucher by now increasingly isolated, on 10 February allowed Napoleon fall on him at Champaubert. Grouchy's horsemen were decisive in the French success. The next day the French turned on the corps of Yorck and Sacken at Montmirail, and again Grouchy's horsemen in the fore gave them a severe mauling. Blucher, failing to come to the support of his colleagues, meanwhile had moved towards Marmont's isolated force. To teach the Prussian a lesson, Napoleon ordered Marmont to fall back on Montmirail so drawing the Prussians closer to his main force, which included Grouchy's cavalry and the Guard.

On the morning of 14 February Marmont halted and allowed Blucher to attack his position west of Vauchamps. At the height of the action Grouchy's cavalry appeared and crashed into the Prussian right flank. Ziethen's division was all but destroyed on the spot. Blucher wisely started to withdraw when he saw the Guard approaching in the distance. Grouchy meanwhile hovering on the flank found a road that ran parallel to Blucher's line of retreat and managed to get ahead of the hard pressed Allied columns. He then swung his men across Blucher's path east of Champaubert. The Prussians appeared hopelessly trapped,but were saved by the mud, which had made it impossible for Grouchy's horse artillery to keep up and join the action in time. After a furious fight Blucher forced his way through the trap and headed for Chalons to regroup. Blucher lost that day 7,000 men and 16 guns, besides large quantities of equipment for the loss of some 600 French. Napoleon's esteem soared as did Grouchy's. Confidence in Paris rallied and memories of La Rothière vanished. There was talk of making Grouchy a marshal there and then.

Napoleon now turned on Schwarzenberg and in two days completed a fifty mile march to bring the Army of Bohemia to battle. In the preliminaries to the Battle of Montereau fought on 17 February Grouchy with Gérard* in support destroyed a force of 4,300 cavalry under Pahlen at Mormant. He then drove Wrede's advance-guard through Nangis in disarray after a sharp encounter at Valjouan. At Montereau the devastating charge over the Yonne and Seine by Pajol* left Grouchy with the task to harry the enemy's retreat. He followed them to Troyes where during the capture of the city on 23 February he was wounded in the arm.

Blucher made a fresh move against Paris at the beginning of March, and Napoleon reacted vigorously by heading north to support Marmont's and Mortier's forces. Grouchy with Napoleon seized the bridge over the Aisne at Berry on 5 March which enabled the army to move on Craonne. Napoleon tried to pin the Allies' attention by a frontal attack while Ney with Grouchy outflanked them from the north with a strong cavalry force. The timing went wrong, Ney attacked prematurely and the cavalry ran into deadly artillery fire. Grouchy hit by a shell splinter in the leg was carried from the field and took no further part in the campaign.

THE RESTORATION AND NAPOLEON'S RETURN

Grouchy found employment under the Bourbons when on 19 July 1814 they appointed him Inspector General of Chasseurs and Chevauléger Lancers. Shunned by Royalists for his apparent betrayal twenty years before it was not surprising he found comfort in news of Napoleon's return. No wide eyed devotee, he bided his time and waited for the imperial summons, which came on 31 March 1815 when offered command of the Army of the South. With troops from the 7th, 8th, 9th and 10th Military Divisions he set out to put down a last ditch attempt to keep the Bourbon flag flying by the Duc d'Angoulème. At Montpellier on 7 April he caught up the remnants of Angoulème's followers and gave them a last chance to acknowledge Napoleon. He expelled those who refused, including the Duc, bundling them on a ship for Spain. On 11 April he took charge of VII Corps, which formed the nucleus of the Army of the Alps. On 15 April 1815 Napoleon made him his twenty sixth and last marshal. The most senior cavalryman in the army, on 8 May Napoleon recalled him to Paris where he helped to organize and equip the cavalry regiments scattered throughout France. On 2 June he became a Peer of France, followed the next day with the appointment as commander of the Cavalry Reserve of the Army of the North. It was a post generally accepted by all within the army since, with Bessières dead and Murat in disgrace he was the obvious choice to head the cavalry.

THE WATERLOO CAMPAIGN

As the campaign started Napoleon did away with the Cavalry Reserve as a separate operational formation and made Grouchy commander of the Army's right wing. Almost immediately things started to go wrong for Grouchy. After the capture of Charleroi as senior commander present Napoleon ordered him to pursue the Prussians with vigor up the Fleurus road. In an oversight the change command structure was not explained to Vandamme, which resulted in a row with Grouchy when he refused to take orders from a mere commander of cavalry. Napoleon resolved the matter but it left the hot headed Vandamme seething, as he had no time for Grouchy. At Ligny on 16 June Grouchy resumed his post as cavalry commander and spent most of the day distracting the Prussians on the French right while Napoleon controlled the movements of III and IV Corps.

On 17 June he once more took charge of the Army's right wing with orders to pursue and engage the enemy falling back from Ligny. He had good troops and a number of distinguished subordinates including Vandamme, Gérard, Exelmans and Pajol. The French, slow to set off, a combination of false optimism and fatigue had led him to believe the Prussians were retiring in the direction of Liège. Contact was soon lost, and amidst a flurry of conflicting reports no one knew the direction of their retreat. To put it mildly, not helped by the carping of Gérard and Vandamme, who showed him scant respect, Grouchy lost his confidence, vacillated then bungled. Opportunities to bring the Prussians to battle during the day were lost. By evening they had placed themselves in a fine strategic position behind the Dyle ready to move either on Napoleon's flank or join Wellington if he chose to fall back towards Brussels. At the same time Grouchy's troops found themselves strung out for several miles on the road

between Walhain and Gembloux still unsure where the main element of the Prussian army was heading. The moves that day tipped the balance in favour of the Allies and ultimately cost Napoleon the campaign.

On the fateful morning of 18 June only Thielemann's III Corps of 17,000 men occupied the left bank of the Dyle between Wavre and Limale and watched as Grouchy's forces approached in the distance. The corps of Ziethen, Bülow and Pirch had regrouped and were heading in the direction of Waterloo and Mont Saint Jean. The French approach was again ponderous and when Napoleon's cannonade opened at Waterloo around 11.30 A.M. Grouchy was taking a brief rest at the notary's house in Walhain seven miles from Wavre. Gérard rode up and urged him to march immediately towards the sound of the guns by striking north-east cross country towards the Dyle. Had he done so there was a chance he could have still disrupted the Prussian advance on Plançenoit. Or even if too late, they could have provided Napoleon with a strategic reserve to fall back on. Whatever the case, had Grouchy heeded Gérard's advice, the damage done to the French at Waterloo certainly could have been minimized. The overall outcome of the campaign in Belgium I believe would have been no different, simply the magnitude of the French defeat. Grouchy had lost the strategic game for Napoleon the day before and had no real hope of reaching the Waterloo before the Prussians. He could disrupt but not prevent their intervention and instead doggedly kept to the strict interpretation of his orders, which were to locate and engage the Prussians.

Around 3.00 P.M. Grouchy started a series of heavy attacks with his 33,000 men and 80 cannon. The Prussians held grimly to their positions despite the odds, with Vandamme engaged in bitter fighting around Wavre before being driven back. At 6.00 P.M. the first terrible truth began to dawn when he received a message from Napoleon sent at 1.00 P.M. urging him to march with all haste towards Mont saint Jean. There was no chance he could do anything that day, but it did prompt him to turn his attention to the bridge at Limale, which once seized could disrupt the passage of troops between the Wavre and Waterloo battlefields. With Hulot's division he personally led an attack on the bridge after Gérard was struck down. Driven back, a charge by Pajol's cavalry carried the bridge and by dusk most of Gérard's troops were on the heights behind the left bank ready to roll up Thielemann's line the next day.

News that confirmed the disaster at Waterloo reached Grouchy around 10.30 A.M. as the renewed attacks on Wavre were under way. He assembled his general officers and held a short council of war giving the news and admitting he was wrong to have ignored Gérard's earlier advice. The humility he displayed rallied the officers behind him including Vandamme. With their co-operation and support he showed decisiveness and generalship not seen before. He withdrew his troops intact from Belgium via Namur and into France through Dinant and Rethel gathering stragglers and reinforcements on the way. By the time he reached Laon on 26 June he massed 45,000 men when he took command of the army from Soult. He started to fall back to Paris by stages when Davout arrived and in turn took over on 28 June. He did his duty and remained with the army till after it retired to the Loire after the Armistice.

LATER LIFE AND CAREER

To expect mercy as an aristocrat who served in the Revolutionary Armies was one thing, but then to return to the service of the King and then desert him for Napoleon was asking too much. Proscribed by the Second Restoration on 24 July 1815 Grouchy made his way to the coast and sailed to Guernsey. From there he took refuge on an American vessel and on arriving in the United States settled in Philadelphia. The Bourbons were surprisingly forgiving and on 24 November 1819 with the help of the Duc d'Angoulème who remembered his treatment after his capture at Montpellier; Grouchy was given an amnesty. His rank of *lieutenant général* restored he returned to France on 20 June 1820. Unattached he took no part in military affairs and retired from the army on 1 December 1824.

The accession of Louis Philippe resulted in the restoration of Grouchy's title as Marshal of France on 19 November 1831. He became a Peer of France 11 October 1832. His life was not easy as he was reviled by Bonapartists and Royalists alike, by the former as being the man responsible for the defeat at Waterloo and the latter for betraying his class. He spent much of his thirty two years after the event trying to justify where he was on 18 June 1815. In 1843 he published a five volume set of memoirs, which in the last he tried to justify his conduct but merely reminded everyone of what he might have achieved. His arguments were convincing and he may have been right, but no matter what he said or did he could not shake the "Curse of Waterloo". He was the proverbial, "unlucky general", and to this day his reputation has had to bear the consequences. Active till the end he died at Saint Etienne on 27 May 1847 while passing through the town on his way to Italy for a holiday.

A FINAL ASSESSMENT

Take away his non appearance at Waterloo, in Grouchy you have a fine leader who was grossly underestimated both during and after his time. Up to 1804 his career certainly put him in the frame for a marshal's baton. Exclude marshals such as Perignon, Serurier, Lefebvre and Kellermann, who were largely political appointees, you are left with men like Ney, Davout and Bessières who up to that time did not merit a baton before him. His association with Moreau certainly marred his prospects. The elevation of lesser men in his eyes dented his pride, resulted in bouts of despair that culminated in lengthy periods of leave and placed him further away from prospects of gaining the coveted baton.

Whilst cavalrymen such as Murat, Bessières and Ney were inspirational, they lacked Grouchy's detached disciplined approach. Among the marshalate they were the only experienced cavalry men. As the French armies increased in size there was a need to raise someone from the cavalry to the dignity. Had it been done earlier it would have given Grouchy the impetus his career needed. A baton would have jolted him out of the periods of self pity and lethargy that often overcame him at the end of campaigns when no recognition came. To have become a marshal any time after the Russian Campaign or for that before the First Abdication would have been justified.

Much has been written condemning his conduct during the Waterloo Campaign, yet his orders were clear - they were to pursue the enemy. Admittedly lethargically, this he did. In the end he made the wrong decision, failed to use initiative and take a risk once the situation changed. That was the mark of a cautious general, but not a bad one. With few exceptions if any, all of Napoleon's commanders were guilty of the same at some time in their careers. Grouchy was unfortunate that he was present during the final act. To enter into a lengthy debate on the rights or wrongs of his actions is outside the scope of this tale other than in the writer's opinion to say the odium he received was not altogether justified. In the few days he led his men from Belgium and at the head of the army in France he showed, admittedly to late, that he was a capable commander of a force of all arms.

GUILLEMINOT - ARMAND CHARLES, COMTE
(1774 - 1840)

BACKGROUND AND EARLY CAREER

Guilleminot was born in Dunkirk (Nord) on 2 March 1774. He came from a solid middle class background, his father being a local shopkeeper. Barely three months past his fifteenth birthday he responded to the National Assembly's call for volunteers and on 23 July 1789 enlisted with the 9th Battalion of the Dunkirk National Guard. Soon bored, the following year he left his unit and travelled to Belgium where he joined the uprising against Hapsburg rule. The poorly organized rebellion was quelled by December 1790 and Guilleminot soon after returned to France. He next emerged as a fusilier with the 4th Battalion of the Volunteers du Nord serving with the Army of the North. The furore that arose after the failure of the first French invasion of Belgium in April 1792 caused his career prospects to improve. Radical elements within the Army exerting influence resulted in his men electing him their leader, which was followed by his promotion to *sous lieutenant* on 23 July 1792.

He campaigned with the Army of the North for the next two years in Belgium and Holland. Transferred on 2 October 1792 to the 12th Demi-Brigade he then spent a short time attached to the staff at Dumouriez's headquarters till 26 December 1793 when he joined to the 24th Demi-Brigade. When Souham took command of the 1st Division, Guilleminot on 1 April 1794 joined its division staff. He took part in Jourdan's invasion of Belgium and fought at Tourcoing on 17-18 May. It was at this time he fell under *adjudant général* Deplanque, a leading topographical expert of the time. Introduced to the intricacies of cartography, a close rapport developed between the two and in June 1795 he moved with Deplanque to the general staff of the Army of the Sambre-Meuse. Promoted *capitaine* on 3 April 1796 he returned to the Army of the North in July 1796 and served under its chief of staff *adjudant général* Musnier. Under Musnier he worked well and with him secured a transfer to the Army of Italy's general staff in December 1798. He fought at Verona on 26 March 1799 and was promoted by Schèrer on the battlefield to *chef de bataillon*. In Italy he caught General Moreau's eye and on 31 December 1799 joined him as an aide de camp on his move to the Army of the Rhine. The following year he was at Moreau's side throughout the campaign in Germany and distinguishing himself at Hohenlinden on 2 December 1800.

BERTHIER'S GENERAL STAFF 1802-1808

From 9 September 1802 till 22 March 1804 he worked in the Topographical Section of the Department of War gaining a detailed knowledge of terrain and fortifications throughout Europe. The aftermath of Moreau's trial in June 1804 brought a check to his career when as a known admirer and supporter of the general on 20 January 1805 he was relieved of his post without any real explanation.

The Grande Armée was in need of topographical experts of his calibre, and he was recalled on 9 September 1805, joining the Historical and Geographical Section of Berthier's general staff. He served in the Austrian Campaign, was present at the Capitulation of Ulm and made careful studies of the terrain before Austerlitz on 2 December 1805. In the Summer of 1806 he took an active part in preparations for war against Prussia. He established a network of spies in Saxony and Prussia to survey possible routes of march. Then with war imminent he personally led a secret mission into Saxony during August - September of 1806. He completed a detailed survey of the routes to Dresden and then compiled a detailed report on the preparedness of the Saxon and Prussian armies.

He was with Berthier's staff throughout the Prussian and Polish Campaigns and fought at Jena, Eylau and Friedland. During this period he gained promotion to adjudant commandant on 9 January 1807. He was at Tilsit during the peace conference, and after the signing of the treaty Napoleon sent him to Constantinople to convey its terms to the Ottoman Turks. In March 1808 he took another semi diplomatic role when he joined the party that tricked Prince Ferdinand into coming to Bayonne to negotiate the overthrow of his father Charles IV of Spain. On 17 March 1808 as a reward for his work of the previous three years he received an annuity of 10,000 francs a year drawn on the revenues of Westphalia
.

SPAIN, AUSTRIA AND ITALY

In June 1808 as the French Army spread its tentacles over Spain Guilleminot replaced Lefebvre-Desnoettes as Bessières's chief of staff with II Corps in Old Castile. He was present on 14 July at Medina del Rio Seco when Bessières routed the Spanish Armies of Castile and Galicia. Promoted *général de brigade* on 19 July 1808, the title *baron de l'Empire* followed on 25 October 1808. But Bessières, though a fine cavalry leader, was not a particularly gifted corps commander and Guilleminot had few opportunities to shine.

When Soult arrived from Germany and replaced Bessières on 9 November 1808 things began to pick up. Guilleminot's efficient staff work enabled Soult to quickly take command, pull II Corps together and the next day rout the Spanish at Gamonal, opening the way to Burgos. He joined Soult's pursuit of Blake through Old Castile, then spent a brief time with Bonnet's division which occupied Santander on 17 November. He rejoined Soult in early January 1809 and shared the hardships during the pursuit of Moore's Army to Corunna. He was one of few that had a good rapport with Soult and was genuinely missed by him when recalled on 28 March 1809 to rejoin the Grande Armée.

Back with Berthier's general staff on 5 May he became head of its Topographical Section. His specialist knowledge, map-making skills, and ability to reproduce hundreds of maps on a small mobile printing press were invaluable during the advance on Vienna, and also at Aspern-Essling and Wagram. After the campaign he briefly served as chief of staff to Eugene's Army of Italy from 16 July till 3 September 1809 before resuming his duties under Berthier.

On 26 April 1810 Guilleminot returned to Spain as Macdonald's chief of staff with VII Corps in Catalonia. A tortuous time followed as Macdonald manfully tried to maintain a French presence in the region after Augereau's period of neglect. It took months to break the stranglehold the guerillas had around Barcelona. Catalonia needed to be secure before Suchet could think of moving on Tortosa and Tarragona. Macdonald with Guilleminot at his side tried to help Suchet on the lower Ebro but his resources were too limited. As soon as his troops were there, they had to rush north again as Gerona was threatened. This pattern repeated itself again and again with them making little contribution to the French war effort in the Peninsula. By the time Guilleminot received his recall to Berthier's staff on 18 April 1811 he had become a much chastened and disillusioned man after this second spell in Spain.

THE RUSSIAN CAMPAIGN OF 1812

The build up of the Grande Armée for the campaign against Russia saw a reorganization of Berthier's staff. Sanson became head of the Topographical Section while Guilleminot on 5 February 1812 took the post of Commander of the Lesser Staff. The post covered the administrative responsibility for the smooth running of Berthier's headquarters. After Spain to be back with Berthier and the Grande Armée was a godsend for him.

On 19 August 1812 at Eugene Beauharnais's request he moved to IV Corps as his chief of staff replacing Junot who left to take over VIII Corps from Jérôme Bonaparte*. Present at Borodino on 7 September he was wounded during the battle, which was ironic since his careful study of the terrain before the battle was responsible for saving thousands of lives during Eugene's assaults against the Great Redoubt. Unlike Davout and Ney who both rushed headlong against the enemy positions, Eugene's commanders heeded Guilleminot's observations and took advantage of the shelter the limited terrain features offered whilst mounting their attacks.

At Malojaroslavets on 24 October he took command of 13th Division during the battle after the death of Delzons, showing to many that a gifted staff officer could make a good fighting soldier. He played a key role in the struggle for the bridge over the Lusha won by the dead Delzons. During the retreat he formed Eugene's rearguard. At Viazma on 3 November he fought a hard action against 12,000 Russians under Miloradovich when the Russians cut between his division and Davout's forces following. Outnumbered, he held his position till the arrival of Compans's 5th Division signalled the arrival of Davout's Corps.

He again fought Miloradovich on 9 November and suffered heavily when the Russians disputed the French passage of the Vop. After another desperate struggle at Krasnoe on 15 November, his troops decimated more by the elements than the Russians, virtually ceased to exist. He joined the remnants of Ney's rearguard where he served as chief of staff during the final stages of the retreat to the Niemen.

THE GERMAN CAMPAIGN OF 1813

Recalled to France he returned to the general staff of the Grande Armée on 13 February 1813 with responsibility for the Topographical and Historical Sections. On 14 March 1813 he moved to Mayence to set up the advance elements of the Grande Armée's headquarters before Napoleon's arrival at the end of April. With the Army during the Spring Campaign in Saxony he fought at Lutzen and Bautzen. On 28 May 1813 he gained long overdue promotion to *général de division*.

During the Armistice on 17 July 1813 he replaced Lourencz who had not recovered from wounds received at Bautzen as commander of the 14th Infantry Division with Oudinot's XII Corps. The renewal of hostilities saw his formation take part in Oudinot's advance on Berlin. He fought at Gross Beeren on 23 August when his troops arrived late during the struggle and helped stem a disorganized French withdrawal when his battalions formed square and held off the pursuing Prussian cavalry. He fought again at Dennewitz on 6 September when Oudinot's behavior that day severely compromised him and caused the destruction of his division. At the start of the battle Guilleminot supported Reynier's advance on Golsdorf that put Bulow's Prussians under extreme pressure. Then Oudinot still smarting from his replacement by Ney as the army's commander, not using his head contradicted his original order. He ordered Guilleminot to leave Reynier and immediately march to Ney's aid at Dennewitz. Protesting, Guilleminot withdrew just as a counter-attack by Borstell carried away Reynier's Saxons. In the resulting confusion the fleeing Saxons overturned his troops as they were marching away. He only managed to restore order once they reached Torgau after a forty mile retreat.

Napoleon disbanded XII Corps on 17 September 1813 and in the reorganization Guilleminot took over Pacthod's 13th Infantry Division after it merged with his. For a short period he served with Reynier till 28 September when he passed to Bertrand's IV Corps. At the Battle of Leipzig he was on the Lindenau sector with Bertrand's corps covering the western approaches to the city against Gyulai's Austrians. On 18 October he spearheaded Bertrand's assault that drove Gyulai back and secured the French line of retreat. In the retreat his division was with the Army's rearguard as it passed through Weissenfels and Erfurt on its way to the Rhine.

On 31 October, with Bertrand, he occupied Hanau after Napoleon had driven Wrede's Bavarians across the Kinzig and re-opened the route to the Rhine. In the afternoon after most of the Army had passed, he beat back a Bavarian assault on the Lamboy bridge before retiring with IV Corps to Frankfurt. On 9 November at Hochheim with Morand* in support he halted a tentative Allied thrust to secure a bridgehead over the Rhine at Mayence before withdrawing into the city. At Mayence he served under Morand who took over IV Corps and stayed to fortify the city. Then using his topographical knowledge to good effect he helped Morand maintain a solid defence with the 15,000 troops that formed the garrison. The effect was to deny the Allies a valuable foothold across the Rhine and win Napoleon valuable time to prepare for the campaign in France. Mayence held out until after Napoleon's fall and only surrendered on 12 April 1814 when ordered to do so by the new government.

THE RESTORATION AND NAPOLEON'S RETURN

The stubborn resistance put up at Mayence and the unwillingness at first to surrender irritated the Bourbons. Considered a die-hard, Guilleminot held no command under them, though they did award him the title *Chevalier de Saint Louis* on 27 June 1814. With nothing to lose he rallied to Napoleon and on 21 April 1815 took the appointment as chief of staff of III Corps. Vandamme on becoming III Corps commander was not happy with Guilleminot preferring his old colleague Revest who had served many years as his chief of staff. Honor was satisfied on 2 June when Jérôme Bonaparte appeared and was given command of 6th Infantry Division with Reille's II Corps. Someone needed to watch Napoleon's wilful young brother Guilleminot moved across to act as his deputy. It was with the expressed hope that as one of the foremost tacticians in the Army he would curb any rash ideas the young Bonaparte might have.

THE WATERLOO CAMPAIGN

Initially Guilleminot's influence had the right effect and Napoleon was pleased the way Jérôme handled himself at Quatre Bras. It resulted in the request to Reille to entrust Jérôme with the attack on Hougoumont. From then on Guilleminot's guiding hand, his warnings to exercise restraint and keep to the given battle plan were ignored. Jérôme's assaults on Hougoumont, initially meant as only a diversion against Wellington's right, turned into a full blooded assault, expressly against Napoleon's orders. Apart from Ney's cavalry charges it was one of the most futile acts of the day.

In the confusion that followed Waterloo, he remained with the Army and rallied the division after Jérôme left for Paris. As the Allies neared Paris he replaced Bailly de Monthion as Chief of Staff on 28 June when Bailly also thought it time to make himself scarce. On 2 July after a temporary cease-fire was agreed he led a party to open negotiations with Blucher. The Prussian, still smarting from a sharp reverse at the hands of Exelmans's cavalry the previous day instead took the party prisoner. The incident nearly resulted in a resumption of hostilities; however sense prevailed, the Prussians released Guilleminot and he was a signatory to the Capitulation of Paris signed the next day.

LATER LIFE AND CAREER

Being a party to the surrender of Paris somehow allowed Guilleminot to survive the purges within the army after Napoleon's fall, and though unattached he remained on the active list on full pay. The Bourbons had revised their judgement of a year before recognizing him more as a military thinker, theoretician and organizer than a rabid, uncompromising Bonapartist. Such attributes were needed for the new role the army would play in France. He had a good rapport with Saint Cyr who as Minister for War on 13 May 1818 appointed him a member of the Committee of Defence. On 27 May he was readmitted to the General Staff where one of his tasks in August 1818 was to settle demarcation lines along the Rhine. On 17 December 1818 he became Inspector General of Geographical Engineers and Director General of the Department of War. Awards followed, Commander of the Order of Saint Louis on 1 May 1821 and then renewal of his appointment as Director General on 23 January 1822.

He received the appointment as Chief of Staff of the Army of the Pyrenees on 12 February 1823 and campaigned successfully in Spain under the leadership of the Duc d'Angouleme. He was present during the occupation of Madrid on 23 May and set up the army's headquarters in the capital. Hostilities concluded on 28 September 1823 after the capture of Cadiz and the release of Ferdinand, he returned to France the following month. Showered with further awards he received the Grand Cross of the Legion of Honor, the Spanish Order of Saint Ferdinand and became a Peer of France on 9 October 1823.

Guilleminot's career now appeared at an all time high when an Ultra backlash against the rising influence of former Imperial officers in the Army rapidly gained momentum. As a result he found himself moved aside and on 30 November 1823 took an inferior posting as ambassador to Constantinople. A forgotten man he languished there for several years, only making a brief appearance on the international stage when his pleas for restraint in 1827 went unheeded as Turkey was dragged into war with England and France over the issue of Greek independence. An Ottoman sympathizer and considered out of step with French policy, it resulted in his recall as ambassador on 30 October 1829. The Ottomans understood his position and further awards came his way, including on his departure the Ottoman Knight of the Order of the Crescent.

He returned to the General Staff but with the July Revolution he was considered a Bourbon supporter, and on 16 July 1831 he was placed on half pay. He was recalled on 25 March 1833 when his topographical skills were needed to head a commission created to settle a demarcation dispute of the border between France and Baden. On 13 August 1839 he retired to the First Section of the General Staff. He died of a stroke on 14 March 1840 during a private visit to Baden.

Guilleminot was foremost a military theorist and tactician rather than a great inspirer of men in the field. An efficient organizer, he was a born staff officer and throughout his career was more at ease holding staff positions than a line command. In the field he proved a capable commander, but found theory more difficult to apply in practice. Vandamme ungraciously thought he was all wind and theory, not his type of general. Vandamme wanted fighters and not persons who he thought may try to second guess him with theory.

THE FRENCH TAKE CADIZ AT LAST, BUT LED BY GUILLEMINOT IN 1823

GUYOT - CLAUDE ETIENNE, BARON (1768 - 1837)

A native of Eastern France the distinguished Guard cavalry general Claude Guyot was born in the village Villevieux (Jura) on 5 September 1768. He enlisted as a trooper with the Bretagne Chasseurs (10me Chasseurs à Cheval) on 1 November 1791. Serving with the Army of the Rhine, promotion in his early days was steady, *brigadier fourier* on 10 July 1792 and then *marechal des logis* on 1 April 1793. He received his commission as *sous lieutenant* on 16 May 1793 after his regiment passed to the Army of the Moselle. In 1794 a transfer followed to the Vendée where he spent two years engaged in the vicious civil war in the region. He then saw a period of service in Italy during the Rivoli Campaign ending with his promotion to *lieutenant* on 30 May 1797.

The outbreak of war against Austria in 1799 saw him with the Army of Germany serving in Württemberg and Bavaria where on 8 February 1799 he gained promotion to *capitaine*. The next year he was with Moreau's Army of the Rhine and took part in the Hohenlinden Campaign distinguishing himself in the pursuit after the battle.

Guyot's career took a dramatic turn in 1802 when Napoleon as First Consul increased the size of the Consular Guard. The Guard Chasseurs à Cheval increased their strength to 600 men by drawing 120 officers and men from line chasseur and hussar regiments. Guyot applied and as a former veteran of the Army of Italy was considered the right material for this elite formation and joined them on 13 October 1802. From this date his fortunes were closely tied to those of Napoleon and the Guard. In the period of peace that followed his steady progress continued with promotion to *chef d'escadron* on 31 January 1804.

THE AUSTERLITZ CAMPAIGN

In the Austerlitz Campaign Guyot led one of the four Guard Chasseurs à Cheval squadrons that provided Napoleon's personal escort. Present at the capitulation of Mack's Army at Ulm on 22 October, Napoleon then sent him to Munich with a small Guard detachment to organize reinforcements arriving from France. His administrative skills were pushed to the limit as many cavalrymen, particularly the dragoons, arrived without mounts. He organized the purchase of mounts, established temporary depots, obtained shoes, clothing, food and forage. Once the army's administrative services had caught up and were in place he was able to rejoin the Chasseurs. He fought at Austerlitz on 2 December 1805 where he took part in an epic struggle against the Russian Chevalier Garde, which had overrun the 24th Légère and threatened the French center. He held back the Russian horse till Bessières arrived at the head of the Horse Grenadiers, and together they turned the tide and drove the enemy cavalry back. The part Guyot played greatly enhanced his reputation and that of his regiment, which had overcome some of the finest heavy cavalry in Europe. The cost however was fearful, for the Chasseurs lost 153 men including Morland its colonel. On 5 December 1805 Guyot was promoted *lieutenant colonel* while Dahlmann the regiment's former second in command became *colonel* and commander.

PRUSSIAN AND POLISH CAMPAIGNS 1806-1807

During the brief period of peace that followed the Chasseurs returned to Paris with Napoleon, and being his favoured escort performed numerous ceremonial duties in the city. When Napoleon left Paris for the campaign against Prussia in October 1806 the Chasseurs again formed his escort. So anxious was Napoleon to reach the front that he abandoned them half way and used a relay of coaches to join the Grande Armée on the Main sooner. Guyot had only just crossed the Rhine when news of the Prussian defeat at Jena on 14 October reached him. So rapid was the pursuit that he did not manage to catch up with Napoleon till he reached Berlin. As a result Guyot with arguably the best cavalry regiment in the Grande Armée ironically missed the major successes of the campaign in Prussia.

He followed Napoleon into Poland where the first major battle he fought was Eylau on 8 February 1807. Napoleon ordered the Chasseurs à Cheval forward to help rescue Augereau's VII Corps from the Russian cavalry as it fell back on Eylau. Murat with the Cavalry Reserve followed and Bessières came up with the remainder of the Guard cavalry in support. Twice they broke through the Russian cavalry and knocked out their artillery. Losses were again heavy, Dahlmann was mortally wounded, and Guyot took over the regiment, which suffered 21 officers and 224 men killed or wounded in the battle. The news of Dahlmann's death on 16 February 1807 resulted in Guyot's promotion to *colonel* and deputy commander of the regiment. Present at Friedland on 14 June, his men held in reserve suffered no casualties during the day. Fresh they were able to exploit the victory during the relentless pursuit to the Niemen. Based at Tilsit during the conference his horsemen participated in the great reviews that heralded the Treaty of Tilsit.

SPAIN AND THE DANUBE CAMPAIGN 1808-1809

Guyot shared in the awards that followed two years of successful campaigning. On 8 March 1808 he received a pension of 10,000 francs a year from Westphalia, followed by elevation to the Imperial nobility on 1 May with the title *baron de l'Empire.*

Under Lefebvre-Desnoettes he headed a detachment of Chasseurs that provided Napoleon's duty squadron during the campaign in Spain. After the capture of Madrid he joined Napoleon's hard march north, which resulted in Moore's retreat to Corunna. Present at Benevente on 28 December 1808 he was fortunate to be on the right side of the Esla when English cavalry fell on the Chasseurs and took Lefebvre prisoner. The French misfortune worked in Guyot's favour for he was made the regiment's commander.

The political situation deteriorating, he was recalled to France soon after Napoleon's departure. He led an advance echelon of two squadrons of Chasseurs from Spain that literally completed their 1,200 mile journey to the banks of the Danube as the Battle of Aspern-Essling was raging. At Wagram on 6 July he covered Macdonald's right flank during the attack on Süssenbrunn. Then at the end of the day as the Austrians retired from the field he overran three battalions of infantry.

Promoted *général de brigade* on 9 August 1809 Guyot soon became a rich man with further pensions of 20,000 francs paid by Swedish Pomerania and another of 10,000 from Galicia in January 1810. On 15 March 1810 Napoleon appointed him to the Imperial household with the post of Chamberlain to the Emperor. The post entailed numerous ceremonial duties, including the task of providing an escort for Marie Louise on her arrival in France to marry Napoleon. On 30 June 1811 he became a Commander of the Legion of Honor.

He returned to Spain in January 1811 where he served on Bessières's staff with the newly formed Army of the North. Whilst the army on paper numbered some 70,000 men its actual numbers were nowhere near that. The vast territories to be covered included Navarre, Biscay, Burgos, Valladolid, Salamanca and the Asturias, which was an impossible task even without being called to provide support to Massena's beleaguered Army of Portugal. He was present at Massena's defeat at Fuentes de Onoro on 3-5 May 1811. After Bessières's recall in August 1811 Guyot was supportive of his replacement Dorsenne. It undoubtedly helped his promotion to *général de division* on 16 December 1811 but there was little glory to be gained from campaigning in Northern Spain.

THE RUSSIAN CAMPAIGN OF 1812

Soon after his return to Paris, he left with the first echelons of the Guard Chasseurs à Cheval for the long march to the Niemen on 20 February 1812. Lefebvre-Desnoettes after escaping from England joined Guyot in Poland on 6 May and resumed command of the regiment. Guyot as a result throughout most of the campaign took the secondary role of providing Napoleon's mounted escort. He was present at the capture of Smolensk on 18 August and likewise at Borodino, where always close to the Emperor's person he took the role of an observer rather than a participant. He remained with the Grande Armée throughout the retreat returning to France in January 1813 to help rebuild his broken regiment, which came out of Russia without a single horse.

GERMANY 1813

With Lefebvre-Desnoettes he rebuilt the Chasseurs à Cheval and at the end of March they were in Saxony with 750 chasseurs formed from the depot squadrons and veterans of line chasseur à cheval regiments. He fought at Lutzen on 2 May where his men suffered heavily from enemy artillery. In the battle they covered the Guard batteries between Starsiedel and Kaya that slowly moved forward and blew away the Russian center. Wounded by a shell splinter he recovered in time to be present at Bautzen on 20-21 May. He saw little action as Napoleon wished to spare his cavalry and did not use his regiment.

During the Armistice he took the title *comte de l'Empire.* Actively engaged in the expansion and reorganization of the Guard cavalry the Chasseurs à Cheval were brought up to brigade strength. Five additional squadrons of Young Guard were added, to give a regimental total of 2,544 officers and men when hostilities began.

While the Chasseurs à Cheval distinguished themselves at Dresden on 25-26 August 1813, for some reason which remains unclear he himself was with Vandamme's forces at Pirna. Later captured with Vandamme* at Kulm on 30 August he was fortunate to be one of the last officers exchanged. He returned to his regiment and in September spent time engaged against partisans in the rear areas. At Altenbourg on 28 September he encountered a large force of Platov's Cossacks plus 5,000 Austrians under Klenau. He nearly achieved a spectacular success till Prussian cavalry led by Thielemann surprised his rear, causing him to flee with heavy losses. Present at Leipzig he fought in the cavalry struggle before Wachau on 16 October 1813. During the retreat to the Rhine Platov's Cossacks caught him at Weimar on 22 October and gave him another mauling. At Hanau on 30 October he played a key role in the victory when his men routed the Bavarian cavalry disordered by foolishly trying to charge Drouot's cannon.

FRANCE 1814

On 1 December 1813 he moved from the light to the heavy cavalry when appointed colonel major of the Grenadiers à Cheval. He served with Nansouty's division made up of his Grenadiers à Cheval, the Empress Dragoons and the Elite Gendarmes. He fought at Brienne on 29 January 1814 when Napoleon checked Blucher's westward march on Paris and again at La Rothière on 1 February when, after managing a haphazard concentration, the Allies turned on him. Guyot posted behind the Guard Horse Artillery placed between La Rothière and Petit-Mesnil had orders to protect the guns against enemy horse. The cavalry battle reached a climax when the Empress Dragoons under Letort* were roughly handled when they tried to assist Colbert's Lancers, which had been overrun by two divisions of Russian dragoons. Guyot then joined the melee, enabling the cavalry to extricate itself. But in the process he left exposed the Guard Horse Artillery, allowing the enemy to overrun them for the loss of 24 guns. He later moved to the French left where he supported Marmont's VI Corps in the struggle for the woods before Chaumesnil. Then with the woods clear he headed off the attempt by Wrede's Bavarians to envelop the French left. Napoleon was furious at the loss of his guns; he was prepared to dismiss Guyot there and then but on hearing the part he had played in the battle took the matter no further.

At Montmirail on 11 February Guyot lent valuable support to Friant's division. Friant's battalions in square broke the cavalry charges, then Guyot in support fell on the disordered enemy driving them back to the ridge of La Meulière. At Vauchamps on 14 February he did not have a good day. He failed to dislodge a Prussian battalion from the farm of Bouc-aux-Pierres and allowed a battery to be overrun by a band of Cossacks. Remembering La Rothière Napoleon in a fit of wrath unfairly stripped him of his divisional command and handed it to Exelmans. Upset Guyot kept his head and four days later distinguished himself at the head of the Grenadiers à Cheval during the Battle of Montereau on 18 February. Napoleon on further enquiry had also found that his treatment of Guyot at Vauchamps was hasty and made amends by appointing him commander of his personal escort, a position of great prestige within the Guard during these dangerous times in France.

For the remainder of the campaign he was at Napoleon's side, was present at Craonne on 7 March, the liberation of Rheims on 13 March and the Battle of Arcis-sur-Aube on 20 March 1814. When news arrived of the Allied advance on Paris he headed a scratch force of 1,000 cavalry that accompanied Napoleon in the dash west to save the city. He reached Troyes on the night of 29 March and then at 4.00 A.M. the next day set off on the final eighty-seven mile ride to Paris. Napoleon left him at Villeneuve and continued by coach as Guyot's men were unable to keep up the punishing pace and then be expected to fight a battle. He rejoined Napoleon at Juvisy on the outskirts of Paris on 31 March, but with the city in Allied hands he fell back to Fontainebleau with Napoleon.

THE RESTORATION AND NAPOLEON'S RETURN

When he abdicated, Napoleon as a mark of his esteem he awarded Guyot a pension of 50,000 francs; the Bourbons never honored it. On 14 November 1814 he was made a *chevalier de Saint Louis*. He remained with the Grenadiers à Cheval, which on 19 November became the Royal Cuirassiers of France with him as their colonel. By the time of Napoleon's return they looked anything but a royal regiment, being poorly uniformed and equipped, and they never received their cuirasses.

Guyot did not have to face any agonizing decisions whether bring his troops over to Napoleon or not. His regiment considered a security risk had moved from Paris to Tours the previous Autumn and, away from Napoleon's route, he decided it prudent to sit tight and await events. Once Napoleon was in the Tuileries he submitted and on 25 March 1815 joined Pajol's command in the pursuit of the Duc de Bourbon's forces in the east. The potential campaign soon fizzled out and recalled, he helped to organize the Guard heavy cavalry regiments. The task was not easy, scattered throughout France, their appearance and equipment was in a pitiful state as they arrived in Paris. He worked hard and by the beginning of June when he assumed command of the Guard Heavy Cavalry Division they had resumed some of their original splendor. Their equipment and appearance meant little, as it was their morale that was important and that was as strong as ever. He had fine commanders in Letort and Jamin, who headed the Empress Dragoons and Horse Grenadiers. Including two Guard Horse Artillery companies attached, his division numbered over two thousand men and included some of the finest heavy cavalry in Europe.

THE WATERLOO CAMPAIGN

On the campaign's first day, in the evening of 15 June his troops fought the Prussians at Gilly. Here Letort leading the Empress Dragoons was mortally wounded. Present at Ligny on 16 June 1815 Guyot's cavalry remained in reserve till the late afternoon. Then as the Old Guard joined the assault on Ligny itself he moved up in support with Delort's cuirassier division. Engulfed by the Old Guard and then Delort, the Prussians broke, while Guyot's horsemen were left behind and failed to cross swords with the enemy that day.

At Waterloo Guyot's cavalry formed up in the morning in two lines behind Kellermann's horse. From left to right stretching from the Nivelle road were the Empress Dragoons under Hoffmayer who had replaced Letort, Jamin's Horse Grenadiers, Dyonnet's Gendarmes and the two horse batteries. They remained static during the opening stages of the battle waiting for the breakthrough. At 3.30 P.M. he joined Ney's folly that led to the cavalry's destruction. When Ney ordered Milhaud's cavalry forward he instinctively followed. Lefebvre Desnoettes with the Guard light cavalry did the same. Moving forward at the trot they threaded their way around broken carriages and debris of every sort that littered the field. Unable to build up any momentum his units became separated and lost all cohesion by the time they reached the enemy squares. Five times they charged up the slopes, but met by a wall of bayonets were unable to pierce the enemy line. Had the timing been right and Foy's and Bachelu's infantry been able to support the charges a break through would have resulted. He lost two horses in the charges and was hit twice but managed to remain with his men and lead them from the field at the end of the day.

After the Army retired behind the Loire he took command of all the Guard cavalry regiments, and on 1 August was given the task of presiding over their disbandment. On 25 November 1815 on the mall at Preuilly in Touraine he was present when the last squadron of Grenadiers à Cheval handed their colors to the Inspector General. It was the last regiment of the Imperial Guard to go. Unattached he retired from the Army at his request on 11 September 1816.

LATER CAREER AND AN ASSESSMENT

Still fit to serve he accepted recall by Louis Philippe on 11 August 1830 as commandant of the 10th Military Division based at Toulouse where he remained till his retirement on 1 October 1833. He died in Paris on 28 November 1837.

Guyot's career was tied to that of the Imperial Guard. From his early days with the Consular Guard in 1801 he had no other master. He was the epitome of the Guard general, loyal and brave. As to his ability there was no question, only men who possessed the highest qualities of leadership and courage could survive within the elite body of men of the Imperial Army. Under Napoleon's direct control he was at his best. If he did display any weaknesses it was the same as most other Guard generals, who often lacked initiative when holding independent commands. This however should not detract from the fine service records of the regiments he led and this must be due largely to his fine leadership qualities.

GUYOT'S HORSE GRENADIERS RETIRE AS THE OLD GUARD ADVANCES

HABERT - PIERRE JOSEPH, BARON (1773 - 1825)

The tough Peninsula veteran Pierre Habert was born in the town of Avallon (Yonne), Central France on 22 December 1773. A natural leader, his men elected him their *capitaine* the day he enlisted with the 4th Bataillon of the Yonne Volunteers on 1 September 1792. Two days later he was promoted *lieutenant colonel* and commander of the regiment's 2nd Battalion. He served with the Army of the North from 1792 till 1796. He fought at Ost Chapelle on 8 July 1793. With the *Amalgame* the Yonne Volunteers were merged into the 107th Demi-Brigade and he became *chef de bataillon* of its 2nd battalion on 22 September 1794.

The fighting days of the *Nord* largely over he moved to Brest in November 1796 as commander of the 3rd Regiment d'Etranger, better known as the Irish Legion. He joined Hoche's first unsuccessful attempt to invade Ireland that left the port on 19 November 1796 and returned on 7 January 1797 after storms had scattered the fleet. On the second expedition he was captured on the frigate *La Coquille* when a British squadron under Sir James Warren intercepted the French fleet on 12 October 1798.

He remained a prisoner in England till May 1800 when he was exchanged. On his return he joined the general staff of the 17th Military Division in Paris. In July 1800 he was sent by the First Consul to deliver despatches to Kléber in Egypt. After a lengthy journey evading the British blockade he reached Alexandria on 23 October 1800. Since he was unable to return to France, Menou commander of the army since the death of Kléber, made him an aide. He fought at Canopus near Alexandria on 21 March 1801 when the British expeditionary force under Abercromby defeated Menou. During the battle Habert gained promotion to acting *chef de brigade*. He retired with Menou to Alexandria and served during the siege till its surrender on 2 September 1801 when he was repatriated to France.

AUGEREAU AND VII CORPS 1803-1807

His rank of *chef de brigade* confirmed by the First Consul on 1 May 1802 Habert joined the 105th Line Regiment at Perpignan as its commanding officer. In May 1803 his regiment moved to the camp at Bayonne as tensions rose between France and Portugal. This was followed by a move to Brittany and a stay at the camps of Saintes and Brest in 1804 preparing for the invasion of England.

In August 1805 when Napoleon abandoned the invasion plan and the Grande Armée was set in motion towards Austria, Habert's 105th Line was with Desjardin's 1st Division of Augereau's VII Corps. Having to march from the far western edge of Brittany VII Corps had to cover the longest distance to reach the main theater of operations. As a consequence Habert with the rest of the corps missed the dramatic sweep around Ulm and its capitulation. Likewise being the last formation to pass through the region VII Corps was left with the laborious task of keeping communications intact and maintaining order in the rear areas. This also resulted in Habert serving in one of the most dramatic and hard fought of Napoleon's campaigns without firing a shot.

During the Prussian Campaign of 1806 he fought under Desjardin at Jena on 14 October. Here his troops played a prominent part in storming Isserstadt, a key position held by the Saxons that secured the Prussian Army's left. Further success came in Poland when his regiment crossed the Ukra at Kolozomb on 24 December and breached Bennigsen's defences along the river. On 26 December he fought at Golymin where after a fierce struggle the Russians gave up their positions along the Vistula and Ukra.

At Eylau on 8 February 1807 with the rest of VII Corps, Habert's troops met disaster. Faced by the whole Russian Army, Soult's IV Corps at the beginning of the battle centered on Eylau deployed on its own, suffered heavy losses from the superior enemy artillery. To minimize his losses, Napoleon decided to seek a quick decision and ordered Augereau with VII Corps to pass through Soult's battered IV Corps and attack the enemy center. Desjardin and Heudelet deployed their two divisions abreast as they cleared Soult's line. A sudden snowstorm then came down blinding men and wetting muskets. Augereau's formation ran out of control in the blizzard. The attack strayed off to its left presenting its flank to seventy-two Russian guns, which raked it mercilessly. The error also brought Augereau's men under fire of French artillery, which continued a blind bombardment through the storm. The storm cleared on the Russian side first and Bennigsen cut loose his cavalry against the floundering French. Moments later the French side suddenly saw the wreckage of VII Corps pouring back to Eylau. It had lost in half an hour, 929 killed and 4,271 wounded, including Augereau and his two division commanders. Habert was the most senior officer of the division left standing, and his own regiment down to fewer than 500 men.

Napoleon disbanded VII Corps, and on 21 February the remnants of Habert's regiment were transferred to a new division formed under Legrand with Soult's IV Corps. He fought at Heilsberg on 10 June, where shot in the head and shoulder he was rendered hors de combat for the remainder of the campaign.

SPAIN 1808-1809

On 18 February 1808 promoted *général de brigade*, he left Soult's command in Germany and moved to Bayonne where he headed a brigade of Merle's 1st Division of the Army of the Ocean Coast. A nomadic existence followed as he wandered from division to division with no permanent command, but the French were facing a crisis in Spain. In June he moved to Verdier's division of Bessières's II Corps where he served at the first siege of Saragossa. A period then followed in August with Grandjean's division of Moncey's III Corps on the Ebro. He took part in the opening attack on Lerin on 25 October that signalled the crossing of the Ebro by Napoleon in his march to Madrid. On 23 November he fought at Tudela when Lannes routed Castanos. He spent the next three months under Grandjean at the siege of Saragossa. He gained great respect for the way his men took the suburb of Valence after particularly savage street fighting, which opened the way for the final storming of the city on 20 February 1809. With Grandjean he then moved into southern Aragon capturing Alcaniz and penetrated into Valencia province as far as Marella. Faced by continuous partisan resistance and a large Valencian force from the south he fell back into Aragon. The situation grew more difficult with the withdrawal of further troops for the campaign against Austria.

In early May Leval replaced Grandjean and Habert was ordered to march on Monzon to relieve the garrison cut off there. His first semi-independent command turned into a disaster as he tried to reach the beleaguered garrison. He rashly crossed the Cinca during a thunderstorm; then the river came down in spate and cut off half his force of 1,200 men. Unable to recross they had to surrender when guerillas emerged from everywhere. Encouraged by this French reverse the whole populace in the region rose against Habert and his remaining troops. Unable to hold on, he had to abandon the whole of north-east Aragon as his troops fell back to the safety of Saragossa.

SERVICE WITH SUCHET 1809-1813

Suchet took command of the badly demoralized III Corps on 19 May 1809 and ordered a concentration at Hijar. He at first fared little better as Habert's troops were at the receiving end of another defeat at Alcaniz on 23 May. The Spanish drove back Habert's assault against their right while Musnier's division attacking their center was decimated by artillery. Suchet fell back to Saragossa, regrouped and resumed the advance. Habert fought under him on the French left at Maria on 15 June and with Wathier's cavalry in support drove Blake from the field. Habert followed up the success with another at Belchite on 18 June. Luck was on his side when he engaged the Spanish right flank; a stray shell blew up Blake's ammunition and caused the enemy to panic. With Blake's army destroyed and the threat to Saragossa removed, he spent the rest of the year engaged in the northern sierras where he recorded successes against guerillas at Pelengua and Pons.

In April 1810 he joined Suchet's invasion of Valencia province and occupied Marella. Just as he began to make real progress, Suchet called him north to face a new threat from O'Donnell's Army of Catalonia. He fought at Lerida on 23 April when Suchet routed O'Donnell and then took part in the siege of the fortress, which fell on 13 May 1810. He played a significant role in Suchet's siege of Tortosa on the lower Ebro from 10 December till its fall on 2 January 1811. His troops had become very accomplished at siege warfare, the storming of fortresses and street fighting. At the siege of Tarragona his troops led the assault through the breach in the walls of the upper city on 25 June 1811. The success won him the long overdue promotion to général de division and formal command of the 3rd Division of the Army of Aragon on 28 June 1811. Tarragona's fall also won Suchet his marshal's baton and Habert became a *baron de l'Empire*. He was made governor of Tortosa on 18 July 1811.

Habert joined Suchet's invasion of Valencia and received a check when he failed to storm Saguntum on 27 September. When Blake tried to relieve the fortress he fought in the battle to the south of the city on 25 October 1811. Habert's division formed the French left anchored against the coast. He had a tough battle as Royal Navy sloops came close inshore and subjected his men to a vigorous bombardment. This was accompanied by an unusually determined Spanish infantry assault on his front that forced him to retire. The French faced by a major breakthrough, only a well timed attack by Harispe's division stabilized the situation. Later in the day as the Spanish fell back he stormed the village of Puzzol, causing heavy casualties as the enemy found themselves caught in a bottleneck as they tried to cross the only bridge over the Picador.

When Suchet renewed his advance on Valencia Habert stormed across the Guadaliviar at the Mediterranean end of the lines before the city on 24 December and linked up with Suchet's force that had crossed upstream. Valencia cut off, surrendered with 17,000 troops on 8 January 1812. The southward advance continued and he captured Denia before halting on the Xucar.

Little action took place in the next year. Drafts recalled for the Russian campaign and the need to assist in other theaters of Spain hampered French operations in the region. Suchet chose to go on the offensive in April 1813 when the vacillation and incompetence of General Murray, who headed a combined Anglo-Sicilian force, became too much of a temptation. Facing nearly four times his number Habert crossed the Xucar and drove 2,200 Allied troops from Biar on 12 April before he rejoined Suchet's main force at Castalla the next day. He then led a diversionary attack against the Allied center, but when the main French assault failed he retired behind the Xucar with the rest of the army. The British decided an expedition against Suchet's rear would be more beneficial and withdrew the bulk of their forces for a sea landing near Tarragona further up the coast. Suchet moved north to face the threat while Harispe and Habert responded by opening a new offensive in the south. Habert crossed the Xucar and fell on Del Parque at Carcagente on 13 June, inflicting over 1,500 casualties.

Habert formed Suchet's rearguard when the army withdrew from Valencia in July 1813 after news of King Joseph's defeat at Vitoria made the French position in eastern Spain untenable. The English followed till surprised by Habert when he led an assault on their encampment at Villafranca in Catalonia on the night of 13-14 September, which panicked the Allies into falling back to Tarragona. On 2 November 1813 he became commander of all troops in Lower Catalonia and was appointed governor of Barcelona. After Suchet retreated behind the Pyrenees he remained at Barcelona and only reluctantly surrendered the city on 28 May 1814 after being ordered to do so by Louis XVIII, well after Napoleon's abdication.

THE RESTORATION AND NAPOLEON'S RETURN

On his return to France Habert received the title of a Grand Officer of the Legion of Honor on 23 July 1814 and then became a Chevalier de Saint Louis on 13 August 1814. It meant little and on 1 September he was placed on half-pay. He showed little hesitation in rallying to Napoleon and on 22 March 1815 Napoleon approved his appointment as commander of the 2nd Military Division at Méziéres. Suchet through Davout recommended him for a command with the Army of the North and on 30 April he was given the 10th Infantry Division with III Corps.

THE WATERLOO CAMPAIGN

His division was lightly engaged at Gilly on 15 June as it tried to push its way up the Fleurus road. At Ligny the next day he was held in reserve during the early stages of the battle, while the other divisions of III Corps, Lefol and Berthezène plus Girard's from II Corps assailed the twin villages of Saint Armand. When Girard's division on the extreme left failed to make headway he moved across in support.

Concealed by high corn and the undulating terrain his skirmishers caught Tippelkirsch's brigade in detail as it tried to hit Girard in the flank. The Prussians panicked and Habert seized his chance driving them back into Wagnelée before being halted by reinforcements. The village changed hands several times as Blucher poured in further reserves and it was only the attack by the Old Guard on Ligny itself that broke the deadlock and caused the enemy to break.

Over the next two days Vandamme became increasingly frustrated by Grouchy's tardy leadership and pursuit of the Prussians. Ordered to take Wavre, in a fit of pique he literally hurled Habert's division to its destruction on 18 June. With a reputation for overcoming difficult objectives Habert was given the task to storm the bridge over the Dyle. Penned in by the narrow streets he was unable to deploy when he reached the river flowing through the center of the town. Raked by guns and sharpshooters positioned on the high ground on the opposite bank he lost over 600 men in the first five minutes. His men for a short time gained a tenuous foothold on the left bank but were unable to dislodge the Prussians from the upper floors of buildings. Shot in the stomach his men carried him to safety as they retired in disorder to the shelter of the nearby buildings on the opposite bank.

FINAL DAYS AND ASSESSMENT

He survived the rigors of the retreat to Paris in the days that followed and took no further part in the campaign. On 1 August 1815 he was placed on the non active list. The Bourbons never recalled him to duty and he retired from the Army on 1 December 1824. He died at his home at Montreal (Yonne) on 19 May 1825.

Habert was a fine general with no obvious or easily recognizable faults. After a meteoric rise in the earlier part of his career, it stagnated for a while till he moved to Spain. In Spain he acquired a reputation as a tenacious leader, who seemed ideally suited for the task of storming towns and fortresses that involved combat in the narrow confines of streets and buildings. As a result he was a commander who received more than his fair share of difficult and "dirty jobs" in Spain. It was ironic that even at Wavre he was called to throw himself into the cauldron of the Dyle crossing. But then there were few men like Habert who knew how to storm a well defended town and at the same time also cross a river. This time the task was too much, for the Prussians were of different mettle to the Spanish. They defended the town in depth, and occupied buildings that stretched from the river up the heights that dominated the town. His failure at Wavre was a sad end to an otherwise very fine military career.

BACKGROUND AND EARLY CAREER

François Haxo, the engineer and Guard general, was born at Lunéville (Meuthe) north-east France on 24 June 1774. From military family he was the nephew of *général de brigade* François Haxo, who died in action in the Vendée on 20 March 1794. He completed his schooling at *l'Ecole de Navarre* before entering the Chalons Artillery School where he gained a commission as a *sous lieutenant* on 1 September 1792. He served with the 6th Engineer Company based at Strasbourg and was promoted *lieutenant* on 1 June 1793. He then spent time at the defence of Landau from October till its relief by Hoche on 27 December 1793.

On 29 January 1794 he joined the Army of the Rhine-Moselle and was at the siege-works of Germersheim. Promoted *capitaine* on 25 April 1794 he led a column over the Queich that destroyed the enemy's positions. In November 1794 he was at Mannheim where he distinguished himself during the defences of the bridgehead over the Rhine. January to June 1795 saw him on the Rhine, this time at the blockade of Mayence. The reputation he gained in siege work and fortifications resulted in his recall to Paris in January 1796 where he lectured to aspiring engineer officers on the subject at the prestigious l'Ecole Polytechnique. In May 1797 he returned to active duty as commander of the engineers based at Bitche and then moved to Geneva in October 1798.

ITALY, DALMATIA AND CONSTANTINOPLE 1800-1807

He joined the Army of the Reserve under the First Consul as it passed through Geneva on 12 May 1800. During the crossing the Alps he played a key role in clearing a path through the Saint Bernard Pass, and later during the siege at Fort Bard, a path that by-passed the guns of the fortress. In Lombardy he was at the blockade of Pizzighettone on 6 June and the crossing of the Mincio four days later before the army's entry into Milan. After Marengo he remained in Northern Italy with the Army of Italy and on renewal of hostilities fought at Monzembano and Caldiero on 26 December 1800.

On 6 March 1801 promoted *chef de bataillon* he took the appointment of Assistant Director of Engineers in the Cisalpine Republic. During his tour of duty he spent several months improving the fortifications at Rocca d'Anfo and in February 1802 commenced a detailed survey of the valleys between lakes Garda and Iseo. Made Director of Engineers in December 1802 he was given the task of completing the reconstruction of the fortress of Peschiera.

On the outbreak of war with Austria he served at Massena's headquarters with the Army of Italy. On 14 February 1806 he returned to his post at Peschiera where he remained till early 1807. He then moved to Dalmatia to complete a topographical survey of the region and compile a report on the preparedness of the fortresses. In May 1807 he joined the military mission under Sebastiani sent to Constantinople to assist the Ottoman Turks in improving the fortifications of the city and the approaches to the Dardanelles. He advocated strong French support for the Ottomans and in particular discouraged French ideas to encourage Greek independence. For his efforts he received from the Ottomans on 8 November 1807 the title, Chevalier of the Turkish Crescent.

HAXO - FRANÇOIS NICOLAS, BARON (1774 - 1838)

SPAIN WITH LANNES AND SUCHET 1808-1810

He returned to Italy in January 1808. Based in Milan he served as Chief of Staff to Chasseloup Laubat, who was Director of Fortifications for the Army of Italy. In September 1808 he left for Spain. Attached to Lannes III Corps he was present at the Siege of Saragossa in December where he received a musket ball in the leg. Promoted *colonel* on 2 March 1809 he remained with III Corps when Suchet took over. He fought at Maria on 15 June 1809 and Belchite three days later. Appointed engineer commander with Suchet's Army of Aragon he served at the sieges of Lerida in April and then at Mequinenza from 20 May till its fall on 8 June 1810. These successes were followed by his promotion to *général de brigade* on 23 June 1810. In September 1810 he began to prepare the siege-works at Tortosa but before the city fell was recalled to France on 17 October 1810.

IMPERIAL AIDE DE CAMP

After a period of leave he joined a committee set up to study and report on the state of fortifications within the Empire. The favorable impression he made resulted in his appointment as an Imperial Aide on 1 March 1811 followed by him being awarded the title *baron de l'Empire*. During this period he spent much time in Germany and Danzig with the brief to inspect and report to the Emperor directly on the state of the fortresses. On 31 December 1811 he was appointed Chief of Engineers for the Army of Germany with special responsibility for the preparedness of the fortresses on the Elbe, Oder and Vistula. He returned to Paris in January 1812 before being appointed Davout's chief of engineers with I Corps of the Grande Armée on 27 January 1812.

THE RUSSIAN CAMPAIGN OF 1812

In Russia Haxo received credit from Davout for his efficient planing at the outset of the campaign that enabled I Corps to cross the Niemen without interruption. He fought at Mohilev on 22 July, Smolensk on 17 August and Borodino on 7 September 1812. During the retreat from Moscow with Davout's rearguard he tried desperately to hinder the enemy by destroying bridges, blocking roads and booby trapping abandoned equipment. On 5 December 1812, the day after Napoleon abandoned the Grande Armée at Smorgonie, he received news of his promotion to *général de division*.

GERMANY AND PRISON 1813-1814

At Konigsberg he fell ill with typhus and on 9 January 1813 Rogniat* replaced him. Allowed to return to Paris, after a brief period of recuperation on 6 March he was appointed governor of Magdeburg, the key to the French defences on the Elbe. The Allies never seriously threatened him at Magdeburg and during the Spring Campaign he was able to strengthen the fortress further. His presence needed elsewhere, in June he was called to move to Hamburg to assist Davout, but within days this was cancelled. Instead he joined the Imperial Guard as Commander of Engineers in place of Kirgener, who died at Reichenbach on 23 May.

Haxo went to great lengths to ensure the preparedness of his men for the coming campaign. The irony came when Napoleon ordered him on 26 August to join I Corps at Pirna with the Guard bridging train. He helped I Corps cross the Elbe and moved on Teplitz with Vandamme* to cut the Allied retreat. Present at the defeat at Kulm on 30 August, he was struck down by a shell splinter in the chest and taken prisoner. Incarcerated in Hungary with Vandamme, he only returned to France on 8 June 1814.

THE RESTORATION AND NAPOLEON'S RETURN

Haxo's nine months in prison had an opposite effect on him than Vandamme and made him obsequiously eager to collaborate with the Bourbons after Napoleon's fall. On 25 June he became a *Chevalier de Saint Louis*, with further appointments following as a member of the Committee for Fortifications and Inspector of Engineers for Neuf-Brisach, Besançon and Grenoble on 13 August 1814.

He tried to disassociate himself completely from Napoleon when he heard the news of his return. On 11 March 1815 he accepted command of the engineers with de Berry's force assembled to bar Napoleon's approach to Paris. His loyalty to the Bourbons was such that when Ney defected he immediately left the army and returned to the King. Protestations of loyalty followed and he chose to accompany Louis XVIII into exile. However by the time he reached the Belgian border Haxo's ardor for the Bourbon cause had taken a knock and Louis allowed him to take his leave. He then had the nerve to offer his services to Napoleon, who surprisingly accepted him. The need for good engineer officers was paramount. On 24 March 1815 he accepted command of the engineers with Reille's II Corps, a rather derisory posting, normally reserved for a colonel. He spent May and June improving the fortifications around Paris. Then literally in circumstances very similar to Ney's recall, he was at the last moment on 11 June 1815 given his old post as chief of engineers with the Imperial Guard.

THE WATERLOO CAMPAIGN

The new found trust Napoleon placed in him at first appeared to pay off. On 15 June at Charleroi he showed great personal bravery when he led the attack by his engineers and marines that cleared the barricades thrown across the bridge over the Sambre. Present at Ligny on 16 June his men then did great work during the day to keep the army moving over terrain that had turned into a quagmire as a result of the thunderstorm that engulfed the battlefield.

A FRENCH OFFICER ASSISTING THE TURKISH BATTERIES AT CONSTANTINOPLE

At Waterloo Haxo carried out a detailed reconnaissance of the enemy lines with Napoleon in the early morning. He then left Napoleon to take a closer look at the obstacles and fortifications put up by the Allies and report his findings. His report concluded there were nothing of consequence beyond the barricade and abatis on the road to La Haye Sainte. It was a fatal summation of the situation. Had Haxo looked more closely the weakness of the Allied left as opposed to the strength of their center would have been patently obvious. He should have noticed that from his position the treetops behind the Allied Center on the northern side of the plateau descended to a pocket where reserves could be hidden. This failure to point out the strengths of the Allied Center was a major factor why Napoleon chose to launch his main assaults in this direction. He genuinely did not believe the Allies had defended this area in depth. Haxo gave further input with fatal results when he sided with Drouot in recommending to Napoleon that he delay the start of the battle till the ground had dried further.

His men meanwhile did great work to ensuring that all guns and troops were in position in spite of the muddy conditions for the start of the battle. He was with Napoleon throughout the day and after the battle returned with him to Paris. On 23 June Davout made him responsible for strengthening the defence works around Paris. When the Allies arrived before the city he made himself unpopular by being one of the first generals to favour capitulation. After the signing of the Armistice, Haxo with Kellermann on 7 July opened negotiations for the army's submission to Louis XVIII and the King's re-entry into Paris. His readiness to deal with the King at first did not appear to save him and on 22 September 1815 he was placed on half pay.

CAREER UNDER THE BOURBONS AND LOUIS PHILIPPE

His recall on 27 February 1816, coupled with his willingness to negotiate with the Allies so soon after Waterloo attracted criticism. Whilst there was no evidence of disloyalty during the campaign, as a former member of the Guard the stance he took was unacceptable. Apart from the traitor Bourmont he was the first general who served in Waterloo Campaign that the Bourbons recalled before Saints Cyr's reforms in 1818 heralded a short period of reconciliation. Undeterred he took an appointment as a member of the Committee of Fortifications. His career was further tainted when he served as a member of the Council of War that on 11 May 1816 tried Lefebvre-Desnoettes in absentia and sentenced him to death.

In the years that followed his career flourished under the Bourbons. He held many appointments and sat on several commissions, including that of Inspector General of Fortresses for the Northern Frontiers and chairman of a boundary commission to settle the borders between France and Savoy. In 1821 he was on a committee to recommend improvements in training of artillerymen and engineers by establishing separate schools at Metz. He remained as a member of the Committee of Fortifications and of the Committee of Engineers throughout the period of Bourbon rule.

Within the army there were calls for his dismissal after the fall of the Bourbons but his stature as an engineer and fortifications expert made him unassailable and as a result he survived. Louis Philippe made him a Councillor of State in January 1831. He returned to the field as commander of the Army of the North's engineers during the French invasion of Belgium in August 1831. He returned to Belgium the following year when the Dutch in breach of the Treaty of London refused to give up Antwerp. He directed the siege of Antwerp's citadel, which capitulated after twenty four days on 23 December 1832. He became a Peer of France on 11 October 1832.

Dogged by ill health during the campaign, he resigned his command and retired from the Army on 1 January 1833. In recognition of his service on 9 January 1833 he received the Grand Cross of the Legion of Honor and the Belgium Grand Cross of the Order of Leopold. He died in Paris on 25 June 1838.

A FINAL ASSESSMENT

As an engineer officer Haxo had the ability and genius of contemporaries like Bertrand, Chasseloup and Elbé. Being able to effect breach after breach in Spanish fortresses, he with Rogniat was another responsible for Suchet's rise to the Marshalate.

As a man there was a dark side to his character. His career after the First Abdication in 1814 became riddled with instances of self interest taking precedence before any bonds of loyalty or honor towards Napoleon. His eagerness to secure a post under the Bourbons when he returned from Hungary in 1814 was to say the least nauseating. He was one of the first to seek terms with the Allies after Waterloo. His recall by the Bourbons within six months of his dismissal had all the markings of being stage managed for him to save face. He also had the temerity to condemn to death a fellow general of the Guard within nine months of Waterloo. Such behavior could be expected of generals who had a grudge to bear against Napoleon - Haxo did not. As a member of the Guard his behavior was reprehensible and put him beyond the pail. It tarnished and made a mockery of an otherwise distinguished career.

JACQUINOT - CHARLES CLAUDE, BARON (1772 - 1848)

EARLY CAREER

Born on 3 August 1773 at Melun (Seine-et-Marne), the cavalry leader Charles Jacquinot received his education at the Pont-à-Mousson Military Academy. On 21 August 1791 he joined the *1er bataillon de la Meurthe* as a *lieutenant* with its grenadier company. On the outbreak of war his first active service was with the Army of the Center and then the North. Wounded in the shoulder by a shell splinter during a skirmish at Crois-au-Bois on 14 September 1792 it put him out of action for several weeks. After his recovery he took part in the invasion of Belgium and fought at Jemappes on 6 November 1792.

On 15 February 1793 his career in the cavalry began when he left the infantry to join the 1st Chasseurs à Cheval of the Army of the Moselle with the rank of *sous lieutenant*. He fought at Fleurus on 26 June 1794 and then passed to the Army of the Sambre-Meuse. Promoted *lieutenant* on 12 September 1794 he fought at the Roer on 2 October 1794.

He left the 1st Chasseurs à Cheval on 15 March 1796 when appointed an aide de camp to General Beurnonville, commander of the Army of the North. The move offered few opportunities as the center of operations had moved to the Rhine rather away from Holland and Belgium. He soon returned to his regiment with the Sambre-Meuse when Beurnonville moved over as army commander in September 1796. Promoted *capitaine* on 6 October 1796 he remained with the Sambre-Meuse till its disbandment in September 1797 when it merged with the Rhine-Moselle to form the Army of Germany. He passed to the Army of the Danube in 1799 and then to Massena's Army of Switzerland where he was present at both battles of Zurich. In January 1800 the 1st Chasseurs à Cheval moved to Souham's division of the Army of the Rhine and Jacquinot served in that theater in the Spring Campaign, gaining promotion to *chef d'escadron* on 5 June 1800. In the Winter he fought at Hohenlinden on 3 December 1800 receiving a sabre wound in the arm.

THE GRANDE ARMÉE 1805-1807

Peace with Austria saw his regiment returned to France where it spent two years in garrison at Verdun. On the resumption of hostilities he moved in May 1803 to the Camp of Bruges in Belgium. Promoted *major* on 29 October 1803 he joined the 5th Chasseurs à Cheval as part of Bernadotte's occupation force in Hanover. The outbreak of war against Austria saw Jacquinot's regiment join Bernadotte's I Corps of the Grande Armée. Serving with Kellermann's 3rd Light Cavalry Division he was at the fore during Bernadotte's march to the Danube. Heading across Germany by the most direct route possible, his troops entered the Prussian enclave of Ansbach on 3 October 1805 nearly causing war with Prussia. Jacquinot was the first to reach the Danube at Ingolstadt on 9 October and cutting south, three days later his was the lead regiment to enter Munich amidst pealing bells and great rejoicing.

At Austerlitz on 2 December 1805, detached from his regiment Jacquinot acted as an aide for Duroc, who led the wounded Oudinot's Grenadier Division during the battle. He acquitted himself well and was with Duroc during the storming of the Pratzen Heights and the great envelopment of the Russians against the Satschen and Menitz ponds. Promoted *colonel* on 13 January 1806 he received command of the 11th Chasseurs à Cheval, which were attached to Soult's IV Corps.

During the Prussian Campaign Jacquinot served under Margaron who led Soult's cavalry brigade that helped screen the Grande Armée's advance as it moved into Saxony. At Jena on 14 October 1806 his regiment covered the extreme right of the French line as they arrived on the field and deployed. While Soult's infantry cleared Closwitz and the adjoining woods, his cavalry collided with a Prussian force under Holtzendorf and drove them from the field. In the afternoon he joined the cavalry charges by elements of IV, V and VI Corps against Tauenzien on the Prussian left that finally broke their will to resist and resulted in the Prussian Army's disintegration that day. Active in the pursuit of the enemy he was present at the capture of Lübeck on 6 November 1806 that ended the campaign.

After the drama of the preceding weeks Jacquinot found the march into Poland was uneventful till he reached the Vistula. The sight of Cossacks on the opposite bank at Dobrykow on 22 December caused him to be more cautious as he screened the crossing of IV Corps. He fought at Pultusk on 26 December. On 31 December he realized an ambition when his regiment was chosen to join Lasalle's newly formed light cavalry division.

The expected glory under Lasalle did not come did not come and present at Eylau on 8 February Jacquinot's troops were only lightly engaged on the French left. During the Siege of Danzig he spent time in Frische Nehrung, the narrow strip of land that stretched north east from the city. He was with the cavalry on 11 May that scattered a force of 4,000 Prussians under Bulow marching from Konigsberg trying to relieve the port by this route. After Danzig's fall he rejoined Lasalle for the campaign that led to the destruction of the Russians at Friedland on 14

June 1807. The French objective being Konigsberg he was with Lasalle's cavalry spearheading Soult's and Davout's march on the city, and so missed events at Friedland thirty miles away.

THE DANUBE CAMPAIGN 1809

Jacquinot shared in the awards that followed, a pension from Westphalia of 4,000 francs a year in March 1808 followed by the title *baron de l'Empire* in October. Trouble brewing with Austria, on 1 January 1809 as the build up of French troops began he joined Montbrun's 1st Light Cavalry Division in Bavaria. Montbrun's division increased to three brigades resulted in Jacquinot's promotion to *général de brigade* on 10 March 1809 with command of the 1st, 2nd and 12th Chasseurs à Cheval.

The beginning of hostilities in mid April saw his troops posted near the Bohemian frontier. They soon fell back as masses of cavalry from the corps of Bellegarde and Kollowrat crossed the border. The tardiness of the Austrian advance enabled him to join Friant's division at Bayreuth on 18 April and together they retired to Ratisbon where they found the rest of Davout's Corps concentrating. Jacquinot fought at Thann the next day when the Austrians tried to drive Davout's forces into the Danube as they fell back to link up with the Bavarians. On 20 April involved in the fighting around Abensberg he surprised and routed an Austrian column led by the veteran General Theirry. In the pursuit supported by Nansouty's and Saint Sulpice's cuirassiers they together took between 3,000-4,000 prisoners. The next day the Austrians on the run he fought in the cavalry action outside Landshut when the Austrian horse under Hiller tried to save the army's artillery and baggage train as they sought the safety of the city. Attached to Bessières's Cavalry Reserve he was at Neumarkt on 24 April when Hiller turned on the overconfident French horse and gave them a mauling to bring the chase to a halt.

When he reached Vienna on 13 May Jacquinot rejoined Montbrun and was posted down river opposite Pressburg. Under Montbrun he was a thorn in the Austrian side keeping communications severed between Archduke John's army retiring from Italy and Archduke Charles's north of the river. On 6 June with Montbrun he moved south and joined Eugene's Army of Italy. He fought at Raab on 14 June where with Montbrun's and Grouchy's divisions outnumbered two to one, they kept over 6,000 cavalry under Mescery in check on the Austrian left. The success enabled Eugene's infantry, which had made heavy weather against the enemy center around Kis-Megyer time to rally, renew the attacks with vigor and drive the enemy infantry from the field.

Present at Wagram, on 6 July Jacquinot with Montbrun supported Davout's great sweep by III Corps to the north of Markgrafneusiedel, which threatened to envelop the Austrian left. Archduke Charles saw the threat and launched the cavalry divisions of Schwarzenberg and Nostitz, plus Lederer's brigade of cuirassiers, to stop the advance. Over 5,000 cavalry descended on Davout's infantry and brought the advance to a halt. Montbrun intervened, but was overwhelmed, Jacquinot's brigade suffered severely. Only when Pajol* entered the fray was the situation stabilized. Then with the arrival of Grouchy's dragoons the Austrians gave way and Davout was able to resume his remorseless advance.

After the war Jacquinot held several posts. On 21 July he became commander of the 6th Light Cavalry Brigade with Davout's III Corps, then rejoined Montbrun on 10 August 1809 as a brigade commander. Ill health caused him to return to France on 23 June 1810, resuming his command the following October. When Davout began to form the Corps of Observation of the Elbe, on 19 April 1811 Jacquinot's formation became its 2nd Light Cavalry Brigade. During 1811 he also held posts as governor of fortresses on the Elbe, first Kustrin till 1 November and then Glogau.

THE RUSSIAN CAMPAIGN 1812

As preparations for the campaign against Russia intensified, on 9 January 1812 Jacquinot moved to Bruyère's 1st Light Cavalry Division with Nansouty's I Cavalry Corps of the Cavalry Reserve as commander of the 3rd Light Cavalry Brigade (7th Hussars and 9th Chevaulégers). His fellow brigade commander was Piré* and together they operated well. He was in action early in the campaign when at dawn on 28 June he engaged the Russian rearguard outside Vilna and was the first to occupy the city. On 1 July he moved to the 1st Light Cavalry Brigade (2nd Chasseur à Cheval and 9th Polish Uhlans) with Davout's III Corps. He fought at Ostrovno on 25 July where his Uhlans on the French right surprised three Russian battalions as they debouched from a wood and took 200 prisoners. At Vitebsk on 28 July he had a near success when he helped pin a large force under Ostermann as elements of III Corps approached, but Russian superiority in cavalry enabled them to slip away.

Before Smolensk on 17 August he was praised for the way he chased away an enemy horse battery positioned on a ridge that dominated the bridges over the Dnieper. From there, French guns were able to pour shot across the river into the city. At Borodino he supported Ney's assaults that carried the Fleches in the morning. He spent the rest of the day in action against the Russian cavalry around the position as they changed hands several times. The occupation of Moscow saw him posted south of the city on the Kaluga Road. On 18 October he fought at Vinkovo. Posted on the French left the Russians overwhelmed his troops and he fell back to Chernischnia. After that his brigade ceased to be a cohesive force and soon disintegrated during the early stages of the retreat. He ended the campaign serving in Grouchy's *le bataillon sacre*.

GERMANY 1813

On 1 March 1813 Jacquinot joined II Cavalry Corps reforming at Mayence, a formation overloaded with officers and no troops to lead. An example being the 2nd Light Cavalry Division he joined, which numbered only 322 men. Prospects limited he was happy to move on 25 March to Lorge's 5th Light Cavalry Division with Arrighi de Casanova's III Cavalry Corps also forming on the Rhine. Commander of the division's 1st Brigade, the chronic shortage of horses resulted in delays and he missed the Spring Campaign in Saxony.

During the Armistice he worked feverishly to ready his formation, but by the start of the Autumn Campaign his preparations were still far from complete. Squadrons were still en route from depots in France. His regiments, the 5th, 10th and 13th Chasseurs à Cheval each fielded only two squadrons numbering some 1,140 officers and men excluding a further 385 en route. Nevertheless he took the field with Lorge's division and was seriously wounded at Dennewitz on 5 September during Ney's advance on Berlin. The campaign took its toll on commanders, which resulted in his promotion to général de division on 26 October 1813 rejoining his troops as they fell back to the Rhine from Leipzig. Across the Rhine he took stock of the situation and for a second time in a year found his command decimated. Down to 545 officers and men his regiments had lost two thirds of their strength after ten weeks of campaigning.

FRANCE 1814

On 15 December Jacquinot replaced Fournier as head of the 6th Light Cavalry Division with Arrighi's III Cavalry Corps. The morale of his new command was hopeless. Napoleon had ordered Fournier's dismissal because he lacked vigor and this had affected the formation throughout. Since the previous August they had suffered more than most, their strength had fallen from 1,370 to 325 men. On the Rhine he joined a mixed force of 4,500 men under Sebastiani that covered the crossings between Coblenz and Lippe. When the Prussians crossed on 25 December he fell back into Champagne. He won an action at Ober-Winter on 1 January 1814 before returning to Arrighi's command on 25 January with Macdonald's wing of the French Army.

A surprise posting occurred on 19 February when he took over a provisional division of heavy cavalry comprising the 3rd Dragoons and 4th Cuirassiers with Kellermann's VI Cavalry Corps. He fought at Bar-sur-Aube under Kellermann* on 27 February when Oudinot's Corps rashly crossed the river and

suffered a serious defeat at the hands of Schwarzenberg. The fine charges he led enabled Oudinot to extricate himself and retire. Jacquinot then covered the retreat as the Army fell back by stages to Troyes and then to Nogent-sur-Seine. He was present at Napoleon's final victory of the campaign at Saint Dizier on 23 March 1814 when Wittgenstein was crushed.

THE RESTORATION AND NAPOLEON'S RETURN

Not a fervent Bonapartist, the Bourbons at first treated Jacquinot fairly after Napoleon's abdication. In May 1814 he headed a commission sent to Vienna to arrange the release of prisoners of war. Uneasiness amongst the Bourbons as to the number of generals who had served Napoleon that retained commands, resulted in him being placed on half pay on 1 January 1815. News of Napoleon's landing resulted in his recall by the Bourbons on 15 March with orders to lead twenty squadrons ordered to concentrate at Langres. When Napoleon's march on Paris became unstoppable he went to ground to await events.

On 23 March the new government called on him to command the cavalry in the 3rd Military Division, but refusing the post, once more he was placed on half pay. He changed his mind when it became apparent that war was inevitable and on 8 April took command of the 3rd Cavalry Division with the Cavalry Reserve. Another moved followed on 23 April to the 5th Cavalry Division before he replaced La Houssaye on 1 June as head of the 1st Cavalry Division with Drouet d'Erlon's I Corps. His command was a fine mixed cavalry division, brigaded together under Bruno and Gobrecht. The first included the 3rd Chasseurs à Cheval and 7th Hussars (led by Colonel Marbot - the memoirs writer) while the second comprised the 3rd and 4th Chevaulégers numbering in all some 1,664 effectives and a battery of six guns.

THE WATERLOO CAMPAIGN

On 16 June, with the rest of I Army Corps, he missed both the battles at Ligny and Quatre Bras. At the end of the day he was with Durutte poised to fall on the enemy's flank at Ligny, but had to obey the paralyzing orders Durutte received to be prudent. Instead his involvement was limited to supporting Durutte's cautious push towards Wagnelée and shadowing the Prussians as they retired from the field in the evening.

Fresh, the next day he took up the chase of Wellington's rearguard as it fell back from Quatre Bras. In atrocious conditions, thunderstorms and muddy roads which made movement difficult, his men performed well in several skirmishes trying to disrupt the retreat. In the evening as he approached the ridge of Mont Saint Jean a crashing volley of twenty cannon brought his men to a halt and unmasked Wellington's position before Waterloo.

At Waterloo Jacquinot occupied the extreme right of the French line and in the battle's early stages supported Durutte's advance on Papelotte. When d'Erlon's assaults against La Haye Sainte faltered around 2.00 P.M. he returned to give support. His arrival coincided with the rout of d'Erlon's divisions as they broke before charges by the Union and the Household Brigades. The triumphant Allied cavalry failed to notice the approach of his lancers and Farine's cuirassiers on their flanks and swept on regardless of consequences. After the infantry they tore into the great French battery cutting down the gunners and horses. Jacquinot and Farine from right and left then crashed into the Union Brigade's blown horses. In the melee that followed they routed the enemy horse, and the regiments of the Royals, Scots Greys and Inniskillings left over a third of their number dead and disabled. The finest cavalry of the Anglo Allied Army, but also the worst led, had gained a spectacular victory against d'Erlon's infantry before it ran out of control. In turn the fine charge by Jacquinot and Farine did much to stem the rout and rectify a serious reverse.

In the afternoon Napoleon placed Subervie's cavalry division under Jacquinot to help counter to the Prussian threat to the French right. With the two divisions he moved across the field to the heights near Fichermont. Here he held the ground till around 4.30 P.M. when Bulow's two leading brigades (15th and 16th) of IV Corps supported by cavalry and artillery debouched from the Paris Wood in force and pushed his cavalry back. Lobau's VI Corps split from Jacquinot by the Prussian onslaught fought heroically as they fell back on Plancenoit. The Prussians, more anxious to cut Napoleon's escape route largely ignored his cavalry during the closing stages of the battle. It helped him to keep his formation intact and lend support to Durutte till all was lost. Then able to extricate his division from the field, he was able to keep some semblance of order during the retreat. This was reflected in the returns gathered at Laon, which showed he brought back from Belgium over 800 men.

POST NAPOLEONIC CAREER

When the Army of the North disbanded the Bourbons retained his services, which at first raised questions as to his loyalties. On 11 October 1815 he organized the disbandment of the cavalry units based at Strasbourg and Belfort. His task completed he was finally placed on half pay on 1 January 1816. Recognized as a true professional cavalryman and no rabid Bonapartist, Saint Cyr as Minister for War agreed to his recall and on 25 July 1816 appointed him Cavalry Inspector General for the 1st, 12th and 22nd Military Divisions. That was followed in 1818 with responsibility for the 1st and 4th Inspectorates. On 31 December 1818 he passed to the General Staff as an Inspector General where he remained till 1820.

He then became Inspector General of the 15th and 16th Military Divisions on 21 April 1820, a post that was renewed the following year. In 1823 he was Inspector General for the 4th, 10th, 11th and 12th Military Divisions. On 6 January 1826 he became commander of the 2nd Cavalry Division based at the Camp of Lunéville. He received the title of Commander of the Order of Saint Louis on 4 October 1826. From 27 May 1827 he served as an Inspector General in the 4th, 11th, 12th and 21st Military Divisions. In 1829 he was Inspector General for the 1st and 4th Military Divisions.

Considered a sympathizer after the fall of the Bourbons, he found himself briefly without a command on 7 February 1831. Recalled the next month, he took the appointment as Inspector General of the 3rd and 6th Military Divisions. He moved to the Camp of Lunéville on 25 November 1831 as commander of the Cuirassier Division. He was at Lunéville for several years, first as commander of the Cavalry Reserve Division from 8 May 1833 and then as camp commandant from 18 June 1834. In March 1835 he returned to the Cavalry Inspectorate with responsibility for the 3rd Military Division before appointed the military division's commander the following August. He remained there till 31 December 1837 when he was placed on the non active list.

Made a Peer of France on 3 October 1837 he then passed to the Reserve Section of the General Staff on 15 August 1839. A final award for a distinguished career came when he received the Grand Cross of the Legion of Honor on 14 April 1844. He retired from the Army on 12 April 1848 after the reorganization that resulted after the fall of Louis Philippe. He died twelve days later at Metz (Moselle) on 24 April 1848.

ASSESSMENT - A FINE PROFESSIONAL

Jacquinot was another case of the classic professional soldier. He was no Bonapartist and his duty was foremost to his country and the Army, hence his refusal at first to adhere to Napoleon in 1815. As a matter of honor, he first remained loyal to the Bourbon Army since he had sworn allegiance to the King. When the return of Napoleon became a constitutional fact and France was threatened he threw in his lot with Napoleon. As a division commander he showed no real faults or weaknesses. He had the misfortune of more often than not of being on the losing side, rather due to factors out of his control than to his own failings as a leader. In his career after Napoleon's fall, he proved to be a very able peace time commander with a fine career in the Cavalry Inspectorate. A man who wisely kept out of politics, he was able to adapt to the changing role the Army played in a radically changing world after the fall of Napoleon.

JEANIN - JEAN BAPTISTE, BARON (1769 - 1830)

EARLY CAREER

Jeanin was born in the village of Laneyriat in the commune of Epy (Jura), eastern France on 22 January 1769. He responded to the National Assembly's call for volunteers in July 1791 and enlisted with the 10th battalion of the Jura Volunteers. On the outbreak of war he served with the Army of the Rhine and after steady promotion through the ranks gained a commission as *lieutenant* on 5 August 1792.

His early active service included taking part in the assaults on the lines before Wissembourg on 13 October 1793 and then in the action near Brumpt on 27 November 1793. During 1794 his unit merged with the 170th Demi-Brigade de Bataille and soon after on 31 October he was promoted to *capitaine* and placed in charge of the formation's artillery company. Most of 1795 he spent on the Rhine involved in the reduction of the fortresses along the river. At Mannheim in January the skilful use of his guns resulted in a commendation from his brigade commander.

In November 1795 he passed to the Army of Italy where he joined the 19th Demi-Brigade de Bataille with Sérurier's 4th Division. On 25 May 1796 after Bonaparte's first successes against Piedmont he moved to the 69th Demi Brigade serving in Augereau's 3rd Division. He fought at Castiglione on 5 August 1796 when Augereau's men forming the army's left wing virtually won the battle single-handed. Present at the crossing of the Isonzo on 19 March 1797 and the capture of Gradisca the next day, he took part in the French offensive which continued to Leoben. The cessation of hostilities against Austria in April 1797 resulted in him spending the next year in garrison in Northern Italy.

In May 1798 he left Toulon with the 69th Demi-Brigade for Egypt. Serving in Menou's Division he took part in the capture of Alexandria on 2 July. At the Battle of the Pyramids on 21 July his unit fought under Vial, who had taken over from Menou who remained at Alexandria as governor. A period of garrison duty in the Nile Delta was followed by the French invasion of Syria in February 1799. Present at the Siege of Acre, on 6 May he received a horrific shell splinter wound to the jaw that left him permanently disfigured. The wound infected, he suffered terribly and was fortunate to survive the rigors of the retreat to Cairo. He recovered and fought at Aboukir on 25 July 1799.

After the break down of the Convention of El Arish he fought at Heliopolis on 20 March 1800, when Kléber routed the Grand Vizier Yussuf's Army of 60,000 north-west of Cairo. On 22 March 1801 he gained promotion to *capitaine de grenadiers* at the second battle at Aboukir. Locked up with the French forces in Alexandria from April till September when Menou capitulated, he was repatriated to France at the end of 1801.

In March 1802 he joined the Foot Chasseurs of the Consular Guard. He remained with the Chasseurs for three years before gaining promotion to *chef de bataillon* on 31 January 1804. Further advancement within the Guard at the time appeared limited and as a result he was tempted away when offered the colonelcy of the 12th Légère on 21 August 1805. It was a mistake, and his career for a time lost direction. He did not take part in the Austerlitz Campaign. While Napoleon was deep in Austria and Bohemia his regiment was with Marshal Brune's Army of the North covering the Channel Coast in the event of any intervention by English forces.

In September 1806 his regiment moved to Mayence where it joined Dupas's 1st Division with Mortier's VIII Corps assembling for the invasion of Prussia. He once more missed the main theater of operations as Mortier's corps played a secondary role in the campaign, covering the Grande Armée's communications by occupying Hesse Cassel in October and Hanover and Hamburg the following month. Activity increased when detached from Dupas, his regiment joined Michaud's 1st Division with Lefebvre's X Corps at the siege of Graudenz in March 1807. Lefebvre's corps deactivated in early June after Danzig's fall resulted Jeanin moving to Rouyer's division of Lannes's Reserve Corps. Wounded by grape-shot at Heilsberg on 10 June, he took no further part in the campaign while his regiment went on to cover itself with glory against the Russians at Friedland on 14 June 1807.

SPAIN 1808-1811

In July 1808 he became a *baron de l'Empire* and in October 1808 with the 12me Légère joined the Grande Armée for the second French invasion of Spain. Made a Commander of the Legion of Honor on 16 October 1808, Napoleon followed the award by promoting him *général de brigade* during a review of the army at Burgos on 19 November 1808. He remained with the 12me Légère serving with Dessolles's Reserve Division as part of the garrison stationed in Madrid.

In May 1809 he returned to France due to ill health. Back in Spain in October 1809 he led a brigade (46th and 65th Line) with Bonnet's 2nd Division of Junot's VIII Corps based at Burgos. Detached from Bonnet in March 1810 he served at the siege of Astorga from end of the month till its fall on 22 April. In May 1810 he took command of the 3rd Brigade of Seras's 3rd Division of Junot's VIII Corps, based at Astorga. The posting was fortunate since the division remained on the plains of Leon whilst the rest of the corps marched off to join Massena's invasion of Portugal.

July 1810 saw his appointment as military commander of Astorga province.

A very difficult period was to follow. During his tenure a successful campaign by the Army of Galicia under Santoclides forced him to evacuate Astorga on 19 June 1811. Severely criticized, he simply had insufficient troops to hold down the region against the overwhelming presence of guerillas and the Army of Galicia. Dorsenne in a renewed offensive by the Army of the North reoccupied the city on 1 September, but a disillusioned Jeanin lacked the stomach and the will to implement harsh measures needed to pacify the region and requested recall. He returned to France on 29 December 1811 and found himself unattached and placed on half pay, with a reputation and career in shreds.

ILLYRIA AND ITALY 1813-1814

Jeanin was so out of favour that he was not even offered a command with one of the many foreign contingents accompanying the Grande Armée in Russia. The ultimate insult came on 11 November 1812 when out of desperation he accepted command of the 14th Brigade of the National Guard. A move to the Illyrian provinces followed in March 1813. where for a brief period he was head of the garrison at Karlstadt on the Dalmatian coast. In July 1813 he received command of the 2nd Brigade with Marcognet's Division. The outbreak of war against Austria coupled with the vast superiority of enemy forces in the region forced him to evacuate Karlstadt on 18 August 1813. He fell back to Fiume where on 15 September he transferred to the 1st Brigade of Palombini's Division of the Corps of Observation of Italy. He fought at Weichelburg on 16 September where for a time Eugene Beauharnais stemmed the Austrian advance.

Active during the campaign in Northern Italy he fought at Caldiero on 15 November, San Michele on 19 November and Boara on 8 December 1813 in vain attempts to prevent the Austrians crossing the Adige. A reorganization of the Army of Italy on 5 February 1814 resulted in him rejoining Marcognet with command of its 2nd Brigade of the 4th Division. He fought at Parma on 2 March 1814 and then with the restoration of the Bourbons he returned to France in June 1814 when the Army of Italy was disbanded.

THE RESTORATION AND NAPOLEON'S RETURN

On 14 August 1814 he became a chevalier de Saint Louis. It was soon followed by him being placed on half pay on 1 September. The Bourbons recalled him on 30 December 1814 as Marchand's deputy commander of the 7th Military Division based at Grenoble. An unexpected promotion to lieutenant général followed on 20 January 1815.

When Napoleon returned from Elba he went through the terrible dilemma of whether to support or resist him. Fortunately Marchand's indecision allowed the troops at Grenoble to defect en masse and caught in the euphoria he decided to join the returned exiles in their march to Paris. His rank of lieutenant général given by the Bourbons was confirmed by Berthier and on 23 April 1815 he received command of the 20th Infantry Division with Lobau's VI Corps.

THE WATERLOO CAMPAIGN

His division was very understrenght at the start of the Waterloo Campaign. The 47th Line comprising 1,048 officers and men had been detached earlier to the Vendée and did not return in time to play a part in the campaign. In all he fielded 3,311 officers and men with eight guns. With VI Corps he formed part of Napoleon's reserve and with his troops was well south of Charleroi at dawn on 16 June. As a result with the rest of Lobau's Corps he only arrived on the Ligny battlefield around 7.30 P.M. in time to lend support to the Guard as they broke the Prussian centre at the end of the day.

At Waterloo he was again with the reserve positioned behind La Belle Alliance to the left of the Charleroi road. As the Prussians were observed approaching the field around 1.00 P.M., with the rest of VI Corps he took position between the Paris Wood and Plancenoit to face the threat. He stood his ground till 4.00 P.M. when the first Prussians emerged from the woods and engaged his troops. There followed a concerted attack by 28,000 men of Bulow's IV Corps at 4.30 P.M. against his and Simmer's divisions, which numbered barely 7,000 men. The threat to Napoleon's right was serious, so with Simmer's 19th Division they stubbornly yielded ground inch by inch. By 5.30 P.M. they had been pushed back to the slopes before Plancenoit, and his men finally broke and fell back through the village in confusion. They rallied when reinforced by detachments of the Young Guard led by Duhesme, who led a furious attack that drove the enemy from the village. As evening fell they held onto the village with the Young Guard sharing in much of the credit for keeping the jaws of the Allied vice open as the French Army left the field.

His division scattered and he escaped during the confusion of the retreat. Once across the border he managed to regroup and fell back by stages to Paris. After the armistice was signed on 4 July he led the remnants of his division south to the Loire. On 1 September 1815, with his division disbanded he was removed from his post and placed on the non active list. He was fortunate the Bourbons brought no charges against him since he was one of the first who chose to rally to Napoleon.

FINAL YEARS AND AN ASSESSMENT

Recalled on 1 April 1820 he held no active

command and eventually retired from the army on 1 January 1825 on an annual pension of 6,000 francs. He died at Saulieu (Côte d'Or) on 2 May 1830.

An old veteran from the Italian and Egyptian campaigns, followed by a posting to the Consular Guard, Jeanin's rise seemed assured. It was not to be. He was tempted away by a colonelcy of the 12th Légère, but from there on his career lost direction for several years. He was unlucky that fate placed him with formations that took part in few of the major battles of the Napoleonic wars. His opportunities were few and the story of his postings show that he was rarely at the right place at the right time. The Waterloo Campaign was the only opportunity he had as a division commander. Here his performance was sound and his defence of the French right at Waterloo revealed here was a leader of true merit.

In the end, the Prussians storm Plancenoit

BACKGROUND AND EARLY CAREER

Son of François Etienne Christophe Kellermann, the victor of Valmy and Marshal of France, Kellermann was born at Metz on 4 August 1770. A sickly child, a military career was not at first envisaged. However his father who had a strong influence over his early career on 14 August 1785 secured him a commission as an acting *sous lieutenant* in his regiment, the Colonel General Hussars. Young Kellermann was an excellent linguist and a diplomatic career seemed the answer to the frail youth's ill health. It helped him secure a posting to the embassy staff on the Duc De Ternan's appointment as ambassador to the United States in 1791.

Whilst with the embassy in Philadelphia on 1 May 1791 he gained a commission as a *sous lieutenant* with the 2nd Cavalry Regiment. Thereafter his progress was steady, *lieutenant* on 10 May 1792, and *capitaine* on 31 May 1792 with the Kellermann Legion, a formation raised by his father in France. After his father's victory at Valmy another promotion followed on 29 October 1792, that of acting *lieutenant colonel* in the Kellermann Legion.

On 10 April 1793 the Ministry for War recalled him to France as *chef de bataillon* with the Hautes Alpes Chasseurs. He never took the appointment up, and instead joined his father then commander of the Army of the Alps as his aide. Present at the siege of Lyons in August 1793, he saw for the first time after the city fell the influence the Representatives with the army held, and the depths of cruelty the Revolution had reached. The next month he joined his father's march into Savoy that secured the province for its later incorporation into France.

As the excesses of the Revolution reached new heights, his father soon fell from grace in the eyes of the Representatives when the common sense and humanity he displayed after the fall of Lyons caught up with him. Condemned for exercising moderation and lack of zeal he lost his command on 12 October 1793 and was imprisoned at L'Abbaye in Paris. Young Kellermann, shaken by the dramatic turn of events and also suspended from his post, sort refuge with an uncle in Metz. Before long he too was arrested and imprisoned in the town's fortress, charged with being in secret communication with his father. The mayor of Metz, a family friend, intervened on his behalf and the authorities seeing no threat released him.

He returned to Grenoble to take up his command with the Chasseurs des Hautes Alpes, but turned away by them, he instead on 8 July 1794 enlisted as a trooper with the 1st Hussars serving in the Army of the Alps. A terrible time for him followed. Small in stature, his frizzy red hair and pallid complexion resulted in perpetual bullying. It was here the cruel nickname "the ugliest man in the army", a description his detractors used throughout his career

first arose. He however did have the last laugh when after his father's recall as commander of the Army he appeared before his former tormentors on 9 March 1795 in a new colonel's uniform as his aide.

Kellermann remained with his father another year through the period when the Army was split and Schérer took charge of a new Army of Italy. Then when

KELLERMANN - FRANÇOIS ETIENNE, COMTE (1770 - 1835)

General Bonaparte arrived as commander of the Italian theater and adopted new plans that relegated the Army of the Alps to a supporting role, he decided to make a change. Encouraged by his father on 25 March 1796 he accepted an offer to join Bonaparte's staff with the rank of *adjudant général chef de brigade*.

SERVICE IN ITALY 1796-1799

Present at the crossing of the Adda at Lodi on 10 May 1796, he then joined the siege of the Milan citadel. He took part in Vaubois's punitive expedition across the Po in June to deal with the Italian Duchies. In charge of the small cavalry force, he worked well under Vaubois, who as a veteran of the Army of the Alps knew him well. They seized Modena on 20 June and then Pistorga on 24 June. After crossing the Arno they wheeled around and headed for Leghorn where on 26 June Kellermann's horsemen, first in the port, captured a vast quantity of supplies unloaded from English ships. He then helped suppress a revolt in Parma that threatened Bonaparte's communications before he rejoined the main army north of the Po in July.

On 16 July 1796 General Bonaparte posted him to Massena's division. With Massena's cavalry brigade he led the 4th Chasseurs in operations against Wurmser as the Austrians pushed down the Adige Valley to relieve Mantua, he fought at Castiglione on 24 July. In the months that followed he also fought at Bassano on 8 September and Arcola 15-18 November 1796, building up a reputation as one of the foremost regimental commanders in the Army. He joined Leclerc d'Ostein's cavalry brigade on 13 January 1797 and distinguished himself at Rivoli the next day. On 20 February his cavalry attached to Dugua's command were the first to force the Piave. The crossing of the Tagliamento on 16 March again brought him to Bonaparte's attention. Here he crossed the river at the head of the 1st Cavalry and the 4th Chasseurs, drove back the Austrian cavalry that threatened Bernadotte's division as it deployed, and seized five guns. Sabred several times in the action, his exploits were enough to win Bonaparte's approval to be sent to Paris on 21 March with twenty four captured Austrian and Papal colors, to present them to the Directory.

Bonaparte's despatches to the Directory also included the highest praise for Kellermann, in particular mentioning his bravery during the crossing of the Tagliamento. The Directory received the young colonel with due ceremony, and Kellermann overcome by the occasion gave a glowing address of Bonaparte's virtues as a man and a leader. The Directors, their popularity at a low ebb and not keen to hear such public utterances about a potential rival, thought it expedient to promote him *général de brigade* on 28 May 1797 and then pack him back to the front.

On his return to Italy Kellermann received command of the 3rd Brigade (1st and 7th Hussars) of Dugua's 2nd Cavalry Division based at Trieste and a pleasant period of rest and recuperation followed. Then as the Army broke up, on 5 August 1797 he moved to the 3rd Cavalry Division under Rey at Castelfranco and thereafter a series of strange postings followed for a man who was primarily a dashing light cavalry leader. On 16 September 1797 he joined Massena's division as commander of a brigade of light infantry. That posting not a success, he returned to Rey's division as head of its 2nd Brigade (22nd Chasseurs) on 4 October 1797. A transfer to the Army of England followed on 12 January 1798 in charge of a brigade of dragoons. Another posting came with the deteriorating political situation in the Papal States after the murder of Duphot in Rome in December 1797. On 12 February 1798 he joined Massena's Army of Rome that occupied the Papal States and seized the city. Then as the Army increased due to threats from the Kingdom of Naples he took charge of the advance guard with Macdonald's division. When the Coalition forces invaded the newly formed Roman Republic on 22 November he inflicted a sharp reverse on the Neapolitan forces at Nepi on 5 December. Two days later he then scattered a second demoralized force under the *emigré* Roger de Damas at Toscanella near Rome, which Macdonald had earlier defeated.

The Army of Rome disbanded in January 1799, he moved to the Army of Naples, a force built up in central Italy under Championnet to remove once and for all the meddlesome rulers from Naples. With the vanguard he crossed the border and took part in the brief struggle for the city that lasted from 20-23 January. From May to June he served with Rusca's 2nd Division. He fought at Trebbia 17-19 June, when Macdonald was defeated by Suvorov. The Army of Naples then merged with Joubert's Army of Italy on 4 August 1799, but he missed its destruction at

Novi on 15 August. A few days earlier suffering from a nervous collapse he was sent to Aix en Provence to recover.

THE ARMY OF THE RESERVE AND MARENGO

After his recovery he was posted to a back-water command as commander of troops in the department of La Manche on 29 December 1799. The victim of a breakdown, his long term prospects looked bleak. Again his fortunes changed dramatically, for Bonaparte back from Egypt and now First Consul, remembered him and included him in the plan to use the Army of the Reserve as the force to destroy the Austrians in Italy. In the weeks that followed he found himself by chance drawn into the events that led to the greatest exploit of his career. The cavalry under him was first incorporated into the Army of the Reserve mustering at Dijon on 29 March 1800. Three weeks later he was given a brigade of light cavalry under Murat that was to be the eyes and ears of the Army as it passed through Switzerland and prepared to cross the Alps. Another posting followed on 14 May when after Napoleon reviewed the cavalry at Lausanne he was given command of a heavy brigade with Harville's dragoon division. It was a move totally out of keeping; his experience had been limited to light formations.

That post led to Marengo, where during the desperate struggle on 14 June 1800 Kellermann could do little wrong that day. Posted on the French left, his first exploit was around 9.00 AM when he routed 2,000 cavalry under Pilati that tried to cross the Fontanone Creek between Sortiglione and Marengo in an attempt to envelop the French left. The Austrians unable to deploy as Kellermann's 500 cavalry descended on them and sabred from all sides were driven into the bed of the muddy stream and destroyed as an effective unit.

As the morning passed overwhelming numbers told and Victor's infantry fell back from Marengo while Kellermann for several hours conformed to the general retirement, withdrawing his regiments by alternate troops. By 3.00 P.M. his brigade was down to a bare 150 men and Victor's division was close to routing. Desaix then arrived on the field at the head of Boudet's division and began to restore order. Victor's and Lannes's formations formed up respectively on his right, and with Kellermann reinforced by another 250 troopers on his left to provide the cavalry support. As Desaix led Boudet's division forward he was hit in the chest by a musket ball and died instantly. The French attack faltered for a moment as his men saw him fall. A column of Austrian grenadiers appeared through the dust and smoke. A blast of canister from Marmont's guns on their flank, plus a deafening blast from a caisson that exploded nearby, caused the enemy column to hesitate.

The critical moment had arrived. A few moments earlier Kellermann's cavalry had been in a line formation behind the French infantry and seeing the Austrian column, he seized his opportunity. With the two squadrons from the 8th Cavalry and 1st Dragoons that brought his numbers up to 400 men, he formed his troops into column, cut across the front of Boudet's line and charged the enemy's flank and rear. Lauttermann's grenadiers forming the head of the Austrian column were tired, their muskets empty, and convinced they were pursuing a beaten foe, failed to see Kellermann's approach that was hidden by a vineyard. In addition, dismayed by Marmont's unexpected grape-shot, the huge explosion of a caisson, Kellermann's cavalry hit them before being able to recover their composure. The timing was perfect, the psychological impact immense. Utterly demoralized the Austrians began to throw down their arms. The front ranks of Zach's column of 6,000 men began to shred away towards the rear. With that the whole formation began to disintegrate into a mass of fugitives heading for the Bormida bridges pursued by the jubilant French infantry. Not finished Kellermann turned his men on the Liechtenstein dragoons to Zach's left that had missed their opportunity to fall on Kellermann's exposed flank. Quick witted, before the enemy saw their mistake, he swung around on them and with Bessières's Consular Guard cavalry drove them from the field.

Napoleon paid tribute to the dead Desaix, Lannes, Bessières and Kellermann, in that order for the part they each played in snatching victory from the jaws of defeat. For Kellermann it was a bitter pill to swallow since in his eyes and the considered opinion of others the bulletin played down his role. The least he expected was immediate promotion to *général de division* and that did not happen. Angry, he openly showed his contempt by publicly claiming it was he, a mere *général de brigade* that had saved the Consulate at Marengo.

THE ARMY OF ITALY 1800-1804

Napoleon at first appeared to ignore the slight and promoted him on 7 July 1800. The damage however was done and Kellermann paid for it in the years to follow where the postings he received gave him little opportunity to excel. His comments showed the selfish side of his character and revealed him in poor light compared to the heady way he praised Napoleon before the Directory in 1797. His posturing resulted in a transfer to head a hussar brigade with a light cavalry division under Quesnel. It was a calculated insult because Quesnel's reputation was questionable, a nondescript and uninspiring leader who at the time was not even a *général de division*! The slight was partly rectified when Davout took command of the cavalry and Quesnel moved on.

With the Army of Italy he served at the crossing of the Mincio 24 December 1800 and the advance to the Adige that resulted in the Convention of Treviso signed on 16 January 1801. After the signing of the Peace of Lunéville he was given a heavy cavalry division sent south to join Murat who was about to invade the Kingdom of Naples. A treaty signed in Florence on 18 March 1801 halted the operation and his future prospects in a new campaign. He remained in Italy, with his career in a rut serving as a Cavalry Inspector General for another three years.

THE AUSTERLITZ CAMPAIGN 1805

On 1 February 1804 he moved to Germany as cavalry commander of the small Army of Hanover, a posting with another malcontent general, this time Bernadotte. The creation of the Grande Armée saw this army become its I Corps and Kellermann's cavalry the 3rd Light Cavalry Division (2nd, 4th & 5th Hussars; 5th Chasseurs). During the march to the Danube that started on 17 September 1805 he was soon involved in a controversy that nearly precipitated war with Prussia, when he rashly cut through the Principality of Ansbach, which was part of neutral Prussia. The threat of possible Prussian intervention remained throughout he campaign and only disappeared after Austerlitz. He reached the Danube on 9 October and heading south entered Munich amidst a rapturous welcome two days later.

Present at Austerlitz on 2 December 1805 he had mixed fortunes on the day of Napoleon's greatest triumph. Detached to the Cavalry Reserve under Murat he fought on the French left supporting Marshal Lannes against Prince Bagration. At 7.00 A.M. his regiments moved forward to screen Suchet's division as it advanced to meet the enemy. As Kellermann came forward in column of squadrons he was suddenly bombarded and his men disordered by a Russian battery unmasked by Cossacks. Another body of cavalry then fell on his flank as he tried to change formation to meet the threat. Caught in mid manoeuvre, his first brigade was thrown into confusion and the division hurriedly retired behind Suchet's infantry.

Despite the setback, he returned to the fray through intervals in the French infantry and deployed in echelon to catch the Grand Duke Constantine's lancers as they milled around the French squares. He routed the ten squadrons, which formed one of the finest regiments of the Russian Army. A regiment of Russian dragoons then hit his flank and his 4th Hussars were badly mauled. The brief action was costly with his division losing over 400 men. He made several more vigorous charges against the enemy horse, till badly wounded he was carried from the field. His part in the battle was invaluable and severely disrupted the much vaunted Russian cavalry. It in particular paved the way for the great charges by Nansouty's and d'Hautpoul's heavy divisions that later routed the Allied horse and secured the reputation of the French heavy cavalry for the next decade.

THE INVASION OF PORTUGAL

After his recovery his career stagnated once more as new cavalry leaders emerged to receive the plum commands. For a while he served under his father leading the small cavalry force with the Army of the Reserve, whilst his peers gained fresh laurels with the Grande Armée in Prussia and Poland. After Tilsit he joined the Corps of Observation of the Gironde, under Junot on 2 August 1807. It was another backwater command that was preparing to invade Portugal. The troops he led comprised depot squadrons of dubious quality from the 1st, 3rd, 4th, 5th, 9th, 15th Dragoons and the 26th Chasseurs à Cheval, formed in two brigades under Margaron and Maurin totalling 1,754 effectives.

When Junot crossed the Bidassoa on 7 October the formation became the Army of Portugal. It covered the 300 miles to the Portugese border at a leisurely pace, only reaching Salamanca on 12 November. News of a possible English landing at Lisbon caused Junot to suddenly quickened the pace and the army including Kellermann's inexperienced horsemen began to suffer. They cut across inhospitable country to Alcantara and then marched directly down the valley of the Tagus. On a map it was the most direct route to Lisbon, but the roads turned out to be non existent. Impassable mountains and valleys were encountered, and the country in that part of Portugal was a virtual desert devoid of any provisions. Kellermann's troops forced to range far and wide to find food for the army, his formation soon fell apart. In the end he suffered the humiliation of seeing Junot with a small advance guard of 1,500 infantry complete the march to Lisbon on 30 November 1808, three days before his broken horsemen limped into the city.

The occupation of Portugal completed after Lisbon's fall, Kellermann became military governor of Alemtejo province based at Evora. When the uprising arose in June 1808 he took the prudent course and fell back to Lisbon picking up the outlying garrisons. In August the main problem became the English landing at Montego Bay. Before the French could deal with the threat he sallied from Lisbon with two battalions of infantry and a squadron of dragoons to clear the region of insurgents. His small force had moderate success and scattered 5,000 rebels at Alcacer do Sal near Setuval on 11 August.

On his return he hurried after Junot to face the English. On reaching the army Margaron took over the cavalry, while Junot placed him in charge of the army's reserve made up of four grenadier battalions numbering some 2,100 men. At Vimiero on 15 August he held the second line as Loison's and Delaborde's divisions formed the first attack. After their failure Kellermann sent forward two of his battalions under Saint Clair, but they in turn were driven back. Finally he personally led another assault against the English centre with his remaining battalions that too failed. The French driven back on all fronts, their situation became desperate. Junot appointed Kellermann, who able to speak English, to open negotiations for a French evacuation from Portugal. Using brashness and bluff he convinced Dalrymple, Burrand and a sceptical Wellesley of the futility of storming Lisbon, and the problems faced by a pursuit of the French to the Spanish border. He signed the Convention of Cintra on 30 August 1808 that allowed the French to leave Portugal for France with all their equipment and booty - on English ships! It was a brilliant performance on Kellermann's part.

SERVICE IN SPAIN 1808-1811

Back in France, Junot's army was renamed the VIII Corps of the Army of Spain and set in motion to return to Spain. In charge of the corps cavalry Kellermann crossed the Bidassoa on 4 December 1808. Soon after the corps was dissolved and its formations parcelled out amongst the army. Kellermann assumed command of the 2nd Dragoon Division stationed at the large base at Valladolid on 9 January 1809. When Bessières left to join Napoleon for the Austrian Campaign on 9 March 1809, Kellermann having proven an able administrator in Portugal took on the additional responsibility for all garrisons in Leon and Old Castile.

In May 1809 he received orders from Madrid to join Ney's forces mustering at Lugo to put down the insurgents in Galicia. Then ordered to retrace his steps, he moved north into the Asturias and crushed a force of 3,000 Asturians at Pajares. Apart from this single success the other converging columns under Ney* and Bonnet achieved little. They failed to bring the enemy to battle, who on their approach fled into the mountains. Ney abandoned the campaign and Kellermann left on his own to hold the region, faced problems. His force was overstretched, comprised of odd units requisitioned from other hard pressed commanders, and in no state to hold down a hostile populace. In mid June he pulled back into Leon and the French lost control of north-western Spain after Ney withdrew from Galicia.

He remained in Leon and Old Castile but achieved little. Apart from his division of dragoons scattered throughout the region numbering around 3,000 sabres he only had another 3,500 infantry available. The situation deteriorated further when Ney's VI Corps under the temporary command of Marchand was defeated by Del Parque at Tamanes on 18 October 1809. The reverse was the most serious inflicted on the French by a Spanish force since Baylen, Kellermann hastened to his assistance with his *column mobile*. He met Marchand on the Douro at Medina del Campo and on 6 November 1809 took command of the demoralized corps and whipped it into shape.

Del Parque, still nearby withdrew when he heard of the Spanish debacle at Ocana on 19 November and headed for the sanctuary of the sierras. Kellermann's cavalry in pursuit for a while lost him, then on 28 November they found him encamped at Alba de Tormes, with three of his five divisions still on the river's eastern bank. Kellermann realized that the whole enemy army might escape beyond the Tormes at any moment if he did not act fast. His options were limited, for he only had his cavalry with him since his infantry were many miles to the rear. Noticing the poor siting of the Spanish pickets before their main position, he decided to risk a surprise assault to try to pin the enemy until his infantry arrived. At the given moment four waves of French cavalry poured down on the startled enemy as they tried to form a defensive line. In a few minutes Del Parques's cavalry was destroyed and Kellermann's hussars and chasseurs cut through the infantry. Leaving 3,000 dead, wounded and prisoners, half the Spanish infantry fled towards the bridge over the Tormes and safety, while the remainder formed square.

Kellermann refused to throw away his advantage by engaging the formed infantry and instead for nearly three hours launched repeated feints that glued the Spanish to their positions and prevented them escaping to the far bank. Unable to move reinforcements across the bridge, Del Parque watched helplessly as in the distance French infantry and guns loomed ominously into view. The Spanish infantry also realized their fate, made a dash for the bridge and in the growing darkness the bulk of them reached the other bank after losing another 1,000 men. For the loss of barely 300 men, Kellermann inflicted over 4,000 casualties against a force which at the start of the engagement was ten times the size of his own. In addition he took nine cannon, five standards and a large quantity of baggage. It took Del Parque's army six weeks to regain any form of cohesion and dare to venture from the safety of the sierras.

When Ney returned from leave on 12 February 1810, Kellermann resumed command of the 2nd Dragoon Division attached to VI Corps with additional responsibility for the garrisons in northern Leon and Old Castile. He also became military governor for the provinces of Toro, Palencia and Valladolid on 4 June 1810 when Ney was ordered to join Massena's Army of Portugal.

After the fall of Ciudad Rodrigo and Almeida when Massena plunged into Portugal Kellermann had the task to cover the army's northern flank in Leon and keep open its communications. For the task he had some 20,000 men of the Army of the North, of which he took command of in September 1810 when Bessières returned to France on leave. As an army commander he failed to apply himself to the task at hand. Power went completely to his head and soon gave himself absurd airs of independent authority. He took little or no heed of orders from Massena who

was in desperate straits, or for that matter King Joseph, to whom he showed supreme contempt. His high handed reputation in Castile won him few friends. Reports of his excesses did not endear him the King. His reputed plundering of the countryside and imprisonment of influential locals was a continual embarrassment. The opportunity was there for Joseph to insist on his recall. When Bessières returned on 11 January 1811 Kellermann's days were numbered. Not to be humiliated, his health was poor, Kellermann conveniently requested sick leave and returned to France on 20 May 1811.

ILLNESS RESIGNATION AND DESPAIR 1811-1812

On leave till 9 January 1812 he then received orders to proceed to Verona in Northern Italy to take command of the 3rd Light Cavalry Division with Grouchy's III Cavalry Corps. On the journey his attacks of neuralgia recurred, which forced him to give up his command on 15 January. It took him till 26 April 1812 to recover and when fit for duty he reported to the cavalry depot at Mayence. There no command available he remained unattached passing throughout periods of deep depression. He eventually took an appointment on 21 October 1812 as Cavalry Inspector General for the 5th Military Division at Strasbourg. Totally disillusioned at not being offered a suitable command after the Russian Campaign he resigned from the Army at his own request on 18 March 1813. He however changed his mind a few days later when on 8 April he received orders to join Ney's III Corps in Saxony.

GERMAN CAMPAIGNS OF 1813

He was in action against Winzingerode's cavalry on 30 April as the Allies tried to oppose the French crossing of the Rippach. At Lutzen the next day his cavalry failed to observe the approach of the Allies under Wittgenstein that fell on Ney's widely dispersed infantry. It justifiably resulted in censure first from Ney, then Napoleon. There was little he could do after the battle since Napoleon refused to risk his cavalry in a pursuit. In the slow march towards Dresden he fought at Koenigswortha on 19 May. The next day he halted Dolff's cavalry after it routed Peyri's division at Klix. Severely bruised by a sabre blow to the chest during the action, he managed to stay in the saddle at Bautzen on 21-22 May. Here he again was injured when his fifth horse of the campaign was shot from beneath him.

During the Armistice he was made commander of IV (Polish) Cavalry Corps on 7 June 1813. He fought at Dresden on 26 August and then in September operated with Poniatowski's VIII Corps as Napoleon's armies gradually fell back on Leipzig. Present during the struggle for the city on 16-18 October he took part in the great cavalry battle before Wachau on the southern sector. Seriously ill for several days before the battle, he handed his command to Sokolnicki on 18 October and took no further part in the campaign in Germany.

THE 1814 CAMPAIGN IN FRANCE

Cavalry leaders of his stature at a premium, he was recalled on 13 February 1814 to lead the VI Cavalry Corps made up of veteran dragoon regiments recalled from Spain. The combination of his leadership and the French dragoon temperament nurtured by years of hard campaigning in Spain worked wonders. It resulted in a carnage inflicted on the Allies similar to that by the great cuirassier generals d'Hautpoul and Nansouty in the halcyon days of 1806 in Prussia eight years before. With the divisions of Treilhard and L'Héritier* he was in action for the first time at Mormont on 17 February as Napoleon launched a new offensive against Schwarzenberg.

The Allies had made a leisurely advance towards Paris then lost their nerve after news of several Prussian reverses. Pahlen, a hero of Liebertwolkowitz in charge of the advance guard covered their withdrawal to Nangis. Together with Milhaud* Kellermann fell on this force of 8 battalions and 24 squadrons of hussars, uhlans and cossacks. Piré* of Milhaud's corps dispersed the uhlans and hussars while Kellermann accompanying Ismert's dragoon brigade routed two infantry battalions before driving off the cossacks. Then with Milhaud they broke the infantry squares one by one. Trelliard and Piré continued the pursuit for seventy kilometers till sheer exhaustion forced them to halt. Pahlen lost 2,000 men and 10 cannon, but Victor following up with the infantry failed to exploit the victory. The casualties were not a major loss to the Allies, but the effect on morale for a while was devastating.

On 19 February, Kellermann's two divisions led by Trelliard and Jacquinot* were placed at the disposal of Oudinot, a marshal who carried a reputation for skill in fighting his way out of tight corners only exceeded by the number of times he got himself in them. At Bar-sur-Aube on 27 February Oudinot crossed the river and attacked a superior force under Wittgenstein. He realized his mistake too late, and for five hours with only Leval's 7,000 infantry he faced 26,000 Allies. At Spoy five miles away Kellermann heard the sound of guns and rode towards the action. He crossed the river at Saint Esprit and descended like a thunderbolt from the blue on the surprised Russians. Trelliard engaged the infantry while Jacquinot routed a regiment of cuirassiers and then the Lubny Hussars. Kellermann at the head of the 4th and 16th Dragoons boldly charged the Russian batteries on the Vernonfays Heights. The Russians kept their nerve till his men were a hundred yards away before opening fire. The effect was too much for Kellermann's men and they fell back to reform. Undeterred they twice charged the batteries in fifteen minutes and lost over 400 men. Foolhardy the attacks were, but they did distract the enemy sufficiently to enable Leval to withdraw across the river.

Kellermann then covered Oudinot's and Macdonald's retreat as the army fell back by stages to Troyes and Nogent-sur-Seine. Sheer exhaustion allowed his men to be caught by Bavarian cavalry at Malmaison on 3 March and in the ensuing action he lost another 400 men. He later joined Napoleon's last eastern drive against the Allied communications in late March. Present at Wittgenstein's defeat at Saint Dizier on 26 March 1814, he led a ten kilometer pursuit of the enemy after the battle. He joined the desperate dash back to Paris to save the city but with its fall on 31 March moved with the army to Fontainebleau.

THE RESTORATION AND NAPOLEON'S RETURN

Weary of war, he readily accepted the return of the Bourbons and declared for them on 6 April once he heard of Napoleon's abdication. Sceptics claim the date was fortuitous since he would have defected anyway. The story had some credence since in the months that followed he ingratiated himself with the Bourbons sufficiently to secure a post as a member of the Council of War for the Royal Guard. That was followed by appointment on 1 June 1814 as a Cavalry Inspector General for the depots at Lunéville and Nancy. Further awards came his way, *chevalier de Saint Louis* on 2 June, the Bavarian Iron Cross and the Grand Cordon of the Legion of Honor on 23 August 1814.

When news of Napoleon's return broke, he joined the Duc de Berry's hastily assembled force as its cavalry commander. Anxious to justify the King's trust in him he rode ahead of the main force in charge of the advance guard. At Mélun on 16 March 1815 he faced a revolt by his men who refused to march against Napoleon. Faced with no alternative, he left his command and retired to his estates to await events. For two months neither he or Napoleon made any moves towards a reconciliation. Napoleon was reluctant to give any important command to a man who had served the Duc de Berry, but cavalry leaders of his calibre were few. The vacancy for commander of III Cavalry Corps remained unfilled. Napoleon wooed him to Paris and made him a Peer of France on 2 June. Moved by the great ceremony on the Champ de Mars, when command of III Cavalry Corps was offered by Davout the next day he could not refuse. III Cavalry Corps was the most impressive formation of the Cavalry Reserve, comprising two regiments of dragoons, four of cuirassiers, an elite brigade of carabiniers and two batteries of horse artillery. Formed into two divisions under the veterans L'Héritier and Roussel d'Urbal, it numbered 3,858 officers and men. On 9 June he arrived at

Vervins to take up the finest command he ever held.

QUATRE BRAS 16 JUNE 1815

When the Army of the North entered Belgium, Kellermann's cavalry was delayed by the traffic jam that occurred before Charleroi and only crossed the Sambre at Chatelet late in the afternoon. In the evening he bivouacked north of Chatelet, in an ideal position to join the battle expected the next day before Ligny. Ney's left wing weak in cavalry, Napoleon ordered Kellermann the next day to move to Frasnes to lend support. The move took time, and when he arrived he received the message from Ney to move up to Quatre Bras. He only had Guiton's brigade of L'Héritier's division with him, the rest of the corps was still strung out on the road between Frasnes and Gosselies. Arriving on the field around 5.30 P.M. with only two regiments he found Ney in a very agitated state. The non arrival of d'Erlon's corps, coupled with Napoleon's declaration that the fate of France depended on his capture of Quatre Bras, haunted him. A life-line offered, he immediately ordered Kellermann to seize the crossroads with his single brigade.

Few tasks would normally have daunted Kellermann, but this time he must have mistrusted his ears when he heard the order. Wellington's infantry loomed like a long red wall barring the way to Brussels. He was to lead under a thousand men against a force of over 20,000 strong, holding a dominant position, ready to receive a charge, and leaven with a good spread of British infantry. That it would be a death ride mattered little to Kellermann, but he felt bound to make sure that Ney was in earnest before he led the brigade to destruction. He pointed out to Ney the difficulty of the task and that he had only one brigade out of four present. Ney took little heed and angrily ordered him to go. The rebuke infuriated Kellermann, who could do little but obey. Unfortunately he never waited for Ney to make the necessary arrangements to support him with Piré's squadrons, or for that matter with infantry or artillery support. He returned to his cuirassiers and at once led them forward at a sharp trot up the southern slope of the plateau where the British stood. Reaching the crest he ordered his men into a gallop, so to prevent them from shirking once they saw the full extent of the danger that faced them.

Like a whirlwind this mass of steel with Kellermann at their head approached the Allied lines. The high rye prevented Halkett's brigade from seeing much of the French movements, or even realizing until the last moment that the cuirassiers were advancing to charge. The 30th and 33rd Line promptly formed square, whilst the 69th believing no cavalry was nearby started to deploy in line. The cuirassiers appeared and closed with the 69th. The regiment stood firm, showed a fine front and held their fire till the last moment. The 8th Cuirassiers received the first volley at thirty yards. The fire failed to stop the French horse as they burst through the line and seized the regiment's color. The 30th formed square and beat off the attacks as the waves of horsemen flowed past its sides. The 33rd suffering severely from artillery fire by a nearby horse battery, fled to the safety of Bossu Wood, while the cuirassiers sabred all in their path as they reached the Quatre Bras crossroads.

Kellermann had achieved the impossible. His two regiments had torn a breach in the centre of Wellington's battle-line. Fortunately for the Duke two guns of a Kings German Legion battery had just arrived before Quatre Bras and saw the French carrying all before them. They unlimbered and when the cuirassiers were a few paces away opened fire with a salvo of round shot and grape. The effect was devastating; the ground strewn with dead, it was enough to halt Kellermann's victorious thrust.

Kellermann paid for his impetuosity. The close support he needed before the Allied infantry rallied was way behind. His cuirassiers having driven into the heart of Wellington's position, found themselves reduced to half their strength, broken by their furious ride, and their chargers exhausted. Subjected to artillery and infantry fire from all sides, his regiments suffered heavily. Kellermann's horse was shot, fell on him, a signal for a wild rush back to the French lines. He scrambled to his feet, vainly tried to rally his men who took no notice as they galloped back the way they had come. Out of control they collided with Foy's infantry nearing the English positions and carried them to the rear. Kellermann hatless and bruised, left the field clinging to the bits of two troopers' horses. By 6.30 P.M. the dramatic episode was over; Ney had lost his chance as the tide turned and the outnumbered and demoralized French were pushed back to their start lines.

WATERLOO 18 JUNE 1815

At Waterloo Kellermann's two divisions were posted to the west of the main Charleroi-Brussels road, with L'Héritier to the right and Roussel d'Urbal to the left. For three hours they stood their ground watching as the French artillery and then the infantry assaults failed to dislodge the Allies from their positions. At 3.30 P.M. the battle took a dramatic turn as the Prussian approach was visible in the distance and Napoleon demanded a quick decision. Then followed the train of events that effectively lost the battle when the French cavalry were hurled to their destruction. Without receiving any orders from Napoleon, Ney apparently mistook an unimportant enemy movement on Mont Saint Jean as the first signs of an Allied retreat. Overreacting he sent an aide to order up another heavy cavalry brigade to turn the retreat into a rout. Farine's brigade of Delort's division was slow to comply, with the result that Ney had lost all self control and in a fury ordered forward Milhaud's entire corps. That not enough, the Guard light cavalry instinctively followed.

For an hour they poured onto the ridge desperately trying to break the squares. The charges achieved little, but the die cast with nothing to lose Napoleon then chose to commit his last cavalry reserve, III Cavalry Corps. Flahaut carried the order to Kellermann who sceptical of its wisdom complied and moved forward, followed by the Guard heavy cavalry. To reach the slopes he deployed in lines but as he passed between Hougoumont and La Haye Sainte the regiments closed up in order to avoid the flanking fire from these points. Bunched they became further disordered as cannon fire from the crest ploughed into them. They then encountered the muddy slopes littered by the debris of Milhaud's assault that stretched up and around the Allied squares behind the crest. Four times his formations charged and were halted by the Allied squares amidst the mud and debris left by their colleagues. By 7.00 P.M. his force was spent, only thoughts of self preservation remained. Too weak to support the last attack by the Guard, when that failed there was little left for him but to lead the remnants of his men from the field.

Without adequate infantry and artillery support, any hopes of achieving another success akin to Marengo disappeared in the mud and debris on the ridge of Mont Saint Jean. It was a supreme irony that for the first time ever the French heavy cavalry had reached a battlefield in prime condition, unaffected by weeks or months of hard marching and inadequate or poor forage, only to be needlessly thrown away once the decisive battle was joined.

POST NAPOLEONIC CAREER

He remained with the Army as it fell back to Paris and on 3 July Davout nominated him, Gérard and Haxo, to negotiate its submission to Louis XVIII. The Armistice signed, he withdrew with the Army to the Loire and once it was disbanded he found himself on 1 August 1815 without a command. Then on 4 September 1815 he was suspended and placed on half-pay. A partial reconciliation with the Bourbons came on 31 August 1817 when his title, marquis de Valmy, previously removed in 1815 was restored. The passing of Saint Cyr's Army Reforms on 30 December 1818 resulted in his recall. The gesture was largely futile, since in the years that followed the Bourbons never trusted him with a command. The death of his father on 12 September 1820 saw him become Duc de Valmy and a Peer of France.

His reputation as one of the foremost cavalrymen in France was belatedly recognized by Bourbon politicians on 1 February 1828 when they invited him to join the Council of War. The body made up of professional soldiers to advise the government on military matters soon became too strident in its demands for reform. It lost much of its influence and by 1830 was totally ignored by an increasingly reactionary government who rarely bothered even to convene it.

Kellermann's behavior after the fall of the Bourbons soon showed he was devoid of any moral scruples when he was one of the five Peers who voted for the death of Polignac, Charles X's former chief minister. He had conveniently forgotten it was Polignac's government that convened the Council of War he belonged to. The Orleanists had uses for him and he headed a commission formed in October 1830 to examine ways to reorganize the cavalry. That task completed he joined the General Staff Headquarters on 7 February 1831. The Minister for War (Soult) by now had the measure of this elderly self-opinionated officer and seeing he was of no further use, on 1 July 1831 placed him on half-pay. Kellermann in his remaining years became an active participant on the lecture circuit recalling his experiences of past wars. Devoid of any modesty, he continued to publish pamphlets portraying that, but for his part at Marengo there would never have been a First Empire. He died of a stroke in Paris on 2 June 1835.

ASSESSMENT - THE FINEST OF NAPOLEON'S WATERLOO CAVALRY LEADERS

Kellermann had his share of bad luck. More often than not ill health side-lined him during several of Napoleon's major campaigns. He was one of the most capable cavalry generals of the period, his coup d'oeil, a term used to express the timing of the vital charge, such as at Marengo and Tormes was impeccable. The care and devotion shown to his men was praiseworthy. Yet as a person there emerged a dark side to his character. Throughout his career he felt he had been cheated and whilst he had sympathy in some quarters (Marmont was an example) and was generally respected by his peers, it won him few friends. He was also an incorrigible looter, few churches and wealthy homes he passed through escaped untouched. A classical comment attributed to him when asked of the fortune he amassed in Spain attracted the response, "I did not cross the Pyrenees merely for my health". He was undoubtedly a great cavalry leader, in fact Napoleon's finest at Waterloo. His biggest failing however was that he had few if any moral scruples. Modesty and honesty were virtues unknown to him, and not surprisingly he was missed by few when he passed away.

BACKGROUND AND EARLY CAREER

The elder son of a wigmaker, the cavalry general François Lallemand was born at Metz (Moselle) on 23 June 1774. His younger brother Henri also had a distinguished career and rose to head the Chasseurs à Cheval of the Imperial Guard. François first enlisted on 1 May 1792 with the 1st Light Artillery Company formed at Strasbourg from the remnants of the disbanded royal regiments. With the Army of the Moselle he served the French guns at Valmy on 20 September 1792. This was a unique distinction, since he was the only general who fought at Valmy in the first triumph of the Revolutionary Wars, at Marengo when the Consulate was secured, at Austerlitz when the authority of the Empire was established and then at its final demise twenty three years later at Waterloo.

After Valmy he joined the triumphant French troops as they drove the enemy back to the borders before he settled down to serving at the siege of Trier in November 1792. A change in career direction occurred when he joined the 1st Chasseurs à Cheval on 10 March 1793. He served with them at Fleurus on 26 June 1794 and then with the regiment after its incorporation into the Army of the Sambre-Meuse. In March 1795 he joined General Elie who commanded the Ardennes Division as an aide de camp. After Elie was cashiered in June 1795, he moved to the 17th Military Division in Paris where he found his way on to General Loison's staff as an aide.

ITALY AND EGYPT 1796-1800

Lallemand's first encounter with Napoleon came during the Vendemaire Coup d'Etat on 5 October 1795. During the defence of the Tuileries the coolness he displayed was pointed out to Loison, who next day formalized his appointment as an aide by promoting him *sous lieutenant*. He remained in Paris with the Army of the Interior till Loison's dismissal by the Directory in December 1796 when implicated in a plot supporting a resurgence in Jacobin activity. Fortune again smiled on Lallemand as he secured the same month a posting to General Bonaparte's Guides à Cheval with the Army of Italy. From there he was frequently in close contact with Napoleon since the unit provided his personal escort. He distinguished himself during a charge by the Guides at Rivoli on 14 January 1797, resulting in his promotion to *lieutenant* on the battlefield.

He remained in Italy till May 1798 when with the Guides à Cheval he joined the expedition to Egypt. Present during the march to Cairo he fought at the Battle of the Pyramids on 24 July 1798. The next year he joined the expedition to Syria where he fought at Mount Tabor on 14 April. Promoted *capitaine* on 25 May 1799 he became aide de camp to General Junot. He left Egypt for France with Junot in October 1799

but while at sea fell into the hands of the Royal Navy under Sir Sydney Smith. Fortunately having acted as an interpreter during the peace negotiations with Sir Sydney earlier in the year, he secured his release with that of Junot and Desaix, landing at Marseilles in May 1800. Forced to spend a period in quarantine with Desaix, they then rushed to Northern Italy to join Napoleon. Lallemand served on Napoleon's staff while Desaix perished at Marengo on 14 June 1800 as his forces intervened and saved the battle for Napoleon.

LALLEMAND - FRANÇOIS ANTOINE, BARON (1774 - 1839)

CONSULAR AIDE DE CAMP 1800-1804

After Marengo he rejoined Junot as his aide; a post he held for five years. They were good years, Junot was one of the rising stars amongst the First Consul's entourage and the duties Lallemand fulfilled were numerous and varied. At this stage there was no sign of the instability and madness that was to dog Junot's later career and result in his premature death. On 16 October 1802 he was promoted *chef d'escadron*, then he was sent to Santo Domingo early the next year to report on the conditions on the island. In 1804 on Junot's appointment as ambassador to Portugal he accompanied him to Lisbon.

THE GRANDE ARMÉE 1805-1807

After his return from Portugal, on 14 May 1805 he was promoted *major* with the 18th Dragoons. As the Grande Armée was set in motion his regiment formed part of the Cavalry Reserve with Bourcier's 4th Dragoon Division. He was present at Ulm on 20 October and fought at Austerlitz on 2 December 1805.

He campaigned in Prussia under Sahuc who had taken over the division from Bourcier, fought at Jena on 14 October 1806 and was at the capture of Lubeck on 6 November. The conclusion of the campaign in Prussia saw him promoted *colonel* on 20 November with a move to the 27th Dragoons, still under Sahuc. In Poland attached to Bernadotte's I Corps with the rest of Sahuc's division he pushed across the Vistula and engaged Lestocq's outposts at Biezun on 23 December. In the Friedland Campaign he served at Spanden on 5 June and then at Friedland itself on 14 June 1807 where he fought under La Houssaye who had the previous month taken over from Sahuc.

SPAIN 1808-1813

On the creation of the Imperial Nobility he became a *baron de l'Empire* on 28 June 1808. In charge of the 27th Dragoons he entered Spain on 7 September 1808 with La Houssaye's 4th Dragoon Division. He was in the march to Madrid and took part in the brief struggle before the city's walls on 4-5 December. In January 1809 he joined the pursuit of John Moore's army as it fell back to Corunna. With La Houssaye's dragoons and Colbert's light cavalry that formed the van of the French advance guard he was active against Paget's rearguard at Cacabellos on 3 January 1809. In the skirmish the dashing Colbert de Chabanais took one chance too many, he ignored the steadiness of English infantry and rushed the bridge over the Cui, was killed and his troops repulsed. At the same time Lallemand forded the river at several points but the rocky terrain and vine groves obstructed their charge. Forced to dismount, his troops acted as skirmishers, which they found difficult against the veterans of the 52nd Foot who held them back with little trouble till the lead elements off Merle's division came up in support.

At Corunna on 16 January Lallemand was again heavily engaged. He joined La Houssaye's and Mermet's sweep to the west of Monte Mero trying to turn the English line. When Paget advanced from his position between Monte Mero and San Cristobal with the Reserve Division to counter the French move, Lallemand was detached and had to face Paget's assault on his own. His dragoons moved against the advancing infantry, but were hindered by the terrain cut up by rough stone walls between the villages of San Cristobal and Elvina, and could do little to halt the English advance. Knowing the French could not effect an orderly charge Paget swept across the low ground and the volley fire of the 95th Rifles and the 28th Foot soon drove Lallemand's men back. In vain he tried to halt the advance by dismounting his dragoons and ranging them along the lower slopes as tirailleurs. The deadly fire from Paget's infantry soon thinned his ranks and forced him to retire while the rest of Paget's division dealt with Mermet's flanking move.

In the van of Soult's invasion of Northern Portugal he was present at the action at Braga on 20 March 1809 and then a week later at the fall of Oporto. Detached to Loison's command in the Tras-os-Montes region he occupied Amarante. He later covered the French retreat to the Spanish border when Beresford's forces outmaneuvered Loison and precipitated a hurried retreat from the region which left Soult's retirement from Oporto dangerously exposed.

Ill health forced his return to France in September 1809. On his return in January 1810 he rejoined La Houssaye's division based at Talavera. Notable actions he fought were at La Rocca on 21 April and Cuenca on 14 October 1810. From January till May 1811 he was at Cordoba and joined Latour

FRENCH CHASSEURS IN SPAIN

Maubourg's division with Soult's Army of the South for the offensive into Estremadura to relieve Badajoz. Present at Albuera on 16 May his regiment took part in the great cavalry sweep around Beresford's flank, which promised much but yielded little in that fierce struggle. At Elvas on 23 June 1811 he was in the fore during the action that mauled the 2nd Kings German Legion Hussars and the 11th Light Dragoons.

Promoted *général de brigade* on 6 August 1811 he continued to serve under Latour Maubourg till December when he moved back to Cordoba as brigade commander with the 2nd Cavalry Division. In January 1812 he moved to Jaen as military governor. He was not long in the post when on 7 February he was recalled to Estremadura to strengthen the 1st Cavalry Division operating under d'Erlon* who was covering the Portugese border against Hill's corps.

Lallemand on 11 June 1812 had a spectacular success at Llerena when he routed the 1st Royal Dragoons and the 3rd Dragoon Guards under Slade. Flushed by an initial success against his horsemen, the English cavalry ran out of control and pursued him for several miles. Lallemand led them straight into a trap with supporting squadrons falling on the exhausted horsemen from all sides. The numbers involved were small, but the effect on morale was considerable since they were two of Wellington's finest regiments and foolishly lost some 500 out of the 700 men engaged. The action had a profound effect on Wellington's attitude towards his cavalry. He likened their performance at to that of a disorganized rabble engaged in a fox hunt on Wimbledon Common. Throughout the remainder of the Peninsular War very rarely did Wellington allow his cavalry to be used as an offensive arm, because he no longer had faith in them. From then on they were used primarily for reconnoitering, a role they carried out with good effect.

After the French withdrawal from Andalusia he joined Soult's advance on Madrid in October and the pursuit of the Allies to the Portugese border. The disasters in Russia resulted in his recall to France on 5 February 1813.

GERMAN CAMPAIGN OF 1813

On 21 March 1813 Lallemand was posted to the 2nd Cavalry Division of III Cavalry Corps reforming at Metz. From there on 4 April he moved to Mayence on the Rhine as a brigade commander with the 1st Cavalry Division de Marche under Milhaud*. After the troops were distributed to their regiments within the Army on 15 June 1813 he was appointed corps cavalry commander with Macdonald's XI Corps. A strange appointment followed on 25 June when he became Macdonald's chief of staff, a post that totally ill suited him. Fortune favoured him and on 1 August he moved to Davout's XIII Corps as commander of the corps cavalry.

After the outbreak of hostilities he took charge of the garrison at Lubeck, then in early December 1813 as the Allies closed in he fell back to Hamburg. Locked up with Davout in the city he won great praise for the way he kept the cavalry in such good condition throughout the six month siege.

THE RESTORATION AND NAPOLEON'S RETURN

He returned to France in May 1814 with the rest of XIII Corps. The Bourbons treated him fairly and on 27 June 1814 he became a *Chevalier de Saint Louis*. After a period of leave he received on 31 August 1814 the appointment as commander of the 2nd Subdivision (Aisne) of the 1st Military Division.

A Bonapartist fanatic, he showed few outward signs of his undiminished devotion to Napoleon. As a result he kept his command, yet became deeply involved in plots to support his return. On news of Napoleon's landing, with Lefebvre Desnoettes he raised the garrisons of Guise and Chauny in favour of the Emperor on 6 March 1815. They then moved to La Fère to join his brother who, also part of the conspiracy, planned to seize the arsenal. But his brother failed to raise the garrison and had to hurriedly flee to avoid arrest. The three were then forced to wander from town to town with the few troops they had, failing to pick up any further support. In despair they sent their troops back to barracks and after a few days dodging the gendarmes were arrested at Ferté Milon on 12 March. Taken to Soissons a royalist mob nearly lynched them en route before they were hurled into prison. With Napoleon installed in the Tuileries, their release followed on 21 March and an appreciative Emperor hearing of their escapades ordered Lallemand's promotion to *lieutenant général* on 30 March.

Immediately given a command Lallemand briefly took charge of a brigade made up of the 4th Hussars and the 13th Dragoons till 5 April. Then on the recommendation of Lefebvre Desnoettes he became deputy commander of the Guard Chasseur à Cheval replacing Lion who initially refused to rally to Napoleon. On 2 June he became a Peer of France.

THE WATERLOO CAMPAIGN

Lallemand's Guard Chasseurs à Cheval formed part of the Guard Light Cavalry Division under Lefebvre Desnoettes. Comprising five squadrons of veterans it numbered 1,197 officers and men at the start of the campaign. Attached to the Army's Left Wing with Lefebvre he passed through Gosselies around 3.00 P.M. with orders to push on to Quatre Bras. A brush with Saxe Weimar's 2nd Nassauer Battalion supported by artillery outside Frasnes caused him to halt. Lefebvre ordered infantry support while Colbert with the Guard Lancers skirted the village and moved on. After a battalion of infantry arrived an hour later Lallemand moved on to join Colbert. A mile short of the Quatre Bras the enemy barred their approach with a force of 4,000 men with eight cannon astride the road. With orders not to commit the Guard Cavalry, Ney merely allowed a few probing charges before they fell back to Frasnes at 8.00 P.M. for the night.

During the struggle for Quatre Bras the next day his regiment remained idle as the Allies gradually built up their strength in full view of his horsemen. Napoleon's paralyzing orders given to Ney the previous day remained in effect after Colbert's horse received unnecessary casualties.

At Waterloo during the early stages of the battle he was positioned in the second line behind d'Erlon's I Corps. In the afternoon he took part in the charges by the Guard cavalry between Hougoumont and La Haye Sainte. The attacks sent in by an impetuous Ney without artillery or the support of Foy or Bachelu's infantry fared little better than Milhaud's corps repulsed earlier. Slightly wounded he did not remain with his troops after the battle and instead chose to return to Paris with Napoleon.

YEARS OF EXILE 1815-1830

He was at Napoleon's side during the second abdication crises and retired with him to Malmaison. As the Allies closed in he moved with Napoleon to Rochefort and was involved in the negotiations with Captain Maitland of the Bellerophon to allow Napoleon to seek exile in England. While Napoleon's fate was decided he remained with him at Torbay till 8 August when the Northumberland set sail for Saint Helena. He offered to accompany Napoleon to the island but was turned down. Classified as a prisoner of war the authorities arrested him in Plymouth when he tried to leave the country and return to France. He was shipped to Malta for his own safety as there was a price on his head and arrived on the island on 26 September. Whilst imprisoned at Fort Manuel the Council of War convened in Paris on 26 August 1816 tried him for high treason and condemned him to death in absentia. Released soon after he spent a brief period in Turkey and Egypt before he settled in the United States.

The next year he moved to Texas to found the Bonapartist colony of Champ d'Asyle for fugitives from the White Terror in France. These fanatics set about clearing the virgin land, dreamt of past glories and hatched mad-cap schemes to spirit Napoleon away from Saint Helena. Within a few months yellow fever got the better of these intrepid pioneers, the project collapsed and the bulk of them settled in the United States, Lallemand in Louisiana.

A follower of lost causes, in May 1823 he returned to Europe landing in Lisbon with a plan to raise a mercenary force to help the junta in Cadiz that had risen against the Bourbons in Spain. When the French armies crossed the Pyrenees, the revolt soon collapsed and his hopes to strike at the soft underbelly of the Bourbons came to nought. From there he moved to Brussels where he fell on hard times before he settled in New York as a French teacher.

After the overthrow of the Bourbons he returned to France in September 1830. His rank and privileges restored on 7 January 1831, he joined the Army General Staff. He then served as Cavalry Inspector General for the 5th Military Division from September 1831 till October 1833 when he transferred to the 17th Military Division in Corsica. On 30 April 1835 he became a Grand Officer of the Legion of Honor and on 13 October 1835 moved to Toulouse as commander of the 10th Military Division. Active till the end he died in Paris of a heart attack on 9 March 1839.

CAREER ASSESSMENT

An inspirational and brave leader, respected by all who served under him, Lallemand also became one of those tragic punch drunk Bonapartists who in later life took loyalty to a ridiculous extreme. His part in the attempt to raise the northern garrisons in 1815 was also foolish. The way the plans were implemented showed reckless and irresponsible behavior akin to that of a later day colonels from a banana republic. Had the revolt turned to bloodshed the peaceful march on Paris by Napoleon could so easily have gone he same way. It instead gave him a reputation and profile greater than his accomplishments in the field ever merited.

EARLY CAREER

Henri Lallemand, the younger brother of the cavalry general François Lallemand was born in Metz (Moselle) on 17 October 1777. After completing his studies at the Ecole Polytechnique he received a commission with the 1st Foot Artillery Regiment on 2 May 1797. His first active service was with the Army of the Rhine followed in 1798 by a brief period with the Army of England.

General Bonaparte was the man of the moment, and Henri's brother managed to secure them both a place with the expedition to Egypt. He travelled to Genoa where he joined Baraguey d'Hilliers's division on 19 May 1798. D'Hilliers was left at Malta as governor, and Lallemand was posted to the 3rd Company of the 1st Artillery Regiment under Menou, who took over the division. Present when the army landed at Aboukir on 2 July, the next day he took part in the capture of Alexandria. Based for most of the next three years at Alexandria he had few opportunities to excel and his career prospects were very much eclipsed by the exploits of his elder brother. He fought against the English in the second battle of Aboukir on 22 March 1801 when Menou's forces were driven back into the city. He then served throughout the siege of Alexandria till its surrender on 31 August 1801. His return to France resulted in promotion to *lieutenant* on 21 January 1802, nearly five years after he was first commissioned.

The period of peace that followed saw him at the 1st Artillery Regiment's main depot at Metz. Then as the French Army prepared for the invasion of England he remained at Metz training and organizing the batteries before they moved off to join the armies. He built up a sound reputation as a trainer and organizer as well as one of the foremost up and coming logistics experts in the Army. Promotion came as *capitaine* on 2 June 1804 and then *capitaine commandant* on 16 April 1806.

CAREER WITHIN THE GUARD 1806-1814

He missed the Austerlitz Campaign, but his skills did not go unnoticed and on 1 May 1806 he joined the Imperial Guard as *capitaine* in charge of the new artificer company. He served in the Prussian Campaign and then the next year in Poland where his guns were in action at Hof on 6 February, Heilsberg on 10 June and Friedland on 14 June 1807.

The rapid expansion within the Guard Foot Artillery helped his career and on 28 August 1808 resulted in further promotion to *capitaine 1er* with command of the 4th Company of the Guard Foot Artillery. He served in Spain and then with the Army of Germany where he received promotion to *chef de bataillon* on 22 June 1809. He distinguished himself at Wagram on 6 July where under Drouot[*] his company was with the massed battery formed by Lauriston between Breitenlee and Süssenbrunn that

LALLEMAND - HENRI DOMINIQUE, BARON (1777 - 1823)

took on the Austrian center. Suffering horrific casualties his gunners were put to the test as they gradually disabled the enemy's guns faster than they lost their own, enabling Macdonald's infantry to smash through Liechtenstein's corps to secure victory.

On 13 August 1810 he was made a *baron de l'Empire*. He fought at Smolensk on 18 August and at Borodino on 7 September. Promoted Guard Artillery Chief of Staff on his promotion to major on 22 September 1812, he later won great praise for the way he tried to save the guns during the retreat. His organizational skills much in need after the disasters in Russia resulted in his recall to France where he set about helping to reconstruct the Guard artillery arm. On 29 March 1813 he became chief of staff to Dulauloy who took command of the Guard Artillery. He served in Saxony, fought at Lutzen on 2 May, Bautzen on 21 May and Leipzig on 18 October 1813. He proved a very effective chief of staff and due largely to his untiring efforts the Guard was able to recross the Rhine intact with some 166 cannon, well prepared for the coming campaign in France. He fought at Brienne on 30 January, at Montmirail on 11 February and Laon on 6 March. Promoted *général de brigade* on 12 March he was with the Guard at Fontainebleau when Napoleon abdicated on 6 April 1814.

THE RESTORATION AND NAPOLEON'S RETURN

Like so many of the Guard he found it impossible to adjust to the Bourbons, often referring to them as, "that canaille of princes and emigrés". It did his career prospects no good, and when plans were mooted to merge the Guard Artillery with the Royal Guard his vociferous opposition resulted in him being placed on half pay on 1 September 1814. A positive result was the plans were temporarily shelved, till his later recall on 14 October with new orders to implement a new plan to disband the Guard batteries by 1 January 1815. He saw no need to hurry, for a secret Bonapartist conspirator, he was confident Napoleon would return.

He was in the plot with his brother, d'Erlon and

Lefebvre Desnoettes to raise the northern garrisons and march on Parijs after they heard of Napoleon's landing. At La Fère on 9 March when the news broke he tried to persuade his superior General D'Aboville to join his fellow conspirators. Not impressed, the wily veteran wanted no part of it and ordered Lallemand to leave the depot. Shaken by the unexpected opposition he did so without winning over a single gunner to the cause. After a few days wandering around the countryside trying to gather support he was arrested and imprisoned with his brother.

Released on news of Napoleon's installation at Tuileries, Davout the new Minister for War recalled him to the Guard on 23 March 1815. Appreciative of the loyalty he had shown Napoleon backed his promotion to lieutenant général on 11 April 1815. Under the watchful eye of Drouot and Desvaux Saint Maurice he spent the next two months feverishly reorganizing the artillery at La Fère. At the beginning of June his formal appointment as commander of the Guard's twelve foot batteries totalling ninety six guns was confirmed.

THE WATERLOO CAMPAIGN

Present at Ligny on 16 June his batteries did great damage to the Prussians massed around and behind the town. He then played a decisive support role in the final assault by the Old Guard in the evening. As the chasseurs and grenadiers advanced, his guns also moved forward by stages blowing great gaps in the enemy lines. At Waterloo in the morning he directed the bombardment against Hougoumont on the Allied right. Here he made the serious tactical error by not directing howitzer fire on the position and instead used ball before the infantry went in. Had he done so, with the chateau aflame sooner the English holding the position in all likelihood would have found it untenable. After Desvaux's death in the afternoon he took overall command of the Guard artillery and spent much time directing fire against the Allied positions behind La Haye Sainte. Again he was responsible for another serious mistake when so absorbed trying to destroy he Allied Center he refused the request to allow two batteries of Guard Horse Artillery to support the cavalry charges. With Wellington forced into square the effect would have been devastating had twelve cannon joined the struggle. His men also had the equipment available to disable the enemy guns once they were overrun, a task the cavalry failed to do.

The suddenness of the Army's rout caused most of his pieces to be lost. By the time the Guard reached Laon only 30 pieces remained. He stayed with the army as it fell back to Paris and was a strong advocate in favour of fighting the Allies before the city as he felt their position was very exposed. A battle not to be, he retired with the Army to the Loire once the Armistice was signed.

EXILE AND DEATH 1815-1823

He realized retribution would be swift, and well before the issue of the Ordinances of 24 July 1815 he had fled the country. On 21 August 1816 a Council of War condemned him with his brother to death *in absentia*. A wandering exile he spent time in Hungary and Turkey before he joined his brother in Texas where they set up the unsuccessful colony of Champ d'Aigle near Galveston in 1817. Here they dreamt of past glories and made plans to spirit Napoleon away from Saint Helena before most of the intrepid settlers succumbed to fever. The more tranquil surrounds of Bordentown, near Philadelphia beckoned and the following year he joined the exiles there, who included Joseph Bonaparte, Vandamme, Grouchy and his brother. He married and settled down there. He caught dysentery and died after a short illness on 15 September 1823.

A BRAVE LOYAL BONAPARTIST - BUT LIMITED

Lallemand was typical of the latter generation of younger general within the Guard, who showed great bravery and possessed a fanatical loyalty towards the Emperor that certainly helped his career. He was also a fine administrator, organizer and trainer of artillerists, but such attributes did not necessarily make a great general. In fact his handling of the guns at Waterloo revealed a certain lack of tactical common sense. To go back any further and judge Lallemand is difficult since the personalities of Desvaux and Drouot were so dominant within the Guard Artillery. The role of the Guard Artillery in the closing years of the Empire was so much their preserve that Lallemand's influence as a subordinate commander on the great events of the day is very very difficult to determine.

His last promotion was certainly not due to his prowess on the battlefield, but rather to an ill judged attempt to rally the northern garrisons to the Emperor. Its failure was turned by the Napoleonic propaganda machine into a brilliant and heroic diversion. The conspirators at the time were labelled as foolish opportunists, which was largely correct since they risked thrusting France into an unwanted civil war that not even Napoleon was prepared to countenance. As events moved on, the criticism turned to acclaim, and as a result he with his brother Henri received credit, which they did not readily deserve. For this reason Henri Lallemand must rank as a leader of only moderate ability.

BACKGROUND, EARLY LIFE AND CAREER

From a modest middle class family, Charles Lebrun was born in Paris on 28 December 1775. His father Charles François Lebrun, a lawyer by profession and once secretary to Louis XVI's minister Maupeau, survived the turmoil of the Terror to become one of the most astute politicians of the Directory. Casting his lot with Napoleon he became a Consul of France and later under the Empire its Finance Minister or Arch Treasurer.

Younger Lebrun's early life took the classic course of a spoilt young man from an influential political family. Women and gambling were more his style and after a period of idleness he avoided being disinherited and helped his father's cause by securing a politically useful marriage to the daughter of Director Barbé Marbois. Then as one of the three Consuls with his political power at its peak after Brumaire, his father sensitive to public criticism insisted his wayward son don a uniform. That done, on 28 December 1799 Charles obtained one of the softest postings he could wish for as a *sous lieutenant* in the 5th Dragoons serving with the Army of the Interior based in Paris. His next career step had definite marks of patronage, an art form his father became renowned for, when Lebrun senior arranged for Napoleon to appoint him as an aide de camp for the coming campaign in Italy.

Never having fired a shot in anger Lebrun was hurled into the Marengo Campaign. In spite of it, carrying the stigma of being a Consul's son, he showed surprising pluck and came through the experience with credit. In the Battle of Marengo itself, Desaix shot by his side is reputed to have died in his arms. It was the type of incident legends were made of; the young aide de camp comforting the dying hero as the battle raged around them. Desaix's heroic sacrifice passed into history, while the episode gained Lebrun a mention in the Marengo Bulletin. It also secured him a permanent place on Napoleon's staff and removed all criticism from his colleagues as to the circumstances of his appointment and ability to serve as an aide.

As with most of Napoleon's aides at the time, he returned to his regiment after the campaign. He remained with the 5th Dragoons till 5 March 1801 when on promotion to *lieutenant* he moved to the Consular Guard. In the period of peace that followed, while most of his colleagues' careers remained static his took off, *capitaine* on 17 March and then *chef d'escadron* on 31 December 1801. All this in spite of his record showing no real trace of brilliance or talent above that of colleagues left behind in the promotion race. The influence his father wielded undoubtedly had its effect.

It was further illustrated when on 1 February 1804 with the experience of only one campaign behind him he was then promoted *colonel* and commander

of the 3rd Hussars based at the Camp of Montreuil. For eighteen months under the watchful eye of Ney* and Auguste Colbert he prepared his regiment for the invasion of England. Then on 30 August 1805 with the Grande Armée set in motion for the campaign against Austria Napoleon recalled him as an Imperial Aide. With Napoleon throughout the campaign he

LEBRUN - ANNE CHARLES, COMTE (1775 - 1859)

distinguished himself at Austerlitz on 2 December 1805 when during Murat's pursuit of the Russians at the end of the day he personally took a standard. Napoleon not to miss a sense of occasion then gave him the honor of carrying the official bulletin heralding the victory to Paris. He arrived in Paris on 1 January 1806 and with due ceremony handed over forty captured standards amidst rapturous applause.

For the campaign against Prussia he returned to the 3rd Hussars serving with Ney's VI Corps. Present at Jena on 14 October 1806 he distinguished himself during Murat's charge by the Cavalry Reserve that split Hohenlohe's front. In the pursuit, which went on for several miles till it reached Weimar he rode down several Saxon battalions and seized their colors. The rest of the campaign he spent tied down before Magdeburg covering the siege till its surrender on 11 November. He then he joined the army in Poland and fought at Eylau on 8 February 1807 where detached from Ney he was posted on the French left during the battle. After the destruction of Augereau's VII Corps he supported Murat's counter charge that broke the Russian infantry and saved the battle as they tried to retake Eylau.

IMPERIAL AIDE 1807-1812

On 1 March 1807 Lebrun was promoted *général de brigade* and rejoined Napoleon's staff. With Napoleon in the campaign that followed he was present at Heilsberg and fought at Friedland on 14 June. In the struggle against Gortschakoff's cavalry he was badly wounded forcing him to return to France to recover. Back to duty on 6 October 1807 he became Cavalry Inspector General for the 16th and 24th Military Divisions based in Paris.

When Napoleon invaded Spain he rejoined him in November 1808 and served at his side during the march to Madrid and then over the Guadarramas in pursuit of Moore's army. He returned to France with

Napoleon at the end of January 1809 and followed him to Austria. He witnessed the carnage at Aspern Essling on 21-22 May and then fought at Wagram on 6 July 1809. Here he led a successful charge by Bessières's cavalry against the Austrian centre that disabled a large number of Liechtenstein's guns before Macdonald's column broke through. After the Peace of Schoenbrünn he remained in Austria till the end of the year as it took the Army of Germany several months to withdraw.

In 1810 Napoleon mindful of the English invasion the previous year ordered Lebrun to Holland to prepare a report on the preparedness of the defences along the Scheldt. That done, as the Emperor's man on the spot he then took time to supervise the strengthening of the defences at Antwerp, Breda and Berg op Zoom. In March 1811 he moved to the 14th Military Division based at Strasbourg where a general apathy and war weariness within the region had caused the problem of desertion to become acute. To overcome the problem and make Strasbourg an example for the rest of the country Napoleon gave Lebrun wide powers in order to find a solution. He introduced mobile columns to hunt down deserters and conscripts who had failed to report to their depots. The effort was considerable and for a while it checked the desertion until later reverses coupled with lack of manpower caused the problem to return. In August 1811 he moved from hunting these men to taking charge of the depot at Strasbourg set up specifically to bring these troops to heel. His work was successful and went a large way to earning him his promotion on 23 February 1812 to *général de division*.

SENIOR AIDE DE CAMP 1812-14

The outbreak of war with Russia saw Lebrun back on Napoleon's staff as a senior aide de camp serving at his side throughout the campaign. He was present at Smolensk, Borodino, Malojaroslavets and the crossing of the Beresina before returning to France with him in December.

In 1813 the assignments he carried out were varied. The early months he spent at Mayence where he helped organize newly formed cavalry regiments into provisional divisions de marche before sending them into Saxony. On Napoleon's arrival at the front he returned to the Imperial Headquarters. He fought at Lutzen on 1 May and then at Bautzen on 21-22 May. In the second battle he headed one of the columns of Young Guard that broke through the Russian centre at the end of the battle. With Napoleon during the campaign in the Autumn he was present at Dresden and Leipzig. He briefly headed I Corps from 30 November till 7 December as it rallied behind the Rhine. The posting as military governor at Antwerp followed till Napoleon recalled him and Lazare Carnot took his place on 25 January 1814. He served on Napoleon's staff throughout the campaign in France and was with him till the final abdication at Fontainebleau.

THE RESTORATION AND NAPOLEON'S RETURN

He acknowledged the return of the Bourbons and they in turn treated him well. The influence of his father whose financial expertise they wanted to tap undoubtedly had some effect. On 22 April 1814 he became the the King's special commissioner in 14th Military Division (Calvados, Manche and Orne), the

DESAIX FALLING FROM HIS HORSE, LEBRUN TRYING TO SAVE HIM (DETAIL FROM A LARGER CANVAS OF MARENGO)

Bourbon equivalent of an Inspector General. In spite of advice to the contrary Louis XVIII realized he needed men who had served Napoleon to maintain continuity within the army and on 14 July 1814 Lebrun became Inspector General for the Hussar regiments. The award of the title *Chevalier de Saint Louis* followed on 29 July 1814.

Lebrun's future apparently assured under the Bourbons as a result was slow in rallying to Napoleon and waited till the new government was firmly established before offering his services. Napoleon took little heed of his hesitancy and genuinely fond of him on 24 March gave him the important command of the 2nd Military Division at Mézières on the Belgian border. Then as prospects for peace evaporated Napoleon thought he would make a good head of III Army Corps as it gathered at Mézières since he was the man on the spot. Davout more aware of Lebrun's limitations persuaded the Emperor otherwise and to only assign him as acting commander till a more experienced man was found.

Lebrun tried his best to carry out the Emperor's instructions. After his arrival he at once set to work bringing his infantry and cavalry regiments up to strength, to mobilizing the National Guard, drilling the artillerists and to inspecting the divisions. He failed to win the respect of his subordinates and his reputation for a "notorious lack of competence" was illustrated in a letter by General Ameil, a brigade commander with the 6th Cavalry Division to a colleague. He just did not have what was required to be a corps commander. It was much to the relief of everyone when Vandamme took over on 20 April and he resumed his functions as an Imperial Aide. It also gave him some spare time and following his father's footsteps he opted for a career in politics, gaining election on 10 May to the Chamber of Deputies as representative for Seine-et-Marne.

THE WATERLOO CAMPAIGN AND ITS AFTERMATH

Serving in Belgium Lebrun was present at Ligny and Waterloo. There is little record of what he did exactly during the campaign. One can only assume as part of Napoleon's entourage when directed he helped ensure the Emperor's orders were carried out. Not like other imperial aides such as Dejean, Flahaut and de la Bédoyère whose individual exploits had some impact on the campaign, his deeds seem to have attracted little attention. It was notable that he was not among the group of aides that returned to Paris with Napoleon. Nor was he in the Chamber of Deputies to support Napoleon when they voted themselves in permanent session and then for his abdication. When all appeared lost, like his father he acted out of prudence and disappeared from the scene till the hue and cry died down.

LATER LIFE AND CAREER

The Bourbons did not give him a second chance and on 1 August 1815 placed him on the active list. Saint Cyr's Army Reforms brought about his recall on 30 December 1818 with the Headquarters Staff in Paris where he remained for several years unattached. He became a Peer of France and took the hereditary title of the Duc de Plaisance after the death of his father on 16 June 1824. The July Monarchy did not bring about a change in fortune either and he never received another command. He was not completely ostracized by the establishment and on 29 April 1833 he received the Grand Cross of the Legion of Honor. At sixty five he was placed in the Reserve Section of the General Staff on 30 December 1840. The overthrow of Louis Philippe saw him retired on 12 April 1848 with a pension of 7,200 francs a year.

The return of Bonapartism in the form of Louis Napoleon gave fresh impetus to Lebrun's standing and on 26 January 1852 he became a senator. That not good enough, he insisted on his recall and on 1 January 1853 was re-admitted to the Reserve Section of the General Staff. Further honors came as Louis Napoleon began to deify former relics of the First Empire with him becoming a Grand Councillor of the Legion of Honor on 26 March followed by being awarded the Medale Militaire on 16 April 1853. He had the satisfaction of seeing the Second Empire reach its peak and died peacefully in Paris on 21 January 1859.

ASSESSMENT - A FINE IMPERIAL AIDE

Lebrun as an Imperial Aide rarely held an independent battlefield command where his talents could be subject to close scrutiny. He was different from other imperial aides in that when he first joined Napoleon's staff he had no record of proven bravery nor was he a specialist in any arm of the service. The appointment was not due to ability, but nepotism. To his credit he very soon proved himself, because there was no place for aides who were not devoted, brave and commanded the respect of others.

As an imperial aide he had authority to act in the Emperor's name. He was a man able of carrying out any mission, to lead an improvised task force, to meet an emergency, mass artillery to support a decisive attack, clear a snarled supply line or conduct large or small scale reconnaissances. He had the authority to require even marshals to present their troops for review. Isolated subordinate commanders could expect his trust to take Napoleon a factual account of their problems. He was loyal, spoke the truth as he saw it and did not flatter. Napoleon in turn gave him his trust, accepted his often outspoken advice and comments, not without occasional anger, but counted on him as a friend. To possess such attributes was the mark of a very talented man.

LEFEBVRE DESNOETTES - CHARLES (1773 - 1822)

BACKGROUND AND EARLY CAREER

The son of a prosperous draper, the distinguished Guard cavalry general Charles Lefebvre Desnoettes was born in Paris on 14 September 1773. He completed his education at Grassins College in Paris in spite of running away several times to enlist. His distressed parents on each occasion had to purchase his discharge. He enlisted as a *chasseur* with the Paris National Guard on 1 December 1789. Life in the National Guard not exciting enough, he quit in June 1791 and next emerged on 15 September 1792 as a *chasseur* in the Allobroge Legion at Lyons. He served with the Army of the Alps during Kellermann's invasion of Savoy. Discharged on 21 November 1792 after the campaign, he travelled north and joined the 5th Dragoons with the Army of the North as a *sous lieutenant* on 15 February 1793. From there he spent time with Armies of the Sambre-Meuse and the Rhine-Moselle gaining promotion to *lieutenant* on 4 January 1797. He passed to the Army of Italy in January 1797 serving on Bernadotte's staff. It was during the crossing of the Tagliamento that his bravery came to the notice of General Bonaparte. After the Peace of Campo Formio he remained in Italy on garrison duty gaining promotion on 11 July 1798 to *capitaine*. A period followed in Belgium and Holland followed in 1798-99 where he served on Brune's staff with the Army of Batavia.

Recommended by Brune, on 1 February 1800 he became an aide de camp to the First Consul. He fought at Marengo on 14 June 1800 and on 28 July was promoted *chef d'escadron*. He served for a brief period on the staff of the Consular Guard before transferring to the Elite Gendarmes on 19 September 1801. On 30 December 1802 promoted *chef de brigade* with the 18th Dragoons he joined Soult's forces with the Army of the Ocean Coasts concentrating at Boulogne for the invasion of England. With the Grande Armée as it set out for the Danube his regiment was with Bourcier's 4th Dragoon Division of the Cavalry Reserve. He distinguished himself at Elchingen on 14 October, breaking a square and was at the Capitulation of Ulm and later Austerlitz on 2 December 1805. He was made a commander of the Legion of Honor on 25 December 1805.

AIDE DE CAMP TO JÉRÔME BONAPARTE

Promoted *général de brigade* on 19 September 1806 he served as senior aide de camp to Jérôme Bonaparte*. When Jérôme took charge of the Bavarian IX Corps with the Grande Armée Lefebvre led the corps cavalry during the Prussian Campaign. Engaged principally in Silesia, he was at the capture of Glogau on 9 November, Kalisch and then the siege of Breslau from 9 December to 7 January 1807. He seized Brieg on 11 January and distinguished himself at the defeat of Kleist's relieving force near Frankenstein on 17 April and drove it back to Glatz. He fought another sharp action at Canth on 13 May and then rejoined the siege of Glatz until its capitulation on 24 June 1807. An appreciative King Ludwig made him a Commander of the Lion of Bavaria on 27 July 1807 for the fine leadership of Bavarian troops he displayed.

When Jérôme's became King of Westphalia Lefebvre transferred to the Westphalian Army on 29 December 1807 with the rank of *général de division* and became Jérôme's Grand Equerry. A split with Jérôme came on 18 January 1808 when he returned to France as colonel of the Guard Chasseurs à Cheval. Part of the circle of dashing young officers that surrounded Napoleon, his rise was helped by a well placed marriage to a great-niece of Madame Mère, who also happened to be Napoleon's second cousin. On 19 March 1808 he joined the imperial nobility with the title *comte de l'Empire*.

SPAIN AND CAPTIVITY 1808-1812

Appointed Bessières's chief of staff on 19 March 1808 he moved to Northern Spain. On his arrival he found confusion and chaos as the revolt that had started in Madrid on 2 May had spread country-wide. Bessières's plan in the north was to keep a firm hand on communications between Madrid, Burgos and Bayonne, while Lefebvre-Desnoettes with a division moved on Saragossa to crush the insurgents who had raised the standard of revolt there. On 3 June he set out from Pamplona with a force of 5,000 infantry, 1,000 cavalry and two field batteries. At Tudela on 8 June he overwhelmed a force of 2,000 ill equipped Aragonese under the Marquis de Lazan sent ahead from Saragossa to meet him. To show he meant business, he sacked the town and executed a number of notables for inciting rebellion. It was an ill judged move since as events turned out it hardened the resolve of the Saragossans to resist. Five days later Lazan again offered battle at Mallen. In a poor position the Aragonese had little protection against the French artillery and when the Polish lancers charged they fled. They kept their spirit and on 14 June this time under Palafox they faced Desnoettes a third time at Alagon 17 miles from Saragossa. The 6,000 infantry (of whom only 500 were regulars), 150 dragoons and four guns were no match for the fourteen French cannon and soon gave way with the first assault by the Polish infantry.

Elated by three easy victories, Lefebvre thought there was little more to do than enter the city in triumph. The next day he arrived before the walls and drove in the pickets. So confident was he, he decided to take the place by storm without making detailed plans. The city appeared defenceless and the French soon broke through the outer walls. Once inside the attack soon broke down as the French and Poles encountered the narrow streets where 60,000 citizens lived. The packed houses, churches and convents were mostly solid, lofty structures of brick and stone, with heavily barred windows and doors, usual to Spain, each served as a small fortress that had to be taken by storm. They were defended by a populace imbued with a fanaticism never before encountered in Spain and Desnoettes was repulsed. His ill prepared assault on 16 June cost him over 700 men.

No sooner had Desnoettes settled down to prepare for a long siege when a relief force under Palafox of 2,500 regulars and several thousand levies threatened him. He took a risky but courageous decision and split his forces, despatching Colonel Cholopiski with the Polish 1st Vistula Regiment, a French battalion, a squadron of lancers and four guns to face Palafox. At the same time with his remaining 3,000 men he made several demonstrations against the Saragossa's defences. The plan was a complete success. While the Saragossans warded off the imaginary assaults, the Poles made a forced march and fell on Palafox at Epila during the night 23-24 June scattering his force. Verdier, a more senior commander, arrived before Saragossa with reinforcements on 2 July and took over operations. Desnoettes was very relieved since sieges were not the forté of cavalrymen, gave him his full support and helped direct a new assault by Verdier on 4 August that cost over 2,000 casualties. Severely wounded he withdrew to Tudela when Verdier abandoned the siege and from there he returned to France to recover.

The reverses in Spain did not affect his career and on 28 August 1808 he was promoted *général de division*. On his recovery he rejoined the Guard Chasseurs à Cheval as its *colonel* based in Paris. He led Napoleon's escort squadron that left Paris on 30 October to join the Grande Armée gathering behind the Ebro, rejoining the rest of the regiment at Bayonne on 13 November.

No sooner had he crossed the Ebro when he once more left his command and joined Lannes's operations, which culminated in the Battle of Tudela on 23 November 1808. He played a notable part in the struggle by leading a decisive cavalry charge as the Aragonese reeled back from the French infantry assaults. With three dragoon regiments of Wathier's division he broke through the centre of Castanos's Army and wheeling outward, attacked both wings in the flank. The enemy dissolved into the olive groves, irrigation ditches and stone fences that

BRITISH CAVALRY SURPRISED LEFEBVRE DESNOETTES AT BENEVENTE

covered the plain south of the city.

He rejoined his regiment before Madrid and on 30 November took part in the attack on the Somosierra Pass. He spent a few days in the city while the army reorganized and then joined the pursuit of the English army. He caught up Moore's rearguard at Benavente on 29 December. Napoleon anxious to force a decision when he saw the English retiring in the distance took a risk and ordered Desnoettes to cross the Esla with a few squadrons, to try and pin the enemy. The Emperor wanted revenge, for events of the preceding days had angered him when Moore's cavalry had surprised and mauled several isolated detachments. Fording the Esla, Desnoettes came across pickets of the 18th Light Dragoons who fell back before him. He followed at the gallop and in the pursuit his men soon became breathless and disordered. The trap was sprung when the 10th Hussars lying in wait behind Benavente swooped down on him. His troops swept back the two miles to the river, while Desnoettes in difficulty with a wounded horse that refused to ford it was taken prisoner.

Bundled back with the retreating army to Corunna, he was then shipped to England. Granted parole he lived in Cheltenham where his charm and good looks soon made him a celebrity amongst the local society. Napoleon genuinely missed him and felt partly responsible for his well-being having issued the fatal orders. He showed his concern by ensuring that while in captivity, the Desnoettes's family received Charles's pay in full. Equally anxious to return to France, Charles was encouraged to break his parole. In the Spring of 1812 his wife joined him at Cheltenham and with her help disguised as a Russian count he escaped to France.

RUSSIA AND GERMANY 1812-1813

He resumed command of the Guard Chasseurs à Cheval on 6 May 1812. His regiment formed Napoleon's escort throughout the Russian Campaign and was present at all the major battles, Smolensk on 17 August and Borodino on 7 September 1812. Wounded at Vinkovo on 18 October he recovered sufficiently to be allowed to join Napoleon's small party that left the army at Smorgoni on 6 December to return to Paris.

During January and February 1813 he worked feverishly to rebuild the Guard light cavalry regiments. He built up the Chasseurs à Cheval from veterans of line regiments added to the small nucleus of 120 men that survived the retreat from Moscow. In March his regiment at brigade strength was at Frankfurt ready for the new campaign. Once again he was not to lead it on campaign and on 1 April 1813 took command of the 1st Guard Cavalry Division of the Army of the Main, which numbered some 1,400 effectives made up of the Guard Lancer regiments. His Chasseurs went to Walther's 2nd Division with the rest of the Old Guard cavalry regiments. He was present at Lutzen on 2 May and then Bautzen and 22 May, but saw little action since Napoleon wanted to preserve his cavalry. After Bautzen on 23 May his Polish lancers did fine work clearing Russian cavalry off the heights behind the Neisse near Reichenbach.

The increase French forces in Germany during the Armistice resulted in his division increasing to over 4,500 lancers. On 19 August he led a strong reconnaissance into Bohemia and seized the pass at Rimberg. After the Battle of Dresden on 26-27 August he was south of the city with Napoleon's main army ready to move through the mountains into Bohemia. The chance never came, and reverses elsewhere forced Napoleon to detach him with 4,000 cavalry to clear the rear areas behind the Elbe between Dresden and Wittenberg where Allied partisans under Thielemann, Mensdorf and Platov were causing chaos and threatened communications. He was initially successful and defeated Thielemann at Merseburg on 24 September, harrying his force as it fell back towards Altenbourg where linked up with 5,000 Austrians under Klenau. As he attacked the combined force at Altenbourg on 28 September when another under Platov took him in the rear. He suffered a sharp reverse, losing over 1,000 men. It was a set-back Napoleon could ill afford, as his rear areas became untenable and hastened his decision to withdraw from Dresden to Leipzig.

In the vicinity of Leipzig during the battle for the city from 16-18 October he did not engage the enemy as he was in the rear covering Napoleon's communications. After the defeat he formed the army's vanguard as it cut its way through to the Rhine. On 22 October he had the indignity of another defeat at the hands of Platov when caught at Weimar and put to flight. At Hanau on 30 October he went a long way to restoring his tarnished record and was instrumental in the rout of Wrede's Bavarian cavalry after Drouot's guns had first caused great damage.

THE 1814 CAMPAIGN IN FRANCE

Behind the Rhine on 25 November 1813 he took charge of a Young Guard Cavalry Division forming in Holland. Trapped at Breda in December he fought his way out to Antwerp in January 1814 before rejoining Napoleon's army at Chalons on 25 January. Serving under Nansouty he fought at Brienne on 29 January and at La Rothière on 2 February, Montmirail on 11 February, Chateau Thierry on 12 February and Vauchamps on 14 February. He returned to Paris to organize drafts for the front and later rejoined Napoleon on 17 March. He served under the Emperor at Arcis sur Aube on 21-22 March. He joined Napoleon's dash to save Paris and then retired with him to Fontainebleau where he was during the Abdication Crisis. He provided Napoleon's escort during the first stages of his journey into exile, taking his leave at Raonne on 19 April 1814.

THE RESTORATION AND NAPOLEON'S RETURN

He swore allegiance to the Bourbons and in turn received the title *Chevalier de Saint Louis* on 19 July 1814. He assumed command of his old regiment renamed the Royal Chasseurs à Cheval. Resentful of their reduced status within the Bourbon Army, his troops were moved from Paris to Cambrai in January 1815 as their presence was considered a threat.

An active plotter against the Bourbons, when he heard of Napoleon's return he sprung into action and with his deputy Lallemand* urged his troops to rise against the monarchy. On 9 March with Lallemand they set out for the artillery depot at La Fère where Lallemand's younger brother Henri headed the former Guard gunners. Their plan was disrupted by the presence of General D'Aboville who had unexpectedly arrived from Paris on an inspection tour. He tried to harangue the troops but D'Aboville threatened to fire on him if he did not leave the depot. Shocked by the upset they moved on to Compiègne where they were equally unsuccessful persuading the Chasseurs de Berry to defect. With nowhere to go he ordered the chasseurs to return to their barracks and with a price on his head went into hiding. General Rigau, who headed the troops in the department of the Marne, helped him and when it became clear that the monarchy was doomed he re-emerged and confronted Marshal Victor at Chalons and rallied the 12th Line and 5th Hussars to Napoleon. On 14 April he resumed command of the Guard Chasseurs à Cheval and on 2 June 1815 he was made a Peer of France. The formation of the Army of the North saw him appointed commander of the Guard Light Cavalry Division comprising his chasseurs and the Dutch and Polish lancers, numbering some 2,077 officers and men.

THE WATERLOO CAMPAIGN

Attached to the left wing of the Army he crossed the Belgian border on 15 June and made rapid progress towards the crossroads at Quatre Bras. His advance elements drove a battalion of Nassau infantry from Frasnes but met tougher resistance at the crossroads. Under strict orders to use his horsemen sparingly he did not press home an attack and as no support was at hand failed to secure the crossroad. Present at the battle for Quatre Bras he remained in reserve all day as Napoleon's orders were still in effect. The decision had dire consequences; had he taken the position the previous day there would have been a very different sequence of events on 16 June. The situation still begged to be retrieved and had he charged earlier in the day, or when Kellermann and Jacquinot made their charges Wellington for all his skill would have been hard pressed to hold the position.

At Waterloo his cavalry was posted to the right of the Charleroi Road alongside the Guard infantry. In the afternoon he joined the second wave of cavalry charges against the Allied positions between Hougoumont and La Haye Sainte heading the Chasseurs while Colbert led the Lancers. When the battle started to turn he tried hard to get himself killed and went down wounded in its final stages. It was understandable, since he was known to be a marked man as he was one of the first to put his foot in the stirrup on Napoleon's return. François Lallemand also wounded, persuaded him that he owed it to his men to lead them from the field. Scarcely able to stay on his horse he remained with his men during the retreat to Paris. His spirits picked up when Soult made him commander of all Guard Cavalry and he became determined to fight a last battle north of Paris. He felt with the army regrouped under Davout they could fall on the over-extended Allies and at the very least be able to negotiate terms.

EXILE 1815-1821

When the Armistice came into effect he retired behind the Loire with the army. His future looked increasingly bleak and proscribed on 24 July, Davout warned him to leave the country as his life was in real danger. He shaved of his moustache and disguised as a commercial traveller made his way to the coast and caught a ship to the United States. On 16 May 1816 a military court sentenced him to death in absentia. He opted for a frontier life and settled in Alabama where like his colleague Lallemand, he dreamed of plots to free Napoleon from Saint Helena. His hopes shattered by Napoleon's death in May 1821 he made plans to return to Europe and settle in Belgium. The Bourbons feeling a major threat had passed and enjoying a brief period of popularity offered him a pardon. He sailed for Holland in the *Albion* but was drowned when the vessel was wrecked in a storm off the coast of Ireland on 22 April 1822.

ASSESSMENT OF CAREER

Desnoettes was the typical die hard Bonapartist. His whole career was dependant on the fortunes of the Emperor and the Bonaparte family. Examples of his undying loyalty and bravery were endless. As a leader he revelled in the glories of battle. In the end he became rather like a punch drunk boxer who refused to give up. In battle his dash and style was always an example to younger leaders, who tried to emulate him. He had the great advantage of leading some of the finest cavalry regiments in the French army. He was a great cavalry tactician, but certainly no great strategist or leader of combined arms. Likewise no person could ever accuse him of being a coward or lacking bravery. Napoleon was very fond of him and thought highly of him to the extent that he was included as a beneficiary in his last will and testament.

LEFOL - ETIENNE NICOLAS, BARON (1764 - 1840)

EARLY CAREER

Lefol was born in the village of Giffaumont near Vitry-le-François (Marne) on 24 October 1764. In June 1786 he enlisted as a trooper in the Colonel General Dragoons (later the 5th Dragoons in 1791). He became disillusioned with army life after failing to gain promotion and he left his regiment on indefinite leave in October 1788. He next emerged with the 3rd Battalion of the Marne Volunteers after his men elected him their *capitaine* on 4 September 1791.

He served in Belgium and on 26 November 1792 joined the Ardennes Chasseurs with the Army of the North. In May 1794 he passed to the general staff of General Fromentin's division just before the latter's replacement by the Representatives. On the formation of the Army of the Sambre-Meuse in July 1794 he joined the staff of Grenier's division. He remained with the army for three years including a time from March 1797 under General Tilly, who was the Sambre-Meuse's Chief of Staff at the time.

A fine staff officer Lefol's career started to take off when he gained promotion to *adjudant général chef de bataillon* after joining the Army of the Lower Rhine on 5 February 1799, followed by that of *adjudant général chef de brigade* on 25 July 1799. His prospects came to an abrupt halt on 18 September 1799 when carelessness on his part while reconnoitering the Austrian positions before Mannheim ended with his capture. He remained a prisoner till 20 March 1800 when he was exchanged and returned to France. On 30 June 1800 he returned to the Army of the Rhine as chief of staff with Hardy's division. He fought at Hohenlinden on 2 December 1800. Like Hardy he was a man with strong republican sympathies and made it known he did not favour the Consulate. As a result, after the conclusion of peace with Austria and a large part of the army disbanded, he found himself unattached and on 23 September 1801 placed on half pay.

THE GRANDE ARMÉE 1805-1807

The increase in size of the Army after the breakdown of peace with England resulted in Lefol's recall on 21 September 1803. His first posting as chief of staff with the 22nd Military Division based at Tours was not encouraging as it offered few opportunities. He then got his chance when he joined Ney's staff on 31 December 1803 at the Camp of Montreuil preparing for the invasion of England. He did well under Ney and when the Grande Armée officially came into being on 25 August 1805, with Ney's forces becoming its VI Corps, Lefol served as chief of staff with Mahler's division. On 9 October during the crossing of the Danube he fought with distinction at Leipheim near Gunzberg when he led one of the assaults that secured the bridges over the river. As a result of his efforts the Austrian force under d'Aspré withdrew and he secured a vital bridgehead on the southern bank close to Ulm.

He was in the fore during the maneuvering before Ulm by Mahler's division and fought at Elchingen on 14 October, the result of which largely determined the capitulation of Mack's Austrians on 20 October 1805. He was not with Napoleon's advance to Vienna nor at Austerlitz on 2 December as Ney's corps followed Archduke John's forces into the Tyrol.

In October 1806 he became Marchand's chief of staff when Mahler was replaced. He served throughout the Prussian and Polish campaigns, fought at Jena on 14 October 1806 and, was at the siege of Magdeburg from 20 October till 11 November. In Poland he struggled through the mud and snow trying to reach Eylau on 8 February 1807. He was with VI Corps during its finest hours when it issued from the Sortlack woods and smashed through the Russians before Friedland on 14 June 1807. After the Treaty of Tilsit he spent a period with the army of occupation in Silesia.

VICTOR'S I CORPS - SPAIN 1808-1811

In March 1808 he received an annual annuity of 4000 francs drawn on Westphalia. He was with Marchand's division during the long march by VI Corps from Silesia to Spain during August and September 1808. On 12 November at an army review by Napoleon at Burgos he was promoted *général de brigade* and made chief of staff with Victor's I Corps of the Army of Spain. On 22 November he became a *baron de l'Empire*. He took command of a brigade with Ruffin's division on 15 December 1808 after the death of Labruyère from wounds received during the storming of Madrid. He remained with Ruffin as part of the force that covered Madrid. He greatly enhanced his reputation at Ucles on 13 January 1809 when I Corps defeated Venegas after the Spaniard tried to overwhelm Victor's weaker force before the capital. Apart from the good fortune of facing a far inferior Spanish general, the battle could have gone horribly wrong but for a stroke of luck for all concerned including Victor, Ruffin and Lefol. Victor's forces were widely dispersed and Ruffin planned to join the rest of Victor's force at Ucles, but lost its way over the roads. Victor was confident he could defeat the Spanish with just Villatte's division but was soon in trouble when his assaults were driven back. The triumphant Venegas inexplicably lost his nerve and began to withdraw. As he was doing so, suddenly Lefol at the head of one of Ruffin's brigades still trying to find his way to Ucles appeared in his rear. Thrown into confusion Venegas lost half his army, with about 1,000 killed or wounded and a further 5,800 rank and file captured.

In March 1809 Lefol took part in Victor's operations against Cuesta's Army of Estremadura. On 29 March he was present at the Battle of Medellin on the banks of the Guadiana where his troops formed the reserve as Victor routed the Spanish. On 12 July 1809 he moved to Villatte's 3rd Division and joined in the French advance against Wellington. At Talavera on 27-28 August his brigade comprising the 94th and 95th Line took post on the Cerro de Cascajal as the French artillery bombarded the English positions across the valley. Then when the French launched their assaults against Hill's position centered on the Cerro de Medellin he remained in reserve as Villatte's and Ruffin's divisions tried unsuccessfully to oust the enemy. At the end of the day he presented a bold front covering the French withdrawal from the field and suffered heavily from the English artillery.

In January 1810, moving from Estremadura with Victor's I Corps, he joined Soult's march into Andalusia. Much of the next two years he spent in the vicinity of Cadiz covering the siege. He fought against Graham's forces in the brief campaign that led to the French defeat at Barossa on 5 March 1811. In the battle itself his involvement was minor, skirmishing with La Pena's Spaniards along the line of the Almanza Creek three miles from the main action. At first criticized for not coming to Victor's help sooner, Victor soon had to retract when it became apparent that by keeping La Pena's force occupied he had probably saved Ruffin's and Leval's divisions at Barossa. Ruffin's death at Barossa should have resulted in a division command, but Soult was never fond of him and recommended Barrois' who got the nod. Instead for his service in Spain Lefol was decorated as a Commander of the Legion of Honor on 6 August 1811.

THE RETREAT FROM SPAIN 1812-1813

His remaining time in Andalusia was spent alternating between the siege lines at Cadiz or fulfilling garrison duties at Seville and Cordoba. After news of Marmont's defeat at Salamanca reached Soult, his brigade with the rest of Villatte's division abandoned the siege lines of Cadiz and fell back eastwards by stages through Andalusia to Valencia. He took part in the renewed advance on Madrid by King Joseph in October, which led to Wellington's forces being pushed back to the Portuguese border by winter. He spent the winter at Salamanca, but in the spring with Villatte's force increasingly isolated they abandoned the town on 26 May 1813 as overwhelming forces led by Hill approached the town. As they fell back to Zamora on the Douro Lefol won acclaim for the way he held the division together. Faced by days of continuous threats from English cavalry that tried to break his march and pin his formation his dogged defence and cajoling of stragglers won the respect of all.

He fought at Vitoria on 21 June 1813 where with the rest of Villatte's division he was held in reserve on the heights behind Arinez during the early hours of the battle. The division then moved on the Puebla Heights on the extreme left of the French line. Here it drove back Morillo's Spanish division and the British 71st Line to the outskirts of Subijana de Aliva before halted by fire from Cadogan's brigade of Hill's 2nd Division. Before able to resume the assault news arrived of the rout of Leval's division at Arinez, which herald a breakthrough by Picton and the collapse of the French line, forcing Villatte to retreat after a fine performance.

In the days that followed as the French withdrew from Northern Spain Lefol received news of his promotion to *général de division* with effect from 30 May 1813. Leval's poor performance at Vitoria resulted in his removal and Lefol replaced him as commander of the Army of the South's 1st Division. Soult's arrival as overall commander of the shattered French armies had an adverse affect on Lefol. The three defeated armies were merged into one and Lefol's division was disbanded on 17 July. He had difficulties with Soult before and now without a command was ordered back to France four days later.

SAXONY AND FRANCE 1813-1814

On 17 August 1813 he joined Augereau's XVI Corps concentrating at Mayence. With no specific command to take over on his arrival, he was made Augereau's deputy on 1 September. He spent the next six weeks with Augereau in Bavaria watching Austrian movements before moving into Saxony in early October. As the Allied armies closed in around Leipzig he was placed in charge of the city's garrison on 14 October, which largely comprised a motley collection of regimental fourth battalions. As the Allies threatened the southern approaches his 4,000 men joined Poniatowski's Poles south of the city on 16 October and helped halt the Austrian advance under Klenau that tried to turn the French flank anchored on the Elster. Engaged in the struggle for the southern suburbs in the final stages of the battle on 18 October he put up a valiant defence before laid low by a bad head wound. His men somehow carried him to safety and he was fortunate to survive the rigors of the retreat to the Rhine, taking no further part in the campaign.

On 25 December 1813 he received orders to head the 2nd Division of Marmont's VI Corps but four days later they were countermanded and he instead he joined Napoleon's staff serving in all the major battles of the campaign. He eventually received a division when he joined a corps under Ney, replacing Janssens wounded at Arcis-sur-Aube on 20-21 March 1814.

THE RESTORATION AND NAPOLEON'S RETURN

After Napoleon's abdication he transferred to Oudinot's VII Corps on 9 April. As the armies disbanded he found himself placed on half pay on 17 May 1814. He received the title chevalier de Sainte Louis on 29 July 1814, which he accepted as a hollow gesture to placate unemployed disgruntled generals formerly in Napoleon's service. Recalled on 15 January 1815 he accepted an appointment in the 19th Military Division as commander of the 2nd Sub-division based at Aurillac (Central Puy de Dôme).

On the news of Napoleon's landing at Fréjus the comte d'Artois ordered Lefol to march his troops to Lyons. By the time he reached the city on 12 March he faced a turmoil of emotions. Napoleon was at the outskirts and his own troops were close to mutiny with the prospect of having to oppose the Emperor. Totally compromised when his troops broke ranks and rallied to Napoleon, Lefol at first refused to join the march on Paris and returned to his home. Like many others he watched events, then changed his mind and on 31 March 1815 accepted command of the 8th Infantry Division with II Corps of the Army of the North.

THE WATERLOO CAMPAIGN

By the beginning of June, the build up of III Corps had not proceeded to plan so his division transferred to Vandamme's command. His division was not at full strength when it crossed the Belgian frontier on 15 June as several battalions were still en route from their depots. In spite of this the division itself was still a fine one, comprising the 15th Light Infantry, the 23rd, 37th and 64th Line Regiments with eleven battalions totalling 5,272 officers and men plus eight guns. Involved in the skirmishing before Charleroi on 15 June, one of his brigade commanders, Billard was badly injured when his horse fell, which placed an additional burden upon him for the rest of the campaign.

Present at Ligny the next day in the morning he deployed on the extreme left of the French line opposite the villages of Saint Armand. The weather

THE BRITISH 71ST LINE TRYING TO STOP LEFOL AT VITORIA

was insufferably hot and still. At 2.30 P.M. a battery of the Guard artillery opened the battle with a salvo of three rounds in quick succession. As the echoes died away Vandamme launched Lefol's division against Saint Armand. With bands playing and colors spread the division advanced in three columns covered by skirmishers. The first round of the battle fired from the Prussian guns fell amongst a company of his 15th Légère and killed eight men, but undeterred by this grim salute they remorselessly moved forward. Point blank discharges of musketry and cannon failed to stop the rush of French, and Lefol at the head of his men forced their way into Saint Armand. After fifteen minutes of desperate fighting Jagow's Prussians were driven from the orchards, the houses and had to evacuate the church. Steinmetz in reserve behind the village sent forward four of his battalions to support Jagow's hard pressed brigade and the village changed hands as Lefol was pushed back.

Vandamme renewed the struggle by deploying Berthézène's division on Lefol's left and then Girard's. A long and bitter struggled followed for the two villages, and once more Saint Armand Le Hameau was wrested from the Prussians who lost nearly 2,500 men in its defence. Blucher poured in more reserves as Vandamme threatened to envelop the Prussian right and by 5.00 P.M. both villages had changed hands a further four times.

Lefol's exhausted troops with the rest of Vandamme's corps panicked and lost the villages around 6.00 P.M. when the unannounced approach of d'Erlon's corps in their rear was mistaken for that of a hostile column. As the triumphant Prussians began to debouch south of the villages the panic became so serious that Lefol turned his cannon on the French fugitives to turn them back. Then when Duhesme's Young Guard joined the fray he rallied his regiments and joined the French counter offensive. By the end of the battle with Berthézène's division they drove Pirch II out of the two villages and regained the ground they had previously won.

WAVRE 18-19 JUNE 1815

He fought at Wavre on 18 June where in the afternoon he joined Vandamme's headlong rush to seize the town. After Habert's troops were brought to a crashing halt his men were also fed into the struggle for the town with little effect. He tried to force the Dyle with his second brigade before the mill at Bierges but well placed Prussian sharpshooters

drove back the first battalion he sent forward. Gérard was angered by what appeared to be lack of vigor on Lefol's part and joined the attack by personally leading elements of Hulot's division into the fray with equal lack of success. As night fell the Prussians were still holding their positions before Bierges, Gérard for all his pains was carried away wounded while Lefol made preparations to renew the assault the next day.

When he moved forward he found the town deserted as the Prussians had withdrawn in the night. News of Waterloo arrived soon after and his troops with Habert's formed the advance echelon in the retreat to Namur. He fought a small action outside Namur on 20 June when his troops drove back overeager elements of Prussian cavalry and that night crossed into France. He remained with Vandamme as the army fell back to Paris and then retired with it to the Loire after the Armistice. As the senior commander on 2 August 1815 he assumed command of III Corps when Vandamme fled the country.

Placed on the non active list on 4 September he later retired from the Army on 7 December 1816. The overthrow of the Bourbons led to his recall on 7 February 1831 and he served with the Headquarters Staff in Paris. He retired on 1 May 1832 and settled at Vitry-le-François (Marne) where he died on 5 September 1840.

AN ASSESSMENT

Lefol was another commander who consigned to Spain had very mixed fortunes. In his earlier days he showed a marked talent as a staff officer and had he continued on that path could have ended as a corps or even an army chief of staff. Instead in Spain he opted for a line command as the way ahead, and in the end this nearly proved his undoing through no real fault of his own. His record as a brigade general was a fine one and he deserved promotion long before he did, but years of defeat in Spain was a tough place to seek promotion. As a général de division he proved a tough opponent. He was one of the heroes of Leipzig where his defence of the southern suburbs saved thousands and perhaps many more had the bridge not blown. Not afraid to take harsh measures, few men would turn their cannon to stem a rout and live to tell the tale. He did what was expected of him during the Waterloo Campaign. A fine career soldier whose characteristics could be best described as cool, reliable, dogged and tough.

**LETORT - LOUIS MICHEL, BARON
(1773 - 1815)**

EARLY CAREER

Letort the imperial aide and Guard cavalry general was born at Saint Germain en Laye (Seine et Oise) on 28 August 1773. He enlisted as a volunteer with the 1st bataillon of the Eure et Loire on 1 November 1791. Popular amongst the men, they chose him as their *lieutenant* on 1 December 1791. A veteran of the Army of the North he fought in Belgium and Holland at Jemappes on 6 November 1792 and then at Neerwinden on 18 March 1793. Taken prisoner in a skirmish on 8 June, he was promoted *capitaine* while still captive on 29 July 1793. Later exchanged on 20 September 1793 he joined General Huet's staff as an aide de camp with the 1st Division of the Army of the Moselle. He fought at Kaiserlautern on 28 November and was then out of action for a year after a bad sabre wound to his left arm received on 26 December 1793 during the relief of Landau, failed to heal.

He rejoined Huet with the Army of the Coast of Cherbourg and served for two years in the departments of Seine Inférieure and Eure. Unhappy with the turmoil in Western France he proved a difficult subordinate, which led to his suspension on 7 September 1796. To gain a posting elsewhere at any price at his request he took a demotion to *sous lieutenant* and on 16 December joined the 9th Dragoons with the Army of Italy. He missed the major actions of the campaign and as a result promotion came slowly with him only rising to *lieutenant* on 13 February 1799.

The renewal of hostilities in Northern Italy saw his left arm shattered by a musket ball on 30 March 1799 during the retreat to the Adige. Encouraged by promotion to *capitaine* on 20 April 1799, he insisted on returning to the saddle with his arm still in a sling. He conducted a heroic but fruitless defence at Lecco on 26 April with a detachment of dragoons before fighting his way back to the Ticino a week later. He was at Joubert's defeat at Novi on 15 August 1799 and later fell back with the army to the Apennines.

On leave in Paris during the events that led to the Coup d'etat of Brumaire, he joined the band of young officers that rallied to Murat and helped eject the deputies from the Assembly convened at Saint Cloud. Back with the 9th Dragoons in Northern Italy, he took part in Brune's offensive in December 1800 after the breakdown of the Convention of Alessandria. He served in Delmas's division and distinguished himself at the second battle of

Montebello on 7 January 1801. After the Peace of Lunéville a period of garrison duty followed in Italy during which he gained promotion to *chef d'escadron* on 24 August 1801. Further promotion came on 29 October 1803 when he became *major* with the 14th Dragoons.

THE CAVALRY RESERVE 1805-1807

In the Austerlitz Campaign his regiment was with Klein's 1st Dragoon Division of Murat's Cavalry Reserve. A person who had significant influence over him was his brigade commander, the legendary Lasalle. After crossing the Danube at Donauworth he fought at Wertingen on 8 October 1805. He later joined the pursuit of Archduke Ferdinand's force that slipped past the French net cast around Ulm and was in the action at Neresheim on 18 October that destroyed his rearguard. With Klein's division detached to Mortier's command he fought at Dürrenstein on 11 November. Part of the forces that covered the Grande Armée's communications, he remained near Vienna and missed the confrontation at Austerlitz.

THE EMPRESS DRAGOONS 1806-1812

Promoted *colonel* of the 14th Dragoons on 8 April 1806, his regiment soon earned a reputation as one of the finest dragoon formations in the army. It was not a surprise that after the creation of the Empress Dragoons he joined them as their *major* on 8 October 1806, the day the Prussian Campaign opened. He was not too happy with the appointment since his new formation was a week's march behind the 14th Dragoons, and not wanting to miss any action attached himself to Klein's staff till it caught up. The speed of the French advance caused that never to happen and his regiment was still lagging a few days march behind when he was wounded in Murat's charge by the Cavalry Reserve at Jena on 14 October 1806. He continued with Klein during the pursuit of the Prussians and only took up his new post after the army halted at Berlin. Eager for action, he found the Prussian and Polish campaigns very frustrating. The Guard Cavalry was not very active and the only time his men drew their sabres was on 14 June 1807 during the pursuit of the Russians after Friedland.

The following year he was with the dragoons in Spain where he distinguished himself before Burgos on 10 November 1808 with a charge that scattered Belvedere's disorganized army. His time in Spain was cut short by the threat from Austria at the start of the new year resulting in his recall. He won acclaim for the fine bearing of his troops after completing a 2,800 kilometer march from Valladolid to Vienna in sixty three days without losing a man or horse. His arrival on 22 May 1809 was a great boost for morale as the Battle of Aspern-Essling reached its climax.

At Wagram on 6 July after a series of confused

orders arising after Bessières was wounded his horsemen stood idle as the Guard heavy cavalry failed to support Macdonald's assault against the enemy centre. Bessières carried from the field, command devolved on Walther who refused to advance much to the annoyance of Letort and Guyot*, whose men suffered at the hands of the Austrian artillery. His dragoons still unblooded in a major action, began to be ridiculed by the rest of the Guard for always being missing.

Two years of garrison duty in France saw Letort build the dragoons up to one of the finest cavalry formations in the army. In the Russian Campaign they never saw action till Borodino on 8 September 1812 when again they were held in reserve. At Ghorodnia on 25 October they proved their worth by saving Napoleon after Cossacks surprised and overran his service squadron. During the retreat he distinguished himself at Viazma on 3 November helping Davout to rejoin the rest of the army.

THE GERMAN CAMPAIGN OF 1813

Recalled to France after the campaign to help reorganize the broken Guard cavalry regiments he was promoted *général de brigade* on 30 January 1813. He then moved to the large depot at Frankfurt on Main to rebuild his regiment. At the end of April when Napoleon took the field with a new army his regiment was considerably understrenght, numbering only some 500 men drawn from the depot squadrons and line regiments. His biggest problem was a shortage of suitable horses, which caused Napoleon not to risk the Guard cavalry. Present at Lutzen and Bautzen he again remained in the rear.

The situation improved in the summer, and when hostilities were renewed on 15 August new drafts had increased his regiment to over 1,200 men. The chance to come to grips with the enemy with a formation he had so patiently built up again eluded him. Attached to Latour Maubourg's corps, he replaced Berkheim at the head of a cuirassier brigade with Bourdesoulles's 1st Heavy Cavalry Division. Here he joined the great cavalry sweeps in the area west of the Elbe against marauding Cossacks and Freikorps that had dogged French communications.

He resumed command of the Empress Dragoons in October and led them at Leipzig. On the battle's first day, it was his turn to head the duty squadrons that formed Napoleon's escort. While his regiment joined the great cavalry battle before Wachau, he remained with Napoleon and the 800 horsemen drawn from the Guard Cavalry regiments. The battle reached a crisis near Napoleon when Oudinot's Young Guard threatened by Austrian cavalry was forced into square. Napoleon ordered a counterstroke and Letort charging with only the service squadrons scored a brilliant victory against the Austrian horse, including the capture of 190

officers and men of the famous Vincent Chevaulégers. Oudinot was able to resume his advance and Letort's action went a long way ensuring a French success that day.

On 18 October after the defection of the Saxons he distinguished himself in the struggle around Kohlgarten that helped plug the breach in the French line. During the retreat to the Rhine his dragoons fought at Hanau on 30 October. In the battle they supported Drouot's bombardment of the Bavarian left and broke three infantry squares before driving them over the Kinzig.

FRANCE 1814

Across the Rhine Letort fell under Mortier's command. With a mixed force of four battalions and his dragoons, he played a key role in Mortier's delaying actions, in particular at Rouvré on 24 January 1814, that kept the Army of Bohemia bogged down in the Champagne. On 1 February he rejoined the main body of the Guard at La Rothière where he helped cover Napoleon's retreat after the battle. Engaged at Montmirail on 11 February his dragoons supported Ney's successful attack against Yorck's Prussians as they issued from Fontenelle. Then wheeling south he caught Sacken's Russian infantry as they fell back from Marchais after Lefebvre launched a furious assault. The next day flushed with success Letort hit the Prussians positioned on the Caqueret Hills before Chateau Thierry. Emerging through the mist on the enemy's right, he rode over three squares causing them to flee across the Marne. Covered with praise, Napoleon the next day 13 February 1814 made him a *comte de l'Empire*, colonel in chief of the Empress Dragoons and promoted him *général de division*. He fought at Craonne on 7 March where after the battle his dragoons floundering on the icy roads were unable to exploit one of Napoleon's costliest victories of the campaign.

On 12 March Letort took command of the 3rd Guard Cavalry Division replacing La Ferrière wounded at Craonne. He was at the recapture of Rheims and joined Napoleon's great sweep east that fell on Schwarzenberg's communications and paralyzed the Army of Bohemia. At Saint Méry-sur-Seine on 19 March he destroyed a Württemberg force of 5,000 men. The next day at Arcis-sur-Aube the situation was reversed when Colbert's Guard Lancers routed, also carried away Letort's cavalry before Napoleon personally intervened to rally them. Undeterred he made up for the humiliation at Saint Dizier when on 26 March leading the Guard heavy cavalry he fell on Winzingerode's corps causing 500 casualties and taking a further 2,000 prisoner.

THE RESTORATION AND NAPOLEON'S RETURN

He was with the army during its frenzied march to save Paris and then moved to Fontainebleau till

Napoleon's abdication. He acknowledged the Bourbons, took the title *Chevalier de Saint Louis* on 19 July and then was decorated a Commander of the Legion of Honor on 23 August 1814. Under the Bourbons his position looked reasonably secure as retained command of the Empress Dragoons. On 19 November dramatic changes came when the regiment was renamed the Royal Dragoons of France and Ornano took command with Letort his deputy. Nervousness amongst the Bourbons as to large concentrations of Imperial Guard around Paris resulted in the regiment moving to Tours. The move much resented by officers and men alike, Letort did his best to keep discord to a minimum.

While sympathetic to Napoleon, on news of his return Letort kept his head and made no overt moves to rally to his cause. His stand was easy to take since at Tours he was far from Napoleon's line of march and the Bourbons never ordered him to march. He returned to Paris and on 21 April took the post as an imperial aide. Ornano, meanwhile wounded in a duel was forced to relinquish his command, which enabled Letort to resume his post as head of the Empress Dragoons for the coming campaign.

THE WATERLOO CAMPAIGN

When Napoleon crossed into Belgium Letort's dragoons took duty as his escort squadrons. It was a duty he cherished but unfortunately that day it led to the chain of events that resulted in his death. While the French troops poured through Charleroi he waited nearby as Napoleon bathed in the glory of seeing his armies once more on the march. Around 4.00 P.M.. the Emperor grew restless as messages came in from the direction of Gilly that the advance had become bogged down. He resolved to find out himself what was the cause and with Letort crossed the Sambre and hastened to Gilly.

Once there, Napoleon found Vandamme and Grouchy arguing over how the attack should proceed. Both had differing views on the orders from Napoleon, Grouchy believed Vandamme was simply being difficult while the former refused to accept orders from a mere, "general of cavalry". Napoleon furious at the delay assumed command and directed Vandamme's III Corps artillery to blow away the Prussians. Vandamme formed his infantry in three columns to attack the Abbey of Soleilment, while Exelmans's cavalry worked their way around the Prussian left. The attack began at 6.00 P.M. and Pirch overwhelmed by the intensity of the French bombardment gave orders to retreat on Lambusart and Fleurus. Napoleon exasperated at the sight of the Prussians slipping away urged Letort to send a couple of duty squadrons ahead to keep them pinned to their position. Ever loyal and ever willing to obey an Imperial order Letort chose to lead an attack. It was an unnecessary risk, not the type of attack at the start of a campaign a general needed lead from the front, more the task for an eager squadron commander to carry out. The attack initially against retiring skirmishers ended with his cavalry engaging two battalions of the 6th Regiment forming square. As Letort's men swept over the two squares to the shouts of "Long live the Emperor" he was hit by a musket ball in the chest.

Letort was brought back to Charleroi, Napoleon visited him that night and was genuinely distressed to see one of his aides dying. He lingered on for two days in great pain, dying during the night of 17 June.

HIS DEATH - AN ASSESSMENT

Letort's loss was a misfortune for the French. His presence and expertise at Waterloo could have made a significant difference to the outcome of the great cavalry charges. He was a favorite, who was the first general to fall in the campaign. The extent of Napoleon's esteem was shown in his will where Letort's young family were included as beneficiaries.

As a leader he was able, courageous, much loved and respected by his men. His exploits in France in 1814 were virtually without parallel, when the odds the Guard cavalry faced are considered. Promotion to *général de division* would have come much earlier had he not remained with the Guard. It mattered little to him, for such pride and affection for his regiment, coupled with a personal devotion to the Emperor meant that he preferred to remain where he was.

BACKGROUND AND EARLY CAREER

The son of a baker, Samuel L'Héritier was born in the village of Angles (Vienne) Western France on 6 August 1772. He enlisted on 23 September 1792 as a grenadier with the 3rd Battalion of the Indre-Loire Regiment and saw service with the Army of the Rhine. Promoted *corporal* on 1 May 1793 he soon after joined the army's headquarters staff as an orderly. In possession of some education he adapted well to staff work and on 17 May 1794 was commissioned a *sous lieutenant* followed by *lieutenant* on 3 April 1796. He caught the eye of General Bellavene and joined him as his aide with the army's topographical staff on 2 January 1797. A successful campaign followed with promotion to *capitaine* on 5 October 1797.

In April 1800 he left the Army of the Rhine when transferred to General Boudet's staff with the 5th Division of the Army of the Reserve. He fought at Marengo on 14 June 1800 and took a musket ball in the thigh as Desaix ordered Boudet's charge at the end of the battle.

After his recovery L'Héritier's career took a change in direction when attached to the 6th Dragoons where he gained his first experience handling cavalry. In November 1800 he moved to the Army of the Grissons in Southern Switzerland as aide to General Laboissiere, who headed the army's small cavalry force. He worked well under him and took part in Macdonald's epic march over the Alps by way of the Splügen Pass north of Lake Como in December 1800, which turned the Austrian flank as the French descended through the Adige valley. He was promoted *chef d'escadron* on 19 September 1801 before continuing to serve Laboissiere on his appointment as Cavalry Inspector General in October 1801. Laboissiere retired from the army after election as a Senator in August 1803. For a short period L'Héritier found himself unattached before securing a command with the 11th Cuirassiers on 14 December 1803.

SERVICE WITH THE CAVALRY RESERVE 1805-1809

He served with the Grande Armée in the campaign against Austria, his regiment operating with Saint Sulpice's 1st Brigade of d'Hautpoul's 2nd Heavy Cavalry Division of the Cavalry Reserve. He distinguished himself at Austerlitz on 2 December 1805 when Murat's charge routed the Russian heavy cavalry before the Santon mound.

Promoted *colonel* on 5 October 1806 he moved to the 10th Cuirassiers. Command continuity was retained for the Prussian Campaign as his new command was part of Saint Sulpice's brigade under d'Hautpoul. Present at Jena on 14 October 1806 he took part in the great charge led by Murat that drove through the Prussian Army's centre. Engaged in the pursuit of the broken armies he later entered Poland in November. It was not till af-

ter crossing the Vistula that he fought his first action against the Russians at Biezun on 23 December 1806. Present at Eylau on 8 February he was with the charge that saved the army, receiving a sabre blow that broke his hand. He missed Friedland on 14 June 1807 as his regiment was with Saint Sulpice before Konigsberg moving against Lestocq's Prussians. After Tilsit he shared in the distribution of awards, receiv-

L'HÉRITIER - SAMUEL FRANÇOIS, BARON (1772 - 1829)

ing an annuity of 4,000 francs a year drawn on Westphalia and was made *baron de l'Empire* on 5 April 1808.

THE DANUBE CAMPAIGN 1809

The renewal of hostilities against Austria in April 1809 saw L'Héritier's regiment still with Saint Sulpice's 2nd Cuirassier Division of the Cavalry Reserve. His regiment helped cover Davout's retirement to Ratisbon at the start of the campaign. He later saw action at Abensberg on 20 April and at Eckmühl two days later. At Aspern Essling on 22 May he was severely wounded in the shoulder as his regiment made valiant attempts to stem the Austrian advance. He had not recovered in time to be present at Wagram on 5-6 July, but managed to rejoin his regiment for the Austrian pursuit and was in action at Znaim on 11 July where he was wounded by a shell splinter to the head. Having shown fine leadership qualities in the campaign he gained promotion to *général de brigade* on 21 July and moved to the 2nd Brigade (7th and 8th Cuirassiers) of Arrighi's 3rd Cuirassier Division replacing Bordesoulle.

After the army's withdrawal from Austria a period of inactivity followed for the cuirassiers since they were unsuited for the type of campaigning in Spain. L'Héritier as a result found himself unattached and in May 1810 took a lengthy period of leave. Recalled to the Cavalry Inspectorate in March 1811 his responsibilities included the maintenance of the cavalry depots in the 1st, 15th, 21st and 22nd Military Divisions. A change in function occurred in May 1810 when he became inspector in charge of procuring remounts for the depots in the 2nd, 3rd, 4th and 5th Military Divisions.

THE RUSSIAN CAMPAIGN OF 1812

The rebuilding of the Grande Armée in the Autumn and Winter of 1811-1812 saw his recall on 23

October 1811 as commander of 2nd Brigade (2nd and 9th Cuirassiers) of the 3rd Cuirassier Division led by Doumerc. The Cuirassier regiments were increased to a size never seen before. It led to the division being split into three brigades, with his formation left with only the 7th Cuirassiers, which at the start of the Russian Campaign numbered 769 officers and men. Doumerc's division was attached to Oudinot's II Corps and covered the Grande Armée's northern flank as it was drawn into the Russian interior. L'Héritier took an active part in countering the threat by Wittgenstein's Army in operations on the Dvina. He was present in actions at Jacobovo on 28 July, Svolna on 11 August and covered the retreat of II Corps as it retired on Polotsk. Engaged in the battle before the city on 17-18 August his brigade formed part of Oudinot's reserve covering the French centre. Late in the second day he did valuable work stemming the Russian cavalry after they had routed some Bavarian cavalry and threatened to enter the city.

He remained around Polotsk for a further two months. He was present at the second battle for the city from 18-20 October as Wittgenstein again tried to cross the Dvina and cut Napoleon's communications. He played an important part keeping the Russian cavalry at bay as the French and Bavarians slowly fell back in early November to cover the Beresina crossings. At the Beresina on 28 November his cuirassiers with the remnants of Doumerc's division, which numbered barely 400 men did valiant work to keep the Grande Armée's lifeline open. After crossing to the west bank they passed through the brush and woods before reforming in the open to fall unexpectedly on the 18th Russian Division under Prince Tchervatov advancing in columns. By Russian reports they sabred over 600 men, took over 2,000 prisoners as well as several guns and so made a significant contribution towards opening the way to Vilna for the rest of the army.

THE GERMAN CAMPAIGN OF 1813

L'Héritier's command was non-existent by the end of the campaign in Russia and he remained with the remnants of the army as it struggled back through Poland and Prussia to the Elbe. On 15 March 1813 he was promoted *général de division*, but there was no formation to lead. He remained with Doumerc during the opening campaign in Saxony where the 3rd Cuirassier Division was such a paltry formation, numbering less than 800 men, that it could do little.

During the Armistice he helped reorganize the cavalry and on 1 July he received command of the 4th Heavy Cavalry Division comprising 3rd and 4th squadrons drawn from the veteran dragoon regiments in Spain. No sooner had he whipped that division into shape when he handed it over on 8 August after receiving orders to join Milhaud's command to organize provisional brigades formed from

depot squadrons into a new V Cavalry Corps. The material he had to work with proved most unsuitable for a man who was a heavy cavalry commander being a mixture of hussars, chasseurs, and dragoons. Assigned to watch the Army of Bohemia as it entered of Saxony he played an important part in delaying its march on Dresden. At Bergieshubel on 22 August a series of skilful feints and counter feints kept a jittery Schwarzenberg in the dark as to Napoleon's true dispositions and won time for Saint Cyr's XIV Corps to prepare Dresden's defences. Not present at the great battle for the city on 25-26 August his division crossed the Elbe to form a screen that covered the area between Blucher's and Bernadotte's armies advancing from the north and east.

On 5 October his dragoon brigades became the 5th Heavy Cavalry Division of V Cavalry Corps under the leadership of Pajol*. He fought at Liebertwolkwitz in the indecisive cavalry battle south of Leipzig on 14 October, where his formation had mixed fortunes. Pajol wounded in the battle, L'Héritier took charge of the corps during the Battle of Leipzig from 16-18 October and then the retreat to the Rhine.

FRANCE 1814

December 1813 saw him behind the Rhine where in several skirmishes he repelled Cossack attempts to cross the river. On 5 January further reorganization of the cavalry resulted in him joining Milhaud's Corps as head of the 6th Heavy Cavalry Division. Then as the French hold on the Rhine became untenable he fell back into Lorraine where he joined Victor's II Corps as commander of the 4th Dragoon Division. He fought at Brienne on 29 January 1814 where under Grouchy their skilful maneuvering pinned down Blucher's Prussians sufficiently for Napoleon to inflict a sharp reverse on them later in the day. Back under Milhaud's orders he fought at La Rothière on 1 February. Here Napoleon flushed by overconfidence was drawn into an unnecessary battle that nearly cost him his army. The tenacity of the Guard infantry regiments and the overall skill exercised by the cavalry in particular L'Héritier enabled the army to be extracted intact.

When Napoleon turned on Schwarzenberg's Army of Bohemia, Victor was responsible for a rather tardy pursuit that failed to take the bridges at Montereau on 16 February. Also involved in the affair, L'Héritier was severely censured when he failed to fall on Lamothe's 3rd Bavarian Division at Valjouan the next day as it fell back from Villeneuve. The Bavarians in an exposed position on an open plain were at his mercy, but they slipped away, his only excuse was he had no orders to advance against them. When Napoleon heard Montereau was not in French hands and an opportunity to destroy a Bavarian division was missed, he flew into a rage and stripped Victor of his command. Also present at the stormy meeting L'Héritier was lucky to retain his, but the incident certainly led to a cooling of mutual respect. He joined Kellermann's command for the rest of the campaign, fought at Troyes on 23 February and was present at the destruction of Winzingerode's force at Saint Dizier on 26 March 1814.

THE RESTORATION AND NAPOLEON'S RETURN

The restoration of the Bourbons and reduction in the size of the army saw him placed on half pay on 1 June 1814. As he was not an ardent Bonapartist, Louis XVIII awarded him the title *Chevalier De Sainte Louis* on 19 July 1814 in an attempt to draw him closer to the old order. A further award came on 23 August 1814 when he was decorated a Commander of the Legion of Honor. Acquiescence paid and on 1 January 1815 he received the post of Cavalry Inspector General for the 16th Military Division at Lille. He was not a party to the conspiracy to raise the northern garrisons and march on Paris on news of Napoleon's return; the likely reason was the conspirators were unsure of his true loyalties. As a result he managed to avoid compromising himself and offered his services once Napoleon had reached Paris. Napoleon recognizing him as the professional soldier he was, on 23 April gave him command of

the cavalry with the Army of the Moselle (IV Army Corps) at Metz. On 4 June when Kellermann took charge of III Cavalry Corps, he paid L'Héritier the supreme compliment by requesting that he join him as commander of the 11th Heavy Cavalry Division.

Outside the Imperial Guard cavalry his division was one of the finest in the army. The cuirassier (8th and 11th) and dragoon (2nd and 7th) regiments, plus a horse battery had some of the finest campaign records in the army, and numbering over 2,000 effectives made it the most formidable formation he ever led.

THE WATERLOO CAMPAIGN

As the Army of the North crossed into Belgium on 15 June L'Héritier's division had a quiet day as it passed through Charleroi and bivouacked between Chatelet and Gilly. Early the next day he received orders to follow Kellermann and join the army's left wing moving on Quatre Bras. His command became strung out on the road as Kellermann in his haste moved ahead with his cuirassier brigade to assess the situation at Quatre Bras. In the meantime he waited at the junction of the Roman road the main Brussels Charleroi highway while Roussel d'Urbal's division caught up ready to act as a reserve to fall on either the Prussians flank at Ligny or join Ney at Quatre Bras. In the end he was sorely needed at

THE CUIRASSIER GENERALS LED THEIR MEN FROM THE FRONT.
JACQUINOT'S LANCERS CAN BE SEEN HITTING THE BRITISH CAVALRY IN THE FLANK.

Quatre Bras. Had Kellermann his extra brigade when he made his valiant charge in the late afternoon, the French could have carried the Quatre Bras position. Instead when he rushed to the field with his dragoons to lend support it was too late, Ney with characteristic rashness had ordered the charge and already Kellermann was streaming back in disarray.

At Waterloo L'Héritier was posted to the west of the main Charleroi-Brussels road. He remained in this position, inactive for the first four hours of the battle watching as first Reille's and then d'Erlon's assaults failed to dislodge Wellington from the heights. At 3.30 P.M. as the Prussians were seen in the distance the battle took a dramatic turn. Napoleon demanded a quick decision and Ney rashly sent in Milhaud's IV Cavalry Corps followed by the Guard light cavalry to overrun the Allied positions. The charges achieved little, because the infantry and artillery support was not there.

The die cast, Napoleon sent Flahaut to with orders for Kellermann to commit the last cavalry reserve. Kellermann in turn urged L'Héritier forward at the head of the renewed assault, followed by Roussel's division and then Guyot with the Guard heavy cavalry. This fresh tempest struck Wellington's centre at 5.30 P.M.. Again the cavalry were not backed by infantry, nor closely supported by horse artillery. Had guns galloped up in the wake of the cavalry when the English gunners left their pieces unmanned and sought safety in the squares then nothing could have saved Wellington's centre from being torn to pieces by case shot and breached.

To reach the slopes L'Héritier deployed in lines and passed between Hougoumont and La Haye Sainte where his regiments closed up to avoid the flanking fire from these points. Bunched they became further disordered as they received the cannon fire and encountered the muddy slopes littered by the debris of Milhaud's assaults that stretched up the slopes. They knew behind the crests lay the Allied squares, for Milhaud's horse to their cost had earlier found that out. The breach in the Allied lines that appeared to the French from their positions was simply a tactical withdrawal behind the crest. In an hour of confused, chaotic strife his regiments charged four times. Halted by the mud and debris left by their colleagues the artillery fire and that from the squares took its toll. Unable to comprehend why no support arrived, by 7.00 P.M. L'Héritier's force was spent, and his troops thought only of self preservation. Shot in the shoulder and unable to continue, his two brigade commanders also wounded, his command was leaderless and too disorganized to support the final attack by the Old Guard. When that failed there was little left for him to do but lead the remnants of his men from the field.

The punishment he took was evidenced by the final returns gathered as the army reorganized at Laon. His brigade of dragoons untouched when it reached Waterloo lost 540 men out of 1,100 at the start of the day. His cuirassiers fared little better, after losing 250 men at Quatre Bras, another 320 were lost at Waterloo, leaving only 230 men from the 800 at the start of the campaign.

Later Life and Career

He remained with his men as they fell back to Paris and then after the Armistice to behind the Loire for disbandment. He was placed on the non active list on half pay on 20 September 1815. Saint Cyr's Army Reforms led to his recall to the General Staff on 30 December 1818 where he remained till 27 January 1819 when appointed a Cavalry Inspector General. He served in the 2nd, 3rd and 16th Military Divisions. He became a victim of the anti-Bonapartist paranoia that gripped the Army and on 1 January 1820 found himself unattached and without a command. He remained unemployed till 27 May 1827 when he took an appointment as Inspector General of Gendarmes. Ill health prevented him renewing it the following year and on 10 January 1828 he again found himself unattached and placed on half pay. He died at his home at Conflans Saint Honorine (Seine et Oise) on 23 August 1829.

An Assessment - A Solid Brave Professional

Typical of the later generation of heavy cavalry leaders L'Héritier never quite possessed the brilliance and outward dash of earlier predecessors such as d'Hautpoul and Espagne, who established the formidable reputation of the Cuirassier Divisions. Although he shared in all the great triumphs of the earlier Napoleonic campaigns his progress was slow and by 1809 was still only a colonel. Nevertheless he was a worthy and redoubtable regimental commander who was equally at home in with either dragoon or cuirassier commands. As the years rolled by, the promotions that came his way were due more to attrition within the cuirassier arm than to his brilliance on the field.

No die hard Bonapartist, he was a professional soldier. As a general of division he was reliable but not outstanding. Then on the other hand the finest hours of the Grande Armée had passed when he led a division. Opportunities were fewer as Napoleon's enemies learnt to counter his art of warfare. Also by 1813 L'Héritier found the quality of his troops was not the same. The scarcity of good horseflesh and of experienced horsemen resulted in the cavalry being used less and less as the army's main offensive arm. As a result cavalry generally was used more cautiously, which led to young leaders like him finding it difficult to emulate the great feats of his predecessors.

BACKGROUND AND EARLY CAREER

From a noble background, Marcognet was born on his father's estate at Croix Chapeau (Charente-Inferièure) south west France on 14 November 1765. Educated at the Ecole Militaire in Paris he then joined the Bourbonnais Regiment (the 13th Line) as an officer cadet on 30 March 1781. The war with England saw his regiment shipped to America where he served with Rochambeau's forces during the War of Independence from 1781 till 1783. Promotion was steady, *sous lieutenant* on 1 July 1782, *lieutenant* on 3 July 1787 and *capitaine* still with the 13th Line on 1 March 1792.

The outbreak of the war in April 1792 saw his regiment with the Army of the Rhine. Notable actions he fought were the capture of Bodenthal on 14 September 1793 where he was shot in the thigh, and Fillingen on 17 November 1793 where he received a musket ball in the arm. A genuine patriot, this did not prevent him running into problems with the Representatives and on 11 December 1793 he was suspended on account of his noble birth. Ironically, prior to handing over his command he showed his patriotism by leading his men in a sortie against the lines at Wissembourg on 25 December 1793, which ended with him hit in the shoulder by a shell splinter.

THE ARMY OF THE RHINE 1795-1801

It was eighteen months before the Directory restored his rank on 5 July 1795 when he joined the 20th Demi-Brigade Légère with the Army of the West. In 1796 he moved to the Army of the Rhine where he joined the 10th Légère on 19 February. He distinguished himself at Etlingen on 9 July 1796, and the next day was promoted *chef de bataillon*. He fought at Neresheim on 11 August and than Geisenfeld on 1 September and Biberach on 2 October. After the army reached the safety of the Rhine he was shot in the arm during the defence of Kehl on 28 November 1796.

He joined the 95th Line as a *chef de bataillon* on 15 November 1798 and then on 15 May became regimental *adjudant général chef de brigade*. He was promoted acting commanding officer of the 108th Line with Grandjean's division on 5 May 1800. He distinguished himself at Hohenlinden on 3 December 1800 when as part of Grandjean's division his regiment played a key part in halting the advance of Archduke John's centre column. At dawn Grandjean and Ney's divisions were drawn up before the village of Hohenlinden itself. Marcognet's 108th Line in line held the centre with the 46th and 57th in close columns on his flanks. As the Austrians emerged down the road through the forest a brisk artillery fire from both sides started the action. Six Austrian battalions attacked the 108th while another eight battalions of Hungarian grenadiers tried to turn the French left by filing through the woods. After offering fierce resistance, against such numbers

Marcognet gave way. The 46th joined the action and after a furious struggle among the pines drove the Austrians deep into the recesses of the forest. The rest of the Austrian assault against the 108th faltered as the 57th joined in and Grandjean's victorious division prevented the main Austrian column from opening out upon the plains of Hohenlinden. Struck down in the confusion of the hand to hand fighting Marcognet was carried away by the enemy as they retreated. Released on 2 February 1801 after the signing of the Treaty of Lunéville, his rank as *chef de brigade* was confirmed on 8 February 1801.

MARCOGNET - PIERRE LOUIS BINET DE, BARON (1765 - 1854)

THE GRANDE ARMÉE 1805-1807

The breakdown of the Treaty of Amiens in 1803 resulted in his regiment moving to the Camp of Montreuil to prepare for the invasion of England. Here on 9 August 1803 he was promoted *général de brigade*. The mobilization of the Grande Armée on 30 August 1805 for the campaign against Austria saw his brigade become the 1st of Mahler's 2nd Division of Ney's VI Corps. His troops had one of the shorter routes to cover to the Danube reaching Hochstadt on 8 October. The next day his orders to take one of the three bridges over the river at Günzberg to cut Austrian communications between Ulm and Augsburg at first appeared a simple operation soon turned into a deadly struggle. Mack at the same time hatched a plan to cross to the north bank and fall on the French communications. In the struggle that followed both sides facing each other tried to secure the opposite banks of the river.

The course of the Danube at this point was not of a regular pattern. The flow was amongst numerous islands, little branches, bordered with willows and poplars. Marcognet's advanceguard went forward resolutely, forded all streams that were an obstacle to them and in the process took General d'Aspré prisoner, who commanded on the spot 300 Tyrolian jager. Marcognet then arrived at the largest arm of the river, over which was lay the bridge at Günzberg. The Austrians retiring had destroyed most of the wooden flooring, which needed to be replaced. At the same time on the opposite bank Archduke Ferdinand began to deploy a force of 15,000 that planned to cross and fall on Napoleon's communications. The Austrians grew alarmed as they began to comprehend the dangers the French operation present to their rear should it succeed, and made a valiant effort to bar this crossing point so close to Ulm. They directed a murderous musket and cannon fire on the French, who with no shelter from the wooded islets, on the exposed gravel banks withstood the fire with great fortitude. Fording the stream was impossible, so they sprang on the piers of the bridge and tried to repair it using joists. The workmen hit one by one by sharpshooters on the opposite bank were unable to succeed. After losing several hundred men he withdrew to the safety of the woods. The sacrifice was not in vain. The determined attack kept Ferdinand pinned and enabled another assault at Reisenburg led by Lacuée's 59th Line to succeed. The Austrians were so demoralized that a force a third their number had effected a crossing, they streamed back to Ulm spreading alarm and confusion.

The action was probably the finest of Marcognet's career, and the tenacity he showed was one of the key factors deciding the fate of the Austrians at Ulm a week later. On 14 October he followed up the success with another at Elchingen. Mack tried to break through the ring tightening around Ulm by driving along the north bank of the Danube in the hope of making contact with Kutuzov's Russians. Marcognet with Mahler having recrossed the river further east fell on the Austrian flank as Dupont engaged them frontally before driving them back through the Grosser Forest towards Ulm. The action was decisive, the French ring around Ulm was complete and a few days later Napoleon received Mack to discuss surrender terms. After the Capitulation of Ulm on 20 October Marcognet took part in Ney's occupation of the Tyrol and then moved into Carinthia to cover a possible intervention by Archduke Charles moving from Italy.

PRUSSIA AND POLAND 1806-1807

He retained his command for the Prussian Campaign of 1806. When Mahler was removed for not showing the necessary vigor during the events that culminated in the Prussian defeat at Jena, he narrowly missed promotion to the division command. Vandamme*, recalled from another of his periods in disgrace, took over the division on 20 October. Marcognet took part in the Siege of Magdeburg from 20 October till its capitulation on 11 November. He led the division for a week before Gardanne took over on 5 December 1806 when Vandamme left to join the Wurttemberg contingent and show Jérôme Bonaparte the arts of generalship.

He served in Poland where he joined Ney's staff at the end of December. On the banks of the Passarge he joined Ney's action against the Prussian rearguard under Lestocq at Waltersdorf on 5 February 1807. With the rest of VI Corps he failed to arrive at Eylau

in time on 8 February to turn the Russian flank. After Eylau another reorganization of VI Corps took place and on 3 March he replaced Roguet* at the head of the 2nd Brigade of Marchand's 2nd Division. In the spring his troops performed well as Ney retired from Deppen on 5 June when Bennigsen with overwhelming force crossed the Passarge. At Friedland on 14 June when Napoleon launched Ney against the town, Marcognet under Marchand was one of the few leaders who failed to perform with merit that day. His brigade issued from the Sortlack woods in echelon on the French right and advanced along the banks of the Alle. Formed in dense columns they stormed the village of Sortlack and drove the Russians into the river. In their enthusiasm the area became constricted and with little room to manoeuvre Marchand's two brigades closed up into a massive single column of battalion columns. As they rounded the projecting loop of the Alle and moved on Friedland a hastily assembled battery of Russian guns on the opposite bank raked them with a merciless fire. Halted by the carnage, they tried to deploy but Russian cavalry caught them in the open and sent them fleeing to the rear. He rallied his men and resumed the assault after Senarmont with Victor's I Corps artillery blew away the Russian guns and cavalry and with the rest of Ney's corps streamed into Friedland. His costly failure to deploy properly was an ill omen, for it was to repeat itself eight years later at Waterloo.

SPAIN 1808-1811

After the Treaty of Tilsit Marcognet spent a period stationed in Silesia till August 1808 when the reverses in the Peninsular resulted in the recall of Marchand's division for operations in Spain. A piece of good news he received after his promotion prospects had diminished the previous year was his elevation to the Imperial nobility as *baron de l'Empire* on 26 October 1808. In Spain he took part in the occupation of Galicia and was present at the capture of Astorga on 1 January 1809. He then joined Ney's pursuit of Moore's Army till mid January when Soult with fresh troops took over, while under Marchand he was detached to hunt down La Romana's force along the Galician coast. In May Ney invaded the Asturias with a force of 16,500 men. Marcognet with the 39th Line was at the capture and sacking of the provincial capital Oviedo on 18-19 May. La Romana avoided battle and the increasingly frustrated French left a trail of destruction when they left the region in July, reflecting little credit on Marcognet or Ney.

Further misfortune came his way in the Autumn of 1809 while Ney was on leave in France. Marchand in charge of VI Corps at Salamanca learnt Del Parque was in the vicinity with a new army and decided to march against him. Del Parque with 20,000 men accepted battle on 18 October, taking up a strong defensive position along a series of hills behind the village of Tamanes facing 12,000 French. Marcognet with his brigade (39th and 76th Line) attacked the Spanish centre east of the village while Maucune's brigade with a force of cavalry in support made progress on the French right. Halted by salvoes and volleys from Losada's division three quarters up the slope, Marcognet's men tired by a full days march could be moved no further. Del Parque ordered a charge and the French recoiled and descended the hill in disorder. Losada's battalions pursued them to the foot of the slope, with Marcognet's men pouring onto the plain in a disordered mass of fugitives, only to halt when Marchand committed his reserves. The French lost around 1,300 men that day, 800 from Marcognet's brigade, opposed to 700 Spanish. The enemy made no attempt to pursue the broken French and Marchand retired to Salamanca. The reverse was unparalleled and Marcognet's 76th Line lost its eagle. No French force had received such a humiliation at the hands of Spanish regulars since Baylen.

Before heads rolled Kellermann* took charge of VI Corps from a demoralized Marchand and in a series of brilliant manoeuvres tempted Del Parque from his mountain retreat. Then in a hectic pursuit that followed the Spaniard was brought to battle at Alba de Tormes on 28 November 1809. Marcognet regained confidence as the prospect of dismissal faded and led a brutal march before the battle that instilled a new sense of vigor in his troops to show they were equal to any in VI Corps.

The following year Marcognet took part in Massena's invasion of Portugal. He was present at the siege of Ciudad Rodrigo from 30 May till 9 July and after its fall moved on with Ney's forces to that at Almeida from 15-27 August 1810. On 27 September 1810 he fought at Bussaco where his brigade suffered badly. In the battle under Marchand he made headway against a stubborn defence by Spencer's division but had to retire when French assaults further along the line failed. He shared in the hardships before the lines of Torres Vedras and his was the first force to meet Drouet d'Erlon's* reinforcements near Ephinal on 26 December 1810.

After Massena's retreat he fought at Fuentes d'Onoro on 5 May 1811 where his brigade drove the Allies from the village of Pozo Bello. The move threatened to turn Wellington's right but with the main assault on Fuentes itself failing, Massena's chance of clinching a victory evaporated. Marmont's arrival as commander of the Army of Portugal resulted in a reorganization of the army with his departure, plus division commanders Marchand, Merle, Mermet and Heudelet all ordered back to France on 11 May 1811.

ITALY 1813-1814

After a short period of leave he was ordered on 13 July 1811 to report to the Camp of Emden in Holland.

Whilst passing through Paris, completely against expectations he found himself promoted *général de division* on 31 July 1811. A possible command with the Grande Armée for the Russian Campaign beckoned but he was to remain disappointed and instead moved to Caen in Normandy as commander of the 14th Military Division.

The chance of an active command came on 30 May 1813, when with rumblings of Austria joining the Coalition, he moved to Italy to join the Corps of Observation of the Adige. On the outbreak of hostilities in August 1813 he took command of the 2nd Division of Grenier's 2nd Lieutenancy with Eugene Beauharnais's Army of Italy. He won a sharp action at San Martino on 22 September as his forces fell back to the Isonzo. He rejoined Eugene's main army and fought at Caldiero on 15 November and then drove the Austrians from San Michele on 19 November. In December he was engaged in the Po Delta where after Grenier's defeat at Rovigo the Allies managed to breach the Adige riverline. Forced back by weight of numbers, in particular Murat's defection with the Army of Naples, he fought in Eugene's last victory on the Mincio on 8 February 1814. Here in a counter blow across the river the Austrians suffered over 7,000 casualties. His last action of the campaign was defending the riverline on 8 April 1814.

THE RESTORATION AND NAPOLEON'S RETURN

After Napoleon's abdication he remained with his command till 20 June when the remnants of the Army were disbanded at Lyons. The Bourbons made him a *Chevalier de Saint Louis* 8 July and then a Grand Officer of the Legion of Honor on 27 December 1814. As a noble who had served Napoleon there was no place for him in a Bourbon army and on 1 January 1815 he was placed on half pay. With such treatment it was not surprising he readily accepted Napoleon's return and on 6 April 1815 received command of 3rd Division of Drouet d'Erlon's I Corps of the Army of the North.

THE WATERLOO CAMPAIGN

His division was part of the misunderstanding that resulted in d'Erlon's corps missing the battles of Ligny and Quatre Bras on 16 June, so he was anxious to make amends at Waterloo. At Waterloo he was posted between Quiot's and Durutte's divisions to the right of the Charleroi road, joining the assaults directed by d'Erlon on the farm buildings of La Haye Sainte around 1.30 P.M.. Inexplicably he advanced to within musket range with his whole division deployed in a dense battalion column. First raked by ball, grapeshot then tore through his ranks as he moved closer. The assault drifted off line and flowed past La Haye Sainte towards the crest of the ridge where a vicious fire fight took place with Pack's brigade. Unable to deploy, he had little chance. When

Ponsonby's Union Brigade fell on his division it broke and ran, and the 45th Line lost its eagle.

He rallied his troops by 3.30 P.M. but instead of rejoining battle with the Cavalry Reserve in support, he watched for two hours as Ney with no infantry support threw the cavalry away against the Allied squares. Around 6.00 P.M. his renewed assaults on the Allied centre pushed past La Haye Sainte, which Donzelot took, while he reached the Ohain road on the ridge's crest. For two hours he held his own and with artillery support did great damage to the Allied line. No cavalry available was the decisive factor not making a break through, since without them the enemy were not forced into square so the artillery could do its greatest damage. Finally when the cry, "La Garde recule," was heard his men broke and fled into the valley in disorder. Sabred by English and Prussian cavalry his division disintegrated. The toll exacted by the assaults and the damage his division suffered is borne out by the returns at Laon. Only 645 officers and men were with their eagles, compared to 4,081 at the start of the campaign, the highest proportion of losses suffered by any division in the campaign, and that after a single battle.

LATER CAREER AND ASSESSMENT

He remained with the army as it fell back to Paris and helped with its disbandment south of the Loire till retired on 9 September 1815. A noble who served Napoleon on his return, the Bourbons never recalled him. A change of fortune came with the July Monarchy and on 7 February 1831 the Ministry of War placed him in the Reserve Section. In his sixty sixth year he retired on 1 May 1832. He spent most of his remaining years in Paris where he died on 19 December 1852 at the age of eighty-nine, the greatest age reached by any of Napoleon's generals who served in the Waterloo Campaign.

Marcognet had a career of fluctuating fortunes. During the Revolution as a noble in the Republican armies he suffered all the prejudices inflicted on his class. With the Empire his career began to take off. Although personally not successful, he did play a key role in some of the momentous events of the Napoleonic Wars. The capitulation at Ulm and the destruction of the Russians at Friedland put him in the frame for a possible division command. However, like many his career faltered when he went to Spain. Ney had a terrible time there, and his immediate superior Marchand was past his best. That could have been to his benefit, but had the opposite effect. The speedy despatch of Del Parque by Kellermann after Tamanes saved him from an ignominious recall and he performed reliably till his recall two years later when Marmont took over. In the Italian theater he regained credibility, which helped him secure a command with the Army of the North, only to throw it away by the poor handling of his men at Waterloo. Bravery and tenacity were qualities he possessed in abundance, but they were of little use when so often he proved to be indecisive and tactically inept. Such failings relegated him to the status of a very mediocre leader.

WATERLOO: MARCOGNET'S 45TH LINE LOSES ITS EAGLE TO THE SCOTS GREYS

MAURIN - ANTOINE, BARON (1771 - 1830)

EARLY CAREER

The cavalry general Antoine Maurin was born in Montpellier (Herault) southern France on 19 December 1771. He enlisted as a trooper with the 6th Chasseurs à Cheval on 23 July 1792 and saw his first active service with the Army of the North. On 15 April 1794 he joined the staff of Goguet's division and remained with the formation when Mayer took over a week later after Goguet's murder by an enraged soldier. He became close to *adjudant général* Mireur the division's chief of staff, who had a strong influence over his early career. At Mireur's side, he fought at Fleurus on 26 June 1794. He served under him with the Army of the Sambre-Meuse. He was at the capture of Nivelles on 7 July and distinguished himself during the siege of Maestricht and as a result gained a commission as *sous lieutenant* with the 25th Chasseurs à Cheval on 25 August 1794. Present at the Battle of the Roër on 2 October 1794 his bravery again came to the attention of his superiors when heading a party of volunteers he swam across the river under heavy fire and led a successful assault against the enemy entrenchments. He received another citation for bravery during an attack on Kreutznach on 30 November 1794.

SERVICE WITH BERNADOTTE 1795-1802

On Mireur's appointment as Bernadotte's chief of staff on 15 July 1795 he followed him to Bernadotte's division. He took part in the crossing of the Rhine at Neuwied on 2 July 1796. At Limbourg on 6 July he again distinguished himself and was promoted *lieutenant* on the battlefield. He took part in Jourdan's deep drive into Germany and during the army's hurried withdrawal received a severe head wound at Burgwindsheim on 1 September 1796.

In January 1797 Bernadotte's division moved to Italy where he took part in the crossing of the Tagliamento on 16 March and the storming of Gradisca three days later. During the day he received another battlefield promotion to *capitaine*. On 4 April a lengthy association with Bernadotte began when he joined his staff as an aide de camp. In February 1798 he moved to Vienna on Bernadotte's appointment as ambassador where during his brief stay he struck up a close rapport with Gérard*, another of Bernadotte's aides. Not at ease with the life of a diplomat, he was another involved in the fracas outside the French embassy when Bernadotte's staff attacked a mob that set fire to a tricolor, which led to their expulsion.

After Bernadotte's return and later appointment as commander of the Army of the Lower Rhine, Maurin was promoted to *chef d'escadron* on 5 February 1799 and continued as his aide. When Bernadotte became Minister for War he followed him to the Ministry on 21 July 1799 with the rank of *adjudant général*. Bernadotte's dismissal two months later, resulted in Maurin's career stagnating as he became one of numerous anonymous officers wandering the corridors of the Ministry. Bernadotte getting back in favour with the Consulate and commander of the Army of the West resulted in Maurin's recall on 3 May 1800 as an aide, followed by serving as his chief of staff from May till November 1801. His relationship with Bernadotte cooled during this period; Bernadotte's strong republican sympathies and jealousy towards Napoleon caused his career to suffer. Maurin mindful of this, was among the first to look elsewhere for better career opportunities; especially as rumors circulated of Bernadotte being sent to America as governor of Louisiana. He readily accepted promotion as *chef de brigade* of the 24th Chasseurs à Cheval on 24 April 1802 serving in the 11th Military Division at Bordeaux.

ITALY 1805-1806

At Bordeaux Maurin missed the Grande Armée's march to the Danube and instead in September 1805 joined Espagne's cavalry division with Massena's Army of Italy. In the Italian Campaign he served at San Michele on 29 October, at Caldiero the next day and was at the seizure of Gradisca on 15 November. His cavalry was in the fore during the army's advance into Austria and was at Laibach when news arrived of Austerlitz and the end of hostilities.

In February 1806 he joined Massena's invasion of the Kingdom of Naples that provided Joseph Bonaparte with a throne. Under Espagne his regiment later took part in the invasion of Calabria and fought at Diavolo on 25 August 1806 before capturing Sora.

PRUSSIA AND POLAND 1807

Russian intervention in East Prussia and Poland saw his regiment hurriedly recalled to the Grande Armée. Within six weeks he traversed the length of Italy and in late January eventually caught up Lasalle's cavalry on the Narew. Here his regiment joined Bruyère's brigade of Lasalle's division and was with the screen covering the retirement by Ney's and Bernadotte's corps after an aggressive reconnaissance provoked Bennigsen to respond with a mid Winter offensive. He fought at Eylau on 8 February 1807 and took part in the great charge by the Cavalry Reserve that saved the day after the rout of Augereau's VII Corps. In the action Bruyère's brigade fell on the rear of a Russian column of 6,000 grenadiers as they followed Augereau's fleeing troops into Eylau itself. Napoleon himself was in extreme danger as his staff tried to stem the onslaught. It was only the timely intervention of

Bruyère's horsemen in the Russian rear with that of the Guard infantry joining from the front that crushed the Russian column and saved him. Bruyère overnight became the new star of the light cavalry arm, while Maurin fresh from Italy also caught the Imperial eye. He again came to notice at Braunsberg when Dupont seized the town on the Passarge on 27 February and played an important role rounding up some 2,000 Prussians prisoner in the closing action of the winter campaign.

Before Napoleon's summer offensive opened in June, Maurin was reconnoitering ahead and confirmed the Russian army was concentrated around Guttstadt. Murat moved out with the Cavalry Reserve and Maurin's men joined the action at Gottau on 9 June as Bagration's rearguard was driven back on Guttstadt. He missed the French triumph at Friedland on 14 June as he had joined Soult's and Davout's march on Konigsberg.

PORTUGAL AND CAPTIVITY 1807-1812

His efforts in Poland were rewarded by promotion to *général de brigade* on 25 June 1807. A large portion of the Grande Armée demobilized after the Treaty of Tilsit he moved to Junot's Corps of Observation of the Gironde at Bordeaux on 30 September 1807 as a brigade commander with Kellermann's light cavalry division. He led the advance guard of Junot's Army after it completed a leisurely march across Spain and invaded Portugal on 19 November 1807. Ordered to make a dash from Alcantara down the valley of the Tagus valley to Lisbon he immediately ran into trouble with his raw troops. The harsh terrain and inadequate forage took its toll on his horses. Days were lost as he spent time wandering around the countryside securing remounts, and he only reached Lisbon on 10 December 1807 ten days after Junot had secured the city with a small detachment of 1,500 infantry.

On 17 March 1808 he received the title *baron de l'Empire* and was made military governor of the Algarve. The subjugation of Portugal a relatively easy operation, he spent a peaceful period in the Algarve till the revolt that started in Madrid on 2 May 1808 spread. On 16 June the small fishing town of Olhao was the first to signal the revolt, followed two days later at the provincial capital Faro where Maurin was in bed suffering from a bout of malaria. The French garrison was overwhelmed by the populace and Maurin was taken prisoner with seventy officers and men. Handed over to the captain of an English ship hovering off the coast, he was to spend the next four years as a prisoner in England till September 1812 when he was exchanged and returned to France.

THE GERMAN CAMPAIGN OF 1813

On 1 March 1813 Maurin became a brigade commander with the 4th Light Cavalry Division forming at Mayence under Exelmans[*]. In April he moved to the Lower Elbe and won a sharp action at Celle on 19 April. Operating in that area in May he missed the major battles of Lutzen and Bautzen. During the Autumn Campaign he continued under Exelmans where he was often compromised by the reckless behavior of his commander that at times put his forces in great danger. A classic case was at the Katzbach on 26 August his cavalry crossed the river, and in the bad weather mistakenly reported Blucher's army was falling back to Jauer. Macdonald in charge of the Army of the Bober interpreted this as the enemy being in full retreat and crossed the river. Blucher at the same time decided to retrace his steps as reports confirmed Napoleon had headed for Dresden with the bulk of the army. The result was the two armies blundered into each other. Maurin's cavalry should have been well ahead looking for the main Prussian army but instead chose to engage isolated detachments on the Jauer Plateau. When the Prussians fell on him, Maurin's regiments hopelessly outnumbered were brushed aside with the rest of Exelmans's cavalry. No cavalry screen left, the Prussians turned on the rest of Macdonald's army catching it in detail crossing the river. It was an avoidable disaster that haunted Maurin for the rest of his days.

During the battles around Leipzig he fought in the cavalry struggle before Liebertwolkowitz on 14 October and then again at Wachau on 16 October 1813. He was with the army during the retreat to the Rhine where his horsemen served with the rearguard. He had a terrible campaign and crossed the Rhine with barely 600 men, a quarter of his former strength. A reorganization of the cavalry followed and his division dissolved Maurin found himself in December unattached and without a command.

FRANCE 1814

On 6 January 1814 Maurin took a post with the cavalry depots at Versailles to help the overworked Bordesoulle raise and train replacements for the frontier under. He led a brigade that formed the Paris Cavalry Reserve under Bordesoulle. He served in Champagne and fought at Champaubert on 10 February, at Vauchamps on 14 February and at Valjouan on 17 February 1814.

The situation very fluid during the campaign in France, Bordesoulle moved on to head I Cavalry Corps, and Maurin promoted *général de division* on 19 February 1814 took command of the Paris Cavalry Reserve. No sooner had he taken over, the formation became the 2nd Light Cavalry Division of II Cavalry Corps under Milhaud[*]. His last month of campaigning saw him at the defence of Troyes at the beginning of March, and then as part of

Napoleon's eastern drive to cut the Allies's communications, he fought at Saint Dizier on 26 March 1814.

THE RESTORATION AND NAPOLEON'S RETURN

After Napoleon's fall he renewed his friendship with Bernadotte and was awarded the Royal Order of the Sword of Sweden on 30 April 1814. A friend of Bernadotte and no radical Bonapartist, the Bourbons gave him with the usual *Chevalier de Saint Louis* on 27 June 1814. It did not however prevent him being placed on half pay on 1 January 1815 as more and more soldiers from the old regime took up posts within an ever decreasing army. News of Napoleon's landing brought his recall and on 19 March 1815 he acted as chief of staff with the Duc de Berry's forces gathered outside Paris to bar Napoleon's entry. The gesture was futile as it was too late, but it did however later save his career since as a point of honor he escorted de Berry to the Belgian border. Then assured the rest of the royal family were safe he took his leave and returned to Paris to offer his services to Napoleon.

THE WATERLOO CAMPAIGN

On 31 March 1815 he became commander of the 7th Cavalry Division attached to Gérard's IV Army Corps at Metz. His cavalry screened IV Corps's advance as it crossed the Belgian border at first light on the 15th June. His pickets however were not sharp enough to stop Bourmont at the head of Gérard's advance guard division defecting with his staff. His role in the campaign was then cut short at Ligny in the afternoon when supporting Gérard's assaults against the twin villages of Saint Armand a musket ball hit him in the chest. The wound serious he was evacuated to Charleroi and took no further part in the campaign.

LATER CAREER AND ASSESSMENT

Suspended on 1 September 1815, he was later reinstated on 30 December 1818. He remained unattached for a short time before on 20 January 1819 he replaced Léger-Belair as commander of the 15th Military Division at Rouen. A Bourbon backlash against former Imperial officers in the Army after the Duc de Berry's murder resulted in his removal on 30 May 1820. He never held another command under the bourbons and retired from the Army on 1 January 1825.

His military career looked set for a revival during the July Revolution when he was recalled to the Ministry of War on 31 July 1830. This was followed by him becoming on 4 August acting commander for the 1st Military Division in Paris. A strong supporter of La Fayette, he commanded the troops in Paris during the delicate period the veteran revolutionary tried to consolidate his power under Louis Philippe. On 18 September 1830 the government confirmed his post, but a week later La Fayette was ousted. Maurin soon followed as he seemed unable to curb the mob, who too frequently were trying to dictate government policy. Replaced by the more politically acceptable Pajol*, who regained his spurs as one of the foremost leaders of the July uprising, the train of events threw him into a state of utter despair. His reputation and honor questioned, let down by his close colleague Gérard*, who as Minister for War was party to the move, he shot himself in Paris on 4 October 1830.

For any man to take his life in circumstances Maurin did, showed a certain instability. It was completely out of character, since all senior commanders, after Napoleon's fall including him went through equally difficult times. In his early career Maurin built up a reputation as a fine regimental commander. Under Lasalle, Bruyère and Kellermann he learnt his trade with some of the finest light cavalry leaders in the French Army. When made général de division in 1814 time was running out for him to show his worth. In the Waterloo Campaign he was unlucky when felled by a musket ball at Ligny. As a brigade general he was reliable but certainly not blessed with brilliance. His promotion to général de division one feels was more due to being in the right place at the right time rather than anything else. For the above reasons and his known instability, often the subject of fits of despair, must place him in the category of rather mediocre leader.

191

MICHEL - CLAUDE ETIÉNNE, BARON (1772 - 1815)

BACKGROUND AND EARLY CAREER

The son of a surgeon Michel was born in the village of Pointre (Jura), Eastern France on 3 October 1772. He responded to the National Assembly's call for volunteers and on 1 October enlisted with the 6th battalion of the Jura Volunteers. A natural leader, within two weeks he rose to *sergent-major*. On 12 January 1792 he received a commission as *sous lieutenant* with the 96th Demi-Brigade.

The outbreak of war saw him with Luckner's Army of the Rhine. Further promotion was rapid, *lieutenant* on 22 August 1792, followed by that of *capitaine* on 6 October 1792. The Prussians cut short his prospects when they captured him at Rhein Durckheim on 30 March 1793. He remained a prisoner in Prussia till 21 June 1795 when he was exchanged. Resuming duty he joined the 93rd Line and served with the advance guard of the Army of the Sambre-Meuse. After a short period he gained promotion to *chef de bataillon* on 1 October 1795. In April 1796 he transferred to the 49th Line with the Army of the Rhine, the reasons for the move unclear.

Peace established with Austria and Prussia in 1797, the main weight of France's military effort turned towards England. Michel's regiment in the Summer of 1797 was with the Army of England in Brittany gathering for the planned invasion. After several false starts caused by the Royal Navy's dominance of the Channel a less ambitious plan emerged. On 6 August 1798 Michel embarked at Rochefort with Humbert's small force of 1,500 men that sailed for Donegal to raise the flag of revolt in Ireland. Contrary winds soon blew the expedition off course. It failed to rendezvous with a larger force under General Hardy that set out the same time from Brest, and instead on 22 August made a landfall at Killala in County Mayo. Once ashore Humbert sent Michel ahead with a small detachment that scattered two companies of the Prince of Wales Fencibles at Balayna two days later. He then fought at Castlebar on 27 August when Humbert's main force supported by 2,000 local rebels defeated a force twice their number under General Lake. As Cornwallis closed in with over 30,000 men and all hope of support from Hardy gone he joined Humbert's last stand at Ballynamuck on 8 September. Prepared to sell himself dearly Michel led a valiant but futile charge against the enemy's center that soon broke under

their disciplined volley fire. The revolt a failure, he took to the hills with the remnants of the Irish rebels, and hunted down was captured on 27 September. Taken to Dublin he spent his captivity with Humbert till exchanged for a second time on 4 December 1798 before returning to France.

Still with the 49th Line Michel spent the next two years with the Army of Batavia. He fought at Egmont op Zee on 2 October 1799 where a musket ball shattered his right arm. The following year as part of Augereau's march into Franconia in support of Moreau's offensive he fought at Nuremberg on 18 December 1800. Then in the precipitous retreat to the Rhine that followed he served well with Augereau's rearguard.

In 1801 the plans to invade England renewed, he moved with his regiment to Rochefort to prepare for another decent. When that was delayed indefinitely he joined Leclerc's expedition for Santo Domingo that left in December 1801. He survived two years on the island and was one of few not only to do that, but also to emerge from West Indian disaster with some credit. On the island he renewed his acquaintance with Humbert and served for a time in a force of 500 led by him that in March 1802 occupied Port-de-Paix and routed over 4,000 rebels.

The imminent breakdown of the Treaty of Amiens precipitated his return to France with the remnants of the 49th Line in May 1803. The regiment virtually wiped out by yellow fever was disbanded, and on 23 September 1803 the survivors merged with the 24th Line at Brest. On 23 December 1803 promoted *major* he moved to the 40th Line at the Camp of Boulogne for another invasion of England.

THE GRANDE ARMÉE 1805-1807

When the Grande Armée broke camp on 27 August 1805 Michel's regiment was with Suchet's 4th Division of Soult's* IV Corps. He took part in the great envelopment of Mack's Austrians at Ulm that ended with its surrender on 22 October 1805. Suchet's division then joined Lannes' V Corps in the advance on Vienna. After the city's fall he fought Bagration's rearguard at Hollabrunn across the Danube on 16 November.

At Austerlitz on 2 December 1805 on the extreme left of the French line his troops fought bravely against Bagration. As part of Suchet's first line of defence he drove back several attempts by Russian cavalry to overwhelm the French left centered on the Santon mound. His coolheaded attitude resulted in promotion to *colonel* of the 40th Line on 27 December 1805.

His return to France saw him make a very advantageous marriage to the daughter of Jean Maret, Councillor of State and brother of Napoleon's Foreign Minister. His elevated status amongst the Imperial elite soon helped his career. When the colonelcy for the 2nd Foot Grenadiers of the Imperial

Guard came up, he applied for the prestigious post and received the appointment as its *lieutenant colonel* on 1 May 1806. In the months that followed he spent much time devoted to ceremonial duties in Paris and ensuring his command attained the exacting standards demanded of it.

The campaign against Prussia unfurled with such dramatic speed that Michel's grenadiers were unable to keep up with Napoleon once he joined the army. Whilst personally present at Jena on 14 October his troops were still several miles from the battlefield and missed the opportunity to show their prowess on the field. In Poland he was present at Eylau on 8 February 1807, and though Napoleon was sorely tested, the Guard infantry forming the last reserve again were not used. For his faultless service in a very difficult campaign he was promoted *colonel major* on 16 February 1807. In the months that followed his troops spent much of their time serving as Napoleon's escort. Present at Friedland on 14 June 1807 they again did little, but went on to play a prominent part in the military pageants that so overawed Czar Alexander during the peace conference at Tilsit. He shared in the rewards distributed after two years of successful campaigning and received a donation of 30,000 francs from Napoleon. His status as one of the elite was confirmed when he marched at Dorsenne's side at the head of the Guard as it made its triumphant entry into Paris on 25 November 1807. He later took part in the great fêtes that signalled the Empire at the height of its glory.

Spain 1808-1809

In March 1808 his regiment moved to Bayonne to support Murat's moves in Spain. He was at Bayonne when Napoleon ousted the Spanish monarchy. In November 1808 he was with Napoleon during the invasion of Spain, his regiment once more forming the Emperor's escort. He was present at the capture of Burgos on 10 November and was with the army as it entered Madrid on 2 December 1808. In mid February 1809 while at Valladolid he received orders to return to France. It was the start of the long 1,750 mile march that eventually led to Vienna.

Austria and Germany 1809-1811

On 11 April 1809 at Courbevoie he was at the inauguration of the 1st Conscript Grenadiers of the Guard, while his grenadiers the same day reached Versailles. The crisis on the Danube forced Napoleon to once more abandon his Guard in order to join the army deep in Bavaria. Michel's men struggled to catch up and at times were carried across France in carts. They missed the early battles around Ratisbon and Eckmühl, but arrived at the Danube as the guns across the river heralded the Battle of Aspern Essling on 21-22 May 1809. As part of Dorsenne's Old Guard Division Michel's men did not see action, but their

mere presence as Napoleon's reserve was a vital steadying influence as the French fell back to the Isle of Lobau. At Wagram on 5-6 July the presence of his men again forming an elite uncommitted reserve had a great psychological effect as the Austrians were driven from the field.

The Russian Campaign 1812

Actively involved in the expansion within the Guard Michel received promotion to *général de brigade* on 24 June 1811. For the Russian Campaign he led the Grenadier Brigade made up of the 1st, 2nd and 3rd Grenadiers of Curial's 3rd (Old Guard) Division. He left Paris in March 1812, and spent April and May marching across Germany by stages to keep the Russians guessing. He arrived on the Niemen only two days before Napoleon crossed the river on 22 June 1812. The early days of the campaign he found was one of ultimate frustration. Days of marching resulted in raised hopes for an expected battle, only to be dashed as the Russians slipped away. Present at Smolensk on 18-19 August, the Russians again slipped from Napoleon's grasp. At Borodino on 8 September came the ultimate disappointment, when with the Russians reeling Napoleon refused to allow the Guard to commit the coup de grace. During the occupation of Moscow Michel's troops were assigned to guard the Kremlin and maintain order in the city. Their task became increasingly difficult as discipline amongst the troops broke down. On 13 October he left Moscow. The retreat to the Niemen took a terrible toll among his men. When returns were gathered at Konigsberg in December only 721 men remained compared to 3,643 at the start of the campaign.

The German Campaign of 1813

Under Roguet* Michel helped reorganize the remnants of the Guard infantry into a single division before his recall to France in February 1813. At Fontainebleau he set about under the direction of Friant* to recruit and train new regiments to form the 4th Guard Division. Napoleon as a mark of his esteem, before departing for Germany decorated him a Commander of the Legion of Honor on 6 April 1813. In May Michel set out for Saxony with the new formation but missed the major battles of the Spring Campaign. On 29 July 1813 he resumed command of the 1st and 2nd Grenadiers as head of the 2nd Brigade of Friant's 1st Old Guard Division.

He was with Napoleon throughout the Autumn Campaign, his troops forming the Emperor's escort. Present at Dresden on 26 August he formed up with the rest of the Grenadiers and Chasseurs of the Old Guard as a central reserve posted behind Marmont's VI Corps at the fourth redoubt. In mid October he retired to Leipzig and took part in the four day struggle for the city from 16-19 October 1813. On the 16th his grenadiers were position at Probthayda as part of the central reserve. Here they supported Victor's II and Lauriston's V Corps that barred the southern approaches to the city. On the 18th he joined the renewed fighting for Probthayda itself, lending support once more to both Lauriston and Victor where they repeatedly drove back the enemy assaults causing over 10,000 casualties.

As the French fell back to the Rhine he fought at Eisenach on 24 October. Three days later he was in action at Hunefeld when one of his battalions guarding the Imperial headquarters drove off a detachment of Prussian cavalry. At Hanau on 30-31 October after Drouot's* guns drove back Wrede's Bavarians, his grenadiers secured the suburb of Neuhof and the bridge over the Kinzig at Lamboy that helped keep the road to Frankfurt open. When he reached Mayence with Morand* he helped restore order in the city that swarmed with stragglers and deserters. His participation in the campaign ended on 7 November 1813 when his troops crossed the Rhine and went into cantonments at Trier on the Saar. A reorganization of the Guard took place on 16 November resulting in his transfer to the 2nd Old Guard Division gathering at Luxembourg to head a brigade of tirailleurs and Italian vélites. On 20 November 1813 promoted *général de division* he took command of the division. On 26 December he received the honorary rank of *colonel major* of the 1st Grenadiers of the Guard.

THE 1814 CAMPAIGN IN FRANCE

When Yorck with 50,000 men seized Trier, Michel abandoned Luxembourg and retired to Metz, which he reached on 14 January. He then fell back to Langres where on 20 January 1814 he joined Mortier's command. He fought at Bar-sur-Aube on 24 January with Mortier where they managed to delay the Army of Bohemia's march through Champagne for two weeks. After Napoleon's reverse at La Rothière on 1 February, Michel's division was the first to go on the offensive when it surprised the enemy at Maison Blanches on 3 February. The next day he drove Liechtenstein's division from Saint-Thibaut and by evening his offensive reconnaissance had convinced Schwarzenberg that Napoleon was mounting a major offensive against him. His communications with Langres threatened, that move on Paris ground to a halt.

At Montmirail on 11 February his series of successes were cut short when a musket ball shattered his right arm. Unable to continue he handed over to Christiani and returned to Paris to recover. On 23 March 1814 was made *comte de l'Empire*. With Paris threatened by the Allies, on 29 March Michel still unfit for duty offered to defend the capital by leading a provisional division of some 3,800 men formed from the Guard depots. The material he had was not promising and included 1,000 conscripts, who only received their muskets

THE GRENADIERS GET THE ORDER TO ATTACK

just before the battle. Outside Paris he first occupied the ground between La Villette and the northern slopes of Belleville. Later he moved forward and positioned his men in strong outposts before Pantin. On 30 March his men stubbornly resisted an assault by 9,000 men of the Imperial Russian Guard. Struck down by a shell splinter in the kidneys, Michel's campaign was over as he was carried from the field. His division reduced to battalion strength fell back to the city's gates as the enemy seized the batteries at Le Rouvroy.

THE RESTORATION AND NAPOLEON'S RETURN

Napoleon's regard for Michel was shown in the list of legacies he drew up on his abdication. He was to receive 50,000 francs, but the Bourbons never honored it. He avoided going on half pay with the disbandment of the Imperial Guard by taking the appointment on 1 July 1814 as Chief of Staff of the Royal Foot Chasseurs of France. Five days later he received the title *Chevalier de Saint Louis.*

Napoleon's return caused him a great deal of anguish when he received orders to head south to intercept him. At the head of the regiment's third and fourth battalions he correctly judged the mood of the men, dragged his feet and judiciously kept them away from the line of Napoleon's march. When met by Napoleon's emissaries he decided to rally to the Emperor and joined him at Sens on 18 March.

THE WATERLOO CAMPAIGN

The Guard's status was restored under Napoleon and Michel on 1 April became deputy commander of the Chasseurs under Morand. On 1 June 1815 the four Chasseur regiments numbering 5,019 effectives and 16 guns were formed into the 2nd Old Guard Division with him as its second in command.

Present at Ligny on 16 June, he was held in reserve till the evening when he led the assault by the chasseur regiments that broke the Prussian center as they stormed Ligny. At Waterloo he remained in reserve around Rossomme till the Old Guard were called forward to lead the final assault. At the head of the 3rd Chasseurs he drove past the English 30th and 73rd Foot as his men moved up Mont Saint Jean. As he reached the crest the massed volleys of the 1st Foot Guards under Maitland cut great swathes through his ranks. It is believed at this point he died instantly when struck by a musket ball. Stories abound that he survived the assault and found his way to the square formed by the 1st Chasseurs and that it was he, and not Cambronne that should be credited with the infamous cry *"merde."* The story has some credence, since Cambronne was not one given to modesty and no one of any standing survived the square who could or dared contradict him. In the aftermath of the battle his body was never recovered. No doubt stripped of its uniform and decorations, it was buried with thousands of others in one of the mass graves.

CAREER ASSESSMENT

A brilliant junior officer, with a reputation for bravery established in the forests of Santo Domingo, coupled with a favorable marriage all helped to secure Michel a command with the Imperial Guard. Like so many fine officers who joined this elite body, the role the Old Guard played allowed few opportunities for such officers to use their talents to the full. The role of the Old Guard was not to fight battles. It was there to stiffen morale and inspire the Army, the fighting was for others and only in a crisis was it to act as a last reserve.

To hold such a prestigious command in the Guard meant a man's bravery and loyalty were beyond repute. In his career Michel never let down the Guard or the trust Napoleon bestowed on him. To the end he was a fine leader. It was men like him that were responsible for the Guard's fine reputation. The charisma of Napoleon helped, but it was Michel and others like him, who were the real heart of the Guard and made it such a formidable cog in the Napoleonic war machine.

MILHAUD - EDOUARD JEAN BAPTISTE, COMTE
(1766 - 1833)

BACKGROUND AND EARLY CAREER

The son of a farmer, the ardent revolutionary and cavalry general Edouard Milhaud was born on 10 July 1766 in the small town of Arpajon near Aurillac (Cantal) Central France. In his early days he studied hard and hoped to become a marine engineer in the navy, but commissions not freely available for men from the land, he was turned down in 1788. Embittered he developed a strong anti-establishment streak that the dark forces of the Revolution were soon to exploit.

From the day the King left Versailles and moved to Paris after the fall of the Bastille, Milhaud's rise was meteoric as he took advantage of opportunities thrown open to all. A colonial regiment gave him a commission as a *sous lieutenant* in 1790. Then as events gathered pace he left to form the National Guard detachment at his home in Arpajon. Highly intelligent and an inspirational orator, he was a natural leader and soon gained influence in local affairs with his election as head of the Aurillac National Guard in March 1791. This led him to stand for election to the Legislative Assembly in September 1791, but up against a very good candidate in the form of Perret preaching moderation he met defeat. The following year when the political tide had turned, he stood as a deputy for Cantal and gained election to the National Convention in September 1792. Associating with radical republicans he sat with the *Montagnards* in the Convention. His military background, coupled with a determination to familiarize himself with the subject, enabled him to appear as a man of great military talent. This helped him secure on 27 September 1792 a place on the Military Committee, a policy making body that played an increasingly important role organizing the war effort. It was through this body that he met such influential personages as Carnot, Prieur, Dubois-Crancé, Gasparin and Chateauneuf-Randon who advocated a policy of total war and a need for politicians to ensure it was conducted with utmost vigor.

On 17 October 1792 he supported the *Montagnard* move in the Convention to impose the death penalty on Louis XVI without plea or mercy. The Girondins, urging moderation won the day and the motion was lost. The King's fate was referred to a committee that resolved the Convention formally try him. Throughout the trial in January 1793 Milhaud urged the King's death and when the verdict was guilty voted against all moves to spare his life. A friend of the wild radical Marat, president of the Jacobin Club, he supported moves to have all deputies expelled from the Convention who had opposed the King's death. When the Girondins retaliated by impeaching Marat on 13 April, he made a fine speech from the floor of the Convention in defence of his colleague. It went a long way to signalling the end for the Girondins and the supremacy of the Montagnards.

REPRESENTATIVE EN MISSION

On 30 April 1793 he was among the first *députés en mission* with Hentz, La Porte and Deville sent to the Army of the Ardennes for a three month period to rally the war effort. Their principle aim was to assist the war effort by helping to stamp out profiteering amongst contractors and ensure supplies reached the armies. However, with morale within the armies at such a low ebb, they soon began to interfere with command decisions in order to turn the tide. Undoubtedly as their influence increased, Milhaud with his colleagues were responsible for the removal and death of many leaders whom in their opinion failed to prosecute the war with sufficient vigor. In effect responsible for sanctioning promotions, while with the Army of the Ardennes, he had its commander Lamarche on 9 May 1793 promote him *capitaine* with the 14th Chasseurs à Cheval.

Recalled by the Convention he then moved to the Army of the Rhine on 19 July 1793. A suitable rank expected with his status, he bullied the army's commander to approving his promotion to *chef d'escadron*, this time with the 20th Chasseurs à Cheval on 22 July 1793. Like a whirlwind the Representatives passed through the army and it is believed Milhaud was involved in the downfall of General Beauharnais, Josephine Bonaparte's first husband.

In November 1793 with fellow representative Soubrany he moved to the Army of the Pyrenees Orientales. Of all the armies he encountered this one was in the worse state. Supplies and morale were nonexistent. The generals were paralyzed by fear, inaction was reckoned the best way to survive, since seventeen of their number had been removed by previous representatives. Influenced by the army's new commander Dugommier, who was a fellow "Montagnard" in the Convention, Milhaud began to realize the underlying morale problem was more the soldiers' hunger than any disloyalty on the part of the officers. He let his fellow representatives deal with the generals while he turned his attention to the army's supply and the organization of the war effort in the nearby Departments. He ranged far and wide to ensure the troops were fed, demanded and obtained forage from far afield as the Army of the Alps, a hitherto unheard of event. A general of high repute, Dugommier in turn praised Milhaud's work in reports to the Committee of Public Safety.

Properly fed, on 30 April 1794 the army went on the offensive. Within two days the Spanish were driven from the Tech back over the Pyrenees. During the crossing at Boulou a potentially ugly situation arose when Milhaud typical of so many Representatives tried to impose his will over Dugommier and urged him to seize the heights of Montesquiou. Dugommier ignored him, and rather than have the general arrested Milhaud seized the initiative, rallied the troops nearby and at their head successfully stormed the position himself. It was a refreshing change, as representatives did not lead troops from the front. Nothing was said of the incident and it certainly won Dugommier to his side and also the begrudging respect of his officers.

With Robespierre's overthrow soon after, and the power the representatives wielded broken, Milhaud disappeared. His deeds as a representative did not come to light for many years. A piece of farsightedness made him take the precaution to use the assumed name Cumin while *en mission*. He re-emerged in the Convention after an absence on 26 August 1794 but soon stepped into dangerous waters when he tried to speak in defence of his colleague, the notorious Montagnard Jacques Carrier. Carrier's trial and later execution on 16 December 1794 by the Revolutionary Tribunal made Milhaud realize the Convention was no longer a safe place for his extreme brand of politics and he quit as a deputy the following year.

He tried to rejoin the army, but few posts were available for discredited politicians and it was not till 28 January 1796 that a post as *chef de brigade* of the 5th Dragoons came up. Two years followed with the Army of Italy where he soon established a reputation as a fine regimental leader. With Augereau's division he distinguished himself at Primolano on 7 September 1796 and the next day joined Murat's wild charge at Bassano, which yielded over 3,000 prisoners and earned him the praise of Augereau. He received a bad head wound during the storming of Saint Georges on 17 September that kept him out of action for several weeks before rejoining the campaign.

BRUMAIRE COUP D'ETAT

Peace concluded with Austria in April 1797 Milhaud later moved to the Army of England in early 1798 where the next blow was expected to fall. He remained on the Channel Coast till Napoleon's return from Egypt. Still politically active and well aware the Directory was on its last legs, he found his way to Paris and joined the Brumaire conspirators. When the coup broke on 9 November 1799, Milhaud took charge of the guard at the Luxembourg Palace to prevent the Directors leaving the building. The next day as Murat's chief of staff, he helped the conspirators at Saint Cloud when the coup ran into trouble. Napoleon facing a hostile Chamber for a moment lost his nerve and was ejected from the building. It was the quick thinking of Lucien Bonaparte, who called on Murat with Milhaud to lead a detachment of Republican Guards that cleared the Chamber and saved the day.

ITALY 1800-1805

An appreciative Napoleon promoted Milhaud *général de brigade* on 5 January 1800. He served briefly in the 14th Military Division before moving to Vaucluse on 2 March as commander of the departmental troops. On 23 May he joined the Army of the Reserve. He led a brigade of dragoons during the Marengo Campaign, but detached missed the battle on 14 June 1800. After the campaign he joined Kellermann's command at the head of a dragoon brigade with the Army of Italy. In April 1801 as the situation deteriorated with Naples he joined Murat's Army of the South concentrating at Florence. The crisis soon passed and he then found himself for a period as commandant at Mantua. In 1802 he moved to the Italian Republic as cavalry commander at

A REPRESENTATIVE EN MISSION

Milan and then the following year to Liguria.

AUSTERLITZ CAMPAIGN 1805

In August 1805 he moved to the camp at Boulogne. Attached to Murat's staff he took part in the events that led to the encirclement of Mack's army at Ulm. On 20 October 1805 he took over a light brigade (16th and 22nd Chasseurs à Cheval) operating with Walther's 2nd Dragoon Division. Involved in the pursuit of Kutuzov's Russians along the Danube he captured Linz on 1 November. Kutuzov's delaying tactics of burning bridges and stripping the countryside of food soon disrupted Milhaud's rapid advances. At Enns on 4 November he narrowly missed cutting Bagration's rearguard before it crossed the river. After Kutuzov crossed to the Danube's north bank at Krems the way to Vienna lay open and with Murat he entered the city on 14 November. On the Danube's north bank he encountered Kienmayer's rearguard at Hollabrünn on 17 November as it fell back towards Brünn. In the action he seized over 180 guns from the Vienna arsenal.

His cavalry did valuable work in the days that preceded Austerlitz. They again mauled the Russian rearguard at Porlitz on 18 November and with Murat occupied Brünn the next day. As he moved towards Olmütz enemy resistance stiffened and he was nearly overwhelmed by a large force of cavalry near the town. As the Allied armies concentrated his patrols kept Napoleon well in the picture as to their movements. It was his patrols that first spotted the movement from Olmütz as the Allies shifted the bulk of their forces south to Austerlitz on 30 November in order to cut Napoleon's communications with Vienna. Present at the battle on 2 December his forces were only lightly engaged. The fine reconnoitering successes he achieved in the preceding days ensured the campaign's successful outcome.

PRUSSIA AND POLAND 1806-1807

On 1 October 1806 for the coming campaign against Prussia he was given an independent brigade made up of the 1st Hussars and the 13th Chasseurs à Cheval that operated with Lasalle. The two formed the light cavalry screen that covered the Grande Armée's march as it pushed through Saxony towards Leipzig. Both his and Lasalle's brigades much to Napoleon's annoyance failed to appear at the twin battles of Jena-Auerstadt on 14 October being drawn away by Bernadotte's I Corps. Their absence later worked in Napoleon's favour, since both formations were fresh and able to conduct a vigorous pursuit of the broken Prussian armies.

At Boitzenburg on 27 October he intercepted Hohenlohe's advance guard and by the skilful use of feints pinned the Prussians to their positions. It won time for Murat to arrive with additional cavalry, then through bluff and guile they convinced

Hohenlohe to surrender his demoralized force of 10,000 the next day. On 29 October Milhaud caught another force under Hagen near Stettin and demanded their surrender after news reached him that Lasalle barred their only other means of escape through Pasewalk. As a result of his threats another 4,000 infantry and 2,000 cavalry fell to his two regiments.

He screened Davout's I Corps march into Poland and on 22 November had his first clash with Bennigsen's Russians at Lowicz. Resuming the advance he entered Warsaw with Murat on 26 November. His next major action was at Golymin on 26 December when exploiting Davout's breach of the Narew-Ukra riverline, he caught Gallitzin's rearguard. Urged on by an impetuous Murat his cavalry fared badly when the superior Russian force turned on him. The action did delay the Russians and forced them to make a stand. It was only Gallitzin's skill as a rearguard fighter that later enabled his force to slip away when darkness fell.

Promoted *général de division* on 30 December 1806 he replaced Beaumont as head of the 3rd Dragoon Division, handing his brigade over to Lasalle who was also promoted the same day. At Hof on 6 February 1807 he fared badly when he came up against the Russian rearguard. Ordered by Murat to charge across a single bridge unsupported, his force was repulsed by hussars and cossacks before they could deploy. Only the timely intervention of d'Hautpoul's cuirassiers prevented him being overwhelmed. At Zieghof on 7 February his cavalry used more discretion and with the support of Soult's infantry drove Bagration's division from the high ground before Eylau. At Eylau the next day posted on the French left he joined the charges by the Cavalry Reserve that saved Napoleon's center after Augereau's VIII Corps had routed. Later in the afternoon his exhausted horsemen supported Davout's attacks on Kutschitten and Anklappen but made little impact.

In the campaign that culminated at Friedland he joined Murat's advance on Konigsberg. After the Russian defeat on 14 June he pursued Lestocq as the Prussian fell back to the Niemen. After Tilsit with a large standing army remaining in Germany and Poland, Milhaud found himself for a time based in Hanover. In the awards distributed after two years of successful campaigning and the establishment of the Imperial Nobility he became *comte de l'Empire* on 10 March 1808 and received an annual pension of 30,000 francs drawn on the revenues of Westphalia.

SPAIN 1808-1811

On 5 August 1808 Milhaud's division was recalled from Germany and moved to Wesel in Holland. Hardly had it arrived when on 7 September he received orders to proceed to Spain to join IV Corps under Lefebvre. When he arrived on the Ebro at the end of October Napoleon detached him to Soult's II Corps for the march on Burgos. On 9 November his dragoons with Lasalle's horsemen caught Belvedere's Army of Estremadura badly positioned at Gamonal. Soult confident of success against 13,000 Spanish ordered the charge before his infantry arrived. His flanks in the air, Belvedere had little to answer when 5,000 French horsemen swept over him. Lasalle was at his destructive best as he tore through the Spanish center. Milhaud's dragoons at the same time skirted the field, formed up on the banks of the Arlanzon and crushed the Spanish right. His men sabred the fleeing Spanish to the gates of Burgos nine miles away.

After the fall of Burgos on 10 November Milhaud moved south-west covering the advance of Lefebvre's IV Corps that formed Napoleon's right wing. Ever conscious that Moore's English army in Portugal could fall on an isolated French corps, his patrols spread west and south in search of them. By 16 November he was south-west of Valladolid and with no contact made was convinced there was no real threat. He ended his sweep and headed for Madrid, arriving on 6 December soon after Napoleon. After a brief rest he headed for Talavera as the advance element of Napoleon's planned march down the Tagus valley to Lisbon. In a brief skirmish outside Talavera on 10 December he drove the remnants of Galluzzo's demoralized army from the town.

When Moore appeared at Salamanca on 5 December with the idea to move against Valladolid and Burgos, it disrupted Napoleon's plan. Whilst the move first threatened Soult's widely spread II Corps and Napoleon's communications, it also left Moore in a very exposed position far from his base. Napoleon gathered every available resource to trap Moore except two key elements, Milhaud's dragoons and Lasalle's light cavalry. It was an unusual oversight, for these formations possessed the mobility and numbers for Napoleon to overhaul Moore's infantry and bring him to battle on the plains of Leon rather than leave Soult an impossible task at Corunna. Instead Milhaud joined Lefebvre's foray across the Tagus against Galluzzo's 7,000 dispirited levies, which achieved little. Then in January 1809 Lefebvre disobeying orders moved north of the Tagus into Old Castile, appearing at Avila where he was of no use to either Napoleon or Jourdan left to guard Madrid with a meagre force.

When Sebastiani became commander of IV Corps, it was soon followed by successes in the field. Milhaud's prospects began to improve. In March the Spanish Army of the Center threatened Madrid and on 26 March 1809 Sebastiani caught it at Ciudad Real. In the battle Milhaud inflicted over 2,000 casualties, took five cannon and three standards before torrential rain put an end to the chase. Present at Talavera on 28 July his dragoons covered the French left flank. The rocky terrain hindered cavalry movement and reduced him to the role of spectator as Leval's division, who he tried to help, dashed itself against the English line. He was more successful at Almonacid on 11 August 1809 when Sebastiani crushed Venegas's Army of La Mancha after it made a half hearted lunge towards Madrid.

In November 1809 Madrid faced a more serious threat as Areizaga the new head of the Army of La Mancha renewed the offensive. With the French troops around Madrid outnumbered, Milhaud carried out a skilful withdrawal to win time. Aggressive maneuvering halted the Spanish at Ocana on 12 November, and the delay enabled King Joseph to first concentrate his army then hurl it at the Spanish lines on 19 November. While Areizaga's attention was diverted by the French advance against his center, Milhaud's cavalry concealed by olive groves on the French left moved unseen towards his line. Bursting from cover, they overthrew the cavalry before them and wheeled onto the exposed flank of the Spanish infantry. The onslaught was catastrophic for the Army of La Mancha; grappling with hostile infantry to their front, the Spanish foot soldiers suddenly found themselves assailed by cavalry from flank and rear. Unable to form square, entire divisions were rolled up, routed and slaughtered. The charge, its timing and execution, made it arguably one of the most devastating of the Napoleonic Wars. The Spanish lost around 27,000 men including 14,000 prisoners. It was the most comprehensive victory of the Peninsula War.

In January 1810 attached to Sebastiani's corps, Milhaud joined Soult's invasion of Andalusia. After forcing the passes of the Sierra Morena they took Jaen on 23 January. They continued south and at Alcala la Real on 28 January scattered 2,000 horse under Freire from the remnants of Areizaga's army. From there they occupied Granada, then moved on to suppress resistance in the coastal valleys between Malaga and the Sierra Nevada. At Antequerra on 5 February 1810 they met their first serious resistance when they drove 5,000 insurgents from the passes before Malaga. The invasion a success, Milhaud was decorated a Grand Officer of the Legion of Honor on 3 June 1810.

A frustrating period for Milhaud followed. Whilst the French controlled the roads and towns, the insurgents took to the hills and maintained a guerilla war, which never ceased for three years. In September 1810 he joined Sebastiani in a foray into Murcia that produced little. A rare success came when Blake in an ill judged move decided to counter by crossing into Granada with the Army of Murcia. Outside Baza on 4 November 1810, Milhaud leading a mixed force caught Blake's 8,000 men strung out en route and removed any further threat the Army of Murcia posed for another year.

OUT OF FAVOUR - THE WASTED YEARS 1811-1812

As time wore on Milhaud became increasingly disillusioned and critical of Soult's style of leadership in Andalusia. A messy guerilla war in a far flung corner of the Empire, waged by a commander he considered overbearing and self seeking, who allowed little initiative to be exercised was all too much for him. The classic political general and arch intriguer, he won Sebastiani over and then encouraged him to do the dirty work of undermining Soult's authority. When Soult confronted Sebastiani and sent him back to France in May 1811 reportedly on the grounds of ill health, it was only a matter of time before Milhaud followed in July 1811 for the same reason.

Out of favour Milhaud remained unemployed for nearly a year till 20 June 1812 when he took up a "backwater" posting as commander of the 25th Military Division at Wesel on the Rhine. Hardly had he arrived when on 13 July he was summoned to join the Grande Armée in Russia. Napoleon, aware the campaign would take its toll, attached him to the pool of generals serving on the General Staff. Whilst he never led any specific formation in the campaign he was present at Borodino, the occupation of Moscow and shared in the horrors of the retreat.

GERMANY 1813

The Russian Campaign cut such a swathe through the Grande Armée's generals, that from the comparative obscurity of the General Staff on 13 February 1813 Milhaud was catapulted to a new corps command at Mayence. Here he set out to form the nucleus of a cavalry corps for the coming campaign in Germany. The command itself was so vague it never received a formal name, since soon as regiments were formed and given the most rudimentary training they were taken away to join other formations. The arrangement not satisfactory, on 5 April he moved to the *1st Division de Marche* forming at Gotha and Hanau as part of I Cavalry Corps. His past experience as a Representative stood him in good stead as he scoured the countryside for horses and fodder. His hard work and guile brought the strength of the formation to 2,200 effectives by the end of April. The numbers and quality of his men would have been greater had his efforts not been dogged by the scarcity of horses; a problem that hindered Napoleon throughout the campaign.

His efforts recognized, with the Armistice he joined Augereau's Observation Corps in Bavaria as commander of the corps cavalry on 18 June 1813. His cavalry merged with V Cavalry Corps under Pajol' with Augereau's forces to cover the movements of Wrede's Bavarians who had defected to the Allies. Then at the end of September as the crisis deepened in Saxony he moved north. On 4 October 1813 his command become 6th Dragoon Division of V Cavalry Corps. On 10 October he fought a fine action at Zeitz when Liechtenstein and Thielemann tried to intercept Augereau's force marching on Leipzig.

On 12 October he rejoined Pajol in the city. He moved south of Leipzig and fought against the Army of Bohemia at Wachau on 16 October. Pajol was badly injured by a shell that exploded under his horse and Milhaud replaced him as corps commander. In the battle he mauled Kleist's and Eugen's infantry caught in open ground before superior numbers of Allied cavalry forced him back. After the defeat at Leipzig he helped cover the retreat and on 30 October fought at Hanau.

FRANCE 1814, THE RESTORATION AND NAPOLEON'S RETURN

Across the Rhine he was attached to Victor's II Corps in north-east France and served with him throughout December and January. He fought at Sainte Croix on 24 December 1813, at Saint Dizier on 27 January, Brienne on 29 January and La Rothière on 1 February 1814. At Mormont on 17 February he wrecked Wittgenstein's advance guard strung out on the road to Paris. He joined Macdonald's eastern offensive that was halted by Wrede's Bavarians at La Fèrte-sur-Aube on 28 February. Covering the retreat down the Seine valley he was at the defence of Troyes on 4 March. He rejoined Napoleon's army and fought at Saint Dizier on 26 March when Napoleon routed Winzingerode in the last battle of the campaign.

He had little difficulty accepting service under the Bourbons and on 1 June 1814 became Cavalry Inspector General for the 14th Military Division, followed by being made a Chevalier de Saint Louis. His murky past soon emerged, and amidst an outcry that a regicide should serve the king, was promptly retired on 4 February 1815. Not surprisingly he rallied to Napoleon on his return and on 31 March was made commander of the 1st Cuirassier Division. He led a scratch force that put down an attempt by the Duc de Bourbon to raise a royalist force in the West. Recalled to Paris on 10 April after the duke had escaped to England he returned to his division, which became the 1st Cavalry Division of the Cavalry Reserve.

THE WATERLOO CAMPAIGN

In May he was appointed commander of IV Cavalry Corps. The formation comprised two divisions, Watier de Saint Alphonse's 13th and Delort's 14th made up entirely of eight cuirassier regiments. The appointment was not without criticism since he had never led cuirassiers on campaign. In his defence, he had proven in 1814 to be a capable corps commander. The real problem was that cavalry leaders at corps level were at a premium, as evidenced by the more controversial appointment of Kellermann to head the III Cavalry Corps.

At Ligny on 16 June his troops performed well, particularly Delort's division that exploited the breech in the Prussian center made by the Guard at the end of the day. Waterloo was a different matter and saw the destruction his corps through no real fault of his. Early in the battle he was positioned to the east of the Charleroi road behind d'Erlon's lines. His division commanders were both veterans of the Peninsula, and Delort in particular had learnt to treat English infantry with respect. Around 4.00 P.M. Ney ordered one of his brigades to move towards Mont Saint Jean. Delort objected claiming no brigade would leave his division except on the direct order of Milhaud. Ney bristling with impatience rode over and demanded Delort obey, then losing all judgement and self control ordered the whole corps in the Emperor's name to attack the heights. Milhaud faced with no alternative obeyed, though he too shared Delort's misgivings. He however compounded the folly by riding back to Desnoettes and urged him to support his charge.

Thus Milhaud's cuirassier corps joined the battle followed by the Guard regiments. It was a strange ludicrous sequence of events. Instead of the six squadrons that Delort had tried to keep back, Ney with 38 squadrons at his disposal hurled 5,000 of the best French cavalry against the Allied squares. The horses soon tired in the muddy conditions. When they reached the high ground they were blasted by Wellington's artillery posted in front of the infantry. The gunners then either cowered under their pieces or ran to join the infantry. Milhaud's cuirassiers at speeds varying from a quick trot to a canter, which was all they could manage, pressed on to the infantry squares. Unable to break them, Milhaud's squadrons milled about the squares while the ranks of infantrymen fired at the horses. The charge ordered by Ney achieved nothing beyond the capture of the British guns, and that was temporary. Counterattacked by cavalry under Uxbridge, the French were pushed back off the heights. The gunners ran back to their pieces, which had not even been spiked, and when Milhaud's divisions returned to the attack the slaughter was repeated. Milhaud did everything expected but led one of the finest, best trained and equipped bodies of cavalry to its destruction.

BETRAYAL AND LATER CAREER

The battle broke his spirit and his behavior after Waterloo was a disgrace to the army. Immediately after news of Napoleon's abdication he was the first to urge the army treat with the enemy. Many other general officers followed his call as the Allies came closer and closer to Paris but the fact he was the first.

It nearly saved his career and after the army was disbanded he was on 6 August 1815 appointed a cavalry inspector general. Too many old enemies emerged from the past and on 5 September he was suspended and then retired on 18 October 1815. Proscribed as a regicide by the Act Additional of 1816, there was talk of a trial. He pleaded to the King for clemency and on 22 October 1817 obtained a reprieve, which allowed him to remain in France but he never served in the Bourbon army.

He lived quietly at Aurillac till after the July Revolution of 1830. Recalled from retirement he was admitted to the Reserve Section of the General Staff on 7 February 1831. He retired due to ill health on 1 May 1832 and died at Aurillac on 8 January 1833.

CAREER ASSESSMENT

From the outbreak of war in 1792 Milhaud rapidly rose to become the consummate political soldier. The influence he wielded as a representative was considerable. He was however one of the few amongst a very discreditable band of opportunists who did achieve some good. Over the years he developed a keen sense of the mood within the country, in particular amongst politicians and the military. Accordingly the political stance he took drifted with the winds of change. From a republican he became a regicide and later accepted the Empire as a viable alternative to the monarchy. Then when the sands ran out for Napoleon, he twice became a sycophantic monarchist. For most observers it was too much to stomach, yet somehow his career survived and prospered. He achieved this due to a quick mind and supreme cunning rivalled by that of the arch intriguer, Joseph Fouché, Napoleon's Minister of Police. After the first heady idealistic days of the Revolution, soldering became to Milhaud a means to an end in his pursuit of power. Self interest came first, country and patriotism meant little, they were merely weapons used to attain his goal.

As a cavalry leader he was undoubtedly talented and brave. At his best, his exploits were to match those of great cavalry leaders such as Lasalle, Montbrun and Kellermann. The first two became legends because they possessed a style and panache that won them the adoration of their men. Milhaud's personality never allowed him to achieve that. As a leader he possessed the same cold, calculating aloofness of the younger Kellermann, though from complete opposite sides of the political spectrum. Like Kellermann, he too was the complete cavalryman, equally at home with light, medium and heavy formations. In the end, despite the defects in character mentioned he was undoubtedly a fine cavalry leader.

MORAND - CHARLES ANTOINE LOUIS, COMTE (1771-1835)

Morand was the junior member of the triumvirate of generals including Friant and Gudin that earned Davout's III Corps the nick name, "Napoleon's X Legion".

Born in Pontarlier (Doubs), Eastern France on 4 June 1771 he came from a respectable middle class family. His beginnings were unusual for a man who later had such a distinguished military career. His parents certainly had no wish for him to join the army and expected him to enter his father's law practice. At first he did all the right things, studied law at the *Ecole de Droit* in Besançon and qualified as a lawyer in July 1791. Despite prospects of a bright future with his father's firm he threw it all in. Moved by the National Convention's call for volunteers on 9 August 1792 he enlisted in the 7th Battalion of the Doubs Volunteers.

Highly intelligent, articulate and a natural leader within days his men elected him their *capitaine*. As his battalion moved up to join the Army of the Rhine on 5 September 1792 he became the unit's commander with the rank of *lieutenant colonel*. The following year he moved to the Army of the North. He distinguished himself at Hondschoote on 8 September 1793 when his battalion was the first to breach the Allied line and secure the village enabling Houchard to break the siege around Dunkirk. He was again conspicuous at Wattignies on 16 October when Jourdan defeated the Allies before Maubeuge, raised the siege and drove them back to the Meuse.

He fought at Fleurus and after the battle passed to the Army of the Sambre-Meuse in July 1794 serving in Bernadotte's division. His bravery came to the fore during the crossing of the Orthe on 18 September when his battalion led the assault on the narrow bridge at Aywaille. Under a withering fire his men crossed a raging torrent and stormed the steep slopes on the opposite bank. In furious hand to hand struggle they then drove the enemy from their positions in one of the finest regimental actions of the campaign. He was again conspicuous during the crossing of the Roer at Aldenhoven on 2 October 1794.

The *Amalgame* of 29 December 1794 saw Morand's unit merged to become part of the 112th Demi-Brigade (later the 88th Line in 1796) with himself as commander of its 1st Battalion. The new year 1795, was one of mixed fortunes for him. He took part in the crossing of the Rhine at Neuweid in September when Jourdan led another offensive to clear the right bank of the river up to Mayence. The Austrians countered by taking Mannheim and then crossing the Rhine, driving a great wedge between the French armies. Threatened from the rear, Morand formed Bernadotte's rearguard as the division recrossed the river to safety. In the tough campaign that followed he helped contain the Austrian advance on Kreuznach before Bernadotte retook the town on 1 December 1795.

In 1796 he was with the Sambre-Meuse as it plunged deep into Germany spending several days before the fortress of Koenigstein till its fall on 26 July. He then rejoined the army's advance guard and fought at Teining near the Danube on 22 August 1796 when Bernadotte's advance was brought to a halt. Cut off from the rest of Jourdan's army Morand formed the rearguard as Berna-dotte's force struggl-ed back to the Rhine.

In January 1797 he left the Rhine with Bernadotte's 10,000 men sent to reinforce the Army of Italy. Whilst some of the regiments were distributed amongst the army, Morand's remained with a new division formed under Bernadotte at Verona on 6 March 1797. Present at the crossing of the Tagliamento on 16 March and the capture of Gradisca three days later, he showed his troops were a match for any in the Army of Italy.

EGYPT (1798-1801)

In April 1798 more by luck than design, he joined the expedition to Egypt. The 88th Line stationed at Rome joined Desaix's troops gathering at Civita Vecchia for the expedition. He fought at the Battle of the Pyramids on 21 July, gaining promotion to acting *chef de brigade* after his colonel was wounded. He joined Desaix's expedition to Upper Egypt and on 7 October 1798 was at the defeat of Murad Bey at Sediman. When Desaix moved further south in pursuit of the elusive Mameluke, Morand on 23 February 1799 assumed command of the garrison left at Girga. His skills were put to the test on 6 April when he repulsed a Mameluke attack on the town. In August he led a column against Murad Bey and defeated him at El Ganaim on 9 August and then again three days later at Samhud on the twelfth.

On 7 September 1799 promoted *adjudant général chef de brigade* he was chosen as governor of Girga province. In the year that followed a treaty signed with Murad Bey enabled him to maintain a tolerable level of law and order in the province. It also went a long way towards his promotion to *général de brigade* on 6 September 1800. By May 1801 with the French venture in Egypt on its last legs, he returned to Cairo to unite with Verdier's division to form the Cairo garrison that faced the Anglo-Turkish advance up the Nile. A bloody struggle for the city was avoided when on 27 June 1801 with Belliard he signed the city's capitulation as representative for Upper Egypt. Generous terms granted, he marched his troops to Rosetta and on 9 August embarked with the last of his men for the voyage to France.

THE AUSTRIAN CAMPAIGN OF 1805

After a period of leave on 13 March 1802 Morand was posted to Lorient as troop commander for the department of Morbihan. He remained on the Brittany coast till 30 August 1803 when he moved to Soult's camp at Saint Omer. He headed a brigade formed from the 10th Légère with Saint Hilaire's division, which on the formation of the Grande Armée became the 1st Brigade, 1st Division of Soult's IV Corps.

In the Austerlitz Campaign he reached the Danube at Donauworth on 8 October and under Saint Hilaire and Soult brushed aside the Austrian attempts to hold the river. He then joined the great sweep by IV Corps that passed through Augsburg, Landsberg and Memmingen to cut the southern escape routes from Ulm. On 16 October he arrived before the city and a week later was at Mack's army capitulation.

At Austerlitz on 2 December 1805 his 10th Légère led the advance by Saint Hilaire's and Vandamme's divisions as climbed the Pratzen Heights to fall on the Allied center. Concealed by the early morning mist he emerged without warning into perfect sunlight astride the Allied positions. Fire from two battalions of Kollowrath's division was deflected by his skirmishers and he deployed on the summit. Then as Kamensky's brigade, forming the rear of Langeron's column heading for the lower reaches of the Goldbach saw the French pouring over the Pratzen, they at once turned about. Two regiments, the Fanagoria Grenadiers and the Ryazan Musketeers, fell on the 10th Légère. Their numbers too great, the Russians flooded around Morand's small command, which grimly held on before Saint Hilaire in person brought up a battalion of the 14th Line at the run and formed up on his right. Thiébault followed with the 36th Line and secured Morand's right by sweeping two Russian battalions from Pratzen.

The battle intensified as Kollowrath and Miloradovich deployed their columns and became increasingly desperate with Austrians on Saint Hilaire's left and Russians to his right. The arrival of Vare's brigade of Vandamme's division with a battery of 12 pounders personally led by Soult at the moment Kollowrath charged broke the Austrian attack. The crisis reached, 5,000 French against 8,000 Allies, Kollowrath followed up with another attack supported by a brigade from Prschibitschewski's column. As that attack came Vandamme deployed in line and poured a devastating fire into Miloradovich's division that had moved up to support Kollowrat's right. For an hour the struggle raged till the Allied center broke leading to the French triumph. Soult's two divisions, with a major contribution by Morand, were responsible for one of the most decisive breakthroughs achieved by a French army in battle.

On 24 December 1805 Morand was promoted *général de division*. Command of the 1st Division of Davout's III Corps followed on 14 February 1806 when he replaced Cafarelli, who became Minister of War for the Kingdom of Italy. Davout, who knew Morand from his days in Egypt, was pleased with the choice. It was a command not without its ups and downs that Morand was to hold uninterrupted till 1813.

AUERSTADT 1806

The 1806 Campaign in Prussia saw the finest exploits of Morand's career during Davout's triumph at Auerstadt on 14 October. His division was the last to reach the field around 10.00 P.M. as the Duke of Brunswick ordered Wartensleben's division forward supported by Schmettau against Davout's hard pressed men. Their aim was to turn the French left flank and gain control of the main road to Kösen that would cut Davout's line of retreat. Schmettau's attack on Hassenhausen was met by fierce resistance from Gudin, but as Wartensleben began to envelop their left the French began to give way. At this decisive moment Morand arrived on the battlefield and Davout ordered his fresh troops into action at the double. Deployed left of Hassenhausen they first steadied Gudin's wavering line and then absorbed the shock of Wartensleben's attack. They drove back a brigade from Orange's division that came to Wartensleben's relief, and spreading out gained the ground around Hassenhausen forcing, Schmettau to abandon his attempt to surround the village.

Another crisis soon arose when a huge mass of cavalry formed up behind Wartensleben to fall on Morand. The Prussians hoped their 10,000 cavalry operating on the level ground that sloped away from Hassenhausen to the Saale would crush Morand's men under their feet or drive them headlong into the river. Once they succeeded overthrowing Morand on the French left, Gudin surrounded in Hassenhausen would be taken while Friant' on the right would be forced to beat a hurried retreat. Morand's infantry calmly formed square. As torrents of cavalry flowed past they were struck by a deadly fire that brought down hundreds of men and horses with each charge. So deadly was the fire, ramparts of dead formed around the edges, preventing a single Prussian closing to touch bayonets with the squares. The cavalry fell back in confusion behind Wartensleben's infantry. The Prussians spent, Davout seized the offensive. Morand formed his squares into columns and fell on Wartensleben, while Friant on the right moved to the attack. Pivoting on Gudin who had grimly held onto Hassenhausen, first Wartensleben gave way, then Orange before Friant. As the two divisions advanced they caught the Prussians in an interlocking crossfire that swept their entire line. Schmettau's division, caught on its own in the center as it faced Gudin, disintegrated.

By 1.00 P.M. the whole Prussian front broke up as the converging French attacks threatened their retreat. Kalkreuth, who took over from the mortally wounded Brunswick brought forward the reserve divisions of von Arnim and Kuhnheim. Supported by whatever cavalry Blucher could rally they tried to make a stand before Gernstadt. A bombardment by Morand's cannon positioned on the Sonnenberg supported by infantry soon flushed them from the position. For Morand and Gudin who had borne the brunt the battle was effectively over, as Friant continued the pursuit and by dusk drove the enemy from Eckartsberg.

The day was the most glorious in the annals of any French corps during the Napoleonic Wars. Davout with 26,000 men had engaged and beaten a well trained enemy of more than 63,000, suffered some 7,500 casualties but inflicted over 15,000 and taken 115 cannon. Morand, considered by some contemporaries like the jealous Thiébault as a leader of moderate ability, silenced his critics. His role had been decisive when with great coolness he had deployed under heavy artillery fire while Gudin was giving way. Then after repelling massed cavalry charges, he defeated piecemeal three infantry divisions one after the other.

After a brief rest, with III Corps forming the Grande Armée's right wing Morand joined the pursuit of the Prussians. He passed through Leipzig on 18 October and crossed the Elbe at Wittenberg on 21 October. Four days later he took part in the great parade led by III Corps through the streets of Berlin.

EAST PRUSSIA AND POLAND 1806-1807

The campaign continued into East Prussia and Poland. On 1 November he seized Kustrin and crossed the Oder. His division made steady progress towards Warsaw and entered the city on 1 December 1806. Two days later he crossed the Vistula. He supported Friant's crossing of the Bug on 11 December. His turn came when he spearheaded the attack across the Ukra on 23 December and stormed into Czarnovo, which he took after a fierce struggle. The next day he marched on Golymin and joined the battle in the late afternoon, hurling his troops against the Russians without artillery support, having abandoned his cannon due to the muddy roads. He received his first check as a general when his assaults failed to dislodge the enemy.

At Eylau on 8 February 1807, while the main army suffered a major reverse his troops fared better. He joined the battle around mid-day after Friant had forced the Russians from Serpallen. Linking up with the French right formed by Saint Hilaire's division of Soult's IV Corps he acted as a pivot as the remainder of III Corps enveloped the Russian left. He resisted repeated attempts by the enemy to retake Serpallen, at one stage capturing 30 cannon. Wounded, he remained on the field and played a vital role supporting role for Saint Hilaire. At the same time he secured the left when Davout's attacks broke down after Lestocq's Prussians intervened late

A FRENCH GENERAL LEADS HIS TROOPS AT AUERSTADT

in the late afternoon.

When the campaign resumed in summer there were high expectations for III Corps. Morand joined the march on Konigsberg, which Napoleon hoped would draw the main Russian army to battle. As events turned out he played only a secondary role in the events that led to the victory at Friedland on 14 June. The Russians under Bennigsen first wrong footed Napoleon, crossed the Alle and instead fell on Lannes at Friedland. Summoned to the battle, Morand completed the twenty five mile march after it had ended and was left to hound the Russians to the Niemen. After the Treaty of Tilsit his division formed the French garrison at Warsaw. Much of his time was spent organizing and training the new Polish army along French lines. There was no better formation to model itself on than III Corps. On 24 June 1808 Morand took the title *comte de l'Empire*.

GERMANY AND AUSTRIA 1808-1809

In the Summer of 1808 Morand moved with III Corps to Franconia. As relations deteriorated with Austria, from Neumarkt he watched with concern the concentration of Austrian troops on the Bohemian border. When Archduke Charles on 10 April 1809 launched a two pronged offensive from Bohemia and across the Inn into Bavaria Morand joined Davout's risky concentration on Ratisbon ordered by Berthier. Faced by converging forces north and south, Davout soon realized the position was untenable and ordered a withdrawal to Ingolstadt. On 19 April Morand set out along the Danube's southern bank to secure the Saale defile, a key point on the route. Once there, the divisions of Gudin and Saint Hilaire forming on his left drove the Austrians from the woods north of Hausen and Schneidhart that tried to bar his route.

The next day attached to Lannes's command, Morand fought at Abensburg where the Austrian Army split in two, with its left wing was driven in confusion to the Isar. In the struggle for Landshut on 21 April he took Ergolding early in the day and then pushed the enemy back to nearby Seligenthal on the left bank of the river. Storming the suburb, he was checked at the water's edge when he found the bridge into Landshut in flames. The attack lost momentum and Napoleon's aide Mouton' became the hero of the hour when Morand gave him a battalion of the 17th Line to lead the assault. The 17th stormed its way across the burning bridge, hacked its way through the river gate and burst into the town. Once in the town Mouton encountered stiffer resistance and Morand with the 13th Légère quickly passed over to support the embattled 17th Line. Unable to dislodge the Klebeck Regiment positioned in the castle and along the river bluff, Morand was nearly cut off as the Kerpen Regiment tried to cut behind him. It needed all his division to stabilize the situation before the Austrians finally retired. There was no respite and at dawn on 22 April Morand started a twenty three mile march to Eckmühl to help Davout who faced the main Austrian army. By 2.30 P.M. as he neared the field he drove Rosenberg's division, that formed the Austrian left from Stangelmuhle. Exhaustion then took hold, and after four days continuous marching and fighting he was unable to exploit the success as the rest of the Austrian line collapsed.

He rejoined Davout at Ratisbon and with III Corps harried Archduke Charles's army as it fell back into Bohemia. That threat removed, he recrossed the Danube on 27 April and joined the French march on Vienna. Detached at Melk he spent several days guarding against a possible Austrian foray over the river till replaced by Württemberg troops. He rejoined Davout outside Vienna on 20 May as III Corps prepared to cross the river. The pontoon across the river was severed and he missed Napoleon's defeat at Aspern-Essling on 21-22 May 1809. As the French recovered from the reverse Morand moved downriver opposite Pressburg. Here with Davout he was part of a major diversion that drew the Austrian attention away from Vienna. For several days they bombarded the city and gathered a great flotilla of boats to indicate another crossing. Archduke Charles not fooled, at the end of June III Corps moved back to Vienna.

On the night of 3-4 July Morand's division crossed the Danube to the Isle of Lobau. His men led the crossing to the north bank around 6.00 A.M. on 5 July followed by the rest of III Corps. Outflanking the strong entrenchments centred on Gross Enzersdorf he deployed beyond Wittau to face the expected Austrian counter blow. It never came and by evening his skirmishers had steadily pushed back the enemy to the edge of the Russbach Heights opposite Markgrafneusiedl. The next day while Gudin and Friant cleared Markgrafneusiedl and moved up the heights, Morand supported by Montbrun's and Arrighi's cavalry began to envelop the Austrian left. By 1.00 P.M. III Corps were masters of the heights as Rosenberg's corps gave way. Charles faced by Davout about to roll up his line, and his last reserves used to stem Macdonald's advance against his center, realized the battle was lost and ordered a retreat. Morand harried Rosenberg's force as far as Brünn when news of the armistice on 13 July reached him. He remained in the area watching Austrian movements till 14 October when the signing of the Treaty of Schonbrünn signalled the French withdrawal.

DIFFERENCES WITH DAVOUT 1810-1811

In July 1810 Morand moved to Hamburg replacing Molitor as governor of the Hanseatic towns. This period he spent in Germany was not a happy one. Although considered by many the most capable of Davout's commanders, the strong bond with his chief began to sour. Envious of Davout's title and rank, he was resentful that he had never been offered a corps command, which could have set him on the road to win a marshal's baton. Davout out of self interest did not wish to break up a winning team. Tired of Davout's brusque overbearing manner, they clashed frequently. Matters came to a head in July 1811 when in a fit of despair he wrote to Savary, the Minister for War declaring he would resign if he had to serve under the Prince of Eckmühl. The request ignored, their differences were put aside as the war clouds had gathered against Russia in 1812.

THE RUSSIAN CAMPAIGN OF 1812.

For the Russian Campaign III Corps became I Corps of the Grande Armée with Morand's its 1st Division. The start of the campaign on 24 June saw his troops at Kovno on the Niemen. The 13th Légère was the first to cross the river and establish a bridgehead for Eblé's engineers to build pontoons. In the weeks that followed the long march yielded little as the Russians continued to avoid battle. It was not till he reached Smolensk that his division saw its first action on 17 August when they helped clear the suburbs on the southern bank of the Dnieper.

As he neared Borodino on 5 September he became involved in the struggle for the redoubts around Schevardino. At the Battle of Borodino on 7 September Napoleon detached him to Eugene Beauharnais's IV Corps command. In the morning supporting the attack on Borodino itself, he drove back the Russian light infantry posted along the banks of the Semyonovskaya. His 30th Line under Bonamy gained a foothold in the Great Redoubt before driven back by Raevsky. A titanic struggle developed for the position. It was only in the afternoon following a cavalry charge by Caulaincourt, which ended in the general's death, did the infantry of Broussier, Gérard* and his own, secure the position. The cost was fearful and Morand lost over 3,000 men. In the struggle he was badly wounded by a shell splinter that broke his jaw, badly disfiguring his face.

Morand's recovery was slow, and though unfit he led his men throughout the retreat from Moscow. One of the last formations to leave, his division had the unenviable task of escorting the treasures taken from the city. That in itself slowed down his troops considerably, which led to problems for III Corps as the retreat progressed. At Viazma on 3 November Miloradovich fell on him, cut the road and drove him back on the rest of III Corps till Compans with the 5th Division joined the action and pushed the Russian back. Before Krasnoe on 17 November he again fought its way through Miloradovich's forces to rejoin the rest of the army.

It was the cold rather than the enemy that destroyed his division. By the time he reached the Beresina fewer than 500 of his men remained. When he crossed the Niemen on 10 December he was down to 300 men out of the 10,000 that marched into Russia six months earlier.

ITALY AND GERMANY 1813-1814

Exhausted when he reached Poland, Morand returned to France to recover from his wounds. The destruction of Davout's command after so many years of success affected him badly, he became listless and lost the appetite for war. It was reflected in the appointment he took when he resumed duty on 17 March 1813. A move to Italy to head a division under Bertrand at first appeared a soft option. It did not last, as Bertrand gathered a force of 40,000 to march through the Tyrol to join a new Grande Armée in Saxony. When it reached Bamberg a reorganization followed and Morand took command of the 12th Division with Bertrand's IV Corps. A frustrating period followed under Bertrand. The 12th Division comprised French troops that had grown soft in Italy. They were not the same as the veterans of III Corps and neither was Bertrand a Davout or even a Morand. At Lutzen on 2 May Morand's attempt to fall on the Russian left was delayed due to Bertrand's indecisiveness. When he did advance his troops stormed into Poserna and Tauchau, but forewarned the Allies were able to retreat without much difficulty. At Bautzen on 20 May he failed to force the Spree till relentless pressure from other sectors enabled him to cross and take Plieskowitz the next day.

In the Autumn Campaign he fought well at Dennewitz on 6 September driving Tauenzien's forces from the town. He resisted a counter attack by Thumen, but when Franquemont's Wurttemberg division gave way on his right he had to quit the field. It was his first defeat as a division commander. At Wartenburg on 3 October the rot continued when Bertrand's poor dispositions failed to prevent Blucher crossing the Elbe. At Leipzig on 18 October he played a key role in Bertrand's assault from Lindenau that drove Gyulai back and opened the way for the army to retire to Weissenfels. He fought at Hanau on 30-31 October and with Bertrand covered the retreat of the army to the Rhine. Bertrand's limitations were recognized and on 16 November Morand replaced him as commander of IV Corps. Ordered to hold Mayence he worked feverishly to prepare the city's defences. From December 1813 till April 1814 he was under siege and with his garrison of 15,000 men was always a threat to the Allied communications during the campaign in France.

THE RESTORATION AND NAPOLEON'S RETURN

Not prepared to serve the Bourbons he resigned from the army after Napoleon's abdication. He hostility towards them later mollified as he accepted the title *Chevalier de Saint Louis* on 31 July 1814. Then on 15 October 1814 he withdrew his resignation but remained unattached and took no part in military affairs until after Napoleon's return. Made an

Imperial aide on 23 March several important posts followed, namely commandant superior for the 12th, 13th, 21st and 22nd Military Divisions, then colonel of the Foot Chasseurs of the Imperial Guard on 13 April 1815. He was also made a Peer of France on 2 June 1815.

THE WATERLOO CAMPAIGN

In the Waterloo Campaign he led the four chasseur regiments formed into the Chasseur Division of the Old Guard. With his colleague Friant he headed the final assault by the Guard that tore into the Prussians at Ligny on 16 June.

At Waterloo he spent most of the day held in reserve behind La Belle Alliance. In the afternoon as the Prussians started to threaten Plancenoit he moved across with two battalions of Old Guard Grenadiers and Chasseurs to support to Duhesme's hard pressed Young Guard. As Bulow threw more and more troops of the Prussian IV Corps into the struggle he threatened to turn the French right. Morand joined the fight just in time and the ferocity of his assault surprised the enemy. His two battalions against fourteen, flushed the Prussians from Plancenoit and pushed them back to the heights beyond the village. Bulow concentrated every available man against them, and unable to do any more they retired to wait the next assault. Napoleon saw the danger and immediately sent Pelet with two more battalions of Old Guard to reinforce Morand. Their arrival was perfect and occupying the Bois de Chantelet they countered further moves to envelop the French right. In Plancenoit itself Morand's men held back Bulow's renewed assaults and a stalemate developed with half the village in Prussian hands.

The deadlock continued until dusk when the army gave way after the Old Guard's assault against Wellington's centre failed around 8.00 P.M. Morand with a battalion of the Old Guard chasseurs made a valiant stand in the church-yard and were the last to leave the village. Down to 250 men he formed column and cut his way through to the Quatre Bras road. The Prussians anxious not to tackle such a formidable foe in the dying stages of the battle left them alone, and Morand reached the Sambre the next day unscathed. Once across the river he joined a column of 3,000 men gathered by Soult and Jerome Bonaparte. At Laon he helped reorganise the Guard infantry before the retreat resumed. His division down to 1,300 men was formed into two provisional battalions. He remained with the army, and after the Armistice withdrew to the Loire where he helped disband the Guard.

LATER CAREER AND CAREER ASSESSMENT

Morand was placed on the non active list on 1 August 1815. Further humiliation followed on 23 December 1815 when the Ministry for War suspended him without pay or pension and ordered him to leave France. Czar Alexander, one of few leaders prepared to show sympathy and respect towards former adversaries, allowed him to settle in Poland. The Bourbons then committed the ultimate folly when a Council of War convened at La Rochelle tried him in absentia. They found him guilty on 29 August 1816 of being an accomplice to a plot to overthrow the King during Napoleon's march on Paris and sentenced him to death. The verdict was ludicrous, based on the flimsiest of evidence. By June 1819 passions had cooled and with the help of Saint Cyr Morand was acquitted of all charges and allowed to return to France. On 20 September 1820 his rank and privileges were restored, but he never held a command under the Bourbons and retired on 17 March 1825.

The July Revolution saw a revival of his fortunes and on 4 August 1830 he was recalled as commander of the 6th Military Division at Besançon. He was awarded the Grand Cross of the Legion of Honour on 18 October 1830 and became a Peer of France on 11 October 1832. In his remaining years he performed his duties diligently and kept clear of politics, which was the undoing of so many of his colleagues in this period of readjustment. He died in Paris of a stroke on 2 September 1835.

The junior member of Davout's triumvirate of generals Morand was the most ambitious and capable of the three. Friant's career by 1815 was in terminal decline and showed definite flaws during the Restoration and Napoleon's last campaign. Gudin, the closest of the three to Davout, never had his loyalty put to the test as he died in Russia. Conversely Morand's clashes with Davout were frequent and often bitter, for he felt he deserved an independent command. His record showed he had the ability and he proved it during the defence of Mayence. Unfortunately being away from Napoleon's centre of operations it did not receive the recognition it deserved when choosing corps leaders in 1815. The Waterloo Campaign was not a time for Napoleon to experiment with new corps commanders. There were too many others in the picture, who through ties of loyalty, seniority and experience came before him. An interesting prospect would have been how he may have fared if he led VI Corps instead of Lobau, who though a tenacious fighter had limitations. The late arrival of VI Corps at Ligny would not have occurred under Morand, as Davout never allowed such tardiness. Equally Morand's experience and tactical finesse would never have allowed VI Corps to be placed the way Lobau did before the Bois de Paris at Waterloo.

All said and done, Morand still ended the Napoleonic Wars with the Old Guard Chasseurs, a fitting reward for a brave and loyal leader. After 1809 he could have been put to better use and unfortunately became the classic case of a leader who was the victim of his own success.

MOUTON - GEORGES, COMTE DE LOBAU (1770-1838)

Georges Mouton, the imperial aide de camp, hero of Landshut and Aspern Essling, was born in Paris on 21 June 1770. His family originally from Savoy settled in Lorraine at the end of the previous century. As a child he moved to Phalsbourg (Moselle) where his father established a thriving business as a master baker. One of fourteen children there few prospects for him in the family business so his father secured him work with an iron merchant at Lunéville in August 1790.

After the outbreak of war and the Prussian invasion of north-eastern France he enlisted on 1 August 1792 with the 9th Battalion of the Meurthe Volunteers. A strong powerful man and a natural leader, within in two weeks he had risen to *lieutenant*. Serving with Kellermann's Army of the Centre in Champagne he distinguished himself before Treves with Beurnonville's advanceguard, which resulted in promotion to *capitaine* on 5 November 1792.

A reorganisation of the armies in the Winter of 1792-93 saw him pass to the Army of the Moselle. He remained with them till 13 October 1793 when he became an aide de camp to General Meynier, who headed the Army of the Rhine's advanceguard. He established a good rapport with Meynier, which was reflected in his service record where he was described as an excellent republican, who loved his country and was a leader who exercised common sense.

ITALY 1795-1799

Mouton moved to Italy in October 1795 when Meynier led a force of 10,000 men from the Rhine to reinforce the embattled army. On 18 March 1796 he joined the 60th Demi-Brigade as part of a new division under his mentor with Massena's right wing of the army. His first contact with General Bonaparte came during the short sharp campaign that followed in Piedmont. He served at Millesimo on 13 April, Dego the next day and at Mondovi on 22 April. Not up to the exacting standards demanded by Napoleon, Meynier soon took a rear command, while Mouton with his regiment joined Serurier's division at the siege of Mantua from June 1796 till February 1797.

At this time he caught the eye of General Joubert who on 22 May appointed him to his division staff.

After promotion to *chef de battalion* on 30 October 1797 he moved to the 11th Demi-Brigade for a short period before Joubert recalled him as his aide de camp on 21 November. When Joubert moved to Holland he opted to remain in Italy and became acting commander of the 99th Demi-brigade on 26 May 1798. The next year he moved to Rome where the 99th formed the city's garrison as a counter to the threats from the Kingdom of Naples. When the Neapolitans invaded in November 1798 and forced Championnet's small army to withdraw he decided to remain in Rome. Not afraid to take extreme measures, he took the unprincipled step of rounding up several prominent citizens as hostages, including Cardinal Consalvi. Then occupying Fort San Angelo, his presence with hostages paralysed the Neapolitan advance and within a month Championnet retook Rome and Mouton gained the high profile every ambitious young officer wanted.

In the months that followed reverses in Italy caused the army's discipline to deteriorate and known for his toughness Mouton was promoted *chef de brigade* on 14 July 1799 in order to put matters right. A particularly bad formation was the 3rd Demi-brigade of Laboissiere's division, which he soon sorted out. He then led the regiment during Joubert's defeat at Novi on 15 August 1799.

THE SIEGE OF GENOA 1800

After Massena took over the demoralised Army of Italy in December 1799 a reorganisation followed. Mouton with the 3rd Demi-brigade joined Saint Cyr's division with the army's right wing as it took post covering the passes over the Apennines. As the Austrians under Melas closed in on Genoa he played a key role halting the enemy's advance by recapturing the strong point of Monte Faccio near the port on 15 December 1799. Shot in the shoulder as he led his men forward he was out of action for several weeks.

The opening of a Spring offensive by Melas on 5 April 1800 to take Genoa saw Mouton and the 3rd Demi-Brigade with Gazan's division. He distinguished himself on 11 April in an attack on Monte Hermette that enabled Soult* after an expedition behind Austrian lines to scramble back to Genoa's defences. Genoa was cut off from all sides on 18 April and the plight of the garrison became critical. On 30 April an Austrian attack led by Richelieu seized Fort Quezzi, a key strong point to the port's defences. Massena ordered Mouton at the head of the 3rd Demi-brigade to retake the position. Shot in the chest as he scaled the fort's ramparts he was left for dead on the field. It was only the personal intervention of Massena rallying the troops that ended with the fort's recapture and Mouton being carried to safety.

Critically ill, a ball had passed through Mouton's

arm, entered his chest and passed out the other side. A hero, his health was of concern to all the garrison as the surgeons battled to save his life in appalling conditions. When Massena capitulated on 4 June 1800, he was one of the first evacuated to France by boat. This coupled with an incredibly tough constitution, ensured his survival. The high regard his superiors had for him was borne out by Soult's report to Napoleon where he stated there was none braver during the siege than he. Mouton's recovery took time and he did not return to duty till January 1801. He then rejoined his regiment with Murat's Army of the South facing the Neopolitans in Tuscany. In April 1801 he moved to Montpellier and then in June 1803 to the camp at Bayonne with the 3rd Line. While there on 23 September 1803 he was promoted its colonel. He later moved to Compiègne and then to Soult's camp at Saint Omer in March 1804 to prepare for the invasion of England.

IMPERIAL AIDE DE CAMP

Aware of Mouton's reputation, Napoleon wanted the young colonel as an aide de camp. Mouton was not interested, being a sincere republican he was disturbed by the Moreau affair and Napoleon's quest for power. A rough tough coarse soldier he was very similar to Vandamme*, though more disciplined and less likely to cause offence. He also felt he would be out of place amongst the bright young men that made up Napoleon's military entourage. The slight did not affect his career and on 1 February 1805 he was promoted *général de brigade* with command of the 1st Brigade of Legrand's Division, which included his faithful 3rd Line.

Napoleon however still kept after him and on 6 March 1805 he succumbed to the appointment as an Imperial Aide. He was completely won over by Napoleon and during the next two years he became one of his closest and most trusted aides. He learnt quickly, proved to be loyal, a fine organiser, paid meticulous attention to detail, knew exactly what was expected and knew how to handle officers both senior and junior to him. Impartial at all times he became a good judge of men. Napoleon so easily tired of those that tried to flatter him, always appreciated Mouton's frankness and directness.

His tasks over this period, apart from battlefield assignments, were varied and interesting. He joined Napoleon's tour of Italy and attended his coronation as King of Italy in Milan. He was then assigned to Boulogne in July 1805. As part of a deception plan to prepare for the arrival of Villeneuve's fleet, he was one of the few who actually knew that Napoleon planned to march the Grande Armée in the opposite direction. On 14 August he left for Stuttgart where he successfully arranged with the Elector of Württemburg for Ney's Corps of 25,000 men to march across his domain. He was present at Austerlitz on 2 December 1805 and after the treaty of Pressburg was back on the Channel Coast from April to September 1806 arranging diversionary preparations that helped keep the invasion scare alive in England.

The campaigns in Prussia and Poland saw him at Jena on 14 October 1806, Pultusk on 26 December 1806, Eylau on 8 February and Friedland on 14 June 1807 where he was seriously wounded. Numerous awards followed, including the Württemberg Iron Cross and the Order of Military Merit in 1807. This was followed by a cash award of 35,000 francs drawn on the Duchy of Warsaw at the end of June 1807.

SPANISH INTRIGUES 1807-1809

On 5 October 1807 Mouton was promoted *général de division*. The importance Napoleon attached to the promotion was shown the next day when he took the post of Infantry Inspector General for the 1st Military Division in Paris. In December 1807 he moved to Bayonne to organise the French depots at Saint Pied-de-Port near the Spanish border. In March 1808 he moved to Spain as an inspector for the provisional regiments supposedly passing through the country en route to Portugal. His real task unknown to the Spanish, was the compilation of a secret report for Napoleon on the dispositions and preparedness of the Spanish army at Vitoria and Valladolid, which he completed with meticulous detail.

Mouton became involved in the constitutional crisis that erupted between Charles IV and his son Ferdinand. He took responsibility for his protection after the unpopular king was forced to abdicate and placed him under guard in Madrid. He then escorted the king to Bayonne to meet Napoleon, who forced both father and son to renounce all rights to the throne and replaced them with his brother Joseph.

On 15 June 1808 after the outbreak of hostilities in Spain he joined Bessières's corps in Castile as commander of its 2nd Division. He fought at Medina del Rio Seco on 14 July where in charge of the French right wing he pinned down Cuesta's forces while Bessières completed the destruction of the remaining Spanish force under Blake. He transferred to the 1st Division on 8 September 1808 and took part in the crossing of the Ebro in October. When Soult took command of II Corps, he played an important role at Gamonal on 10 November when his infantry completed the rout of the Spanish before Burgos. Napoleon in need of him, he quit his command to resume duties as Imperial aide on 30 November. He remained with Napoleon throughout the rest of his stay in Spain and at the end of January 1809 returned with him to France.

LANDSHUT AND ASPERN ESSLING 1809

As the Austrian Campaign of 1809 unfurled Mouton achieved the finest exploits of his career first at Landshut and then at Aspern Essling, personifying the role expected of an Imperial aide. Soon after his arrival in Bavaria Napoleon fell on the Austrian army and split it in two at Abensberg on 20 April. The main element under Archduke Charles fell back on Eckmühl while Hiller's two corps harried by Bessières's cavalry made for Landshut on the Isar. Napoleon mistakenly thought Hiller's left wing was the main army and directed the bulk of the army to follow. Amidst a retreat that soon became a panic the last Austrians crossed the Isar with moments to spare leaving a huge park of wagons, cannon and equip.m.ent on the left bank. Outside Landshut Bessières's horsemen galloped across the first bridge leading to an island, but could go no further as the second wooden bridge spanning the main channel was in flames. Napoleon needed the bridge, and seeing the only one available going up in smoke called on Mouton to solve the crisis.

A column of Morand's* 17th Line in the meantime had reached the bridge and for a moment hesitated when they saw the enemy sharpshooters and cannon positioned on the opposite bank. Mouton dismounted and sword in hand urged the grenadiers of the 17th Line to follow him. They quickly crossed to the island, which gave them ample cover and room to stage the next phase of the assault. After a pause while the rest of the regiment closed up, the column approached the second longer bridge to the right bank. A frontal attack appeared a desperate thing to attempt. The bridge in flames, smoke

crowned the river bluffs above Landshut as Austrian artillery poured fire on the exposed French. The plunging fire kicked up spouts of spray as the shells hit the water. Directly ahead a stone wall and a wooden gate barricaded the bridge's exit. From windows and roofs along the bank and from the church situated behind the bridge came musket fire. Had steady troops occupied the opposite bank they could have held the position indefinitely.

Mouton, brave but not foolhardy, saw that the position was not as strong as first imagined. The cannon from the heights had to fire at a narrow target. The river swallowed up the misses, and muffled the lethal ricochets that occurred on land. The irregular fire of the musketry on the right bank lacked the stopping power of controlled massed volleys. A column he realised would suffer losses, but provided it kept moving it could cross the river within a minute. If prepared to use the bayonet a column could storm the bridge and enter the town. His stentorian voice feared throughout the army carried the simple words above the noise of battle, "No firing - March!" Striding forward, Mouton knew that some of the enemy fire would strike home and that men in the front ranks would take the brunt of the losses. Sword in hand, at the head of the column he led the way. Surging behind him at a steady jog came the grenadiers of the 17th Line. The Austrian fire struck down a handful as Mouton and the grenadiers reached the barricade. A section of

MOUTON STORMS THE BRIDGE AT LANDSHUT

sappers strode up to the gate and began to smash it with their axes. The wood splintered and the grenadiers shouldered their way into Landshut with the rest of the regiment pounding across after them.

It was here that the celebrated remark by Napoleon, "Mon Mouton est un lion" (My sheep is a lion) occurred. After the campaign the Emperor acknowledged Mouton's conduct when he presented him with a magnificent painting by Hersent depicting his aide leading the attack over the bridge. The next day he was at Napoleon's side when the army turned north and defeated Charles at Eckmühl.

On the second day of the Battle of Aspern-Essling on 22 May 1809 Mouton again covered himself with glory. A crisis developed as Boudet's division driven from Essling took refuge in the village's granary and a great gap appeared on the French right. To meet the crisis Napoleon ordered his aide forward with two battalions of Young Guard tirailleurs to recover the village. With Roguet* in support he swept it clean in one rush and saved the remnants of Boudet's division. Rosenberg's Austrians again closed in around the village and forced Napoleon to commit Rapp with the last two Guard battalions to now extricate his aides. When Mouton and Roguet saw Rapp approaching, instead of abandoning Essling as ordered they counterattacked, scattering Rosenberg's corps.

It was a magnificent feat of arms that had a telling effect on the outcome of the battle. Charles withdrew his shaken infantry from the French right and centre, and replaced them with masses of cannon. For the remainder of the day the French almost without artillery withstood a merciless bombardment, closing ranks as men fell. Lannes was mortally wounded, but the beleaguered French reinforced by the Guard, the enemy never dared renew the attack. By 4.00 P.M. the second crisis passed as firing died away and that night Mouton with Roguet covered the withdrawal to the Isle of Lobau.

Mouton for his outstanding contributions to the campaign on 28 May 1809 was made *comte de Lobau*. The title was after the island on the Danube, from where the battle was launched and after which Napoleon originally wished to name the battle. At Napoleon's side during the Battle of Wagram on 5-6 July for once there was no record that he had any involvement in events that influenced its outcome.

PEACE AND CONTENTMENT

In the awards that followed the campaign a grateful Bavarian government made Lobau a Knight of the Order of Saint Hubert on 23 October 1809. A pension of 27,000 francs per annum from the department of Deux Sèvres followed on 21 November plus another of 50,000 francs from the Roer on 1 January 1810.

Affairs of the heart also came to the fore. Napoleon thought the tough old bachelor, nearly forty, needed a wife. He always encouraged his aides to marry into the old noble families and with Josephine's help they helped Lobau secure the hand of Félicité d'Aarberg one of Josephine's "maids of honour". She came from one of the finest noble families and they married on 22 November 1809. It was a success and three daughters came from the union. Content, Lobau's future was assured, he had become a wealthy man and was among the chosen few of Napoleon's confidants. Ever grateful, he never forgot to whom he owed his good fortune. It however never prevented him speaking his mind and when Napoleon told him in 1811 he would make an ideal Minister for War to replace Clarke he flatly refused to be considered.

THE RUSSIAN CAMPAIGN OF 1812

He openly opposed a campaign against Russian, feeling it was unnecessary and too risky. Yet when the decision was taken and the die cast he threw his full weight behind Napoleon to ensure its success. In the campaign he served as Napoleon's senior aide and was present at Smolensk on 17-18 August and Borodino on 8 September 1812. In Moscow he had long and bitter arguments with Napoleon urging the case for an earlier withdrawal. He was tactful enough not to remind Napoleon of his advice during the retreat. Napoleon took no notice of any offence that he may have caused and showed his regard at Smorgoni on 6 December 1812 when he chose Lobau to accompany him back to Paris.

GERMANY 1813

With Napoleon during the Spring Campaign in Germany Lobau distinguished himself at Lutzen on 2 May 1813. Ney's was corps driven back in confusion at the start of the battle, he took the lead at the head of Ricard's division in the bloody fight around Kaja and is credited with stabilising the wavering French line. On 29 July 1813 with Soult's return to Spain he replaced him as *Major Général* of the Imperial Guard. He held the post till 3 September when he took command of the remnants of I Corps at Dresden after Vandamme's* defeat and capture at Kulm.

With XIV Corps under Saint Cyr they covered Southern Saxony and the approaches to Dresden against Schwarzenberg's army, winning a sharp action on 13 September against its right wing at Giesshübel. He should have withdrawn westward when the Allies breeched the line of the Elbe but Napoleon's fixation with occupying capitals proved a costly mistake. Had the 30,000 troops at Dresden been available at Leipzig the outcome of the battle and the campaign could have been different. As the Allies gradually tightened the noose around Dresden Lobau defeated Ostermann at Racknitz on 17 October. He urged Saint Cyr as senior commander to use the opportunity before winter fell to link up with the

Torgau and Magdeburg garrisons and with this force of over 40,000 men fight their way back to the Rhine. Saint Cyr normally a very clear headed leader, refused to budge without orders. By the end of the month Ostermann, reinforced by the corps of Chasteler and Klenau, completed the encirclement. Then Saint Cyr, belatedly spurred into action, tried to break out along the Elbe on 7 November but soon fell back to the city. Inadequately provisioned the garrison could not withstand a long siege. On 11 November an agreement was reached with Klenau to surrender provided the garrison could return to France with its arms and equip.m.ent. Once outside Dresden's walls, Schwarzenberg refused to ratify Klenau's terms and the French faced with cold and hunger. Offered the alternative to return to the city, Saint Cyr to Lobau's disgust decided to surrender regardless and allowed 30,000 men to pass into captivity. Lobau with Saint Cyr joined Vandamme in captivity in Hungary, only returning to France in June 1814.

THE RESTORATION AND NAPOLEON'S RETURN

Missing the agonies that France went through during the dying days of the Empire, Lobau had a thirst for revenge. He refused to acknowledge the Bourbons till he realised his release and return to France depended on it. The Bourbons in an attempt to win him over on 5 July 1814 made him a *Chevalier de Saint Louis*. Unrepentant and openly hostile to the monarchy, he was placed on half pay on 2 September 1814. Soult on becoming Minister for War rightly believed he was less of a danger when employed and recalled him on 30 December 1814 as an infantry inspector general. It was the right ploy since when news of Napoleon's return broke, Lobau from outward appearances remained staunchly neutral. It was only at the last moment as the Bourbons fled Paris did he throw caution to the winds and join the Emperor at Esconnes on 20 March. The reunion was emotional. Napoleon who had not seen his favourite aide for nearly two years immediately made him an imperial aide. The next day Napoleon appointed him commander of the 1st Military Division and ordered him to Paris to secure the city before his arrival. The new government established, Lobau's was one of the first line appointments for the Army of the North, when he took charge of VI Corps on 3 April. The troops, based in Paris, were veterans from some of the finest line regiments. They comprised the divisions of Simmer (19th), Jeanin (20th) and Teste (21st) and numbered 11,000 infantry with 38 guns.

THE WATERLOO CAMPAIGN

On 2 June he became a Peer of France and attended the ceremony on the Champ de Mars. Soon after he left Paris to join his troops as they concentrated on the frontier. Forming Napoleon's reserve with the Guard his formation was the last on the line of march and took no part in the early skirmishes. On 16 June the day of Ligny, Napoleon used Lobau's formations poorly. At mid morning they were only on the outskirts of Charleroi and still had to cross the Sambre. By 2.00 P.M., half an hour before the battle began he had reached Bois de Soleilment, still three miles from the battle. Here he halted and waited for over three hours for orders, not knowing whether to support Ney or join the Emperor.

When the orders arrived he set out for Fleurus. As Napoleon's plan was to crush Blucher's right, it was strange he was not directed to form on Vandamme's left. So placed, the VI Corps would have been in an ideal position to back the decisive blow, connect with any troops Ney might detach to co-operate at Ligny, and act with them in enveloping and destroying Blucher's right. Also had he come up on the French left, his corps not engaged would have been free to establish the identity of d'Erlon's oncoming troops when they were first sighted. His men fresh, would have been less likely to panic when the strange corps was seen. If the newcomers had been the enemy, he possessed just the undaunted fearless personality, which fitted him admirably for the task of blocking a hostile advance against Napoleon's left at Ligny. On the other hand if the newcomers were friends, contact would have been made quickly between d'Erlon and the main army, and all the later misunderstandings avoided.

Where Lobau deployed was Napoleon's fault, however he cannot go without censure as he was uncharacteristically slow in reaching the battlefield. It took him till 7.30 P.M. before his corps came streaming through Fleurus and began finished forming up on the heights to the east. His late appearance undoubtedly delayed Napoleon's blow against the Prussian centre. It saved the Prussians, who as a result were able to leave the field as darkness fell, and the French were unable to determine the direction of their retreat. It was a delay that in the long run proved decisive, yet it didn't attract the attention it deserved due to the wanderings of the unfortunate d'Erlon.

At Waterloo Lobau's corps with the Guard formed Napoleon's reserve. In the morning his divisions formed up to the left of the Charleroi Road behind Reille in front of Rossomme. Around 1.00 P.M. when the Prussians were first observed in the distance, Napoleon ordered Domon's and Subervie's cavalry to move to the right to delay them. Shortly after he ordered Lobau to follow with his two divisions and take up a good intermediate position. With this mixed force of 10,000 men he was to stop Bulow's 30,000 Prussians. Further, as soon as he heard Grouchy's guns thundering in the rear of the Prussian corps signalling his approach, he was to attack Blucher vigorously. He drew out at right angles to the French front between Plançenoit and the Bois de Paris. He waited for three hours while the Prussians never seriously threatened the French right. It was not till 4.00 P.M. they emerged from the woods and started to deploy in front of him and then were not really dangerous till an hour later.

Had Napoleon three hours earlier exercised a different option and allowed Lobau to deploy in echelon on d'Erlon's right, the battle could have taken a very different turn. The danger of Wellington's left being enveloped from Frichermont would have required him to weaken his centre. The Prussians observing in the distance the Allied right outflanked and overwhelmed in all likelihood would have halted their advance and not joined the battle. Lobau also with the orders he received displayed a tactical ineptitude and lack of initiative that went a long way towards losing the battle for Napoleon that day. It was no time for half measures. He should have followed Domon's cavalry towards the Lasne brook where they were observing the Prussian approach. Domon could have caused considerable problems had he decided to dispute the passage. Even had the enemy crossed in force before he arrived, Lobau could easily have occupied the Bois de Paris. Bulow would have had to deploy his corps to master this mile wide band of woodland. The close nature of woodland fighting would have disordered the Prussians to such an extent that even had they been successful, they then would have had to rally and reform before advancing any further against the French right. Valuable time would have been gained, and Plançenoit would not have come under threat for many hours.

Instead it was not so much the dogged resistance of Lobau that delayed the Prussians, but rather their reluctance to open an action which could have turned horribly wrong, if Wellington in the distance at any time gave way. In the end, the wood was surrendered to the Prussians without a blow. Thereafter it was used by the Prussians to screen their movements as Bulow collected his troops in it before they fell on VI Corps. Around 4.30 P.M. their resolve stiffened by Blucher's presence, Bulow's two leading brigades (the 15th and 16th) debouched from the Bois de Paris. Supported by cavalry they drove away Domon's squadrons shielding Lobau's infantry a mile back holding a line stretching from Frichermont along the heights that ran parallel to the woods.

The 7,000 infantry of VI Corps could not for long expect to hold an attack by 30,000 fresh troops once Bulow's whole corps was committed. Lobau presented a bold front, showed an unconquerable spirit and his troops fought stubbornly and well. For a time he held Bulow off. By 5.30 P.M. the whole of Bulow's corps deployed had pressed him back with Domon's cavalry to the heights before Plançenoit.

From there the Prussians drove him into the village and roundshot started to plough up the Charleroi Road. The main burden of defending the French right then passed from Lobau's exhausted men as Napoleon promptly ordered Duhesme to reinforce him with the Young Guard and recapture Plançenoit at all costs. Duhesme attacked at once, dislodged the Prussians and by 6.30 P.M. completely cleared the village.

The battle for the village continued for a further three hours as Lobau with the remnants of his corps helped keep the closing jaws of the Allied vice open. His troops, which now included the Young Guard stiffened by elements from the Old held firm to cover the retreat of the Army along the Charleroi road. Here as at Essling on the banks of the Danube, he once more held the blazing ruins of a village to save his Emperor's army from annihilation. His courage afforded Napoleon a chance to rally the exhausted Army and conduct a withdrawal had he not committed the Guard in a last desperate throw against Wellington. Then when all was lost, his heroic defence of Plançenoit kept the way open for the Emperor and gave the debris of the Armée du Nord an avenue of escape from the field.

During the rout the remnants of his command was trampled underfoot at Genappe. The next morning he tried to rally stragglers making their way cross country to Charleroi but wounded in the foot was captured by a group of Prussian cavalry near Glabais. Taken to Blucher's headquarters, an inn called the Roi d'Espagne in Genappe, he was cared for with the dying Duhesme by the Prussian staff. Later handed over to Wellington, he was sent to England, which turned out to be fortunate as he was proscribed by the Ordinances of 24 July 1815.

THE JULY REVOLUTION

Released after six months he lived in Belgium till the end of 1818 when he was allowed to return to France. His rank and privileges were restored but he never received a command throughout the period of Bourbon rule. He instead took up a political career and gained election to the Chamber in April 1828 as a liberal republican deputy for Lunéville.

Opposed to the repressive ordinances of 25 July 1830 that tried to curb the Chamber and reassert the power of the monarchy, Lobau with Lafayette became a leading light in the July Revolution that followed. As the government lost control and collapsed on 28 July, fearful of a breakdown in law and order he supported the move to recall the National Guard disbanded in 1827. He became a member of the five strong Municipal Committee that set up in the Hotel de Ville as an alternative government. On Lafayette's appointment as head of the National Guard he helped defend key points of the city as Marmont at the head of the Paris garrison tried to quell the insurrection. It was very much his force of personality and iron will that maintained discipline amongst the insurrectionists and forced Marmont to abandon the capital two days later.

When the question arose as the alternative to the monarchy Lobau supported a republic. It brought him into conflict with Lafayette who had no stomach for the possible re-emergence of Jacobinism and a return to the days of the Terror. Lobau with the republican leadership was outmanoeuvred by Lafayette and Thiers, and on 9 August 1830 Louis Philippe was proclaimed king. In the interest of unity he modified his stance and was won over when on 19 August 1830 he received the Grand Cross of the Legion of Honour from Louis Philippe for his part in the uprising. Officially recalled on 30 September 1830 he served on the General Staff taking on the task to review the plight of former officers who had served Napoleon and wished to return to active duty. His reputation and influence was considerable, and he became commander of the Paris National Guard on 26 December 1830 when the Chamber curbing Lafayette's growing power forced his resignation.

He found the position tough, as Paris in 1831 continued to seethe with unrest. Plots, riots and disturbances hatched by Ultras, Republicans and Bonapartists were not uncommon. Much must be said for his patience as he skilfully managed to avoid bloodshed in the capital. His ingenuity was at its best on 5 May when he personally directed his men to turn half a dozen fire hoses on a mob of Bonapartist sympathisers, who dispersed amidst much laughter to burn effigies of him elsewhere. He also became a Marshal of France at the time of his election to the Chamber as representative for the 10th Arrondissement in Paris on 30 July 1831. The award was one of several made by Louis Philippe that tried to revive the martial glories of the past. Whilst most had the effect of devaluing the dignity of the rank, Lobau's was generally well accepted. On 27 June 1833 he became a Peer of France. He continued his political and military career till the end. Suddenly on 27 November 1838 the old chest wound he received thirty eight years before at Genoa reopened, and within the day he was dead.

ASSESSMENT OF CAREER

A bold, brave and loyal follower of Napoleon, his courage and deeds were an inspiration to all. He was the classic example of an Imperial Aide de Camp. The part he played at Landshut and Aspern Essling showed to perfection the role Napoleon expected them to fulfil on the battlefield. He also was a fine organiser and administrator. The fact Napoleon considered him for the post of Minister of War bore testimony to that. As a strategist and planner he too was amongst the best, being part of Napoleon's inner circle that laid the plans for the campaigns in Spain in 1808, Russia in 1812, Germany in 1813 and Waterloo.

His role as a corps commander was another story. He possessed an abundance of courage, but for some inexplicable reason all strategic and tactical skills he possessed often deserted him during the Waterloo Campaign. It brought out the best and worst in him. His defence of Plançenoit was inspirational, where the order was simple and clear. Earlier he had shown a lack of initiative and tactical common sense by choosing his ground poorly. His excuse that Napoleon's orders never specifically stated he should stand backs up the argument. As an independent commander he was free to exercise his own judgement within the context of the battle plan, which in this case was to keep the Prussians at bay at all costs. Ideally he was the man temperamentally suited to lead the Guard infantry in the absence of Mortier. It certainly would have made life easier for Drouot. On the other hand, Napoleon expected Mortier to recover and once the campaign was underway it was not advisable to switch commanders in mid stream.

Soult after Waterloo grew to dislike him and in private considered him a bit simple, but then he had few kind words for anyone. Lobau's name must have often come up at Saint Helena when Napoleon reflected on past glories. Considering that his exploits at Landshut and Aspern Essling earned Napoleon's praise with the words, "My sheep is a lion," the former emperor must have felt after Mouton's showing at Waterloo and Ligny that the words, "My lion is a sheep," more appropriate. There is no record of such a comment, but Lobau certainly deserved it at the time.

NEIGRE - GABRIEL, BARON (1774 - 1847)

BACKGROUND AND EARLY CAREER

The son of an artillery sergeant with the Metz Artillery Regiment, Gabriel Neigre was born in the garrison town of La Fère (Aisne) on 28 July 1774. Brought up amongst soldiers he was attracted to the way of life, and knowing no other profession followed his father into the regiment's ranks as a gunner on 14 July 1790.

Little is known of his early career apart from that he served with the Army of the Centre and then the Army of the Moselle in the opening campaigns of 1792-93. Promotion to *sergent* came on 31 December 1793, followed by a move to the Army of the Rhine. On 10 January 1794 he became *capitaine* of an independent artillery company known as *la compagnie d'artillerie du Mont Terrible*. Again little is known of this body as it disappeared in the cloud of independent formations that came and went during the early years of the Revolutionary Wars. He came to prominence during the blockade of Mayence from November 1794 manning the guns before the city for over a year. In March 1796 he moved to the 106th Demi-brigade as commander of the regimental guns.

A change in career direction came in April 1797 when still with the artillery he led one of the bridging trains during Moreau's crossing of the Rhine. In 1798 he was with the bridging trains attached to the Army of Mayence in 1798 and then the following year with Lamartillière's artillery serving in the Army of the Danube. The Spring Campaign in Germany in 1800 saw him with Lecourbe's left wing of the Army of the Rhine where he was present at Moeskirsh on 5 May 1800.

In the aftermath of Moreau's victory at Hohenlinden he played a valuable part during the crossing of the Inn on 9 December 1800. Moreau's pursuit had stalled after he failed to batter his way across the river at Rosenheim. As a diversion

Lecourbe moved the bulk of his artillery and Neigre's bridging train upstream to Neubeuern. There his guns on a dominant rise above the town overlooking the opposite bank poured a steady fire on the Austrian positions. At the same time, Neigre leading his *pontoniers* launched his boats across the river and supported by light infantry to protect them, started to build a bridge. Using specially designed portable equipment that gave greater mobility, the construction became a landmark in bridging operations for the French army at the time. It took Neigre two hours to span the freezing river with a 200 yard pontoon bridge Montrichard's division stormed over to the opposite bank. In the extreme conditions this was an unheard of speed. Such operations had often failed in the past because bridges took too long to construct and gave the enemy time to prepare their defences. The Inn forced, the Austrians soon gave up Rosenheim and the advance to the Salza resumed. Lecourbe received the praise while the exploits of an unknown artillery captain went unnoticed.

It was not till 2 October 1802 that Neigre received promotion to *chef de bataillon* and an appointment to the Artillery General Staff. He then moved to Strasbourg as an artillery sub-director, a post similar to an artillery inspector general but with less authority, in January 1803. A move to Boulogne followed in July 1803 when he became sub-director of the artillery park with the Army of the Ocean Coasts. The same appointment followed with the activation of the Grande Armée on 23 August 1805.

THE GRANDE ARMÉE 1805-1811

Neigre took part in the Austrian Campaign and was present at the capitulation of Ulm and fought at Austerlitz. On 11 April 1806 back in France he became major of the 3rd Foot Artillery Regiment based at Strasbourg. For the campaign against Prussia he joined Murat's Cavalry Reserve on 14 August 1806 as head of the artillery train. Able to keep the horse gunners supplied during the headlong dash across Saxony and Prussia he was

FRENCH PONTONIERS AT WORK

promoted *colonel* on 17 January 1807. A move to the Artillery Reserve staff followed, serving with them at Eylau on 8 February and Friedland on 14 June 1807.

With Tilsit and peace he spent a short period in October 1808 as Artillery Director General at Toulouse before moving the next month to Danzig as artillery commander. The campaign against Austria led to his appointment as Director General of the artillery train with the Army of Germany in June 1809. The successful conclusion of hostilities saw his elevation to the imperial nobility as *baron de l'Empire* on 29 September 1809. A few years followed with little active campaigning as he moved to Antwerp as Artillery Director on 1 March 1810 and then in June 1810 was to head the artillery park of Massena's Army of Portugal. The orders then countermanded he remained at Antwerp till March 1811 before becoming artillery commander at Metz.

THE GRANDE ARMÉE ARTILLERY PARK 1812-1814

After despairing of gaining another field command, Neigre's reputation as a fine administrator and logistics expert resulted in the post as head the artillery park for the Grande Armée in February 1812. It became the most challenging post of his career. His responsibilities involved the maintenance and replenishment of equipment to ensure the smooth operation of 1200 artillery pieces. It was a daunting task, and considering the climatic conditions and vast distances covered, he did well. The work by the artillery park at Smolensk and Borodino made an important contribution towards the fine performance by the artillery.

He found his way back to Poland with the *Bataillon Sacre* during the retreat from Moscow in December 1812. Promoted *général de brigade* on 10 January 1813 he was attached to the Artillery General Staff in Germany, which at the time presided over an arm that was virtually non existent. On 11 March 1813 he took charge of the artillery park with Eugène's Army of the Elbe. During the Armistice he worked feverishly to build up the artillery for the coming campaign. On 25 November 1813 he gained promotion to *général de division*. He was present with the army at Dresden and Leipzig. During the campaign in France it found it increasingly difficult for the artillery parks to operate efficiently as the Allies occupied most of the major depots. At the end of the campaign he was at Fontainebleau with a few wagons and forges trying to keep the artillery operational.

THE RESTORATION, NAPOLEON'S RETURN AND THE WATERLOO CAMPAIGN

He acknowledged the return of the Bourbons and on 25 June 1814 took up the appointment as an artillery inspector general. The award of *Chevalier de Saint Louis* came on 29 July, soon followed by his placement on the non active list on 6 August 1814 as the artillery establishment was reduced. It made him a ready convert to Napoleon on the latter's return and on 1 May 1815 he became commander of the artillery park for the Army of the North.

As the campaign unfurled Neigre's work was vital to ensure the artillery was not short of spares and replacement equipment. When the army fought at Ligny and Quatre Bras he was at Charleroi supervising the movement of the hundreds of wagons that made up the artillery park. On the day of Waterloo he was with elements of the artillery park at Quatre Bras. As the trickle of wounded from the battle turned into a flood of fleeing soldiery, he tried to organize a defence. He formed barricades with upturned wagons at the crossroads while he waited the arrival of Girard's division from Fleurus. The infantry he hoped would act as a breakwater around which the army could rally. It never arrived, an act for which its commander Colonel Matis was much criticized. With a mere handful of men Neigre tried to stem the rout but in the crush was knocked to the ground and trampled underfoot.

He rejoined the army at Charleroi and remained with it during the retreat to Paris. After the signing of the Armistice he retired to the Loire where on 6 July 1815 Macdonald made him artillery commander of the Army of the Loire. He co-operated with the Bourbons during the army's disbandment and remained in his post till 18 October 1815.

LATER CAREER

For a short while he remained unattached, but the need for experienced artillery officers was foremost for the new army. The Bourbons overcame their prejudices and on 16 February 1816 Neigre took a post on the influential Central Committee of the Artillery. His loyalty always suspect was tested when he headed a Council of War convened to court martial Drouet d'Erlon for treason. He adroitly side-stepped a conflict of loyalties when after careful consideration he declared there was insufficient evidence to bring charges. The accused was not in the country and could not be called on to give evidence. The case collapsed and d'Erlon was acquitted. In April 1817 he became Inspector of Material and Personnel at Strasbourg. The post of Artillery Inspector General followed in March 1822 in addition to being a member of the Artillery Consultative Committee. In June 1824 he took the post of Artillery Inspector General at Rennes.

The July Revolution brought new opportunities. The declaration of independence by Belgium and the resultant sabre rattling by Holland to restore its authority resulted in Neigre's appointment as inspector general for the artillery in the northern departments as men and equipment moved to the border. After the Dutch invaded Belgium, France countered with the Army of the North, which crossed the border on 4 August 1831 with Neigre heading the artillery. In the awards that followed after the army's return he received the Grand Cross of the Legion of Honor on 9 January 1832. He became a Peer of France on 11 October 1832 in recognition of a distinguished career where he built up a reputation as one of the foremost artillerists in the army. He soon had the opportunity to put new theories into practise during the siege of Antwerp in November-December 1832 when the Dutch refused to withdraw from the city.

From a gunner's perspective the siege was a valuable exercise. New weapons were tried, in particular a mortar that fired a 1,000 lb shell. Whilst the damage it inflicted was minimal, its effect on morale was considerable as it led to a further 20,000 shells of different calibres to rain down on the citadel over a period of six weeks. He recognized the importance of artillery as both a defensive and offensive weapon. He encouraged research and development and by 1840 the French artillery had become the foremost in Europe in an army that had become innovative and yet matured as a result.

In the years that followed he was a member of the influential Artillery Committee that formulated and implemented artillery policy under the Ministry of War. In January 1839 he took up the less onerous post of Director of Powders and Saltpetres, which he held till his death at Villiers-sur-Marne (Seine-et-Oise) on 8 August 1847.

AN ASSESSMENT

Under Napoleon Neigre held one of the least glamorous commands within the Grande Armée. There was little glory to be found within the artillery park. If the artillery performed badly, was not replenished or repaired it was the park that invariably suffered criticism. Writers rarely gave this arm of the artillery much attention, yet Neigre with his mobile workshops, trains of spares and supplies was always nearby to lend valuable assistance. The role the artillery park played was a vital to the continued success of the artillery. It received little recognition or reward and could be likened to the mechanics of a Grand Prix motor racing team of today, vital for the driver who took the public acclaim. Neigre never let Napoleon down. He was a fine leader and a man dedicated to his profession.

FRENCH ARTILLERY AT THE SIEGE OF ANTWERP

NEY - MICHEL, PRINCE DE MOSKWA (1769 - 1815)

BACKGROUND AND EARLY CAREER

Michel Ney referred to as, "the Bravest of the Brave" is arguably the most famous of Napoleon's marshals. He was also the most complex, whose temperament few could predict or understand. German by descent Ney was born in Sarrelouis on 4 January 1769; the same year as Napoleon, his arch rival Soult and Wellington. His father was a barrel-maker who had served in the French armies during the Seven Years War. Michel the second son was highly intelligent and his father expected great things of him. To give him the right start in life he was sent to school at the College of Saint Augustine, a place of learning for children from good families. After school he joined a law office, his father's idea being it was the right step to towards an influential job with the *Procureur du Roi*. Soon bored the attraction of a military life took hold and he soon took the first steps towards his chosen career by running the office at the Apenwarler iron-works. The post of overseer at the Saleck works followed where he had the opportunity to study cannon construction. After eighteen months at Saleck much to his family's dismay he threw the job in and enlisted as a trooper with the 4th Hussars on 1 October 1790.

Ney's rise to fame with the onset of the Revolution can best be described as meteoric. Promoted *brigadier fourrier* on 1 January 1791 as a young man with strong republican ideas his troopers looked up to him. On 1 February 1792 he was promoted *maréchal des logis* and three months later *sergeant major*, followed by on 14 July 1792 *adjudant sous officier*.

In the opening campaigns he served with the Army of the North as the 4th Hussars fell back before the Duke of Brunswick's forces. They then moved to Lafayette's camp at Carignan near the Belgian border before pulling back to the narrow valleys of the western Argonne. Summoned to join the army between Verdun and Rheims he received his first baptism of fire at Valmy on 14 September 1792. He joined Dumouriez's pursuit of the enemy into Belgium gaining a commission as *sous lieutenant* on 29 October followed by that of *lieutenant* on 5 November 1792. In Belgium he fought at Jemappes on 16 November 1792. Noticed by Colonel Bernadotte as he led a squadron against the Coburg Dragoons, it led to Bernadotte's comment, "Ney will go far if he doesn't kill himself first". He was one of the first to enter Brussels and his division commander General Lamarche made him a provisional aide de camp. The promotion promised much but the campaign soon turned sour resulting in Lamarche's suspension. The general, however for what it was worth did find time to write Ney a good report mentioning how he led the 5th Hussars in a gallant charge at Gossancourt that halted the enemy before Brussels. It was enough to secure him a place as Colaud's aide on 3 February 1793, but that general's career was soon in trouble and in May 1794 Ney found himself back with the 5th Hussars and his career soon in a rut. An old captain who was woefully incompetent was promoted *chef d'escadron* before him, which led him to pour his heart out to an old colleague Claude Pajol' who was aide to General Kléber. It was the break he needed, as Pajol spoke to Kléber the general in turn called Ney over. Both German speaking they got on well. As a result Ney received command of a body of 500 horsemen attached to Kléber with the task to guard supply columns and protect the flank of the army against surprise attack.

THE ARMY OF THE SAMBRE-MEUSE

Posted to the Army of the Sambre-Meuse on 28 June 1794 Ney later gained promotion to *adjudant général chef d'escadron* on 31 July after distinguishing himself near Louvain when he captured an Austrian general. As the army pushed its way to the Rhine he fought in several small actions including Aldenhoven on 2 October. His role at the time changed to that similar to a partisan leader attacking convoys far behind enemy lines as far afield as Cleves and Nijmegen. It resulted in further promotion to *adjudant général chef de brigade* on 15 October 1794.

On 22 December 1794 whilst leading a feint against an earthwork during the siege of Maestricht Ney was shot in the left shoulder and carried from the field. The wound was serious and as he grew delirious there was talk of amputation. The surgeons saved his arm but his recovery was slow and Kléber recommended on 5 January 1795 sending him home to convalesce. It was backed by the Representative Merlin de Thionville who was so impressed by Ney's exploits that he also recommended immediate promotion to *général de brigade*. Ney was reluctant to accept since he felt he would lose the semi-independent role he had under Kléber. The Terror over, his refusal was accepted as the days were past where such a gesture was considered unpatriotic. Ironically Ney's display of modesty was a far cry from later years when his ambition took precedence to all else and he proved a difficult subordinate to anyone.

He returned to Kléber's command in August and

on 8 September 1795 defeated an emigré force at Opalden. He took part in Jourdan's campaign on the right bank of the Rhine in the Autumn of 1795 distinguishing himself in several actions including Dierdorf on 14 September. When Jourdan renewed the offensive across the Rhine in June 1796 he was again in the fore. He fought at Altenkirchen on 4 June and made a celebrated charge at Uckerath on 19 June when at the head of the 14th Dragoons checked an Austrian advance on Kléber's positions. The Rhine cleared he was at the fore during Jourdan's advance deep into Germany. On 15 July he took the citadel at Wurzburg by a bold ruse maneuvering a small force of 100 men so well that he convinced the enemy they faced an army. So intimidated were they when he demanded their surrender over 2,000 infantry and 300 horse filed out of the fortress. His daring resulted in promotion to *général de brigade* on 1 August 1795. In action at Forcheim on 8 August he led a mixed column of infantry and cavalry that pinned the enemy till the arrival of Colaud's division completed their destruction.

As the campaign turned against Jourdan, Ney covered the army's retreat after the reverse at Amberg on 1 September. Overconfidence on his part led to his first reverse at Wiselhof when cut off by Austrian cavalry he abandoned two battalions of the 23rd Line. In the days that followed he showed his worth as a rearguard fighter covering the army as it fell back to the Rhine. Jourdan removed and the army reorganized under Hoche, Ney served under Grenier a fellow compatriot from Sarrelouis who replaced Kléber. Leading a Hussar division he was to act as the eyes of the army as it once more crossed the Rhine at Neuwied on 17 April 1797. He did well during the advance to the Lahn pushing the outnumbered Austrians hard. Misfortune struck on 19 April when at the head of one charge too many his horse fell in a ditch resulting in his capture near Giessen. His reputation well known, he was a fine prize. Hoche tried to secure his exchange but it was not till 27 May, well after the news of the Armistice of Leoben on 23 April, that he was released for General O'Reilly taken at Rivoli.

Mannheim and the Campaigns in Switzerland and Germany 1799-1800

An event that made Ney a household name and showed to Paris, the politicians and the people that the armies in Germany produced heroes was his capture of Mannheim in February 1799. He arrived before the city with his cavalry and proposed to Bernadotte they try stealth to take it rather than be involved in a lengthy siege operation. Leaving a few companies of infantry and some guns on the French side of the Rhine in full view he led the enemy to believe they were just a covering force. The remaining infantry and cavalry he kept well out of sight. Then disguised as a peasant selling vegetables

he crossed the river and entered the city unchallenged. He moved about freely gathering information on the garrison and the condition of the defences. The main problem he found was a deep moat that could only be crossed by an ancient drawbridge, which when up sealed off the city. While leaving he encountered a young woman heavily pregnant who knew the officer commanding the drawbridge and had arranged for it to be lowered at night if she needed a midwife in an emergency. With this in mind Ney hurried back to his camp, selected fifty of his best men and after dark transported them across the river in skiffs. They waited in the shadows near the drawbridge till the women in on the plan appeared. The drawbridge lowered, Ney's men surged forward and overwhelmed the guard. Making such a commotion the Austrian commander thought an army had entered the city immediately surrendered before Bernadotte's force arrived.

Bernadotte immediately recommended him for promotion but again Ney was reluctant. Not prepared to accept an excuses Bernadotte lost his patience and after a blunt exchange of letters Ney's promotion to *général de division* went through on 28 March 1799. The fact was the Directory needed a hero and if Ney was not suitably rewarded it would put them in a poor light. He had captured the imagination of Paris who were starved of good news from the armies on the Rhine. It wasn't just the Government who discussed the Mannheim affair but people in wine-shops, restaurants and fashionable salons. Instead of building stories around General Bonaparte's proteges like Murat and Lannes from the Army of Italy they could turn their attention to a new hero - Michel Ney.

On 4 May 1799 Ney moved to Massena's Army of the Danube and Switzerland commanding a light cavalry division before taking over Oudinot's advanceguard. With a force of barely 2,000 men he fought Archduke Charles at Winterhur on 27 May. He received a severe mauling, suffering some 600 casualties and another 100 prisoner before having to give up his position to a force five times the size of his. It was his first experience handling infantry during a retreat and he came out of the affair with credit, Massena personally commending him. Wounded three times by a ball in the leg, a bayonet thrust in the foot and a pistol shot in the hand he was out of action for several weeks recovering at Colmar. The next occasion he came to the fore was during the defence of Mannheim on 18 September. He was nearly killed when his horse was shot from under him and a cannon ball reopened his leg wound. That not all, a spent bullet caused severe bruising when it hit him in the chest and hurled him to the ground. He replaced Muller as acting commander of the Army of the Rhine from 24 September till 25 October when Lecourbe took over.

Both commanders helped to keep that sector of the front active to divert large numbers of Austrians away from Massena in Switzerland.

Ney was indifferent to Napoleon's overthrow of the Directory and the setting up of the Consulate after Brumaire. As a soldier he possessed a contempt for politicians and preferred the idea of a general ruling the country. Brumaire also resulted in the reinstatement of Moreau as commander of the Army of the Rhine in December for giving his tacit support to Napoleon during the coup. The army formed into four corps, Ney joined the Center Corps under Saint Cyr with command of its 1st Division. Compared to the spectacular successes in Italy culminating in Marengo, Moreau was a cautious leader and advancing on a broad front the campaign in Germany turned into a slow drawn out affair. Ney's troops headed the march across the Rhine and were in action at Engen on 3 May and again two days later at Moeskirsh driving Kray back on Ulm. Moving into Bavaria and down the Danube valley he fought at Hochstadt on 19 June and took Ingolstadt on 16 July 1800 before an armistice came into effect.

The renewal of hostilities in November culminating in Hohenlinden on 3 December saw Ney's profile to the fore, and more than anything brought him closer to Napoleon's inner circle and the marshals baton. The battle was one of the most significant of the Revolutionary Wars. It secured the Consulate, led to the end of the Second Coalition and the Treaty of Amiens. The campaign opened with the Austrians crossing the Inn as they tried to turn the French left and envelop their rear so forcing them from the forests and difficult country before Munich. Ney's division now with Grenier's 20,000 strong Center Corps blunted an advance by 40,000 under Archduke John's men at Ampfing on 1 December 1800. The contest was tough, and defending the heights of Ampfing he displayed great energy and courage holding back twice their number. Finally forced to give up his positions he managed to effect a retreat without any serious loss.

John gave up the idea of an envelopment and instead decided to take Moreau head-on in the forests before Munich. His advance in four widely spread columns that stretched across a front of several miles made it possible for Moreau's commanders to defeat the Austrians piecemeal. Ney with the French left fell back through the forests to Hohenlinden where Moreau decided to give battle. The weather conditions and terrain made it difficult to cover the six mile front on either side of Hohenlinden so Moreau with the main body formed up on the small plain before the village. To the right of the village Moreau placed Grouchy's* division, next to him was Ney and then further to the left the divisions of Legrand and Bastoul in the villages of Preisendorf and Horthofen. The divisions of Richepanse and Decaen formed the weakened

French center some miles to the right before Ebersberg.

In the morning of 3 December Riesch's column of some 20,000 men advanced down the Hohenlinden Muhldorf road and hurled itself at the French line. Ney and Grouchy* bore the brunt of the fighting and drove the Austrians back. The Austrians resumed their assaults several times till sufficiently weakened when the French formed into column and drove them through the defile and into the forest. Richepanse at Ebersberg with little opposition before him, had moved forward along forest tracks and emerged in the Austrian rear overrunning their baggage train and artillery reserve. With Ney pursuing from one direction and Richepanse the other the result was a rout. Over 8,000 Austrians fell, another 12,000 were taken prisoner and 80 guns abandoned.

Hohenlinden increased Ney's reputation immeasurably. Its strategy was Moreau's helped by a measure of luck and good timing by his subordinates, in particular Richepanse soon to die of fever in the swamps of Santo Domingo. The detailed battle reports showed Napoleon that Ney was one of the Army of the Rhine's principal fighting generals, a master of all arms who again showed his toughness in extreme difficulties. No longer was he simply an enterprising cavalry general who had made a name for himself capturing towns by stealth and cunning, he was now a commander of first order and an inspiration to all who served him.

A CHANGE OF ALLEGIANCE

The rivalry between Napoleon and Moreau could have caused lasting damage to Ney's career. Whilst he was not active politically, Ney was weary of Napoleon's ambitions as First Consul, considered himself a good Republican and therefore a supporter of Moreau as a counter to Napoleon. He soon became disenchanted with Napoleon as the First Consul sought to isolate Moreau. The Army of the Rhine was broken up and generals he knew as personal friends were passed over or given inferior postings. Ney on the other hand was drawn away from the political rivalries that developed between the two by succumbing to the oldest ploy by falling in love with a member of the opposition - Agalaé Aguié.

On leave Ney was summoned to Paris in May 1801 for a meeting with Napoleon. It was the first time they met and the encounter was frosty, Napoleon in all likelihood would have let him rusticate on leave or given him an inferior command. Ney in turn was not too worried, with France at peace he had already spent several months on leave and was confident his time would come when there was another call to arms.

Just as they were about to part Josephine interrupted the proceedings and with her charm completely captivated Ney. Walking him through the

Tuileries Gardens before he caught his coach she introduced him to Agalaé and her daughter Hortense. Agalaé was one of Josephine's ladies in waiting and a close friend of Hortense. The meeting was brief, but Josephine the match-maker *extraordinaire* noticed the chemistry and in the months that followed worked on it for the benefit of all; Ney, Agalaé and Napoleon.

Ney did not see Agalaé for another six months. He remained on leave till 1 December 1801 when his appointment as Cavalry Inspector General for the whole army came through. Ney showed ingratitude by asking to join the Santo Domingo expedition and on 18 December became head of Leclerc's cavalry, but the expedition had already left. While in Paris he saw Agalaé several times and they were soon deeply in love. On 1 January 1802 he wrote another letter requesting he remain in France. Napoleon had hooked Ney and accepted the sudden change of mind as he knew Ney was his. The marriage took place on 5 August 1802 and was probably the most successful of Napoleon's marshals. Ney was no philanderer, their love was deep and lasting; four fine boys were born.

SWITZERLAND 1802-1803

Through Josephine Ney made a good marriage, which placed him amongst Napoleon's inner circle of favorites. On 29 September 1802 he received his first truly independent command as head of the French forces gathering on the Swiss border at Geneva. His task was to re-establish order in the friendly but precarious Helvetian Republic. It was an important mission, requiring diplomacy as well as soldiering, and signified that not only did Napoleon favour him but trusted him completely.

Switzerland though small and politically insignificant had become a country whose strategical importance was recognized by France and her enemies. The principles of the French Revolution had penetrated the Swiss cantons, swept aside the old feudal federation with its oligarchic privileges and led to the creation of the Helvetian Republic with civil equality and religious tolerance. Following this it remained firmly allied to France. By 1802 however, external influences subsidized primarily by Britain and with the overt support of the Hapsburgs had nearly brought about civil war with counter revolutionary forces in the mountainous cantons rising against the central government in Berne. It was a situation France could not tolerate French forces gathered on the border.

Talleyrand as Foreign Minister urged Ney to use diplomacy rather than brute force. Ney set out on 17 October 1802 with the powers of a Minister Plenipotentiary to convince the government at Berne the need for French intervention. His move appeared too late as several cantons in the hands of insurgents were prepared to declare and fight for their independence. Within hours he ordered his troops to march on Zurich and break the insurgents. Their presence so overwhelming, they occupied the city on 29 October without resistance. It was a clever and bloodless victory where Ney using judgement timed his show of strength to perfection. The anti French faction leaders arrested, he treated them well and persuaded them of the mutual benefits of aligning with France. After patient negotiations he won them over. Peace restored led to the Act of Mediation, the basic constitution still in effect in Switzerland today. A trade and defence agreement was also negotiated with France and four regiments were provided for the French army. On 30 June 1803 he was the official French representative at the opening of the Diet at Freiburg that heralded the new government. His task of bringing Switzerland closer to France completed, on 17 January 1804 he was recalled from Freiburg.

THE CAMP AT MONTREUIL 1804-1805

On 4 March 1804 Ney took command of the Camp of Montreuil at Etaples near the modern resort of Le Touquet to prepare for the invasion of England. Without delay he threw himself into the task of turning the three infantry divisions assembled into a cohesive and effective fighting force. As a commander he was stern but just, a firm disciplinarian with a warm heart. He would never ask officers or other ranks to do anything, which he could not do himself. His corps generals Dupont, Loison and Mahler all remarked on his skill in handling men and his great physical strength. As for his troops they were prepared to follow him anywhere.

Apart from drill and musketry he made his men change their muskets for hammers, saws and shovels to make them proficient builders, earth-movers and trench diggers. His men had to realize that no army could expect to rely on its engineers alone. For landing in England he practised embarkation and disembarkation procedures, claiming his corps was the fastest in the army.

The Army's almost universal wish, except for a few old republicans, for Napoleon to become Emperor led to Ney writing on behalf of all at the Camp of Montreuil urging him to take the Imperial crown. This attracted the scorn of Moreau about to stand trial for his part in the Cadoual plot. On 19 May 1804, the day after the proclamation of the Empire Ney became *Maréchal de France*. He was twelfth in seniority, behind Mortier and before Davout, and the only one from the Army of the Rhine.

Apart from interludes like attending Napoleon's coronation, Ney spent much time at Montreuil. His military thinking became influenced by the theorist and professional soldier Jomini who joined as an aide and was to serve him on and off till 1813. Jomini criticized the outmoded rigid tactics of Frederick the Great and encouraged the maneuverability of formations adopted by Napoleon in his Italian campaigns. Ney helped the Swiss writer to publish his great work, *Traite des Grandes Operations Militaires* that eventually stretched to eight volumes.

THE AUSTRIAN CAMPAIGN OF 1805

When VI Corps received orders on 24 August 1805 to break camp at Montreuil it numbered 24,500 men with 25 battalions and 12 squadrons. The training soon paid as it traversed France via Arras, Rheims and Chalons in less than two weeks. The fact that Ney was on leave in Paris reflected the suddenness of the move and he rejoined his troops at Saint Dizier. VI Corps crossed the Rhine at Seltz where to their credit the corps engineers constructed a pontoon from boats and barges in fifteen hours. Through Württemberg and Bavaria he marched as part of the Grande Armée's great encircling movement, while Murat's cavalry disguised the movement by forming a great cavalry screen through the Black Forest, the traditional route of previous wars. From Stuttgart VI Corps moved steadily down towards Ulm through the Swabian Alps. When Ney reached the Danube at Günzberg on 9 October he heard that with Lannes (V Corps), he was under the operational control of Murat. Murat a great cavalry leader was no strategist. He ordered VI Corps to cross the river leaving only Bourcier's dragoons on the north bank. It left the Grande Armée's communications dangerously exposed if Mack to avoid encirclement attacked in that direction. Lannes and Ney warned Murat of the dangers. An acrimonious meeting followed, blows were nearly exchanged till Ney at least forced Murat to concede that Dupont's division should remain on

the north bank. Subsequent events were to vindicate Ney when what he feared happened. Obeying orders he brought his troops to the south bank while on 11 October the Austrians fell on Dupont. They nearly broke through as Dupont's casualties mounted, but remembering their discipline and training at Montreuil the French fought with amazing courage. Mack convinced he faced a far greater number fell back on Ulm. Napoleon galloped to Murat's headquarters extremely worried by the close call. Without the bravery shown by Dupont's division his entire campaign may have been ruined. At first Napoleon blamed Ney for leaving his subordinate on his own, but as the facts came out he rounded on Murat with the crushing comment, "to stick to his horsemen." From then on individual corps commanders were responsible to him alone.

By 13 October the French had virtually closed all the gaps around Ulm, when on the morning of the 14th Ney led his celebrated attack across the bridge at Elchingen. The Austrian commander Riesch unable to demolish the bridge had fortified the plateau that looked down on it with 9,000 men and many cannon. On one flank the Austrian was protected by a dense wood, on the other by the village of Elchingen and a monastery full of troops. He hoped to keep the French pinned down long enough for Mack to attempt another break-out to the north past Dupont. Napoleon watched as Ney's attack with VI Corps went in at daybreak. Ignoring heavy artillery fire Ney was in the midst of the struggle for the bridge. He helped to throw makeshift planking across and then led the first infantry over with fixed bayonets. The

THE BATTLE OF ELCHINGEN

turning point came when he returned to lead his cavalry over the loose boards in a fine charge that helped effect the break-out from the bridgehead. His infantry went up the slope and cleared the plateau while he led his horsemen forward to take the village by storm. It was a magnificent feat, capturing over 4,000 Austrians and taking over forty cannon. More importantly, the escape route along the north bank was blocked, and the fate of Mack's remaining 30,000 men in Ulm was sealed. Marmont and the Guard had almost reach the outskirts of Ulm along the southern bank, while Soult* had steadily moved up the Iller's west bank from Memmingen cutting any chance of escape to the Tyrol. On 21 October Ney and VI Corps, having endured the most, were given the honor of occupying Ulm. It was their contribution more than any other formation that was responsible for the first major military triumph of the Empire.

Ney was detached to occupy the Austrian Tyrol and took no part in the Battle of Austerlitz. Supported by several Bavarian units the invasion went without a hitch. He drove back Archduke John and took Innsbruck on 7 November and a week later all resistance in the region ceased. After the Treaty of Pressburg he remained as Governor of the Tyrol and did not return to Paris till 13 September 1806, only ten days before his recall for the campaign against Prussia.

PRUSSIA AND JENA 1806

Leading the 24,000 men of VI Corps for the campaign against Prussia Ney at first did not have a very successful time. Having missed the glory of Austerlitz the previous year, the pursuit of further glory resulted in an overeagerness that seriously jeopardized Napoleon's plan at Jena on 14 October 1806. The preliminary action at Jena started at 6.00 A.M as the French advanced from the Landgrafenberg to secure a larger area of ground to help the deployment of troops for the coming battle. The move was entrusted to Augereau on the left, Lannes in the center, and Saint Hilaire's division of Soult's corps on the right. By 10.00 A.M with the help of a slowly lifting fog that helped conceal the initial French movements, these aims were largely achieved in spite of determined Prussian resistance.

Ney's advanceguard numbering barely 4,000 men then made its appearance on the field as the advance element of a central reserve to be formed by VI Corps. Ney had waited several hours for them and anxious not to miss the battle resented the inactivity. Chafing at the delay he launched an unauthorized attack on his own account. With two regiments of cavalry and five battalions of infantry he moved into action between Lannes's left and Augereau's right, covered from everyone's sight by the lingering remnants of the morning fog. Thinking little of the overwhelming odds he faced, he made straight for a powerful Prussian battery. At first fortune favoured his

boldness, the right of the Prussian line crumbled at his approach and he drove the Prussians from their pieces. It caused forty-five squadrons of Prussian cavalry to turn about, which were to launch a charge against Lannes' troops around Vierzehnheiligen. The success was only temporary as the enemy cavalry recovering from the surprise returned to the attack. Ney was forced to form square to withstand the new onslaught. Worse still the impetus of his advance had carried him beyond supporting distance of both Lannes and Augereau, and consequently he was soon cut off.

Ney's insubordination forced Napoleon to launch his attacks early. Bertrand* galloped forward to the rescue with only two regiments, which represented the total cavalry reserve since Murat had not arrived on the field. At the same time Lannes was told to press forward and re-establish contact with him, while Augereau made haste to form a second line to Ney's rear. As a result the steady French advance ground to a halt as attention and resources were diverted to save Ney, giving the Prussians valuable time to reorganize. Fortunately the battle turned out well in the end, due more to poor Prussian leadership than anything else. General Savary, an Imperial Aide, later wryly described Napoleon's reaction to Ney's foolhardiness. "The Emperor was very much displeased at Marshal Ney's obstinacy. He said a few words to him on the subject - but with delicacy." Napoleon's venom was reserved for Bernadotte, whose men didn't fire a shot all day.

Anxious to redeem himself Ney joined the vigorous pursuit spending the first night after the battle at Weimar. From there he moved on to Erfurt where he had to sort out the 14,000 man garrison that surrendered to Murat on 17 October. Rather than interrupt his march on Berlin, Napoleon diverted Ney north to Magdeburg on the Elbe, to take the most strongly fortified city in Germany. The siege started on 20 October after Murat's summons to surrender was ignored. Ney had only 18,000 men with him to face a garrison of 25,000. After the commander Kleist made an unsuccessful sortie from the walls on 4 November, Ney threatened an intense bombardment, which resulted in negotiations. An armistice was concluded on the 7th, a capitulation signed the next day and on 11 November the garrison marched out as prisoners of war. Napoleon on hearing the news sent Ney his warmest congratulations; the indiscretions at Jena were forgotten.

THE WINTER OF 1806-1807

The campaign in Poland and East Prussia was to see Ney at both his worst and best. He crossed into Poland at Frankfurt, passed through Posen, then cut north-east to arrive before Thorn on the Vistula on 6 December. Lestocq's rearguard disputed the crossing. That night some French light companies passed the river in boats and after some sharp fighting in the town the Prussians withdrew. The Vistula took ten days to rebridge; then Ney slowly pushed Lestocq into the depths of the lakeland and forest region of central East Prussia. His corps exhausted after fighting around Neidenburg and Soldau, he set up near the latter on 26 December. With the French across the Vistula and Bug in force, Napoleon ordered no further forward movements of any kind until the spring as he wished to build up his forces.

Ney's impetuosity led him to break ranks and disobey orders, arousing the sleeping Russian army and forcing a winter campaign on Napoleon he dearly wished to avoid. The weather and lack of food forced his men to make a forward move from their bivouacs. He pushed through the Polish lakeland region around Allenstein as far as Heilsberg, before turning south on the 17th. The move opened a great gap in the French front and helped Bennigsen decide to launch an unexpected and unseasonal offensive through northern Poland to press the French back to the Vistula. The Russians fell on Bernadotte's Corps and Ney's scattered troops as they continued their winter migration. Napoleon was furious at Ney for deliberately flouting his orders, and blamed the advance for stirring up a hornet's nest. Then as Napoleon reacted with startling speed Bennigsen lost his nerve and began to retire. The result was the Battle of Eylau on 8 February 1807.

Ney's part in the battle or lack of it, to say the least was unflattering. He had orders to intercept Lestocq's 9,000 men to the north-west but the wily Prussian kept giving him the slip. The deadening effect of the snowstorm made Ney unaware that a major battle was unfolding nearby at Eylau and a summons from Napoleon sent at 8.00 AM only reached him six hours later. Lestocq aware of the battle managed to slip away and nearly won the battle single-handed when his columns fell on Davout's exhausted troops in the late afternoon. Ney once he received Napoleon's message acted fast, ploughed his way through Lestocq's rearguard and arrived at Eylau three hours later in time to restore French fortunes. It wasn't that VI Corps did much fighting, but rather the appearance of fresh troops ready and eager for action that put new heart into the wavering French. Resistance stiffened all along the line and when another Russian attack failed Bennigsen gave up and retired leaving the French possession of the field.

THE FRIEDLAND CAMPAIGN

Eylau was a pyrrhic victory, and the French retired to winter quarters in the towns and cities along the Passarge with Ney in command of the rearguard. At Güttstadt on 1 March he settled down to face Russian troops estimated at 60,000. For the rest of March, through April and May he held the line, frequently at bayonet point. Finally when the Russians opened a new offensive on 5 June he fought a brilliant delaying action at Deppen on 6 June disputing every yard till he retired over the Passarge.

Ney won time for Napoleon to unleash his counter-blow the next day by driving north-east with the East Prussian capital of Konigsberg as the objective. It was nearly stopped by the Russians at Heilsberg on 10 June, but resumed till the decisive battle at Friedland on 14 June 1807. The Russians retreated to the position along the right bank of the Alle to avoid being outflanked in their effort to retire on Konigsberg. Then when Lannes (Reserve Corps) appeared before the town on 13 June they obligingly began to pour troops back over the river to destroy him before the arrival of reinforcements.

The battle started around 9.00 A.M as the Bennigsen launched his 60,000 Russians against the 35,000 of Lannes and Mortier (VIII Corps) supported by cavalry under Grouchy*. The French held as reinforcements including VI Corps strengthened their line. Napoleon mindful that VI Corps had missed Austerlitz and only played minor roles at Jena and Eylau gave Ney the opportunity to lead the main blow against the Russians. By 4.00 P.M Napoleon had massed 80,000 men and considered the battle ripe to launch his blow. He waited a little longer to allow the Russians to commit themselves further, then satisfied gave the signal for Ney's advance when a twenty gun battery fired three salvoes.

Bennigsen had no forewarning of Ney's coming attack. His observers in the church steeple at Friedland had warned him that the French were constantly being reinforced, but they were unable to see Ney's divisions supported by artillery forming up in the woods. As a precaution before withdrawing behind the Alle Bennigsen shortened his front by withdrawing from the Sortlack Woods and resting his left in the village by the river. As Ney's attack burst from the forest, Bennigsen cancelled all orders to withdraw and instead poured his remaining troops on the left bank across the river to stem the new threat. Ney's attack came on in echelon, Marchand's division leading the right along the river while Bisson formed the left. Both divisions were in cramped formations, just as they had assembled in the roads in the forest, but Ney did not pause to deploy. Using the Friedland church steeple as a marker as it was the only part of the town not hidden by the undulating ground, Marchand quickly drove the Russians holding Sortlack into the river. Bagration threw a mass of cavalry against Marchand's left flank, but they were routed by Latour-Maubourg who had moved into the space between Marchand and Bisson. Marchand with ten battalions formed into a single column of battalions rounded the projecting loop of the Alle and moved directly on Friedland but was raked by Russian guns on the left bank. At the same time, Bagration concentrated his own guns against the heads of the French divisions riddling their deep ranks. Staggered by this crossfire Ney's advance halted. On his left Bisson's troops formed in two lines of battalion columns attempted to deploy, but were caught by Russian cavalry and sent to the rear dragging Marchand's infantry with him. Only three regiments managed to form square and stand firm amidst the rout.

Covering all eventualities Napoleon had ordered Victor forward when Ney's advance began. Then Dupont (later of Baylen fame) on his own initiative at the head of the lead division of I Corps, came forward smartly to support Bisson's left. Latour-Maubourg's cavalry charged again and rode down the enemy cavalry still sabering the broken battalions of VI Corps. Behind Latour-Maubourg, Ney's artillery disabled the Russian guns on the opposite bank one by one. Victor's artillery chief Senarmont accompanying Dupont's division with two fifteen gun batteries on either side did fine work blasting a great hole through Bagration's infantry forming a new line. In 25 minutes at a range of 120 yards they inflicted over 4,000 casualties and sent the rest streaming back to Friedland. While Dupont was knocked to pieces by the Russian guns it gave Ney time to rally his men. Together they then surged into Friedland. Amid scenes of growing disorder Bennigsen tried a third time to stem Ney's maddened infantry with the Russian Imperial Guard. With the bayonet, the French cut through them like a knife through butter. By 6.00 P.M. Ney reported the capture of Friedland and the seizure of the four bridges over the Alle. What was left of the Russian army on the left bank was trapped. It was the signal for Napoleon to complete the victory by rolling up the Russian line north of Friedland with Lannes, Mortier and Grouchy's cavalry.

Friedland was decisive in bringing the war to an end. It forced the Russians to open negotiations, led to the disarming and partial dismemberment of Prussia, gave the Poles their freedom and offered hopes of a European peace. The French lost 7,000 men, over 2,500 from Ney's two divisions alone, compared to 25,000 Russians and 80 guns. Lannes was the hero of the day for his part holding the Russians before the Army's arrival. It was Ney without a doubt who was the hero of the hour by turning a successful holding action into a decisive victory. At Friedland he led the finest and most successful attack of his career. Napoleon quick to single him out for praise observed, "that man is a lion."

THE HEIGHT OF GLORY 1807-1808

For Ney after the signing of the Treaty of Tilsit a year of peace followed. On 30 June 1807 besides his pay of 40,000 francs as a *maréchal* he received an annual annuity of 28,000 francs. Westphalia gave him a grant of 100,000 francs while Hanover had to fund another annuity of 83,000 a year. It was followed by a cash award of 300,000 francs plus a further 300,000 in Government Bonds. Such awards made Ney a wealthy man and with other leading lights of the First Empire he led a hectic social life. He was also a national hero. On 6 June he joined the Imperial Nobility with the title Duke of Elchingen in recognition of his first success under Napoleon. He wanted to add something to his life-style. Algae heavily pregnant with their third child they travelled to the Château Coudreaux near Châteaudun in Eure-et-Loire, fell in love with the property and purchased it in May 1808.

THE INVASION OF SPAIN, THE CORUNNA CAMPAIGN 1808-1809

On 8 August 1808 Ney was summoned to join the Army of Spain. He was to spend nearly two-and-a-half difficult years in the Iberian Peninsula during a period of misfortune for French arms. It was to show Ney in particularly poor light as a thoroughly insubordinate, unreasonably touchy, quarrelsome and uncooperative commander. From the heights of glory his reputation was to plummet, ending with his ultimate dismissal from the army.

By 30 August Ney reached Irun on the Spanish frontier. From there he moved to Vitoria to secure a base across the Pyrenees for the arrival of VI Corps and the rest of the army. For two months he reconnoitered and probed in an ever widening arc, gathering all intelligence he could pending Napoleon's arrival. The arrival of Lefebvre (IV Corps) enabled Ney to launch a minor offensive. On 26 September after moving north-west he captured the port of Bilbao. The move surprised Blake's Army of Galicia and drove them east straight into the path of Lefebvre who beat them at Durango on 31 October.

The foray ended when Napoleon arrived at Vitoria on 5 November. With a Grande Armée numbering 150,000 men he outlined his strategy for the conquest of Spain. Victor took charge of the right-wing and defeated Blake at Espinosa on 10-11 November. Moncey (III Corps) was on the French left to deal with the armies of Castanos and Palafox. Ney following Soult (IV Corps) was part of the Grande Armée's main thrust that fell on the Spaniards outside Burgos and from there was to bludgeon its way through to Madrid. When Ney reached Aranda Napoleon started a complicated manoeuvre by ordering Ney to swing east up the Douro and then head for Calatayud via Soria or Almazan as circumstances dictated. He was the southern jaws of a great pincer movement against Castanos which Moncey and Lannes would close from the north.

The plan at first went well, on 23 November 1808 Lannes routed Castanos at Tudela, chasing the Spaniards southwards in headlong flight through the passes of the Sierra Moncaya supposedly into the arms of Ney at Calatayud. Instead Ney was at Soria miles behind his ordered schedule where he had halted for three days on hearing reports that Castanos with 80,000 men had defeated Lannes and pushed him back to the Ebro. As a result of the ruse Castanos escaped virtually intact, Palafox shut himself up in Saragossa with Lannes and Moncey spending many months in a costly siege. Napoleon was furious to hear Ney had been fooled by unsubstantiated rumors and reports. In Ney's defence, the French were new to the area. They lacked maps, and there was no time to carry out a proper reconnaissance. The terrain was appalling, with some of the worst roads in Spain. The population was hostile and the mountains through which he passed were infested with *guerrilleros* who harried every step of the way. Communications with Lannes were nonexistent.

The fall of Madrid restored Napoleon's confidence in Ney and his mistakes were soon forgotten. VI Corps was reviewed by the Emperor as it entered the capital on 14 December. Plans were put in hand for Napoleon to continue the march to Lisbon when on 19 December news broke that Moore's force threatened Soult and his communications in the north. Napoleon spurred into action on 22 December, left Madrid with Ney and 42,000 men to intercept Moore. The two hundred mile march in appalling weather through wind rain snow and mud, crossing the Guadarramas with Napoleon at his side was one of history's epic marches. Not prepared to face such an eventuality in such conditions, Moore's began to fall apart as both armies raced to cut the bridges over the Esla at Benavente. On 29 December Ney was present at Benavente when Napoleon's Guard Chasseurs à Cheval received a mauling from the English rearguard resulting in the capture of Lefebvre-Desnoettes*. A further 60 kilometers on at Astorga on 1 January he called a halt and Soult (II Corps) took over the pursuit to Corunna with his troops providing the supporting role. A personal tragedy occurred at Caceballos on 3 January when Ney's close friend and commander of his cavalry Auguste Colbert was killed by a sniper.

GALICIA AND LEON 1809

While Soult made preparations to invade Northern Portugal after the English evacuation at Corunna Ney was left to police Galicia and the Asturias. The job was well nigh impossible. Unhelpful advice kept arriving from Napoleon to hand over administration to the Spanish allowing him to move out from Corunna and Ferrol to crush the opposition in the provinces. Napoleon simply didn't appreciate that there were no Spaniards prepared to administer the region since to a man they violently opposed any form of French occupation. In early May he brought up Kellermann* from Leon and Bonnet from Santander pulling off a fine converging attack on the Asturian capital Oviedo. While there trouble broke out behind him in Galicia, while Santander was captured to the east. The French retired, Bonnet to retake Santander, Kellermann to Leon and Ney to Galicia.

No sooner was Ney in Galicia when Soult's II Corps routed by Wellesley at Oporto descended on him calling for help and supplies. Ney's surprise soon turned to anger as Soult regarding himself the senior commander demanded Ney meet his every need. Ney in return became difficult and refused any help at all. Not a musket, gun or uniform was set aside. A violent spate of feuding broke out between the two marshals, which became so bad staff officers were duelling. On one occasion at Lugo Ney in fury even drew his sword on Soult before the two were parted. Their behavior was appalling and led to a series of incidents that resulted in the permanent loss of Galicia. First they appeared to bury the hatchet by concluding the Convention of Lugo on 30 May 1809 laying down their respective responsibilities for waging war in Galicia. Ney marched south from Corunna, and was to fall on Romana and capture Vigo. Soult at Lugo agreed to support by moving on Orense, and between them to trap Romana. Ney started the movement and on 8 June met stiff resistance on the coastal road crossing the Oitaben. He summoned help from Soult and on 11 June learnt that the marshal after reaching Monforte had turned about and marched for the safer plains of Leon. Soult's reason was that his priority was to keep an eye on Wellington's movements on the Portuguese frontier where the well provided depots in Leon were better able to keep his troops replenished.

Abandoned by his colleague, the problem of Galicia left to him with barely 18,000 men, Ney refused to risk the safety of VI Corps in such an unequal struggle. He evacuated Corunna and Ferrol and on 22 June concentrated his whole force at Lugo. There he picked up the sick and wounded left by Soult and by forced marches reached Astorga on 30 June. He was harried by guerillas but Romana did not dare to meddle with him. Ney's anger knew no bounds as he sacked every place he passed, from Villafranca and Ponferrada down to the smallest hamlets. Twenty-seven Galician towns and villages were burnt by VI Corps during its retreat. It was conduct unworthy of Ney, he had previously always taken care to separate guerillas from peasants and townsfolk. From his earliest days his sense of honor always dictated that war was for soldiers and not to be waged against the common people. His act served no military purpose at all, his critics considered it simply vindictive spite, a smear on himself and the army. In sympathy, frustrated after nine months in Spain where he had achieved nothing, abandoned by a fellow marshal and harried out of a region by guerillas, his nerve simply snapped with regrettable consequences.

Worse news was to follow when on 1 July Ney and Mortier were placed under Soult's orders. He joined Soult's operations to trap Wellington's army after Talavera when Soult ordered him south from Leon to Plasencia as part of his plan to block their retreat. Wellington saw the danger and crossed the Tagus at

Arzobispo. Soult still thinking there was a chance to intercept Wellington ordered Ney to Almaraz. There he hoped to use a ford near the broken bridge to cross the river, but when he arrived on 8 August he failed to find it. The English able to retire undisturbed through Trujillo, Medellin and Badajoz to Portugal and the campaign fizzled out. Ney ordered to return to Leon by Joseph immediately started his march north to be rid of the influence of his old rival. Ney's departure raised old wounds as it was now Soult's turn to feel abandoned. A great opportunity to follow Wellington into Estremadura and Portugal was lost, as there were no defences at Torres Vedras to overcome. Ney on his northern march had an unexpected success on 12 August when nearing Salamanca he ran into the elusive Robert Wilson's Lusitanian Legion and gave it a severe mauling at Banos.

Ney suddenly received a letter of recall on 4 October 1809 and Marchand took over command. He had fallen victim to Bonaparte intrigue. Joseph not happy with Victor's performance at Talavera had offered Ney command of I Corps without referring the matter to Napoleon. Ney was pleased as it would distance him from Soult, but had wisely not accepted without Napoleon's blessing. When Napoleon heard of Joseph's move he inexplicably rounded on Ney and ordered his recall to France. Recalled for the wrong reasons, Ney accepted it with good grace since Spain appeared to him no place for glory. In twelve months, through no real fault of his own, his career had gone from bad to worse as he had tumbled from one set of mishaps to another.

MASSENA AND THE ARMY OF PORTUGAL

After a brief stay with his family Ney had an interview with Napoleon, who wanted him back in Spain. The situation had changed; while away news came through things had gone badly for Marchand and VI Corps. His deputy had underestimated Del Parque, marched out of Salamanca to challenge him and was himself beaten at Tamanes on 18 October - the first French defeat at the hands of a Spanish army since Baylen. With the threat from Austria removed that had caused Napoleon to leave earlier in the year, the time had come to resolve matters in the Peninsula. Buoyed up because the Emperor in the new year was to take charge personally Ney felt reasonably optimistic about the future when he arrived back at Salamanca in December.

His optimism proved ill founded. Whilst reinforcements poured over the Pyrenees there was no sign of Napoleon. Napoleon meantime had given up all serious thought of returning to the Peninsula and on 17 April 1810 appointed André Massena to command a new Army of Portugal. This army was an autonomous command, made up of VI Corps (Ney), II Corps (Reynier) and VIII Corps (Junot), and Massena apart from invading Portugal would have total responsibility for the province of Leon and Salamanca.

Ney respected Massena, Napoleon's foremost independent army commander, whose reputation went back to the Army of Switzerland where Ney served under him. His leadership of the Army of Italy during the Siege of Genoa was inspirational as was his performance in 1805 when outnumbered by the Archduke Charles his generalship helped by the Capitulation of Ulm forced the Austrians out of Italy. He looked on Massena's coming as at least better than the status quo. Bickering between generals and marshals, and the inept overall leadership of King Joseph he felt were the root of all problems.

That Ney's admiration soon turned to uncooperatveness, disrespect and outright insubordination was due to his inability to come to terms with Massena's greed, avarice and sex life. Massena's wanton plundering of Spanish property and artifacts for personal gain he could not abide and soon made it clear when he returned gifts of stolen property. Ney a happily married man disapproved of Massena's womanizing, but was prepared to go along with provided it did not cloud his chief's judgement. However when the conduct of the campaign was affected by Massena being accompanied by his mistress dressed as a hussar all respect went out the window. He was not prepared to conduct interviews through a closed bedroom door while Massena entertained his lady. Again the failings he previously displayed with Soult and King Joseph appeared, but that was still several months down the line.

Massena faced a formidable opponent in Wellington whose forces were much stronger than they were in 1809. Besides the basic force of 33,000 British and Germans, Wellington could call on 20,000 regulars of the Portuguese Army, well armed and equipped whose training by Beresford had worked wonders. Massena theoretically he had control of over 130,000 men but after garrisons and detachments he had fewer than 65,000 under him at any one time. Ney had the bulk of the forces with direct operational control over some 33,000 centered around Salamanca.

Ney opened the campaign by moving from Salamanca to reconnoiter Ciudad Rodrigo. He found the fortress garrisoned by the Spanish veteran Herrasti and 5,500 regulars. Nearby was Wellington's Light Division led by Craufurd in a thin line along the Agueda river. Wellington's main force was deep inside Portugal at Viseu. Based on this information he urged Massena to mask Ciudad and Almeida with a division and punch a hole through Craufurd's line and launch a strike direct at Wellington's headquarters. Massena opposed it, deciding as Napoleon had suggested to first reduce the border fortresses by regular siege operations. Ney on the spot felt he knew better and sullenly brought up the siege train to Ciudad and began to invest the fortress. It was the first clash of wills and for Ney the start down the slippery slope that ended in his dismissal.

Ney's three divisions were placed in a tight circle around the fortress with his cavalry deployed before Craufurd to prevent any surprises. The sappers started digging and by 15 June their first parallel was completed. Five days later batteries in place, the guns opened up a bombardment that was duly rewarded on 25 June when a direct hit on the town's central powder magazine caused it to explode. Herrasti to his credit after such a shattering explosion held on and it was to his credit that only when a breach was made on 9 July did he agree to surrender.

The siege also was the cause of another incident between Ney and Massena that nearly was to terminate his career in Spain. Colonel Valazé an officer on Junot's VIII Corps staff appeared carrying orders from Massena to take over siege operations from Ney's highly regarded engineer Conche. It appeared to Ney with the fall of the fortress imminent that Junot wanted to take some of the credit. Ney sent him back to Valladolid with the stinging jibe, "I don't want the Duc d'Abrantes to bother me with his protégés. If they are so good let him keep them for himself." Massena lost no time sending the unfortunate Valazé who was a very capable engineer back to Ney. Furious Ney in turn produced a letter that, while protesting his esteem for his commander, was so insolent that it provoked the normally patient Massena to order him to accept the man or resign. Massena got his way, Ney backed down and Valazé conducted the siege operations from 29 June till 10 July when he was severely wounded by a grenade. It was with credit that Ney held nothing personal against the engineer who went on to serve under him in III Corps in 1813.

On 21 July Ney started operations against Almeida. He had a success against Craufurd on 24 July when against Wellington's orders the Light Division deployed south of the fortress. The whole of VI Corps descended on him at Coa driving him across the river leaving Almeida totally isolated. At Almeida Ney was more fortunate. After completing the investment he opened a bombardment on 26 August, which after 14 hours found the defender's powder stored in the cathedral. Seventy tons of powder blew up devastating half the town forcing Brigadier Cox with the remnants of his 1,200 man garrison to surrender.

The way into Portugal now open, Wellington with his 51,000 men rather than retreat to Lisbon and the lines of Torres Vedras chose to make a stand before Coimbra on the eight mile ridge running north to south at Bussaco. Massena superior in cavalry could have outflanked the position but chose to attack it head on. It was a grave mistake by Massena, who like Napoleon compared British infantry's will to fight with the days of the Duke of York in Flanders. So overconfident he didn't complete a proper reconnaissance, it was also reckoned he was more concerned with the well-being of his mistress eight miles to the rear. Hardly on speaking terms with Ney, he also didn't heed his subordinate's advice advising caution and a possible flanking movement towards Oporto.

In the event, on 27 September Ney dutifully accepted Massena's plan. Before dawn his troops moved into position. At 7.00 A.M his attack began as skirmishers cleared the lower slopes. Setting off up the Coimbra road from the village of Moura, Ney divided VI Corps into two separate strike forces, twelve battalions under Loison to the north and eleven more to the south under Marchand. As they toiled up the rock-strewn slopes they met a hellish fire from artillery further up the slopes. Ney's troops believed in him and made steady progress. Loison drove Craufurd's Light Division back, 1,500 Rifles and Caçadores from the village of Sula. Attacking the strong-points further up they came under withering artillery fire, managing to overrun a battery of twelve cannon and Craufurd's command post at Sula Mill. Reaching the crest they were confronted by Wellington's reserve. Seconds later 1,800 fresh troops descended on the battered exhausted French hurling them down the slopes. Ney not surprisingly had a sense of foreboding about the battle and uncharacteristically remained at the start line. With his third division led by Mermet he was able to stem the rout. Loison pursuing a hopeless venture up the slopes also lost over 1,000 men.

Meanwhile, Marchand's division south of the Coimbra road had fought its way to within sight of Bussaco Monastery. Ney for a time led the attack but artillery fire supported by the steadiness of Spencer's Foot Guards forced them back. Ney's men had done all they could. It was a hopeless attack against hopeless odds. The men they faced were the finest in the British Army and they had pressed them hard. The skill and tenacity the enemy showed, left an unforgettable impression on Ney. By midday Ney had abandoned the fight and was back at Moura. Bussaco was the costliest defeat in the Peninsula so far with 4,500 casualties opposed to 1,200 Anglo-Portuguese.

In danger of Massena's cavalry outflanking his position, Wellington withdrew and Coimbra fell to the French on 1 October 1810. Indiscipline amongst the troops resulted in the sack of the city, and weeks worth of valuable supplies were lost. Ney in disgust returned a telescope taken from the University and given to him by Massena claiming he would not accept stolen goods. Ney also had reservations about continuing the march on Lisbon. He urged the army rest at Coimbra, collect provisions and most important of all reopen communications with Almeida and Ciudad Rodrigo, as XI Corps marching from France was due on the Spanish border to join them. Massena chose to press on and without providing a garrison for Coimbra left 3,500 sick and wounded who within days were taken when the Portuguese reoccupied the city.

Two weeks later the French encountered the Lines of Torres Vedras. Ney not one to shirk a frontal assault realized they could not be broken while Wellington lay behind them with an army that now outnumbered Massena. The campaign developed into a stalemate. Massena's only hope was that Wellington would risk an assault on the French before the lines. Communications to Spain severed, the surrounding country swept bare of food and forage, it was the French that became the besieged. The deadlock could be broken if Soult's Army of the South could pass through Alemtejo and approach Lisbon south of the Tagus. Such a step needed Napoleon's direct intervention and that never occurred. By February 1811 the situation had become critical as the army had withered down to 44,000 men. Ney at a conference of senior commanders held at Golegao on 19 February 1811 proposed the temporary abandonment of Northern Portugal by crossing the Tagus to join Soult. On the Guadiana the united armies could complete the conquest of Estremadura and then re-enter Portugal through Alemtejo and attack Lisbon from the south. Massena rightly rejected it, as it meant the surrender of all the previous years gains and allowed Wellington to fall on Almeida and Ciudad Rodrigo. To cross the Tagus before Wellington was too risky and Alemtejo sparsely populated could not hope to support a French army.

Massena took the sensible option and decided to retreat the way he had come and on 3 March 1811 his troops started to abandon their positions. Once the Army of Portugal cleared Santarem Ney's VI Corps formed the rearguard. Wellington took up the pursuit with vigor but his generals met their match when they came up against Ney the rearguard fighter. Repeatedly he delayed the Allied advance to win time for Massena. The qualities he displayed were to repeat themselves in Russia the next year. His first successful action was at Pombal on 11 March when with Mermet's division outside the town he brought the Light Division crudely handled by Erskine to a halt. The next day he fought at Redinha, again using Mermet with Marchand in support. The same tactics were applied. Ney appeared to deploy in strength, let the enemy build up their strength, then once they started to advance he withdrew after offering token resistance.

The tactic was very effective and won time, however it did not always work. On 13 March the enemy turned his position and he had to make a hurried withdrawal from Condeixa. The move exposed Massena's headquarters at Fonte Coberta and a sudden cavalry raid nearly resulted in his capture. The incident caused another blazing row as Massena thought Ney had deliberately endangered him. As rearguard commander he bore the brunt of

the fighting. Another crisis arose when a small Portuguese force occupying Coimbra could not be forced from the city. It delayed Massena's march and Wellington's lead divisions caught up and pressed Ney hard at Casal Novo, Miranda do Corvo and Foz de Arouce on 15 March. A route around Coimbra found, there was a lull as Wellington had outrun his supply trains. On 22 March the Army of Portugal reached Celorico and was able to regroup after a sixteen day retreat. Massena with Ney's invaluable help had extricated himself from a previously impossible situation for the loss of under 1,000 men. The army was safe, only thirty-five miles from the frontier the main depots in Spain were in easy reach.

The immediate danger passed on 22 March Massena inexplicably decided to lead the army in a reckless venture. He astonished and appalled his corps commanders when he decided to adopt Ney's old plan to join Soult. He ordered the retreat on Almeida and Ciudad Rodrigo not to continue and instead the army to strike south through Coria, Plasencia and Alcantara into Estremadura. Such a move was quite beyond the exhausted and emaciated troops who had suffered months of privation. Ney refused to point blank to comply and came up with several excellent reasons for continuing the march to Almeida. When Massena rejected them he stated verbally and in writing VI Corps would not accompany the Army of Portugal on such a venture. It was the supreme crisis of the stormy relationship between the two marshals. Ney's was a foolish act and Massena relieved him of his command and ordered him to leave the army immediately. Ney, right by way of strategy but absolutely wrong in committing such a gross act of military insubordination, Massena had no choice but to act in the way he did. It was a sad departure for his corps, where Ney knew every officer by sight and many of the men. He had been with many of them since the Camp of Montreuil when VI Corps first gathered in 1804. So outraged was VI Corps that they were prepared to overthrow Massena, but Ney baulked from the prospect of declaring himself commander-in-chief though he knew Massena was universally unpopular within the army.

Ney left and Loison took charge of VI Corps. The army headed south for five days then found the country so inhospitable, it retraced its steps. Ney's remonstrances, but not his insubordination, was vindicated. VI Corps went on to fight at Fuentes d'Onoro where it broke itself against Wellington's infantry. After Massena's dismissal in May 1811 the Army of Portugal was reorganized and VI Corps disbanded.

REHABILITATION

For a marshal to be relieved of his command was a serious matter. Napoleon was furious as it brought discredit to the Army. However, as Massena's failings came to light Ney got off with just a severe reprimand and was ordered to cool his heals at Coudreaux. In mid August he received orders to report at the Camp of Boulogne where he took over the establishment from Vandamme on 31 August 1811. He hated the posting, as the camp was drab and run-down. Gone were the days of order and enthusiasm as the troops prepared for the invasion of England. It was a military back-water second to none with no prospect of action or glory.

Relations deteriorating with Russia, Ney was soon back in favour as Napoleon needed his best strike commanders. In January 1812 the two divisions at Boulogne became the *Corps d'Observation des Côtes de l'Océan* with Ney as their commander. That move was followed by orders for his formation to move to Mainz. After the arrival of more troops and a tough period of training the formation on 11 April 1812 was redesignated III Corps of the Grande Armée. Ney's German was soon used to good effect when a contingent of 10,000 Wurttembergers arrived to swell his corps to nearly 40,000 men including 86 guns and 3,500 horsemen. During April and May they moved across Germany and on 29 May 1812 they reached Thorn on the Vistula. From there it was a short march to the Niemen where pleased with the condition of his men on their arrival Ney expressed confidence they would prove equal to VI Corps when called to give battle.

THE RUSSIAN CAMPAIGN

With Davout (I Corps) Oudinot (II Corps) and Eugene (IV Corps) Ney would be called to join the Grande Armée's main thrust. The Grande Armée crossed the Niemen on 24 June with I Corps in the van. Ney had to wait the next day before he crossed with Eugene's Army of Italy following a day later. III Corps in a support role behind Davout did not spare them the difficulties of the march. As the center columns Ney and Eugene suffered the most as anything the Russians left behind was snapped up by the advance elements. Likewise they also went hungry since the transport system soon broke down due to the speed of the march meant the commissariat was unable to keep up whereas the rear echelons were at least fed. The heat followed by rain coupled with poor food and inadequate forage played havoc with Ney, and within a month he had lost a quarter of his strength without seeing a major action.

Ney's first action was at Krasnoe on 14 August, where chances of success were blocked by Murat's futile cavalry charges against the rock solid 27th Division under Neverovski. The Russian with 7,000 men formed in a huge square moved slowly towards the safety of a nearby defile. It caused Murat's cavalry to halt and mill around hoping to make a breach while Ney's infantry supported by artillery would have been more effective.

On 17 August Ney with Davout and Poniatowski stormed the outer suburbs of Smolensk, his troops taking Krasny on the southern bank of the Dnieper. So anxious was he to come to blows Ney recklessly led a bayonet charge. First into the city the next day after the Russians evacuated it he regrouped before heading east on the Moscow road. On 19 August his men pushed Barclay's rearguard from Stabna and then encountered increasing resistance as they passed through Valutino Gora. A rearguard action turned into a major battle as Barclay seemed prepared to make a stand outside Smolensk. Gudin's division attached to Ney fought magnificently as did Ledru's and the Württembergers under Marchand as they pushed the Russians back. With the Russians pinned by Ney's attacks, a spectacular opportunity was lost by Junot who crossing the Dnieper downstream failed to take the enemy in the rear.

The battles around Smolensk had a sobering effect on Ney who suffered over 4,000 casualties reducing III Corps to around 22,000 men. With other commanders he urged Napoleon to consider halting at Smolensk for the Winter to build up strength and supplies to renew the campaign in the Spring. Napoleon who had never failed to complete a campaign in a season apart from the Winter of 1806-1807, refused. After a brief rest he resumed the march, convinced the Russians would offer a final battle before Moscow. The die cast, Ney not wishing to be seen as a faint heart asked if III Corps could form the advance guard. Napoleon mindful of Ney's unwelcome advice, gave Davout with Murat the job.

BORODINO

At Borodino on 7 September Ney's 20,000 men of III Corps formed the center of the French line with Davout's I Corps to his right. Davout's objective were the Fléches behind the Semyonovkaya creek defended by Bagration's Second Army. At 6.00 A.M Davout opened a bombardment of the Russian positions and soon after moved forward. Davout's attacks faltered and at 7.00 A.M Ney led Ledru's division forward. Within an hour at the head of his men, leading them "like a captain of Grenadiers" he took the first two fléches. To his right Davout's battered divisions took the third. Bagration countered and his cuirassiers brushed aside Ney's Württemberg light horse driving him from the fléches. He rallied his troops and with support from Nansouty's cuirassiers who drove the Russian horse back, he led his men forward. By 1.00 P.M the fléches were again in his hands as was the whole ridge south of the Semyonovkaya creek.

Supported by Friant's division Ney slowly pushed the Russians from the village of Semyonovkaya. With Murat's cavalry nearby doing damage to the Russian horse while French artillery mercilessly pounded the massed ranks, to Ney the battle appeared right for the final push. In vain he, Davout and Murat sent messengers urging Napoleon to launch the Guard to exploit the success. The Emperor that day had a heavy cold was listless and uncertain, did nothing. Both sides exhausted the fighting petered out around 4.00 P.M and no breakthrough made, the enemy were able to break off unimpeded. Ney was beside himself with anger, 30,000 French never fired a shot. It led to the celebrated outburst by him, "Have we come so far simply to possess a battlefield? What is he doing so far back? There he hears our reverses, not our successes. He is not a General any more, he is an Emperor. Let him return to the Tuileries and leave the fighting to us." When he later heard Napoleon was ill he regretted the outburst.

THE RETREAT FROM MOSCOW

In Moscow his troops were billeted in the western suburb of Vladimir away from the richest pickings around the Kremlin and central Moscow. Once the fires gripped the city Ney was unable to prevent his men joining the looting short of arresting the whole corps. On 19 October 1812 the retreat from Moscow started, in which Ney was to show his finest qualities. On 25 October he was present at Maloyaroslavets. He passed Mohaisk on 28 October and then encountered the horror of thousands of unburied troops littering the Borodino battlefield as the first snows fell. When he reached Viazma on 31 October III Corps remained reasonably intact still with some 10,000 effectives under his command. Up till this time Davout formed the rearguard and as a result Ney's progress had been uninterrupted by the enemy. While resting at Viazma news came through that Davout's I Corps was cut off by Miloradovich at Fiodorovski. Ney at their head, Razout's division distinguished itself cutting its way back twenty miles to rescue I Corps. With Davout's troops in no condition to continue as the rearguard, Ney with III Corps took over on 3 November.

From here Ney's name went into immortality as he continually drove off cossacks that harried every step the army took. On 6 November he fought a stubborn rearguard at Viazma, which gave the army twenty-four hours to clear the city. For seven days he fought non-stop defending the struggling columns. At times he was forced into square twenty times in a day to repel cavalry. Boldly standing in the open he would jeer at Platov's cossacks daring them to attack. Sometimes it worked, it made his men feel better when they did, but most important it won the columns ahead time. He reached Smolensk on 14 November where he hoped the army may winter, but the depots had been sacked.

On 17 November he left Smolensk with 6,000 men, 12 guns and a squadron of cavalry. Near Krasnoe Miloradovich cut between him and Davout. He attacked the Russians head-on and was repulsed. Miloradovich offered him honorable terms, which prompted the immortal response, "A Marshal of France never surrenders. One does not parley under

fire with the enemy!" The Russians renewed the bombardment. Ney down to 4,000 men plus 1,500 wounded told his incredulous commanders he intended to fall back towards Smolensk to allow the enemy to believe they were heading in that direction. When darkness fell they made campfires to fool the enemy and then cut across country to the Dnieper, which fortunately was frozen. From there they marched across fifty miles of snow drifts to join the army at Orcha on 21 November. Only 900 men survived but the appearance of Ney's column fighting its way through a ring of cossacks caused the army's morale to soar. It prompted Napoleon's remark, "Michel Ney is the bravest of the brave."

Napoleon gave him another rearguard of 4,000 men that covered they army as it made its way to the Beresina. On the east bank he made noisy demonstrations against Tshitshagov forces on the opposite bank at Borisov while preparations to bridge the river took place further north at Studenka. When Oudinot on the west bank fell wounded on 28 November, Ney took charge of operations to keep the route open. Mounting vigorous attacks he won twenty-four hours for the army to continue its march west. In the final days Ney never showed weakness and was an inspiration to all. Each morning he roused the men from their sleep, shook them to their feet and shared their gruel. Fezensac in his diary pays tribute to Ney's ability to keep the army going. "The temptation to lie in the snow and end their terrible suffering, the hunger, the cold and the despair, was resisted only because the leader they loved and trusted never for one moment showed weakness, indecision or even discouragement. His strong body and his strong soul seemed unassailable. Ney not only saved the rearguard, but he saved the whole army".

He shared the rearguard duties with Victor but when that marshal showed neglect he denounced him in front of his troops. Victor never forgot the humiliation and was to exact revenge by voting for his death at his trial by the Chamber of Peers in 1815. On 5 December Napoleon left the army to return to Paris. Ney entered Vilna the next day where he expected the huge depot to give his remaining troops some succor. There were no supplies and on 10 December with less than 1,000 men he headed out on the last stage of the retreat to the Niemen. On 13 December he entered Kovno with 200 men. Stragglers were still streaming over the river so with Generals Marchand and Gérard* and 700 men he held onto the bridge for another day till enemy crossing lower down threatened to cut him off. He was literally the last man to leave the palisades at the Russian end firing the last musket before he led the remnants of his men to Konigsberg.

THE GERMAN CAMPAIGNS OF 1813

Ney returned to Paris in early January 1813. Depressed by the mood of pessimism in the capital he immediately requested leave and after a few days was with his young family at Coudreaux. Napoleon needed his presence in Paris to help restore morale and bring a sense of normality. His return was signalled on 8 February by the award of a huge gratuity of 800,000 francs from "a grateful nation". The bill was in fact footed by Rome, Milan, Westphalia and Hanover. His title Prince of Moscow originally conferred the day after Borodino, Napoleon proclaimed it afresh in an elaborate ceremony at the Tuileries to give maximum publicity and raise public morale. These distractions prevented him getting back to the army till 9 March, in spite of being appointed commander of I Corps of Observation of the Rhine on 17 February (later redesignated III Corps of the Army of the Main).

Much of the preliminary work had been done by the time he arrived at Frankfurt-on-Main to take up his command. His corps was further along in its organization thanks to a nucleus largely comprised of eight regiments of 32 battalions formed from the National Guard cohorts in 1812. When formed III Corps had four infantry divisions numbered 8 to 11, a cavalry brigade, and an artillery reserve with a total strength of some 40,000. The division commanders to lead these raw troops knew Ney, and were no strangers to warfare. Souham led the 8th Division, Brenier commanding the 9th Division and Girard˚ were all Peninsular veterans, while Ricard commander of the 11th Division, was a former division commander with Davout in Russia. Ney's biggest problem during the campaign was his shortage of cavalry, which barely numbered 1,200 horsemen led by the veteran Kellermann˚.

III Corps formed the Napoleon's advance guard as the Army of the Main marched into Saxony. Ney's young soldiers soon proved their worth at Weissenfels on 29 April when a large corps of Russian cavalry supported by Cossacks engaged Souham's division. The Russians tried to panic the young French infantry with a series of sudden charges that were pressed home with great determination. To repulse these horsemen Souham's men formed into battalion squares and with the support of the division's artillery withstood charge after charge. With Souham's infantry standing solid in square, the Russians lacking infantry support and unable to break down the squares with horse artillery due to the superior French gunners drew off after four hours.

In terms of a Napoleonic battle Weissenfels was a small action. Souham praised the troops in his report to Ney with measured optimism. Ney heaped excessive praise on the new troops and let himself be carried away by the victory. A single action did not make veterans and in the days that followed it became apparent his troops may have possessed courage but it was not matched by fortitude under fatigue and privation. March discipline soon deteriorated and III Corps became widely scattered. At Lutzen on 2 May this coupled with a failure by Ney's meagre cavalry to penetrate the Cossack screen resulted in Wittgenstein's army of 73,000 men led by two corps under Blucher and Yorck falling on III Corps.

Heavily outnumbered and outgunned, Ney's troops first put up stiff resistance. The Allies had however learnt the mistakes at Weissenfels and after two hours their combined infantry and artillery assaults became too much for Ney's inexperienced conscripts. Souham stubbornly held onto Gross Görschen but was prised out of the position. His and the other divisions fell back holding formation but soon a collapse in morale associated with a retreat and enemy pressure took its toll; battalion after battalion lost cohesion, broke ranks and fled. The situation became critical as the entire III Corps was on the verge of collapse as Marmont (VI Corps) formed up on his right and later Macdonald (XI Corps) on the left. Napoleon later arrived on the scene around 2.30 P.M. Seeing Ney's troops still wavering he rode forward and joined the marshal rallying his shaken conscripts by voice, boot, and personal example until they began to push Blucher back on Gross Görschen. By 4.00 P.M with both flanks threatened Wittgenstein's advance lost its impetus and a lull developed as both sides consolidated their positions. At about 6.00 P.M Napoleon judged the battle "ripe" and after an intense bombardment by the Guard artillery under Drouot˚, the Guard infantry and III Corps swept forward while Marmont and Bertrand (VI Corps) swept in on their right. Superior in cavalry the Allies able to absorb the blows withdrew from the field intact. French losses were around 22,000 of which 15,000 were on III Corps against the Allies of 20,000.

Lutzen was a close call and had the men of III Corps completely collapsed Napoleon's armies would have been on the Rhine six months earlier. That Ney's raw troops recovered was of great credit to the energy both he and Napoleon displayed in the battle. It however did not conceal the fact that though the army was highly motivated by the infusion of young blood, its morale was extremely brittle when faced by adversity. The failure to notice Wittgenstein's movements was due largely to French cavalry being unable to break through the Cossack screen. As was the inability to follow up the victory with a vigorous cavalry pursuit. With such a handicap the dispositions of his division commanders left room for censure. Ney can take blame for this, but in his defence he was with Napoleon near Leipzig when the thunderbolt fell. Napoleon realized the value Ney's presence had on morale and gave him command of the army's left wing. It included his battered corps, Lauriston's (V Corps), Reynier's (VII

Corps) and Oudinot's (XII Corps) with II Cavalry Corps, in all 77,000 men with 220 guns.

As the Allies prepared to make a stand at Bautzen, Napoleon prepared a battle of envelopment. Ney was to fall on the Allied right flank and rear while Napoleon pinned their front with the rest of his army numbering some 109,000 combatants and 355 guns. With 186,000 men present under arms against 120,000 Napoleon had never achieved such a numerical advantage on the field. It would have meant the complete annihilation of the Allies. Napoleon's plan to pin the Allied left while Ney fell on their flank was botched. Ney the army wing commander showed he was clearly out of his depth in matters of grand strategy where he had to head anything greater than an army corps. The orders he received from Napoleon he claimed were confused. He stopped to attack the village of Porlitz when he should have by passed it. The mistake cost the French dearly. Blucher able to discern the French plan extricated his forces with little difficulty. The scarcity of cavalry prevented a vigorous French pursuit. Ney's critics rightly said he took a village that day - Porlitz - but lost an Empire.

Ney still had Napoleon's full confidence and during the Armistice as the French armies reorganized he held no specific command as Napoleon intended to use him when the need arose. On 25 August Napoleon ordered him to take over from Oudinot who had received a mauling two days before at Gross Beeren as commander of the Army of Berlin. Before he left he distinguished himself at Dresden on 26 August when he led two divisions of the Guard against Schwarzenberg's center, sending the Russians and Austrians reeling back from the city.

On 4 September he took over from a very unhappy Oudinot and immediately rushed into an unnecessary battle against Bernadotte's Army of the North at Dennewitz the next day. In the early hours he drove back Tauenzein's Prussian corps, but didn't conduct a proper reconnaissance that allowed Bulow's corps to ambush him at Dennewitz. Reynier's Saxons gave way followed by IV Corps under Bertrand*. It was a disaster, forced to fall back to the Elbe, he lost 24,000 men and 50 guns. He was so upset by the reverse, he contemplated suicide. In defeats he had suffered before he could lay the blame on others, but this time there was no way out. He had behaved like a lieutenant in a general's cloak personally leading charges and plugging gaps rather than taking notice of the wider strategic implications of the conflict.

It was from Dennewitz that Ney's dislike for Napoleon began. The Emperor's criticism of him was savage as it was against all his commanders. Matters deteriorated further when Napoleon learnt Bernadotte had been in communication with him and other marshals.

At Leipzig Ney commanded the northern sector where things did not go well on the first day.

Outnumbered and furiously attacked by Yorck's Prussians Marmont's VI Corps gave ground retiring slowly and in good order on the suburbs of Mockern and Gohlis. The reverse could have been avoided and Marmont given much needed support had Ney not despatched III Corps under Souham to the threatened sector at Wachau. It was an error of judgement. Souham spent the day marching from one sector to another without ever firing a shot. With his help Marmont could have held Mockern.

On 18 October Ney spent the early part of the day in fierce fighting against Bulow and Langeron, retaking and defending the villages of Paunsdorf and Schönefeld after Marmont lost them. When the Saxon's defected it caused a great gap in his line as he started to retire towards Reudnitz. Two horses killed under him during the day, it was while retiring from Schönefeld he received a musket-ball in the shoulder. He remained with his men but by evening he had become delirious and could continue no longer. Orders for the withdrawal from Leipzig given, Ney barely able to keep in the saddle passed through the city early on the morning of the 19th. From there, placed in a coach, he was soon speeding back to France.

THE 1814 CAMPAIGN IN FRANCE

Ney's wound took longer to heal than expected. It was two months before he was able to raise his sword arm. Napoleon missed him. He was having problems with his commanders, several having lost the will to fight. Macdonald in the north was dynamic and honest, but unlucky. Marmont further south was limited and showed signs of disloyalty, so could not be trusted. Ney still showed resilience and whilst outwardly close to the Emperor his presence was invaluable for raising morale. As a result on 6 January 1814 Napoleon entrusted him with overseeing the formation of two Young Guard divisions under Decouz and Michel* at Thionville and Luxembourg to defend the Vosges. The plan fell apart when Blucher crossed the Rhine and in no time was also over the Marne nearby at Joinville. Ney fell back on Saint Dizier before cutting across to join Napoleon near Brienne on 28 January. The Prussians isolated, Napoleon ordered a concentration and Ney leading the Guard divisions drove Blucher from the town. A severe loss was the death of Decouz. At La Rothière Blucher regrouped and turning on Napoleon inflicted a sharp reverse on 1 February. It was the Young Guard under Ney that went a long way to halting the Allied advance and then covered the retirement to Troyes.

Ney's next major success was Montmirail on 14 February where he showed he had lost none of his old energy. Sword in hand at the head of six battalions of Old Guard he crushed Blucher's left flank and at one stage looked as if he would roll up their entire line and destroy them piecemeal. So formidable was Ney's reputation, Sacken's Russians fled at the mere sight of seeing him with the Guard. As the campaign dragged on he was at all Napoleon's battles, Chateau Thierry on 12 February, Craonne on 6-7 March, Laon on 9-10 March, Rheims on 12-13 March and seized Chalons-sur-Marne the next day.

The turning point for Ney was the reverse at Arcis-sur-Aube on 20-21 March when Napoleon mistook Schwarzenberg's army for Wrede's isolated corps. He fought in the skilful rearguard action with 25,000 men against 80,000 Allies, but soon after realized the game was up as the Allies were between Napoleon and Paris. Ney had become increasingly disillusioned with Napoleon's rejection of the terms of the Treaty of Chaumont on 9 March, which offered France her 1791 frontiers. As Napoleon rushed ahead of the army to try to save Paris he detached Ney on 29 March to gather troops at Troyes. Paris in the hands of the Allies, he made his way to Fontainebleau arriving there on 2 April.

NAPOLEON'S ABDICATION

Ney's role in the events that led to Napoleon's abdication were crucial. He was the first one to take the lead at Fontainebleau after Napoleon gave the Guard a rousing speech declaring they would march on Paris. The gesture he knew was sheer folly and showed on his face while Napoleon harangued the troops. He stood there silent as they broke into shouts of "Vive l'Empereur! à Paris, à Paris." In Napoleon's study afterwards he opened the conversation telling him politely that a provisional government had formed in Paris and the Senate had declared him unfit to rule. He then declared a march on Paris was utter folly, the army was not capable of taking Paris and further military operations would achieve nothing. Macdonald who had arrived with reinforcements backed him and in his accustomed brutal frankness said the affair must be brought to an end, the prospect of civil war too much.

Napoleon not listening, it was only when Ney declared that the Army would no longer follow him but its commanders that Napoleon realized he faced a mutiny and the game was up. They suggested for the good of France he abdicate in favour of his son. Within hours Ney and Macdonald were with Czar Alexander in Paris with Napoleon's abdication letter hoping the Allies would accept the terms. Events moved fast as Tallyrand the arch intriguer rallied support for the recall of the Bourbons. Marmont who hoped he might become a General Monk defected with his troops and with it the prospect of Napoleon marching on Paris collapsed. Meanwhile Alexander was urged by his fellow monarchs that Napoleon must abdicate unconditionally. Napoleon angered at the Allied lack of good faith decided to make a last stand at Fontainebleau with the Guard. Ney's and Macdonald's intervention at this stage was decisive. They forbid his orders be passed onto the troops. Napoleon devastated, he signed an unconditional abdication on 6 April. It was Ney who delivered it to Czar Alexander.

THE RESTORATION

Napoleon's departure for Elba saw Ney as France's foremost soldier. Louis XVIII took an instant liking to Ney and won him over by flattery often referring to his bravery shown in past campaigns in which he showed a keen interest. There was talk of making him Minister for War, but the influence of the Comte d'Artois secured Dupont the job. The Bourbons however did not hesitate to use his talent. He became a member of the Council of War on 8 May 1814 and cavalry commander for the whole army, followed by on 21 May Governor of the 6th Military Division at Besançon.

He was pleased France was at peace and was able to spend time with his family. The goodwill Louis professed soon turned sour as his brother Charles's influence began to assert itself. The reduction of the army from half a million men to 200,000 was regrettable but necessary. Ney's criticism was that Dupont carried out the program with indecent haste. It caused great hardship and ill feeling. The disbandment of the Imperial Guard on 12 May and its replacement by *La Maison Militaire du Roi* came as a profound shock as only officers of noble birth were allowed to lead that body. Ney's suggestion to win over the Imperial Guard by renaming them Royal Guard, at first so well received by the king, went out the window.

Despite the cutbacks he did fine work keeping the cavalry intact and in shape. Right up to the end he was travelling the length and breadth of the country visiting depots and inspecting regiments. What really began to play on his nerves was the sheer military incompetence of the royal princes, Artois and de Berry. They were an embarrassment to him and gave the army little confidence if they should ever be led by them. It was also the social slights, the snide comments and rudeness of the Ultras towards him and his wife that touched a raw nerve ending with him publicly threatening the Duchess of Angoulême. Louis tried to soothe Ney by making him a Gentleman of the Bedchamber, but soon dropped the idea when faced by the hostility of the Duchess.

NAPOLEON'S RETURN

When news of Napoleon's return arrived Ney was one of the first to offer his sword to the King. Convinced such a venture would bring disaster if not stopped, he made the unwise gesture he would return to Paris with Napoleon in a cage. It was a dramatic gesture and no doubt he meant it at the time. Ordered to lead the troops at Besançon he was disturbed to find so few troops there when he arrived on 11 March. Still full of fight he pushed on to Lens-le-Saulnier where he wrote to Suchet at Strasbourg urging reinforcements.

What then caused him to change sides once more reflected his volatile and unpredictable nature. Public humiliation at the hands of the Bourbons, while on the march news of garrisons defecting and the demeanor of his own troops, which numbered some 6,000, all played their part. What struck the right chord was the arrival of emissaries on the night of 13 March with a message from Napoleon's chief of staff Bertrand* urging him to join Napoleon at Chalons. Included was Napoleon's hand-written note, "Execute Bertrand's orders and meet me at Chalons. I shall receive you as I did after the Battle of Moscova." It was not the Bourbons who had made him a marshal or the Prince of Moscova. The next day he told his subordinates Bourmont* and Lecourbe the Bourbon cause was lost and on 18 March joined Napoleon at Auxerre.

Napoleon welcomed Ney, but did not trust him. Ney returned to Besançon to await orders and when none were forthcoming appeared in Paris on 3 April. After kicking his heels around the Ministry for War for a week he realized Napoleon had no place for him and returned to Coudreaux. It was not till 12 June that he received a letter from Davout ordering him to join the Army of the North. Added to it was also a scribbled note from Napoleon to Davout, "Send for Marshal Ney and tell him that if he wishes to be present at the first battles, he ought to be at Avesnes on the fourteenth".

NEY'S RECALL AND ITS REASONS

The message threw Ney's mind into turmoil. Throughout the intense activity of the Hundred Days Napoleon quite deliberately had kept him at arms length. Their meetings were few, invariably cool and once or twice decidedly abrasive. The old comradely ease based on mutual military respect had gone, and apparently as far as Ney was concerned for good. The traumatic scenes of Napoleon's abdication the previous year and Ney's part in them were too much for him to expect the Emperor to again give him his trust. A tangle of conflicting ideas and emotions, he knew that changing sides back to Napoleon had hopelessly comprised his relations with the Bourbons. Should Napoleon fail again it meant his personal fate was sealed. His life depended on Napoleon's survival yet the Emperor up till this time still refused to employ him! Many things had happened to restore his belief in the Empire. As a practical soldier, always close to the troops under his command, he was fully aware of the army's mood, defiantly nationalistic and anti-Bourbon. The social pin-pricks escalating to snubs and open insults, which he and especially his wife had received from returned Royalists were sufficient to convince him that merit no longer counted for much in France.

Most important, he had changed sides because of an ineradicable loyalty to the single man whose personality was almost always a dominant influence upon his life. It was a loyalty that had raised him to the heights of his chosen career. It also suffered from regular disillusionment when Napoleon made mistakes and the Empire's sun began to set. Nevertheless his loyalty had a bedrock quality about it that neither the Bourbons nor the Allies could hope to compete against.

Napoleon on the other hand possessed no such sentiments. His long-standing faith in Ney had been severely shaken and knowing the emotional turmoil his marshal was going through was prepared to exploit it to the full. There were several reasons why Napoleon recalled Ney, none were due to a change of heart but rather due to practical necessity. Napoleon was chronically short of experienced senior commanders. Only three marshals could be spared to join the Army of the North; Soult, Mortier and Grouchy. A further problem arose when Mortier's health began to fail with attacks of gout and sciatica. It was the loss of a tough fighting general whose lack of imagination was far outweighed by him being the perfect subordinate. Ney's inclusion in the army was also a propaganda coup and Napoleon was too adroit a politician to miss the opportunity. It was a calculated blow against Louis XVIII's prestige to employ the former Bourbon commander-in-chief. In addition Ney's recall could help serve to persuade others who deserted Napoleon in 1814 that their folly could be overlooked in return for new tokens of devoted service to his cause. The most important factor in Ney's favour was his personal popularity with the ordinary French soldier. Until Russia he had been a hero essentially to those who had served under him and to a minority who stationed elsewhere had followed the Imperial campaigns in detail. After the retreat from Moscow his bravery became a legend throughout the army. Amongst the conscripts, the Classes of 1814 and 1815, most had never heard the names of Napoleon's senior commanders, whilst Ney's was known to all.

THE WATERLOO CAMPAIGN - OPENING MOVES

Ney reached Avesnes on 13 June where Napoleon gave him a warm spontaneous greeting. The Emperor promptly invited him to dinner and was his old mercurial self, showing all his old charm and optimism assuring all that the campaign would be a success. Underneath Napoleon was relieved, he had gambled and had Ney rejected his offer and word spread, it would have been a blow to morale. Napoleon to his detriment was so carried away by his own exuberance, neglected to tell Ney what his command would be and said absolutely nothing about the overall strategy for the campaign. Ney, on the other hand, hesitated to press him in case he spoiled the convivial atmosphere. The next morning as the French headquarters moved to Beaumont, the last stop before the border, Ney was with neither a command nor a horse. He made his way to Beaumont in a requisitioned peasant's cart and quickly sought out Mortier, offering the sick marshal his sympathy and purchasing his horses.

Already the French had crossed the border, and completing a twenty-six kilometer ride to Charleroi Ney finally caught up Napoleon at 3.30 P.M. in the afternoon of 15 June. The Emperor was sitting outside an inn being cheered by soldiers of the Young Guard as they marched past. Ney dismounted and hurried over to him. He received a short formal greeting and was told,"I want you to take command of I and II Corps. I am also giving you the Light Cavalry of the Imperial Guard - but don't use it without my orders. Tomorrow you will be joined by Kellermann's cuirassiers. Go and drive the enemy back along the Brussels Road".

Napoleon had effectively given Ney nominal command of the French Army's left, a force numbering around 50,000 men. It was not what he expected, on hearing of Mortier's illness he thought he was likely to replace him. It was a great imponderable of the Waterloo Campaign that Napoleon did not inform Ney of his command earlier at Avesnes. Perhaps he hoped that Mortier would make a recovery and he could use Ney as a stand-by. What is known is that the delay placed Ney at an enormous disadvantage. He had no time for proper reconnaissance, nor to become acquainted with his new staff and commanders.

Within an hour he was amongst the men placed under his command. There was no opportunity to call a meeting of the principal commanders as they were already on the move. He knew many of them having fought in Spain with Foy, Kellermann, Lefebvre-Desnoettes and also d'Erlon. Piré he had met in Spain while Reille and Bachelu he knew by repute. He had to acquaint himself with the situation ahead as Reille's advanceguard had encountered strong Prussian resistance on the Brussels Road at Gosselies. He ordered Reille to attack and the enemy, who soon abandoned their position and started to retire across the fields to Ligny. To hurry them on their way, Girard's division followed.

Ney then ordered Lefebvre-Desnoettes to continue probing north with the Guard light cavalry. Four kilometres up the road at Frasnes, musketry halted his Polish lancers. Several squadrons dismounted to engage the enemy on foot while others skirted the village only to see the enemy identified as Nassauers retiring towards Quatre Bras. The cavalry followed at a safe distance but came under heavy fire from a larger force dug in on either side of the road before the Quatre Bras cross-roads. Hearing the news ahead, Ney immediately ordered a battalion of infantry to join Lefebvre-Desnoettes and put the remainder of Reille's corps on full alert. Anxious to be familiar with the situation he spurred north through Frasnes to see for himself and by early evening had reached the French positions. Failing light, high rye fields, the

outline of Bossu Wood and an intense crackle of musketry from concealed positions made it impossible to calculate the number of enemy before him. The noise they made, and where they showed themselves spread across a two mile front convinced Ney he faced a full corps. Had he known that before him was only a brigade of Perponcher's Dutch-Belgian troops numbering 4,000 men with eight guns he would never have backed off. Mindful of Napoleon's order not to risk the Guard cavalry was his first fatal error. A bold cavalry charge up the road supported by a brigade of infantry would have exposed the weakness of the Allied position. The brittle shell broken, Quatre Bras taken on the evening of 15 June, the sequence of events that followed the next day would have been very different.

Ney rode back to Gosselies where he hoped there were further orders from Napoleon. As a wing commander he still had not been briefed on the operational plan. Concerned he returned to Charleroi to speak to Napoleon. After mid-night the Emperor was woken up and Ney had his first briefing. His mind blurred from lack of sleep Napoleon gave him an outline of the operational plan. Napoleon planned the first blow to fall on Wellington and to achieve this end it was essential the French occupy and hold Quatre Bras. Once Wellington was brought to battle the rest of the Army of the North would support him.

Ney returned to Gosselies and had a few hours sleep. He only started to carry out Napoleon's orders when he woke at 7.00 A.M. sending messages urging all formations to hurry their concentration on Quatre Bras. Messages that should have gone out on his return the previous night to reach his commanders at dawn and not mid-morning. That he had to wait for Napoleon's written confirmation showed a total lack of initiative or drive and lost him a valuable six hours.

Napoleon not the man he was, only bothered to despatch Ney's written orders confirming their conversation at 8.30 A.M. Hardly had he done so when Grouchy reported a Prussian concentration at Ligny. Blucher was obligingly advancing towards the French to allow the "knockout blow" to fall on him rather than Wellington. The support Ney expected was at a stroke diverted towards Ligny. More damning, Ney was never told of the events or the change to Napoleon's plan.

QUATRE BRAS - 16 JUNE 1815

In the morning Ney missed his second chance to break through to Quatre Bras as he still outnumbered the Allies three to one. Orders not to risk the Guard cavalry he still interpreted as take no risks at all. Such orders contrary to his hussar temperament, he went completely the other way and proceeded slowly with caution, the psychological presence of Wellington beginning to have an influence his and Reille's thinking. Both Ney and Reille, Peninsular veterans, had learnt their lesson against Wellington the hard way. Ney had suffered defeats since 1812, which he could put down to circumstance or even to his own mistakes. But when he fought the Iron Duke at Bussaco in 1811, he had fought well and knew he had been beaten by a man of extraordinary talent. The more he looked at the meagre force before Quatre Bras the more convinced he became that Wellington was using the Peninsular tactic of concentrating a large force of infantry out of sight on the reverse slope behind the cross-road. Reille endorsed his view and the caution Ney exercised that morning went a long way to losing the campaign.

Ney determined there were to be no lightning cavalry manoeuvres or fancy enveloping attacks. He decided to wait till Reille's troops were fully massed then roll over Quatre Bras in a simple continuous assault. Casualties he expected to be heavy but with d'Erlon's Corps ready to support if II Corps faltered he was assured of success. The French guns opened at 11.45 A.M and one by one the main defence points in front of Quatre Bras fell. Bachelu's division fanned out to capture Piraumont Farm, driving back Perponcher's left, and following this Foy took Gemioncourt. Foy then moved against Pierrepont Farm but a steady stream of Allied reinforcements arriving found the opposition too strong. Ney ordered up Jérôme who captured it and then pressed on to take Bossu Wood. Their position critical by 2.30 P.M. the only defensive positions before Quatre Bras still in Allied hands were the northern fringes of Bossu.

Between 1.00 P.M and 4.00 P.M the Allied position strengthened by the arrival of 15,000 men from Picton's division, the Brunswick contingent, Halkett's brigade and more Nassauers. The numerical advantage Ney possessed a few hours earlier disappeared as both sides faced each other with around 24,000 men. Ney became increasingly agitated at failing to breakthrough. At 4.15 P.M a message (sent at 2.00 P.M) arrived from Soult urging him to drive vigorously all before him and turn east to envelop the Prussians. Apart from the sporadic boom of cannon from the south-east heard through the din of his own battle, this was the first indication Ney had that Napoleon intended to dispose of Blucher first and that he Ney was expected to gain the crossroads as an arterial means of falling on the Prussian flank.

Angry at not being informed sooner, it was time for Ney to redouble his efforts and not to falter. A final push by Reille would open the way for d'Erlon's 20,000 fresh troops to seize the crossroad and pivoting to fall on the Prussians. Reille's troops exhausted, Ney became increasingly concerned of the whereabouts of d'Erlon. The staggering news then arrived that the Imperial Aide La Bèdoyére had in the name of the Emperor intercepted the head of d'Erlon's column and directed it towards Ligny. The unfortunate d'Erlon reconnoitering ahead was not present to question the order and was compelled to follow. Ney, now beside himself with fury, had at a stroke lost the force that could complete the objective given. A few minutes later Forbin-Janson one of Soult's aides

NEY AT WATERLOO

arrived with orders confirming d'Erlon's redeployment and the move to envelop the Prussian right and fall on their rear. Forbin-Janson also carried a verbal message from Napoleon telling Ney, "to look sharp and finish the Quatre Bras business." The aide seeing Ney's demeanor approached nervously and only managed to blurt out the verbal part of his message before a tirade from the marshal rained down on him. So flustered did he become he rode off without handing over Soult's orders, which Ney only received later in the evening. To resolve the Quatre Bras business, Ney immediately ordered an aide to recall d'Erlon who as he was concerned still belonged to his command as he had nothing in writing to the contrary.

The fatal order went out to recall d'Erlon, which resulted in the unfortunate general milling around between the two battles with 20,000 men and not firing a shot all day. At the crossroads the scales had tipped in Wellington's favour. Without d'Erlon's corps plans for rolling over Wellington were finished. Ney made a final effort with his cavalry. At 4.15 P.M he launched Piré's lancers in an audacious attack that rode over the 42nd and 44th Regiments. That was followed by a supreme effort by Kellermann's cuirassiers that broke the 69th and 33rd Foot and reached the crossroads. Had he bothered to allow time for infantry support he may have succeeded.

By 6.30 P.M Wellington had built his strength up to 36,000 when he launched his counter stroke. Had Ney shown proper generalship and watched his opponent's movements more closely he could have prepared a proper defence and used his skill as a rearguard fighter to good effect. Instead when the storm broke he was trying to rally and lead another fruitless infantry assault against Pack's brigade that was sent reeling back to their original start positions. He had lost 4,000 men against the Allies 4,800.

Ney's failure at Quatre Bras had by no means ruined the campaign. Napoleon had won a fine victory at Ligny and Ney by battering Wellington to a stalemate had succeeded in his objective of keeping the Allied armies separated. The events of 17 June were ones that decided the campaign and here Ney and Napoleon must take responsibility. As Blucher fell back Wellington inexplicably decided to hold at Quatre Bras. It gave Napoleon the ideal opportunity of marching up the road from Sombreffe to fall on his flank. Equally Ney joined by d'Erlon and the rest of Kellermann's cavalry could have renewed the struggle. Wellington pinned by a frontal assault would have found it impossible to extricate his troops. Napoleon convinced that Wellington would be well on his way to Brussels did not even think such an opportunity would present itself.

THE CHASE AND WATERLOO 17-18 JUNE 1815

Instead a large part of the morning was wasted at Ligny preparing bulletins for the politicians in Paris and rousing the troops. The truth dawned of Wellington's danger when his cavalry reported at 11.00 A.M the English still in position. Jarred into action he led a cavalry force up the road to Quatre Bras followed by VI Corps and the Guard. There no sound of cannon from Ney's quarter he sent messages to immediately resume the attack. Whether the orders were received or ignored is unclear, but at 2.00 P.M when Napoleon arrived at Quatre Bras Wellington had flown. Unnoticed by Ney he had started to thin his line at 10.00 A.M and all that was left were some cavalry and a few horse guns and rockets. When he moved it was too late. Wellington's cavalry kept the pursuing French horse at bay and when a thunderstorm made it near impossible for infantry to move any hope of catching up was lost. On this day more than any else Ney's failure to pin Wellington lost the campaign.

On 18 June attending a breakfast conference Ney urged Napoleon to start the battle immediately as he had seen signs of a possible retreat. Napoleon agreed but Drouot describing the wet state of the ground was allowed to change his mind. When the battle started Ney had orders to head the assaults against the Allied center. A furious cannonade rained on the Allies around 1.00 P.M but Wellington had moved most of his men to the reverse slope or sheltered them in the sunken road along the ridge. At 1.30 P.M he passed the order to d'Erlon to start the main attack. Mounted and sword in hand to raise morale he led the troops himself the first hundred yards before handing over to d'Erlon. The columns took heavy casualties but cleared La Haye Sainte, with Marcognet cresting the ridge while Durutte took Papelotte. Disaster struck when Uxbridge's two heavy brigades fell on d'Erlon's infantry sending it fleeing into the valley.

Napoleon decided La Haye Sainte must fall and sent a categorical order to Ney to that effect. The Guard moved closer ready to exploit Ney's anticipated breakthrough. Ney gathered two brigades from d'Erlon and around 3.30 PM personally led them back up the road against the farm. The attack was repulsed by a Hanoverian battalion. At this juncture, Ney well to the fore noticed in the Allied center what he interpreted as a sign Wellington was about to retreat when several battalions moved back to take advantage of the reverse slopes.

The time was approximately 4.00 P.M as the next great drama of the battle opened. In an excess of ardor, Ney ordered a brigade of Milhaud's cuirassiers to charge forthwith, hoping to rout the Allies. Delort the commander questioned the order since from his viewpoint there was no sign of a retreat. Ney in a fury then ordered the whole division forward and what started as a relatively small cavalry charge rapidly escalated into a major mounted attack. Division after division of cavalry were drawn into the battle, many without orders. The entire Guard cavalry followed Milhaud's corps and soon 5,000 horsemen, both heavy and light were pounding their way towards the crest, towards the Allied right center. The movement was a disaster for several reasons. First it obscured the fire of the French artillery who were doing more damage than horsemen could hope to do. Secondly there was too little room to manoeuvre with such a large body of horsemen, the threat of flanking fire from Hougoumont and La Haye Sainte reducing their front to 800 yards. Thirdly the ground sodden reduced the heavy horses to little more than a trot. Fourth and worst of all, Ney took no measures to support the cavalry with sufficient

NEY TRYING TO RALLY HIS TROOPS

infantry or horse artillery units. In his haste to seize what he considered the great opportunity of the day, Ney overlooked the importance of sparing a few minutes to lay on a properly co-ordinated attack.

The serried ranks of horsemen made an unforgettable target for the Allied gunners. Wellington's infantry transformed themselves into squares. The gunners manned their pieces till the last moment, then took shelter between the wheels or ran for the nearest squares. The cavalry could not break them and as the cavalry fell back the Allied gunners remanned their pieces. As each wave approached the process repeated itself.

Napoleon reviewed these events with alarm. "This is a premature movement that may lead to fatal results. Ney is compromising us as he did at Jena." The only way to redeem the situation and extricate the forty squadrons was to commit remaining French cavalry. Within an hour 10,000 cavalry had swirled around the Allied squares. Wellington was under severe pressure as horse artillery came into play. It was Ney, not Napoleon, who belatedly thought to use 8,000 men of Reille's II Corps not tied up in the fight around Hougoumont, but it was 5.30 P.M before they came up. Then as they advanced the cavalry too exhausted failed to offer support. Mercilessly cut down by Allied fire they lost 1,500 men in ten minutes.

Around 6.00 P.M Napoleon ordered Ney to make another assault on La Haye Sainte, which was successful as the garrison had almost run out of ammunition. Unable to exploit the success since he had exhausted the cavalry he asked the Emperor for more troops. The Allied line opposite La Haye Sainte was thin and very unsteady. To Ney it was time to launch the Guard. Napoleon faced by remorseless Prussian pressure in the direction of Plançenoit refused. It was around 7.30 P.M when Napoleon decided to commit the Guard. Again he entrusted Ney with leading the assault. Gallant and headstrong as ever the marshal led the last seven battalions forward in two huge columns. Again he blundered, instead of leading them against the buckling line behind La Haye Sainte he led them further left over the ground strewn by the wreckage of the earlier cavalry charges. Their oblique movement across the valley allowed Wellington an extra few vital minutes to reorganize. Its failure was decisive, Ney losing his fourth horse of the day fought on foot. As the Guard fell back Ney's last act was to try to rally remnants of Durutte's division for a last desperate charge. Then as darkness fell he owed his escape to an officer of the Guard Lancers who obtained a horse.

THE AFTERMATH OF WATERLOO

By miracle or miscalculation Ney survived the fighting, for he should have been sabred or collected a bullet at least a dozen times. Leaving the field he entertained no illusions, that the Bourbon Ultras would come hot on his heels in the wake of the Allied armies and the witch-hunts would follow. Ney was without friend or ally. Napoleon did not need him, rather he needed a scapegoat to carry the full odium of the Waterloo defeat. Prepared for a verbal onslaught and public humiliation Ney started to bend the ear of anyone who would listen. Arriving in Paris on the afternoon of 20 June, a day before Napoleon, he immediately saw Davout then Fouché and outlined the hopelessness of the situation. To Ney the reason for the defeat was Napoleon's, it was the Emperor who dragged France into the reckless venture.

It was Ney's visit to Fouché that proved fatal and accelerated Napoleon's overthrow. Fouché in the eighteen hours before Napoleon's arrival quietly gathered support. He offered Ney passports to Switzerland, one in his own name and another with the assumed identity of a merchant. Clearly on top of the Bourbon hit list, Ney chose never to use them. The reason is unsure, as he had plenty of time to make his escape. Most likely he simply did not know how to run, because totally lacking in cowardice, running away was alien to his nature.

On 21 June the Chamber of Peers declared itself in permanent session and debated Napoleon's future. Ney meanwhile wandered around Paris in a state of acute uncertainty, a gloomy disconsolate figure suffering from a mixture of battle fatigue and post defeat depression. The next day he pulled himself together. Washed and in a new uniform, the blood stained bandages removed from his head he dramatically entered the Chamber around 2.00 P.M and took his seat. Carnot as Minister of the Interior completed the reading of Napoleon's act of abdication and then turned to a prepared speech on the state of the nation. In it he referred with optimism to the army claiming it had defeated the Prussians and 60,000 men were coming to the defence of Paris.

Angered by such false optimism he stood up and told Carnot not to mislead the members. He gave an account of the campaign, praising the troops for their bravery and detailed the mistakes in command, which placed defeat beyond repair. He even ventured to reproach Napoleon for putting in the Guard too late and claimed the Prussians would be in the heart of Paris within a week. The Peers listened with stupefaction. They had overthrown an Emperor, now the country was leaderless, but never for once did they believe the situation was so hopeless. His most dramatic statement, though he knew it would seal his fate but could save France, was to urge they open negotiations for the end of hostilities. Many horrified, thought he was suffering from battle fatigue and had lost his head, but they realized this was a man speaking from the heart.

Unable to decide what to do, Ney lingered in Paris till 6 July. He travelled to Lyons where he heard the border at Geneva was closed. Not too concerned on

29 July he reached friends at Roanne, near Aurillac (Cantal). On 3 August the local prefect, hearing of his presence ordered his arrest. He remained at Aurillac till 14 August before starting the journey to Paris. He had a chance to escape when Exelmans* with a column of cavalry intercepted his coach, but refused having previously given his word not to in exchange for not being manacled. He reached Paris on 19 August and was imprisoned in the Concierges.

The King's minister Decazes was one of the first to see him. Decazes a skilful lawyer tried to trick a confession out of him that his defection on 14 March 1815 was a premeditated act and he was guilty of treason. Ney insisted on his innocence, claiming the situation was lost long before and that he had merely gone with the tide. Decazes soon lost interest in Ney, for he was gathering evidence from Ney to bring about the downfall of Fouché and found the prospect of bringing a patriot to trial unappetizing. He deftly passed the problem onto the army for him to be tried by a Council of War.

It took till 9 November 1815 to convene a council as no one wanted to serve. Marshal Moncey was imprisoned for three months for refusing to, while Massena complained his conflicts with Ney in the past compromised him. That excuse was not good enough and Jourdan the council's president got his three marshals, Augereau, Massena and Mortier and three generals, Claparède, Villatte and Gazan. A military trial was Ney's best chance, few soldiers would dare condemn him to death. Whilst they feared the consequences if they acquitted him, a prison sentence for several years was the most likely outcome. In the end Ney was a victim of his own fate and foolishly refused to recognize the court demanding he be tried by the Chamber of Peers. The proceedings ended in uproar and the council retired to consider the matter. Self interest got the better of them and by a majority of five to two (Massena and Augereau) declared they hand the matter to the Chamber who were baying for his blood.

The Bourbons lost little time bring him to trial and new proceedings started on 21 November. Ney again condemned himself when his lawyers tried to use the ploy that he could not be tried for treason as he was not French since his birthplace was no longer French. Ney would not accept it, sprang to his feet and declared aloud he was French and was prepared to die as one. He conducted an eloquent and lively defence but he was doomed. His cross examination of the traitor Bourmont* the principal prosecution witness was dramatic, but the evidence against him was damning; Bourmont had eight months to prepare himself. A principal defence witness to his defection, the old republican general Lecourbe, had recently died. The carefully worded charges ensured his guilt and the Peers voted by large majority for a death sentence. Wellington genuinely felt sorry for Ney and his family, tried to intervene. The French marshals in Louis XVIII's entourage, notably Victor and Marmont, plus the Ultras at Court anxious to see Ney's end caused the King to snub the Iron Duke. Politicians in London did not help; Ney's fate was a French affair and Wellington was under orders not to intervene.

On 7 December 1815 he appeared before a firing squad in the Luxembourg Gardens dressed in civilian clothes. Thus he avoided the indignity of having his badges of rank and epaulettes torn off. His bearing in the words of the General Rochechouart in charge of the proceedings was "noble, calm and dignified beyond reproach." He took charge of the proceedings, refusing a blindfold. Then standing before the firing squad said "Soldiers, when I give the command to fire, fire straight at the heart. Wait for the order, it will be my last to you." Shouting the order to fire as he drew his hat across his chest he fell struck by eleven bullets, one missing and striking the top of the wall.

NEY - A FINAL WORD

Ney was repeatedly at his best when under the direct eye of Napoleon. As an independent commander his worst failings came to the fore. He was often temperamental and changeable, insubordinate and quarrelsome, stubborn and proud. He was certainly no strategist, and at times his tactical skill often deserted him, but he never lacked courage. As an administrator and disciplinarian he was probably on a par with Davout. His ability to rouse troops was formidable and after Napoleon was probably the most inspirational of all his commanders.

His unpredictability and changeability, some attributed it to his experiences in Russia offering the excuse that he suffered from some form of battle fatigue after the campaign. Yet throughout his career he had done the unexpected and more often than not had been successful. That he was temperamental and changeable is fair if one examines his deteriorating relationship with Napoleon, and then his acceptance and later criticism of the Bourbons. That he was insubordinate and quarrelsome, proud and touchy, should not be confused with shell shock. Long before the retreat from Moscow these failings had come to the fore. His dealings with Massena and Soult immediately spring to mind. In the art of strategy Ney was certainly wanting, Poland in 1807, Spain in 1809 and Saxony in 1813 were all examples of lost opportunities. That his tactical skill often deserted him as it did at Dennewitz and Waterloo, as a result altered the course of European history. However in spite of all his failings no one could never accuse him of ever losing his courage and that he truly merited being referred to by Napoleon as the, "Bravest of the Brave".

NOURY - HENRI-MARIE, BARON (1771 - 1839)

BACKGROUND AND EARLY CAREER

From a minor noble family the artillery leader Henri-Marie Noury was born at Cracouville in the commune of Saint-Aubin du Vieil-Evreux (Eure) eastern France on 6 November 1771. He entered the Pont-à-Mousson Military School in 1787 and from there on 1 September 1789 studied at the Douai Artillery School. On 1 September 1791 he gained a commission as a sous lieutenant with the 7th Foot Artillery Regiment. He served on the Artillery Staff with the Army of the North where promotion was steady, *lieutenant* on 6 February 1792, then *capitaine* on 26 July of the same year. On 24 March 1794 he moved to the Army of the West before being recalled on 8 June 1795 to spend a period of duty based in Paris. In 1796 he moved to the Army of the Ocean Coasts and was there when it became the Army of England under General Bonaparte in 1798. He moved to the Army of Italy in April 1799, later becoming aide de camp to General Dulauloy on 8 January 1800. Promoted *chef de bataillon* with the 8th Foot Artillery on 27 April 1802 he moved to Elba on 25 May 1802 as director of artillery on the island. His tour of duty coming to an end, on 18 April 1803 he opted for a further period on Elba and was promoted *chef d'escadron* with the 1st Horse Artillery Regiment.

THE GRANDE ARMÉE 1805-1808

His return to the mainland resulted in promotion to *major* on 23 May 1803 with the 5th Horse Artillery based at Strasbourg. On 8 September 1803 he moved to the Army of the Ocean Coasts to prepare his regiment for the invasion of England. A sound administrator, his forté was with staff work and on 23 September while the Grande Armée marched across France he joined the Artillery General Staff. Present throughout the campaign against Austria he fought at Austerlitz on 2 December 1805. On 9 March 1806 he gained promotion to *colonel* of the 2nd Foot Artillery Regiment. In April 1806 he joined Lannes' V Corps as its artillery chief of staff. He served in the Prussian and Polish campaigns distinguishing himself at Jena on 14 October 1806 and then at Ostrolenka on 16 February when Essen fell on V Corps spread along the Narew. On 17 July 1807 he became *colonel* of the 2nd Horse Artillery Regiment. He was made a *baron de l'Empire* and on 17 March 1808 received an annual pension of 4,000 francs drawn on Westphalia.

CAMPAIGNS IN SPAIN AND AUSTRIA 1808-1812

The war in Spain saw his return on 15 October 1808 to V Corps as Mortier's artillery chief of staff.

He served at the siege of Saragossa from the end of December till the city's fall on 20 February 1809. Promoted *général de brigade* on 23 March 1809, he headed the corps artillery while its commander Foucher was absent ill.

Recalled to join the Army of Germany he became its artillery chief of staff on 26 April 1809. He served throughout the Danube Campaign in that capacity and fought at Aspern-Essling and Wagram. On 2 September 1809 he transferred to the Saxon Corps under Reynier as artillery commander. His period with the Saxons led to major changes within their artillery remodelling it along French lines. The fine performance by the Saxon artillery in the years that followed bore testimony to his hard work.

On 23 January 1810 he returned to Spain as deputy artillery commander with the Army of Catalonia. His time under Augereau was not a success and in August 1810 he was recalled taking the appointment as commandant at the Rennes artillery school. His talents were needed again when the Army of Catalonia became bogged down at the siege of Figueras he served there from April till its fall on 17 August 1811. The appointment as artillery commander of the Corps of Observation of the Reserve based at Madrid followed. The formation was a mixture of ad hoc scraps of unallocated forces in Spain, and it appeared for a time it appeared his career had reached a dead end.

RUSSIA SAXONY AND BELGIUM 1812-1813

The Guard Artillery was going through a period of expansion at the time he unexpectedly received orders to join it on 7 February 1812. He served in Russia throughout the 1812 Campaign, was present at Smolensk and Borodino. One of the last to leave Moscow, he is credited with blowing up a large part of the Kremlin before sharing in the hardships of the retreat.

In the Spring of 1813 he helped reorganize and equip the Guard Artillery before moving to XII Corps (Oudinot) as artillery commander on 7 June 1813. Serving in Saxony he missed Oudinot's defeat at Gross Beeren by Bernadotte on 23 August, being at Dresden helping to prepare the city's defences as the Allies approached. His experience was of great value and on 26 August the skilful sighting of the guns went a long way towards repelling the enemy columns. After the disbandment of XII Corps in October he was attached to the Artillery General Staff. He fought at Leipzig on 16-19 October 1813 and was active during the retreat where under Drouot he helped serve the Guard Artillery at Hanau on 30-31 October.

On 25 November 1813 he was promoted to *général de division*. Command of the artillery of VI Corps followed on 5 December. The need for an experienced artillery commander in Belgium resulted in his transfer to that theater with Maison's I Corps on 22 December where he remained till Napoleon's overthrow.

THE RESTORATION AND NAPOLEON'S RETURN

His return to France saw him involved in the reorganization of a smaller peace-time army. The Bourbons awarded him the title *Chevalier de Saint Louis* on 29 June 1814. The cut-backs resulted in him being placed on half pay on 1 September 1814. Out of work he readily rallied to Napoleon on his return and he worked under Ruty helping to reorganize the Artillery. The formation of VI Corps (Lobau) from troops stationed in Paris resulted in him taking charge of the corps artillery on 24 May 1815.

THE WATERLOO CAMPAIGN

Lobau's failure to arrive promptly at Ligny on 16 June resulted in VI Corps artillery not deploying for the battle. In the afternoon of 18 June he deployed his guns skilfully before the Bois de Paris, delaying the Prussian advance. Numbers in the end told and one by one his guns were disabled as VI Corps was forced back on Plançenoit. No returns were submitted by VI Corps for artillery personnel after Waterloo. It is generally acknowledged that once the army broke, none of his pieces survived from Jeanin's and Simmer's divisions or the Corps Artillery Reserve. For the rest of the campaign he served on the General Staff as it fell back to Paris.

LATER CAREER

At first suspended on 1 September 1815, the acute shortage of experienced artillery officers soon resulted in his recall. On 10 February 1816 he became an artillery inspector general and joined the Central Artillery Committee. In 1817 he became Inspector of Material and Personnel for the Northern Departments. He returned to the artillery inspectorate in 1820 and the following year became a member of the Artillery Consultative Committee.

He steered well clear of politics and during the period of Bourbon rule avoided the purges that cut great swathes through generals that had previously served Napoleon. The July Revolution in 1830 had little effect on his career in August 1830 and continued to serve as an inspector general and was a member of numerous consultative committees that reviewed policies and recommended reform. He was placed on half pay on 1 January 1837 and then moved to the Reserve Section of the General Staff on 15 August 1839. He died soon after at Evreux (Eure) on 25 September 1839.

ASSESSMENT OF CAREER

Noury was like so many other artillery generals of the Napoleonic wars, a loyal, reliable and hard-working professional. The artillery was one arm that rarely let Napoleon down and Noury as with most of his peers served Napoleon and France well. After Napoleon's end he saw the value of his profession, and realizing the artillery was not a popular arm amongst returned Bourbons was able to use his knowledge and experience to good effect. Innovative and hard-working, he encouraged development within the artillery and his support of technological changes was commendable. In the years after Waterloo it resulted in the modernization of that arm to make it the most efficient in the Army, and arguably in the whole of Europe.

FRENCH GUNS IN ACTION

PAJOL - CLAUDE PIERRE, COMTE (1772 - 1844)

One of the finest light cavalry generals in the French Army, Pajol was born in Besançon (Doubs) in the north east of France on 3 February 1772. The son of an advocate, he was destined for a legal career joining his father's practice. He soon gave up law when he joined the Besançon National Guard on its formation in August 1789. Ambitious, he saw better opportunities for advancement in the new volunteer formations, which resulted in him enlisted with the Doubs Volunteers on 21 August 1791 with the rank of *sergeant-major*. The first *Amalgame* saw his unit merged with the 82nd Line and on 12 January 1792 he gained a commission as *sous lieutenant* followed by that of *lieutenant* on 27 May of the same year.

His first active service was with Custine's Army of the Rhine where he was at the capture of Speyer on 30 September 1792. In the final struggle for the city, at the head of a company of grenadiers he scaled the walls and in spite of being bayoneted in the chest and hand showed great bravery and determination. His return to duty saw him at the defence of Mayence. He broke his arm during a sortie from the walls on 10 April 1793. The wound didn't heal and in August he returned home and spent several months on convalescent leave.

KLÉBER AND THE ARMY OF THE SAMBRE-MEUSE

Pajol's bravery had caught the eye of Kléber, at the time an up and coming general who made his reputation during the defence of Mayence. On 12 May 1794 Pajol promoted *capitaine*, joined Kléber's staff as an aide de camp where in the next few years his career was much dependent on the fortunes of that mercurial leader. He served at Fleurus on 26 June 1794 where a charge by Kléber leading the army's left wing was largely responsible for the Austrian defeat. With Kléber during the occupation of Belgium he was present at the capture of Mons on 1 July, Engheim and Ath on the 8th, and Esneux on the 11th where he was wounded taking an enemy standard.

It was at this time that he helped Michel Ney* whose career had run into difficulties. Ney's earlier appointment as aide de camp to General Lamarche with the Army of the North had promised much, but had come to nothing. Unhappy, he was back with his old regiment the 4th Hussars, still a captain. The successes of the new army he served had brought him no promotion, he had seen himself passed over for *chef d'escadron* and in despair considered resigning. Fortunately for Ney, while in this mood he was providing Kléber's mounted escort and was in daily contact with Pajol. He poured out his grievances to his friend, who sympathized and repeated the tale to Kléber. The next day Kléber called Ney over and during a long conversation found they both had much in common. Like Pajol, both Kléber and Ney were German speaking and came from the same border region. The result was Ney's appointment on 8 July 1794 to Kléber's staff with the rank of acting *adjudant-général*. It was Ney's sure step on the ladder that within ten years would lead to the Marshalate.

Pajol in the period that followed spent time at the siege of Maestricht. He was in action against the Austrian relieving force at Orthe on 18 September where he carried away another enemy standard. After the campaign his bravery was recognized when he joined a small party sent to Paris that on 8 November 1794 presented thirty six captured enemy standards to the Convention.

Very loyal to Kléber he turned down several promotion opportunities including that of *chef de bataillon* with the 6th Légère. He took part in the investment of Mayence and was shot in the leg while crossing the Salza on 20 September 1795. Promoted *chef de bataillon* on 9 February 1796, he later distinguished himself at Freiburg on 10 July when he took another standard.

When Kléber resigned from the Army of the Sambre-Meuse in January 1797 his future looked uncertain. A change in career direction followed on 4 February 1797 when he joined the 4th Hussars as *chef d'escadron*. The move benefited him, as Ney was the regiment's former commander and now led the Hussar Division. With influential friends and a touch of bravery, Pajol soon came to notice during the crossing of the Rhine at Neuwied on 18 April 1797 when he was in the fore during the capture of the redoubts on the opposite bank.

The resumption of hostilities in the Spring of 1799 saw him with a cavalry brigade under Soult* as part of Decaen's division of the Army of the Danube. In the advance from the Rhine he fought at Pfullendorf on 21 March and then at Stockach four days later. Here in one of Jourdan's few successes of the campaign, with a squadron of two hundred men he forced two Austrian battalions to surrender. Promoted *chef de brigade* on 21 July 1799 he became commander of the 6th Hussars with the Army of the Rhine. He remained with the army for two years serving with distinction under several great commanders including Leclerc, Gudin, Nansouty and Ney. A notable feat by him occurred at Neuburg on 27 June 1800 when at the head of two squadrons of hussars he routed 500 Austrian cuirassiers. The incident resulted in his first meeting with Napoleon when on 28 July 1800 he was awarded a sabre of honor. In May 1801 he joined Ney's division stationed at Kreuznach. A move followed to Sarrelouis in August 1801 till June 1803, when he moved with the 6th Hussars to the Camp of Utrecht in Holland to prepare for the invasion of England.

THE GRANDE ARMÉE 1805-1807

The formation of the Grande Armée saw his regiment serve with La Coste's light cavalry division of Marmont's II Corps. He took part in the advance to the Danube and the envelopment of the Austrians at Ulm. So fast and decisive were Napoleon's movements that when the city fell on 20 October 1805 his men had still not seen action. The remainder of the campaign he spent moving through the Tyrol, Carinthia and Styria to Leoben, which he reached on 10 November 1805. From there while events unfurled in Bohemia, he watched the movements of Archduke Charles's army as it retired from Italy.

The Treaty of Pressburg signed, in January 1806 his troops moved into Istria as part of the occupying force. He later helped set up a large cavalry depot at Frioul and remained in the region till April 1807. Promoted *général de brigade* on 1 March 1807 he was called to join Napoleon's general staff at Osterode in Poland. On his arrival on 12 May 1807 a change in plan occurred and two days later he took command of Latour Maubourg's brigade with Lasalle's light cavalry division. He fought Bagration's rearguard at Güttstadt on 9 June, and the next day he was in action at Heilsberg coming to the rescue of Murat's heavy cavalry after it received a severe mauling by Uvarov. Then diverted north, he joined Davout's march on Konigsberg and as a result missed Friedland on 14 June 1807. After the battle he joined Murat's pursuit of the Russians, being the first to reach the Niemen at Tilsit on 19 June.

By the end of the campaign, Pajol's dash and bravery had done much to re-establish the reputation of the hussar regiments that had become somewhat tarnished after the heady successes of the previous year. His reputation as a hussar leader soon stood on a par near to Lasalle's. Whilst he did not possess the reckless dash and bravado of the latter, his efforts enabled Lasalle to lead from the front knowing he was had a capable and level headed man behind him. It was reflected in the awards Pajol received, *baron de l'Empire* on 19 March 1808 and the Bavarian Order of Merit on 24 July 1808.

GERMANY AND AUSTRIA 1808-1809

After Tilsit Pajol spent a period of garrison duty in Poland and then Prussia. When Lasalle moved to Spain in October 1808, he remained in the east with his brigade (5th, 7th Hussars and the 11th Chasseur à cheval) under the orders of Montbrun. He later joined Davout's command when Montbrun's division was attached to III Corps on 10 March 1809.

At the beginning of the Austrian Campaign he was near the Bohemian frontier at Amberg. On 15 April when the Austrians crossed the border his outposts at Regen north of Ratisbon were driven in. Conducting a skilful withdrawal over three days, he screened the concentration of Davout's III Corps as it fell back to Ratisbon before crossing to the south bank of the Danube. Still in a dangerous position as the Austrians closed in, Davout was forced to take a risky manoeuvre by marching along the river across the front of five Austrian corps in order to rejoin the army gathering at Ingolstadt.

Covering the retirement, on 19 April Pajol fought a rearguard action around Dinzling that pushed the Austrians away from the rear of the French columns. He fought again at Schierling on 21 April as Davout consolidated his position after effecting a junction with Lefebvre's VII Corps. Then at Eckmühl on 22 April, when the whole weight of Archduke Charles's army turned on III Corps before Napoleon came to its rescue, his brigade put up another fine performance. Under Montbrun he led a series of aggressive feints and manoeuvres that misled Archduke Charles as to the true size of Davout's force. His moves helped protect Davout's exposed flank from the Abbach defile on the Danube to Friant's outposts at Dinzling. Some 30,000 Austrians under Kollowrat and Liechtenstein poised to smash their way into Davout's rear were immobilized as a result. The next day he fought at Ratisbon when the Austrians were sent reeling back over the Danube. He then formed Montbrun's advanceguard that crossed the river and shadowed Austrians to the Bohemian border before breaking contact and returning to the south bank a week later.

Summoned to Vienna to join Napoleon's crossing of the Danube, Pajol was stranded on the wrong side of the river with the rest of III Corps and missed the Battle of Aspern-Essling on 21-22 May 1809. At Wagram, on 5-6 July he supported Davout's crushing assault against the Austrian left centered on the Markgrafneusiedel. Active during the enemy's pursuit he reached Brunn eighty miles away when news of an armistice reached him on 13 July.

He remained in Austria with the army till the last French forces withdrew in January 1810. A period followed in Germany till April 1810 when dogged by ill health he returned to France on leave. He returned to Germany in June and took over Davout's cavalry brigade headed by Jacquinot* till the latter's return in October 1810. In May 1811 he moved to Danzig as commander of the large cavalry depot established there. He rejoined Davout in January

1812 as head of his old brigade with the Corps of Observation of the Elbe, which later became I Corps of the Grande Armée for the invasion of Russia.

THE RUSSIAN CAMPAIGN

Leading Davout's advanceguard he crossed the Niemen with his cavalry on 24th June 1812. He took Kovno and Vilna, then caught Bagration's rearguard at Ochmiana on 30 June 1812, but unable to pin him down before Davout's main force arrived, the Russian slipped away. On 7 August 1812 he was promoted *général de division* and in a reorganization of the cavalry replaced Sebastiani as commander of the 2nd Light Cavalry Division with Montbrun's II Cavalry Corps. Present at Borodino on 7 September his division was badly mauled by Russian artillery in the battle's early stages, suffering most of his 800 casualties before even crossing sabres with the enemy. He later joined the charges that overwhelmed the Great Redoubt in the afternoon and helped scatter the Russian Reserve cavalry that tried to retake the position. Engaged in harrying the Russians after the battle, he was shot in the chest outside Mohaisk on 9 September. His early evacuation from Russia soon after undoubtedly saved his life and spared him the horrors of the retreat from Moscow.

GERMANY 1813

Recalled on 26 February 1813 to rebuild his old division, he was still unfit and unable to take up the post. On 8 May 1813 he returned to active duty and headed a provisional division made up of drafts for I Cavalry Corps in Saxony. Once in Saxony he took command of the 10th Light Cavalry Division attached to Saint Cyr's XIV Corps at Dresden. In the opening days of the Autumn Campaign he gained great credit covering Saint Cyr's withdrawal to Dresden. During the battle for the city on 26-27 August he was attached to Murat on the French right. He did great damage to Eszhe's and Mamb's brigades, forcing them into square near Gorbitz before compelling them to surrender once his horse guns arrived.

The energy he displayed resulted on 29 September 1813 in his appointment as commander of the 5th Cavalry Corps. It was not a very promising command, made up largely of inexperienced and poorly mounted dragoons rushed from the depots in France. With Murat's Cavalry Reserve as the French forces were slowly pushed back on Leipzig, he fought at Wachau on 16 October 1813. A shell exploded under his horse, which fell heavily and crushed him. His arm being broken in several places, there was a fear of amputation. He was evacuated from Leipzig and took no further part in the campaign.

THE 1814 CAMPAIGN IN FRANCE

During his recovery Pajol was made *comte de l'Empire* on 25 November 1813. France in crisis, he was barely able to sit in the saddle when he returned to duty on 17 January 1814. He briefly headed the Yonne National Guard before handing it over to Pacthod in early February. Then returning to head the 2nd Light Cavalry Division at Melun he operated with Allix's and Pacthod's forces against Schwarzenberg's Army of Bohemia. Pajol unable to

PAJOL'S CHARGE AT MONTEREAU

stem the tide, the French were slowly pushed down the Seine Valley towards Paris.

With the capital apparently at the mercy of the Allies Napoleon countered with a brilliant series of moves that led to Pajol achieving the finest exploit of his career at Montereau on 18 February 1814. Blucher's defeat at Vauchamps on 14 February, coupled with reports of a mighty host led by Napoleon about to fall on Schwarzenberg, caused the Austrian to lose his nerve and order a withdrawal towards Troyes. Napoleon saw his chance and early in the morning of 17 February went on the offensive barely thirty miles from Paris.

First Gérard wrecked Wittgenstein's advanceguard at Mormont and then turned on Wrede driving him from Nangis with heavy losses. As the offensive developed, it became a race against Schwarzenberg's columns desperate to reach the safety of the Seine's south bank. The most exposed was the Crown Prince of Württemberg's corps on the Allied left. Montereau with its stone bridges at the confluence of the Seine and Yonne rivers, was important to all, since it offered the best escape route for those units that formed the left flank of the Army of Bohemia. Schwarzenberg ordered Württemberg to hold the town till the night of 18 February. Reinforced by some of Bianchi's corps the Prince held a strong position on the north bank of the Seine with some 8,000 infantry, 1,000 cavalry, and 26 guns.

Pajol was the first of the French to come up at about 8.00 A.M. by the Paris road with some 1,500 cavalry and 3,000 National Guards under Pacthod. His cavalry was mostly raw recruits, and Pacthod's men were also indifferently equipped and trained, so they made little headway against Württemberg's position. Victor arriving over an hour late, attacked piecemeal and suffered a series of repulses. Angered by his performance Napoleon replaced him with Gérard and while he moved up the Guard with Pajol's cavalry close behind. Massing all available guns, Gérard gained artillery superiority, which became overwhelming when the Guard's guns came into action.

At 3.00 P.M. Württemberg attempted an orderly disengagement, but it turned into a wild flight down the steep slopes towards the bridges as Napoleon personally led the artillery to within a short distance of the ridge. With the Emperor nearby, Pajol's men now sensed victory. After days of retreats and reversals they charged with an elan unseen earlier in the day, or for that matter in the entire campaign. Their timing was perfect, and they caught the enemy crossing the main road and scattered them, Then forcing the bridge over the Seine they hurled themselves amongst the fugitives on the other side. Sabering to their right and left, his men continued over the Yonne bridge and clearing Montereau completely of the enemy.

Montereau was a severe reversal for the Allies, they lost nearly 5,000 men, including 3,400 prisoners, and 15 guns. It gave Napoleon the important bridges over the Seine and Yonne, and removed the threat to Paris. For Schwarzenberg the retreat continued till he reached Troyes. It nearly brought the Allies to the conference table, but Napoleon's peace terms were so outrageous that the war continued. For all this, Napoleon was indebted to the magnificent energy of Pajol's charge, which converted a retreat into a rout, and left the enemy no time to destroy the bridges. The cost was heavy, for the French suffered 2,500 casualties, among them Pajol who had another a bad fall from his horse. Unable to continue, he returned to Paris on 20 February to convalesce and took no further part in the campaign.

THE RESTORATION AND NAPOLEON'S RETURN

The Restoration saw Pajol's appointment as a Cavalry Inspector General on 23 April 1814. He became a *Chevalier de Saint Louis* on 1 June 1814, and then commander for cavalry in the 1st Military Division based in Paris. He tried hard to accept the Bourbons, but the martial ineptitude of both Ducs de Berry and Angoulême soon had him seething. An example arose when Dupont as Minister for War arranged wargames outside Paris. It was to culminate in a mock battle for the benefit of the two dukes to show their martial talents. The event was a fiasco. Angoulême the elder was supposed to win, but de Berry was unable to contain his resentment and charged around the field like a man possessed. He struck two of his officers, and threatened to have another shot. Pajol who commanded his cavalry was also insulted and left the field. Keen to get away from such antics, he readily accepted command of the Orleans garrison on 17 January 1815.

He was always a Bonapartist, but when Napoleon landed he kept his feelings to himself as Orleans was not a city sympathetic to the former Emperor's cause. He waited till he thought it safe to do so, then on 20 March 1815 struck down the white banner and raised the tricolor. Gouvion Saint Cyr suddenly arrived in the city to drum up support for the King had him arrested and reversed the flags. The Cuirassiers du Roi (1st Cuirassiers), remembering how earlier that winter they had been turned out of their quarters in Paris, released Pajol and ran Saint Cyr out of town. Pajol then set out after a small force of Royalist sympathizers nearby led by the Duc de Bourbon, which soon scattered when his horsemen appeared.

THE WATERLOO CAMPAIGN

On 2 June 1815 Pajol became a Peer of France and the next day took command of the 1st Cavalry Corps of the Army of the North. His command was one of the finest bodies of light cavalry fielded for many years. It comprised the Hussar Division of Pierre Soult and the mixed Lancer Chasseur division of Subervie, all fresh, newly mounted and eager veterans.

At the start of the campaign he had the task to screen Vandamme's III Corps on its march to Charleroi. With his usual reliability he burst across the Belgian border in the early hours of 15 June and brushed aside the Prussian outposts. Then things went wrong for him as Vandamme's troops didn't receive their marching orders. They set out four hours late and consequently were miles behind Pajol, who reached Charleroi at 9 A.M. and tried on his own to rush the bridge over the Sambre. It was no repeat of Montereau, because the barricades were too strong and with no infantry or artillery support, the attack was driven back. Valuable hours were lost until Napoleon arrived with the Guard Marines, which seized the crossing around mid-day. Able to continue Pajol pushed up the Fleurus road till halted by the Prussians at Gilly. Trying to work his way around the Prussian flank he made little progress until the full weight of III Corps arrived, forcing them to abandon their positions in the evening.

At Ligny on 16 June his cavalry was positioned on the French right. Operating with Exelmans's II Cavalry Corps, by the skilful use of feints and counter feints they managed to pin down Thielemann's 23,000 strong III Corps for most of the day. This prevented the Prussian from supporting Blucher's center as the main battle reached its climax. At the end of the day still comparatively fresh, his men never received orders to harry the enemy's retreat nor did he himself think it necessary to do so. The next day he set out early and established Thielemann's rearguard had evacuated Sombreffe. As his patrols probed forward they found the Namur highway cluttered with fugitives and wagons. He mistakenly thought himself on the heels of the main army when he overtook and captured an ammunition train, some ten guns and overran a squadron of Uhlans near Mazy. All this and other reports convinced him, and more importantly Grouchy, that the Prussians were falling back on Liege. This had calamitous consequences on the outcome of the campaign as Grouchy's wing was drawn further away from Napoleon. By evening Pajol was at Tourinnes five miles east of the Wavre Gembloux road, still unsure where the main Prussian army had headed.

It was only at mid-day the next day that he learnt the Prussians were swarming around Wavre. As commander of the main light cavalry force he had let down Grouchy and Napoleon badly. The thunder of the guns to the west jolted Pajol into action. On his own initiative he immediately headed in that direction, on the way picking up Maurin's division of Gérard's corps. He reached the Dyle at Limale and found a Prussian battalion positioned across the road behind the bridge. Although only wide enough to allow four horses abreast he repeated the bold stroke used at Montereau and charged the bridge with Vallin's hussars. The audacious venture succeeded and the Prussians were ridden down, scattered and dispersed. Teste's division following took Limale, and

as darkness fell, Pajol firmly established on the plateau above the river was confident of rolling up the Prussian line the next morning.

When news arrived of Napoleon's defeat the next day, Pajol's horsemen formed Grouchy's rearguard as they recrossed the Dyle and fell back towards Namur and the French border. Before Paris on 30 June Davout appointed him his deputy and the army's cavalry commander. After the Armistice he retired with the Army to the Loire and presided over its disbandment. On 22 October placed on the non active list, he returned to live in Paris and retired from the army on 31 December 1815.

Later Life and Career Assessment

He went into business and joined a venture to establish a steamship company, which was not a success. In 1823 he founded a metal works at Paraclet near Nogent-sur-Seine, and had it succeeded little more would have been heard of him. The resultant inactivity and a sense of failure brought out in him an underlying hatred of the Bourbons. It led him to becoming a member of the Carbonari, a secret society dedicated to the overthrow of the monarchy. He also started to write his memoirs but events took over and he never completed them. They were later rewritten by his son and published in 1874.

When the July Revolution broke out on 30 July 1830 he took to the streets and helped organize the Paris National Guard. With Lobau he was a key figure in the struggle that forced the Paris garrison from the city. On the night of 2-3 August at the head of the insurgents he marched on Rambouillet. It broke the deadlock, Charles X stopped trying to rally support outside Paris, gave up and fled the country.

A grateful Louis Philippe awarded Pajol the Grand Cross of the Legion of Honor on 19 August 1830. Recalled under the new monarch, Pajol had a distinguished military career. On 3 September 1830 he became commander of the 1st Military Division and Governor of Paris, a post he held with distinction till October 1842. On 19 November 1831 he was made a Peer of France. He died in Paris on 20 March 1844 as a result of a fall from a horse.

Pajol was one of the finest leaders of light cavalry in the Napoleonic era. Before joining Lasalle, he had learnt his trade well under men like Ney. The role model against whom light cavalry leaders were compared, Lasalle had largely gained his reputation in exploits against demoralized and beaten foes. Pajol on the other hand was not so fortunate. By the time he led large formations, the Napoleonic armies were more often than not on the defensive and the quality of men and equipment with the light arm was not the same. For that reason it made his achievements all the more impressive. Not as charismatic as Lasalle, Pajol matched him in courage and ability, and certainly had a cooler head, but was largely eclipsed, more by the legend that grew as a result of the former's untimely death.

PÉCHEUX - NICOLAS LOUIS, BARON (1769 - 1831)

BACKGROUND AND EARLY CAREER

The infantry general and Peninsular veteran Nicolas Pécheux was born in the village of Bucilly, (Aisne), northern France on 28 January 1769. Little is known of his early life including when he first enlisted. He first came to notice with the Army of the North on 19 August 1792 when promoted capitaine of the grenadier company of the 4th Battalion of the Aisne Volunteers. The army was going through a crisis at the time, which resulted in further promotion for him to *lieutenant colonel* on 8 September 1792 when he replaced his commanding officer dismissed for incompetence.

A reorganization of the armies in 1793 resulted in his battalion passing to the Army of the Ardennes where he served at Fleurus on 26 June 1794. He then moved to the Army of the Sambre-Meuse and shared in that army's triumphs as it cleared Belgium of the Austrians and advanced to the Rhine. He took part in the crossing of the river near Coblenz in April 1795 and then the retreat to the Rhine the following October when Archduke Charles nearly cut off and destroyed the army.

On 21 March 1797 he left the Aisne Volunteers and joined the 41st Demi-brigade, which soon after left France in early 1798 to join the Army of Italy. With that army for the next three years, there is little to show that he distinguished himself in any notable way during the turbulent period that followed. The glory won by the Army of Italy under General Bonaparte was squandered and lost by the indifferent leadership of his successors. Pécheux however obeyed orders, kept his hands clean and as a result gained promotion to *chef de brigade* on 7 September 1799. With Suchet's forces in the Spring of 1800 he avoided being cut off in Genoa and during the Marengo Campaign spent his time facing the Austrians across the Var. In the Winter Campaign of 1800 he took part in Brune's crossing of the Mincio that ended with the Austrians being driven from Venetia. He remained in Italy with his regiment stationed in Liguria after the Peace of Lunéville till its disbandment and merger with the 17th Line in August 1803. Pécheux came out of the change rather well with a promotion to *colonel* on 29 August 1803 and a posting as commander of 95th Line with the Army of Gallo-Batavia.

THE GRANDE ARMÉE 1805-1807

He joined Mortier's occupation force in Hanover. When Bernadotte became governor of Hanover and later commander of I Corps of the Grande Armée, Pécheux with the 95th Line served in his 2nd Division led by Drouet*. At Austerlitz on 2 December 1805 in spite of Bernadotte's indifferent leadership, which in turn affected Drouet, he handled his regiment well. Soult successfully secured the Pratzen Heights while his men with the 94th Line and 27th Light Infantry consolidated the breach in the Russian line and pushed the Russian Guard from Kreznowitz.

In the Prussian campaign of 1806 he fought at Schleiz on 9 October when Tauenzien was driven from the town in the first action of the campaign. With the rest of I Corps his regiment failed to make an appearance at Jena or Auerstadt on 14 October. The whole corps was stung into action as a result and conducted a vigorous pursuit of the Prussians. He fought at Halle on 17 October when Drouet's division following Dupont's joined the action that secured the Saale crossing in the face of a vastly superior force of 17,000-18,000 men led by Prince Eugene of Wurttemberg. The momentum of the advance continued till Lubeck on 6 November when his men joined in the reduction of the town that heralded the end of the campaign in Prussia.

He took part in the campaign in Poland and East Prussia where he served under Villatte who took over the division from Drouet in March 1807. He fought at Spanden on 5 June when the new campaign opened and was present at Friedland on 14 June 1807.

SPAIN 1808-1813

In October 1808 he moved to Spain with Villatte's division. He fought at Espinosa on 10-11 November when Victor forming Napoleon's right wing at the head of I Corps scattered Blake's army. Victor rejoined Napoleon's march on Madrid and Pécheux was at the fall of the city on 4 December.

He fought under Villatte at Ucles on 13 January 1809 when Victor routed Venegas's Army of the Centre as it tried to recapture Madrid after the bulk of the Grande Armée was in the North dealing with Moore. The battle itself was won virtually single-handed by Puthod's brigade, to which Pécheux's 95th Line belonged. Whilst Villatte with one brigade presented a bold front, the other including the 95th Line executed a wide flanking movement under the eyes of the enemy, turning their left flank. Then scaling the slopes, they formed up at right angles to Venegas's positions and rolled up the Spanish line. He fought at Medellin on 29 March 1809 when Victor defeated Cuesta's Army of Estremadura near the Portuguese border and the river Guadiana when they threatened French communications.

The run of successes ended against Wellington at Talavera on 28 July 1809. His regiment gave a good account of itself in one of Wellington's toughest battles as it supported the assault by Ruffin's division on the Cerro de Medellin, but failed to effect a breakthrough. At the Battle of Ocana on 19 November 1809 he was in trouble as a result of a mix up in orders when his regiment did not arrive on the battlefield in time.

Soult's invasion of Andalusia saw him at the siegeworks before Cadiz in February 1810. He was promoted *général de brigade* on 23 June 1810 while serving before the city. A transfer followed in October 1810 to Sebastiani's IV Corps in Grenada where he spent time as commander at Xeres. Continually engaged against guerillas in the region he had a success of note at Vanta de Leche in June 1811.

November 1811 saw his return to Villatte's division as commander of its 1st Brigade. He joined Victor's unsuccessful siege of Tarifa that lasted from 20 December 1811 till 4 January 1812. From there he returned to the siegeworks at Cadiz where he was in charge of the left wing of the besieging forces. In August 1812 he formed the advance guard as the French withdrew from Andalusia to Valencia.

GERMANY 1813

He returned to France and after promotion to *général de division* on 30 May 1813 he moved to Hamburg where he joined Davout's XIII Corps on 9 August. With the renewal of hostilities on 17 August he took part in Davout's operations against Wegesack's Swedish corps that after a week took Schwerin and occupied most of western Mecklenberg. When Oudinot's advance on Berlin failed and left XIII Corps exposed on its own in the North, Davout reverted to his secondary mission, the defence of Hamburg and Holstein. Pécheux was given temporary command of Vichery's small 50th Division and detached to the southern bank of the Elbe to secure communications between Hamburg and Magdeburg. Surprised by Wallmoden at Gohrde on 16 September, his 3,500 men outnumbered four to one fought a fine defensive action and then evading the enemy's converging columns reached Hamburg after losing 1,500 men.

The action itself was heralded as a major triumph by the Allies, since so little had happened on the northern front. At one time in order to raise Allied morale they had the temerity to claim that Davout himself headed the defeated force. Pécheux proved to be a reliable and hardy subordinate as the Allies tightened their grip around Hamburg during the eight month siege. In June 1814 Pécheux returned to France with Davout. The delay was caused by Davout, who only agreed to the city's surrender once he had proof of Napoleon's abdication and orders from Louis XVIII of France.

THE RESTORATION AND NAPOLEON'S RETURN AND THE WATERLOO CAMPAIGN

He received the usual treatment on his return, made a *Chevalier de Saint Louis* on 20 August 1814

and then unattached placed on half pay. Not surprisingly on Napoleon's return he had no conflict of loyalties and rallied to him. Being Davout's man, his was one of the first appointments made by the Minister for War. On 31 March 1815 he received orders to leave for Metz to head the 12th Infantry Division with the Army of the Moselle.

Later his preparations complete, as part of Gérard's IV Corps he left Metz on 6 June for the long march to the Belgian frontier. On 16 June his troops played a major role in the struggle for Ligny itself. Soon after Vandamme's first assaults against the Prussian right commenced around 2.30 P.M., Gérard's corps advanced to storm Ligny. The deadly struggle that followed for the village became one of the historic encounters of the campaign.

Pécheux advanced in three columns and at once came under heavy fire from the guns on the high ground above the brook. The two left columns gained the outskirts of the village, but were driven back by the fire from the men of Henkel's brigade posted in the houses and the chateau. The right column however burst in, but then reaching the church came under fire from all sides. The carnage was fearful and in little time some 20 officers and 500 men fell. The head of this column was crushed, and with the others it fell back to reform.

Twice more Gérard ordered Pécheux to renew the attack and twice more he failed. Napoleon sent some 12 pounder batteries of the Guard to co-operate with Gérard's artillery. Shortly after they opened fire the village was aflame and Pécheux was ordered forward again, this time supported by one of Vichery's brigades. After house to house fighting the French succeeded in mastering the upper part of the village, and the Prussians rallied around the church and the cemetery. A final effort was made by Pécheux. His men became disordered by their efforts and were charged by a mass of Prussians. A furious melee resulted, in which no quarter was given or asked. The Prussians retired over the two bridges that spanned the muddy stream followed by the French spurred on by their previous success.

As victory seemed assured, four battalions from Jagow's brigade appeared to lend Henkel support. The Prussians promptly took the offensive and drove back Pécheux's men from the left bank. The terrible struggle continued. Only separated by the small stream the two forces engaged in a heavy and murderous firefight. Above the roar of the flames and the rattle of musketry rose the shrieks and cries of the wounded as they burned to death in the flaming houses. That anyone survived at all was ascribed to the dense pall of smoke, which enshrouded the troops engaged in this terrific struggle in the streets, houses, stables and gardens of Ligny. In one eye witness account of the battle, as fires raged through the village the horror was described like a scene from Dante's Inferno. Great clouds and banks of smoke rose over the Ligny brook, while the fires reflected in the water resembled in ferocity the burning rivers of Hades.

This furious fight for the village never let up throughout the struggle along the five mile front. For Blucher it was the most costly of all, with so little reward. It wore him down; against one French division he was compelled to send reserve after reserve into the fiery furnace. At 8.00 P.M. the Imperial Guard broke the deadlock and enveloped the village from right and left. Pécheux still in possession of that part of the village on the right bank, with Gérard at his side, made one last effort and drove the Prussians before him. Elsewhere the defence collapsed, and the battle was won.

Next to Girard's, Pécheux's division suffered the most at Ligny. He lost over 1,500 men in the struggle and the next day his shattered troops were in no condition to resume the march. On 18 June they were the last to reach the Wavre. In the late evening they took post on the Dyle near Limale before crossing the river with the rest of IV Corps. The next morning on the Limale Plateau he formed the left wing of Grouchy's envelopment of Wavre. After two hours he had made steady progress when the arrival of the news of Waterloo around 10.00 A.M. brought the advance to a halt.

As Grouchy ordered a retreat on Namur, Pécheux's division on the farmost left of the French advance formed the rear of IV Corps, while III Corps withdrawing from Wavre itself fell in behind. The next morning Vandamme's troops broke camp early in the morning without informing IV Corps. As a result as they moved past, Pécheux's division became the French rearguard without realizing it. Prussian cavalry under Hobe with a horse battery caught up with him on the Namur road about five miles from Gembloux. The march came to a grinding halt as shells tore into his ranks and enemy cavalry prepared to charge. Grouchy fortunately nearby with Vallin's hussars of Pierre Soult's division, reacted with the customary coolness of a fine rearguard cavalry leader and drove them off, while Pécheux's march on Namur continued uninterrupted.

LATER LIFE AND CAREER

He remained with the Army during the retreat to Paris and its disbandment behind the Loire in July. On 1 August 1815 he was placed on the non-active list without pay. A man of exemplary character with a reputation for being a fine professional soldier, Saint Cyr's Army Reforms of December 1818 resulted in his recall as an Inspector General of Infantry. On 12 January 1819 he became commander of the 12th Military Division based at Nantes in place of Rivaud. The following year he moved to the 16th Military Division as an Inspector General.

The assassination of the Duc du Berry in February 1820 brought a check to his career, as in the period of reaction that followed the loyalty of most officers who had served Napoleon was suspect. Many lost their posts and Pécheux was no exception and found himself unattached on 1 January 1821. The French intervention in Spain in 1823 to restore Ferdinand VII required commanders of talent who had Peninsular experience. Pécheux possessed both, which resulted in his recall on 25 June 1823 as commander of the 12th Division of Lauriston's II Corps of the Army of the Pyrenees. Ill health caused his early return from Spain on 8 January 1824. He never received another command before his retirement on 1 January 1825. For his part in the Spanish Campaign he became a Grand Officer of the Legion of Honor on 23 May 1825. The fall of the Bourbons saw his recall on 7 February 1831 but he never took up any post and remained unattached till his death in Paris on 1 November 1831.

ASSESSMENT OF CAREER

Throughout his career Pécheux showed he was a loyal and brave professional At no time did he fail to show true leadership qualities expected of him. Equally decisive in both defence and attack, his courage and persistence at Gohrde and Ligny marked him above the average division leader. In Spain, the graveyard for most reputations he came through unscathed, which was an achievement in itself. Unluckily, this was followed by being locked up in Hamburg, which prevented his talents being used to the full. As a general he missed the cut and thrust of Napoleon's later campaigns that could have shown his talents to the full and allowed him to rise higher.

PIRÉ - HIPPOLYTE MARIE GUILLAUME, COMTE DE (1778 - 1850)

BACKGROUND AND EARLY CAREER

Of noble birth, the light cavalry leader Piré was born in the city of Rennes, western France on 31 March 1778. His father was a soldier who left France with his family during the second great exodus of emigrés that followed the King's flight to Varennes in April 1791. When Piré senior, a *colonel* in the old French Army joined the emigré Army of the Princes gathering at Coblenz on 2 January 1792, young Hippolyte not yet fourteen became his aide.

After the outbreak of war Piré saw service in Holland and Belgium. On 15 March 1794 as a lieutenant he served on the staff of the emigré Rohan Regiment that fought with the English forces under the Duke of Cumberland. On 21 July 1795 he took part in the English sponsored landings at Quiberon that supported the Royalist insurrection in La Vendée. Shot in the chest soon after he landed, he was fortunate to be evacuated before Hoche brutally crushed the expedition. He joined the Comte d'Artois's forces that landed on the Ile d'Yeu on 5 October 1795 in the hope of using the island as a base for another descend on Brittany. When forces on the mainland appeared led by Grouchy*, Artois abandoned the venture five days later. Piré after showing his disgust with the whole affair was dismissed. He returned to France in March 1796 and served with the Army of the Vendée under Puisaye till 1 July 1796 when a treaty was signed with Hoche. He left the rebel army, opted for civilian life and spent the next three years wandering Europe as a penniless exile.

A dramatic change in Piré's fortunes came in March 1800 when Bonaparte as First Consul offered an amnesty to all emigrés. Former emigrés were allowed to serve in the army, and knowing no other profession Piré opted to join the Bonaparte Hussars. Made up of volunteers recruited from reconciled aristocrats the formation was not a success. He however remained with it till August 1801 when the formation disbanded, having been promoted *capitaine* on 20 June 1800.

THE GRANDE ARMÉE 1805-1808

He remained unattached for several years till 22 September 1805 when a family connection helped him gain a place on Berthier's General Staff with the Grande Armée. During the Austrian Campaign he helped Berthier's deputy Camus and was cited for bravery at Austerlitz on 2 December 1805. The Prussian Campaign saw him attached to Lasalle's light cavalry with the Cavalry Reserve where his hot-headed brash temperament certainly suited a man like Lasalle. Within a few days he gained a certain notoriety when two days before the Battle of Jena he boldly rode into Leipzig at the head of fifty men from the 10th Hussars and causing great alarm amongst the populace.

With Lasalle during the great dash across Prussia after Jena he gained promotion to *chef d'escadron* on 30 December 1806. He was present at Eylau on 8 February and then returned to Berthier as one of his aides on 17 March 1807 for the rest of the campaign. On 25 June 1807 he gained promotion to *colonel* with the 7th Chasseurs à cheval with Lasalle's division.

On 2 August 1808 he received the title *baron de l'Empire*. He rejoined Berthier as an aide for Napoleon's invasion of Spain and on 30 November 1808 distinguished himself at Somosierra. Lasalle secured his recall on 1 January 1809 and he served briefly with Bordesoulles's brigade till his promotion to *général de brigade* on 10 March 1809.

THE DANUBE CAMPAIGN 1809

Recalled to France as the war clouds loomed with Austria he moved to Bavaria. On his arrival he took command of the 2nd Brigade (8th Hussars and 16th Chasseurs à cheval) with Montbrun's light cavalry division attached to Davout's III Army Corps. On the Bohemian frontier at Amberg he barely had time to familiarize himself with his new command when the Austrians crossed the borders on 15 April. His command made up of fine veterans screened the concentration of Davout's infantry formations in the rear. With Montbrun he successfully covered Davout's retirement to Ratisbon and crossed to the south bank of the Danube on 18 April. The next day under Montbrun he covered III Corps's southerly march down the Danube to close up with Lefebvre's VII Corps. He held up Austrian probing movements around Dinzling earlier in the day and then supported Davout's infantry as they drove the Austrians from the woods around Thann. On 21 April he screened the French left as the bulk of the Austrian army turned on Davout's forces after the fall of Ratisbon in what became known as the Battle of Abensberg.

By evening Davout's plight had became desperate. Repeated messages to Napoleon failed to convince the Emperor that it was Davout and not he that faced the main Austrian army. Davout, knowing Piré had the ear of Berthier and was respected at Napoleon's headquarters, decided to send him to appeal for help. In darkness he made a thirty-seven mile six hour ride across hostile country to Landshut to deliver his message, arriving shortly after midnight on 22 April. Much to his credit he retained his composure as he described the events of the previous day to Napoleon. Most importantly he outlined the dangers that had arisen due to the fall of Ratisbon, which Napoleon was not aware of. It secured Charles's line of retreat but also enabled him to bring two additional corps across the river to fall on Davout. He concluded that Davout would be unable to hold for another day on his own, and that the Emperor should march with all haste to strike the enemy's southern flank around Eckmühl.

After listening to Piré, Napoleon realized he had been mistaken and quickly ordered the army to march to Davout's aid. Then as if he had not done enough, Piré was ordered to return Davout with news that help was on its way. By 3.00 AM Lannes set out with his corps, followed by Vandamme, then Massena. Charles aware of Davout's weakness was determined to roll up his left flank with Kollowrat's newly arrived II Corps supported by Liechtenstein's Reserve. It was not till 10.00 A.M. that Kollowrat began to get his attack moving and even then, his advance was soon slowed to a crawl by aggressive cavalry movements directed by Montbrun. The indecision of the Austrian commanders was not helped by Charles suffering an epileptic attack. Then as the pressure grew it was Montbrun and Piré, now back with his brigade that probably saved Davout's left as the French pounded up the road from Landshut. Unable to bring his full weight to bear, pressed hard by French attacks from front and flank, Charles had little option but to break off and fall back to Ratisbon. Piré took part in the cavalry battle before the city the next day as Austrian cavalry covered Charles's retreat as the army streamed over the Danube.

The Ratisbon phase of the 1809 Campaign was over. In a whirlwind campaign of five days after Napoleon took over the Army of Germany from Berthier, he turned a potential disaster into a brilliant success. Charles's army after losing some 50,000 men was split in two, with its main element on the wrong side of the Danube, and Vienna at the mercy of the French. Whilst Napoleon did not destroy the Austrians, he showed energy comparable with his Italian days of 1796. Davout established himself as the foremost of the fighting marshals and later took the title Prince d'Eckmühl. The pattern that so often emerged later in Spain and in Germany in 1813 where the fruits of Napoleon's victories were lost by defeats of his marshals, could so easily have started on the banks of the Danube in 1809. This time it was not due to the failings of his marshals that disaster loomed, but rather as a result of Napoleon's own intransigence. It is here that Piré must take credit for his role in the events. His valiant ride, the precise and determined way he described events to Napoleon, convincing him it was not he who faced the main Austrian threat but Davout, was critical to the outcome of the campaign. Piré's endeavors

merited a division command. It was not to be, Napoleon did not reward men for pointing out his mistakes, only for winning battles!

Ratisbon retaken on the 23 April, Piré spent the next few days north of the Danube shadowing the Austrian army's movements as it retired towards the Bohemia. He recrossed the river at Passau on 30 April and joined the push against Hiller's forces, fighting at Ebelsberg on 3 May. With Vienna occupied on 12 May, he moved to the Neusieder Lake forming part of the cavalry screen covering the southern approaches to the city.

Piré renewed his association with Lasalle when the flamboyant leader arrived from Spain to head a new light cavalry division that included his brigade. Recalled to Vienna he crossed to the north bank of the Danube on the evening of the 20 May. He played a vital role in the Battle of Aspern-Essling that followed. Fanning out across Marchfeld the next day he failed to pierce the enemy's cavalry screen that hid the Austrian army. Then as the Austrians swung into the attack, his skilful use of cavalry feints won valuable time as the French formed up between the villages of Aspern and Essling. As Austrian pressure grew, Piré's cavalry unsupported by infantry or artillery charged forward to help shield Essling from further Austrian assaults. It gained time for reinforcements to pour onto the field from the Isle of Lobau and deploy. By nightfall his brigade was finished, and it was Lasalle's indomitable spirit kept his men going as the battle continued the next day. When Napoleon realized that his gamble to cross the river had failed, further sacrifices were made as the cavalry covered the army's retreat to Lobau. Piré lost over 600 men in the two day battle.

At Wagram on 5-6 July 1809 Piré was with Lasalle attached to Massena's command. On the first day after crossing from Lobau they pushed forward aggressively and soon discovered apart from a rearguard in Aspern Essling the main Austrian army was well to the north. By 2.00 P.M. Piré's outposts were beyond Süssenbrünn before he encountered serious resistance that prevented him from piercing the Austrian cavalry screen to observe the enemy's movements. Most important, they confirmed that the Grande Armée did not face the prospect of disputing a river crossing, as the corps of Massena, Davout, Eugene and Oudinot deployed and occupied the Marchfeld with little difficulty. The next day the fighting fell largely to the infantry and heavy cavalry while Piré's men were held back to exploit the breakthrough. At the end of the day he harried Klenau's corps as it fell back on Stammersdorf after Charles ordered a withdrawal.

Lasalle's death at Wagram did not result in the division command many expected Piré to receive. Instead he remained with Massena's corps as head of the 4th Light Cavalry Brigade as part of Quesnel's division, a commander well past his best. Relief came in January 1810 when he moved to Holland to head the small cavalry force with the Army of Brabant. In May 1811 he transferred to the Corps of Observation of the Elbe for a time as commander of the 3rd Light Cavalry Brigade.

RUSSIA AND GERMANY 1812-1813

The build up of the Grande Armée for the campaign in Russia saw Piré in January 1812 rejoin his colleague Bruyère's 1st Light Cavalry Division with his old command, the 4th Light Cavalry Brigade. The formation operating as part of Nansouty's I Cavalry Corps, he was at the capture of Vilna on 27 June, fought at Ostrovno on 26-27 July and Smolensk on 17 August. At Borodino on 7 September his troops suffered severely from artillery fire as they supported Ney's assaults against the Fleches.

Piré survived the retreat from Moscow and rejoined Bruyère for the Spring Campaign in Germany where he served at Bautzen on 22 May 1813. After Bruyère's death from wounds at Reichenbach, he served under Corbineau* who took over the division. He fought at his side at Dresden on 27 August 1813. The long awaited promotion to *général de division* finally came on 15 October when he was made commander of the 9th Light Cavalry Division. Hardly had he taken up his new command under Milhaud* (V Cavalry Corps) when he found himself hurled the next day into the great cavalry struggle at Wachau south of Leipzig. He suffered heavily at Leipzig and then again at Hanau on 30 October during the retreat to the Rhine. It showed in his returns, which showed his division down to 500 men when he crossed into France compared to 2,000 at the campaign's start.

FRANCE 1814

The campaign in France saw his division continually switched from corps to corps as the need arose. On 10 January he was with Victor at the action of Saint Die. With Grouchy's cavalry corps at Brienne on 29 January his division was the first to bar the Russian advance. At La Rothière on 1 February he returned to Milhaud's command and served under him at Mormont on 17 February, the recapture of Troyes on 24 February and the action at La Fèrte-sur-Aube on 28 February.

He spent most of March in Champagne with his division operating on its own raiding the Army of Bohemia's communications, causing alarm and confusion far in excess to what it actually achieved. A notable exploit was the destruction of two Russian battalions on 23 March escorting a convoy near Saint Dizier.

THE RESTORATION AND NAPOLEON'S RETURN

The Restoration saw him placed on the non active list on 1 November 1814. The Bourbon expedient of testing the loyalty of unemployed generals by recalling them on news of Napoleon's landing was used on him. He received his marching orders on 10 March 1815 to go to Rennes to help organize troops in the 15th Military Division. He dutifully took up his post, played safe, did little and said little. When news arrived of Napoleon's installation in the Tuileries, to secure his future he quickly declared for him and raised the tricolor over Rennes on 22 April. A busy period followed as commander of troops in the 7th and 19th Military Divisions. On 23 April he accepted command of the 2nd Cavalry Division, a fine formation made up of a brigade of lancers and another of chasseurs à cheval that joined Reille's II Army Corps.

THE WATERLOO CAMPAIGN

As the campaign opened at dawn on 15 June he led the way across the Sambre at Thuin for II Corps. After minor skirmishing he cut across country and reached Gosselies on the Brussels-Charleroi road at 4.00 P.M. At the same time his lancers had pressed on ahead and were at Heppignies, mid point between Quatre Bras and Fleurus. Another formation Colbert's Guard Lancers were further up the road hovering around the outskirts of Quatre Bras after a few volleys from Dutch troops had toppled some men from their saddles. Had he and Colbert's Lancers realized each other's presence, together they could easily have overthrown the small force at Quatre Bras.

The next morning Piré resumed his march and moved up to Quatre Bras. There were too many Allied infantry preparing defences so he dared not risk trying to carry the position until infantry support arrived. It was not till 2.00 P.M. that his chasseurs supported Bachelu's assault on Piraumont and destroyed a battalion of Dutch militia that tried to reinforce the garrison in the farm. His lancers did damage to the Dutch cavalry when they routed a dragoon and a hussar regiment that tried to disrupt Foy's infantry as it advanced on the Quatre Bras cross-roads itself.

In an attempt to break the deadlock that developed on the Namur Road he moved to support Bachelu, then making heavy weather against Picton to the east of Quatre Bras. When Bachelu gave way, Piré turned his lancers on the triumphant enemy. First he charged the 28th Line but was unable to breach the hastily formed square. Another regiment pursuing the broken Brunswick cavalry caught the 42nd Highlanders and the 44th Line also forming square, and after a furious melee was driven back. While Wellington repelled this assault another regiment broke through to the Namur Road where it cut to pieces an unformed Hanoverian battalion. The havoc his cavalry caused that day was out of all proportion to the number involved, and was a fine reflection of his leadership. He tried to join Kellermann's charge that so nearly carried the cross-roads but didn't receive the orders in time to lend proper support.

As Wellington poured more and more reinforcements into the struggle and d'Erlon didn't arrive, none of the early gains were held and the French were driven back to their start lines. At the end of the day Piré's tireless horsemen had one final success when they drove Maitland's Guards emerging from Bossu Wood to seize Piraumont back into the undergrowth.

Piré's division exhausted, he did not join the pursuit of Wellington the next day. At Waterloo on 18 June he was positioned on the extreme left of the French line. His orders were simple, to support the attack on Hougoumont and cover any Allied moves against the French left. He performed these tasks without fault. His horse guns positioned astride the Nivelle Road forced Bull's Royal Horse Artillery battery to retire after they fired on his lancers as they approached Hougoumont from the west. Allied cavalry supported by guns were also active on his flank as they covered the arrival of reinforcements from the direction of Braine l'Alleud. To disrupt their movements he made many feints in that direction. Later in the afternoon the skirmishing between Jérôme and the Allied infantry spread west beyond the Nivelle road as each tried to outflank the other. Piré was drawn into the action having to protect the skirmishers from the occasional rush by enemy infantry and cavalry.

When orders to retreat came in the evening he cut across country to Genappe and formed up in the hope the army could rally behind his horsemen. The army was a swarm of uncontrollable fugitives and swept his men away as they tried to form up. Once across the Sambre he regained control and covered the retreat, picking up stragglers on the way. He reached Laon on 25 June and then fell back by stages to Paris.

He took part in the operations around Rocquencourt on 1 July that resulted in the defeat of Blucher's cavalry south of Paris near Versailles.

LATER CAREER AND ASSESSMENT OF CAREER

There was little sympathy for ex emigrés who had rallied to Napoleon. He was proscribed by the Ordinances of 24 July and took refuge in Russia. He returned to France in 1819 and on 26 May 1819 resumed his rank and served on a commission formed to review the plight of officers on half pay. On 20 February he became a *chevalier de Saint Louis*. His work came to a virtual standstill after the assassination of the Duc de Berry and he retired from the post on 1 April 1820. Unattached he remained with the Army General Staff but he held no post of importance under the Bourbons.

The overthrow of Charles X resulted in his appointment as commander of the 2nd Military Division based at Mezières on 6 August 1830 till 7 February 1831 when he rejoined the General Staff. On 5 January 1834 he became a Grand Officer of the Legion of Honor. He moved to the 3rd Military Division on 12 June 1834, which he headed till 1 August 1835. He then spent a period unattached till 11 June 1839 when he became commander of the 9th Military Division. From 10 June 1841 till 3 March 1844 he was Inspector General of the Gendarmerie in the 5th Arrondissement. Placed in the Reserve Section of the General Staff on 1 April 1849 he retired on 8 June 1848. He lived in Paris till his death on 20 July 1850.

A leader of fine repute, forthright and not afraid to speak his mind, Piré was unfortunate not to receive a division command sooner than he did. His royalist past undoubtedly raised questions as to suitability. Nevertheless it was his fine work at Quatre Bras and Waterloo that upheld the finest traditions of the light cavalry and showed him up as a fine tactician and cavalry leader.

LANCERS SCOUTING FOR THE CUIRASSIERS (BY DETAILLE)

RADET - ETIENNE, BARON
(1762 - 1825)

BACKGROUND AND EARLY CAREER

From a middle class background the gendarme general Etienne Radet was born in the small town of Stenay (Meuse) on 19 December 1762. He enlisted as a *fusilier* with the Sarre Infantry Regiment on 4 April 1780. A fine soldier, his promotion through the ranks was fast for the old army, *caporal* on 20 March 1781 and then *sergeant* on 26 April 1782. He spent a time in the colonies on Santo Domingo. Discharged from his regiment in October 1786, he joined the mounted police as a trooper and in December 1787 gained promotion to *brigadier*. He resigned on 11 August 1789 and entered the National Guard as a *sous lieutenant*, later gaining promotion to *lieutenant* on 10 November 1789. He became interested in guns, which resulted in him becoming an artillery *capitaine* on 1 October 1790. He later moved from Metz to Varenne in May to organize the National Guard in the town. It coincided with Louis XVI's flight to Varenne when the monarch was caught with his family at an inn in the town on the night of 18 June 1791. Whilst not directly in involved in the King's arrest, he kept a cool head in the tense hours that followed and presented a bold front against any persons who tried to release him or do him any harm. This was more admirable since a body of cavalry loyal to the King were nearby and could easily have entered the town and spirited the Royal Family away. He escorted the Royal Family during the first leg of their journey back to Paris and for his part on 9 August 1791 was promoted *major*. He remained at Varenne and on 16 March 1792 with further promotion to *chef de bataillon* became district commander of the National Guard.

He joined the 2nd Battalion of the Meuse Volunteers and on 25 June 1792 became their *adjudant-major*. He was at the defence of Verdun on 2 September, and after catching the eye of General Dillon became his aide de camp on 15 September, serving with the Army of the Ardennes. After Dillon's suspension in February 1793 he cleared himself of any suspicion and emerged as an aide to General Dubois with the Army of the Moselle. He fought at Arlon on 9 June 1793 and served for a period before the lines at Wissembourg.

As the Terror gripped the country, the supreme irony for Radet came on 4 February 1794 when the Committee of Public Safety suspended him. Charges brought against him included being in correspondence with agents of the King in 1791, and a party to the escape attempt foiled at Varennes. Brought before a Revolutionary Tribunal, the evidence was so flimsy they released him after Dubois vouched for his character. He resumed his post under Dubois with the Army of the North, where promoted *adjudant général chef de brigade* on 4 May 1794, he became division chief of staff. He fought at Fleurus on 26 June 1794 and was present at the capture of Mons on 1 July.

He spent the next three years with the Army of the Sambre-Meuse where under Dubois he had the chance to sharpen his skills with one of the finest cavalry leaders of the Revolutionary Wars. When Dubois moved to the Army of Italy he remained with the Army of the Sambre Meuse as Bonnaud's chief of staff in July 1796. A fall from his horse caused him to be out of action for several months and on his return to duty he joined d'Hautpoul's cavalry division as chief of staff. He served at Dierdorf on 18 April 1797 during the dying days of Jourdan's brief campaign across the Rhine before the Peace of Leoben.

THE GENDARME GENERAL

A successful career with the cavalry seemed assured when suddenly he moved to Avignon on 17 February 1798 as head of the Gendarmerie for the 24th Legion. For two years he remained in the Midi restoring order to an area of France that abounded with bands of armed brigands. The speedy pacification of the area owed much to his efforts and resulted in his promotion on 5 May 1800 to *général de brigade de gendarmerie*.

With wide powers under General Moncey the Premier Inspector General of the Gendarmerie, he helped to reorganize the force throughout France. He stressed the importance of proper recruitment and training. In addition he required that men came from Army veterans who had served in at least four campaigns and were a minimum age of twenty five. The reputation of the police was cleaned up. Gentleness, prudence and moderation was required in dealing with law abiding citizens, while forceful and prompt action was in order to combat unrest. In the years to follow Napoleon placed great reliance on his reports to judge public opinion and often referred to his men as his eyes and ears to gauge the mood of the people.

ITALY AND PAPAL INTRIGUES

Radet's attention moved towards civil order as opposed to the Gendarmerie's secondary role, which was similar to that of modern day military police serving with the field armies. His assignments as an Inspector General of Gendarmerie took him to Corsica, Piedmont and Genoa. In March 1806 he moved to the Kingdom of Naples where he organized and headed the Gendarmerie for two years before moving to Tuscany. He gained an extensive knowledge of Italy, the mood of its people and a feel for the political climate.

When the Concordat broke down due to the Pope failing to adhere to the Continental System, Radet was asked for a solution and recommended the invasion of the Papal States. When General Moillis occupied Rome in February 1808, the Pope refused to move and continued to exercise temporal power over the Catholic world, which irritated Napoleon further. Finally on 16 May 1809 the Papal States were annexed to France. When the Pope heard the news, and saw the French flag hoisted above the Castel de Santo Angelo, he excommunicated Napoleon. Napoleon reacted angrily by referring to the Pope as a raving lunatic, who ought to be locked up. Radet in Rome hearing Napoleon's reaction, overreacted, invaded the Quirinal Palace and ordered Pius VII to renounce his temporal power. When he refused, Radet removed him by force from Rome and after a pitiful journey deposited him at Nice.

Napoleon's bluff called by his overeager subordinate, all hope of gradually coercing the Pope into submission with the aim of establishing the Papacy in Paris floundered. Pius finally arrived at Savona, where in spite of the reverent splendor that surrounded him, he behaved as a prisoner, refused any allowance, washed his soutane self on his own and spent his days deep in prayer. As the self proclaimed prisoner of the Emperor, he refused to consecrate bishops nominated by Napoleon. With an impasse reached by two stubborn men, in the years that followed the spiritual well being of France and the rest of the Catholic world declined. Radet's hasty move had widespread and regrettable consequences.

GERMANY AND FRANCE 1811-1814

Radet received the title *baron de l'Empire* on 2 September 1810 and remained in Rome till the next year. A period followed based at Hamburg where his efforts were directed towards enforcing the Continental System. On 30 March 1813 he resumed a field command when made Grand Provost of the Grande Armée reforming in Germany. The change was dramatic, the Empire after the Russian Campaign was in crisis. Moving from a civil police role to that of a military one and with the armies in turmoil, it stretched his resources to the limit. The duties his men performed, providing prisoner escorts, guarding supply trains, rounding up stragglers, supervising the evacuation of the wounded, was all too much for his limited resources.

In addition, the Campaign of 1813 brought the major problem of desertion. The quality of the soldiery was not the same, conscription and the resultant desertion amongst the rank and file was rife. Ranging far and wide in the rear areas the number of deserters was near impossible to control. How effective Radet's measures were in stemming the rot is difficult to judge. Napoleon felt in the circumstances he did a good job and on 3 November 1813 promoted him *général de division*.

THE RESTORATION AND THE WATERLOO CAMPAIGN

He remained in his post till after Napoleon's abdication, then on 1 June 1814 he was placed on half pay. Not surprisingly he greeted Napoleon's return and on 31 March 1815 joined Grouchy at Lyons in charge of the gendarmerie for the 7th, 8th and 19th Military Divisions. Engaged in the operations against the Duc d'Angouleme, he was present at his capture on 14 April and supervised his journey into exile from Cête.

On 1 June 1815 Napoleon appointed him Grand Provost for the Army of the North. He served in Belgium, and was present at Ligny and Waterloo. In Waterloo's closing stages shell splinters struck him in the knee and hip. In great pain he made it to Genappe, where with Neigre he tried to restore order, but was overwhelmed by the fleeing mob.

LATER LIFE AND CAREER

Radet's name was amongst those proscribed by the Ordonnances of 24 July 1815. Dismissed on 3 August 1815, it was not till 4 January 1816 that he was arrested. Brought before a court martial at Besancon on 26 June 1816, the court sentenced him to nine years imprisonment. Incarcerated in the town's fortress the King granted a pardon on 24 December 1818 with his rank and privileges restored. An ill and broken man he retired from the army on 1 December 1819. He died at Varennes-en-Argonne (Meuse) on 28 September 1825.

Being primarily a policeman, Radet's military capabilities as a field commander were never put to the test. His loyalty towards Napoleon was never in doubt. A gifted administrator and highly intelligent man, his organizational skills laid the foundation for the Gendarmerie in France today. He may have been heavy handed at times, but then few men who enforced the law at the time escaped criticism. As a military man he built up a reputation as a fine cavalry leader. As a policeman he certainly never received the odium meted out to such men as Fouché and Savary who served the Empire as Napoleon's Ministers of Police.

REILLE - HONORÉ CHARLES MICHEL, COMTE (1775-1860)

BACKGROUND AND EARLY CAREER

A protege of André Massena, the Imperial aide de camp and Peninsular veteran, Honoré Reille was born in Antibes (Alpes Maritimes) on 1 September 1775. He first saw service as a *grenadier* with the 2nd battalion of the Var Volunteers when he enlisted on 16 September 1791. After the outbreak of war he transferred to the 94th Demi-brigade with the Army of the North. Reille's promotion through the ranks was swift, within a year gaining a commission as a *sous lieutenant* on 15 September 1792. He campaigned in Belgium, was present at the siege of Liege and on 18 March 1793 fought at Neerwinden.

A family connection helped secure him a post nearer to his home, with the Army of Italy as an aide to André Massena, who as a newly promoted *general de brigade* was a rising star. Taking up his new post on 27 October 1793 he was with Massena during his first victory at Castelgineste on 24 November and three days later was promoted *lieutenant*. Present at the Siege of Toulon he took part in the successful assaults on the forts Lartigue and Saint Catherine on 17 December, which later secured Massena a divisional command. After Toulon's fall he was with Massena during the invasion of the Genoese Republic, being present on 29 April at the capture of Saorgio and on 8 May 1794 the Tende Col at the crest of the Maritime Alps.

As Massena's aide, his first direct contact with Napoleon came in March 1796 when he took command of the Army of Italy. Very active during Napoleon's first campaign he fought at Montenotte on 12 April, Dego on 14 April and was present at the crossing of the Lodi on 10 May 1796. Sharing in the promotions and awards after the capture of Milan, on 23 May 1796 he was promoted *capitaine*. In the renewed offensive against the Austrians he distinguished himself at Saint George on 15 September, Caldiero on 12 November and Arcola on 15-17 November 1796. Promoted by Napoleon on 7 January 1797 acting chef d'escadron with the 15th Dragoons he fought at Rivoli on 14 January and was again in action at La Favorite on 16 January 1797 during the pursuit. Cited for bravery during the capture of Tarvis on 22 March, the Directory confirmed his promotion to *chef d'escadron* on 23 May 1797.

Further promotion came on 26 January 1799 when on Massena's staff of the Army of Switzerland he received the rank *adjudant général chef de brigade*. During this time he did valuable work reconnoitering the Rhine crossings above Lake Constance. He shared in the fluctuating fortunes of the campaign and was present at Zurich on 25-26 September when Massena won his crushing victory against the Russians under Korsakov.

On Massena's appointment as commander of the Army of Italy on 17 January 1800, Reille moved with him to Nice. He carried out many reconnaissances of enemy positions between Mont Cenis and Nice, compiling a detailed report that highlighted the dangers faced by the Army of Italy in an exposed position wedged along the narrow coastal strip. Sent to deliver the report to the First Consul he was away when the Austrian offensive began and the bulk of Massena's army was trapped in Genoa. Ordered to Genoa he slipped through the English naval blockade and reached the port on 2 May. With him he carried a much needed war chest of a million francs and a letter from War Minister Carnot outlining Bonaparte's plan to cross the Alps and take the Austrians in the rear. His mission, vital in stiffening the moral of the isolated garrison, no doubt was a major factor in prolonging the siege.

During the remainder of the siege Reille played an active role. On 11 May he led a column from Moillis's division that retook Monte Faccio and seized much needed supplies from the enemy. Two days later he led Spital's brigade against Monte Cretto after the latter took a mortal wound. Then as the siege drew to an end he acted as an intermediary during negotiations for the surrender of the city, finally marching out with the garrison on 6 June 1800 with full honors of war.

A RISING STAR 1800-1805

Returning to France in August 1800 he spent a brief period at Amiens before in November 1800 joining Murat's general staff with the *Armée du Midi*. He remained in central Italy till the army's disbandment in April 1802. A move to northern Italy followed serving with the troops stationed in the Cisalpine Republic. On 13 May 1803 he moved south when appointed chief of staff to Saint Cyr's Corps of Observation of Naples. Promoted *general de brigade* on 29 August 1803 he spent time with the Army of the Ocean Coasts as its chief of staff from 29 December 1803 till May 1804. He then took on a task to prepare a secret study of the preparedness of the Austrian army, which involved frequent visits to Germany and Austria. He was also a member of an enquiry established to examine the reasons for the failure of the Santo Domingo expedition.

On 29 November 1804 he was made deputy to Lauriston, who commanded the troops with the Toulon squadron preparing for the invasion of England. With the fleet when it slipped out of Toulon on 29 March 1805 he was with Villeneuve during the voyage to Spain and the West Indies. On the fleet's return he was in the battle off Cape Finistere on 22 July 1805. Lauriston frustrated by Villeneuve's inactivity, allowed Reille to leave the fleet at Ferrol and return to Paris with a detailed report of events. Fortunately as a result he missed Trafalgar.

THE GRANDE ARMÉE 1805-1808

On his return, there being no Napoleon or Grande Armée in France, the change of events made his reports totally irrelevant. There little else to do, he obtained permission to report to Berthier's headquarters staff with the army in Germany. He joined the army at Ulm and on 7 November 1805 took command of the Linz garrison and all troops in Upper Austria. Soon after on 14 December 1805 he was given the 2nd Brigade of Suchet's 2nd Division (V Corps) after the death of its commander Valhubert from wounds at Austerlitz.

Reille soon made his mark under Suchet early in the Prussian Campaign when his brigade (34th and 40th Line Regiments) distinguished itself at Saalfeld on 10 October 1806. In action after completing a lengthy flank march, his troops debouching through Aue fell on the Prussian right and took Sandberg, completing the rout of over 10,000 enemy, who lost some 2,000 men and 33 guns compared to 172 Frenchmen.

He was again conspicuous during the Battle of Jena on 14 October 1806. The night before the battle he crossed the Saale and with Claparède's brigade in support, gained the wooded slopes of the Landgrafenberg establishing a precarious foothold on the plateau. Then driving the Prussian advance-guard under Tauenzien back to Lutzeroda and Closewitz, Lannes was able to deploy during the night setting the scene for the hammer blow the next day. In the battle itself, Reille at the head of the 40th Line took the key position of Vierzehnheiligen after a stubborn defence by Tauenzien.

In November 1806 he renewed his experiences as a cavalryman, taking over from Savary an independent cavalry brigade (1st Hussars and 4th Chasseurs) to complete the pursuit of the broken Prussian armies. Returning to command his infantry brigade in Poland he fought at Pultusk on 26 December 1806. In the battle he supported Gazan's division against Barclay de Tolly on the Russian right but their assault made little headway against the strong defences.

On 30 December 1806 he was promoted *général de division*. He briefly headed Gudin's division with Davout's III Corps while the latter was recovering from wounds. He then returned to V Corps as chief of staff on 7 January 1807 after Victor's capture in a skirmish. Lannes having been wounded at Pultusk, Reille gained credit for the way he handled the V Corps at Ostrolenka on 16 February when he repelled an attempt by Essen to push him into the Narew. He renewed his acquaintance with Massena who took command of V Corps on 24 February 1807 and he resumed his duties as corps chief of staff. On 13 May Napoleon appointed him an Imperial Aide. He fought at Friedland on 14 June 1807 and was present at the signing of the Treaty of Tilsit on 12 July.

In September 1807 in his capacity as an Imperial Aide, Reille joined Brune during the French occupation of Swedish Pomerania. Keeping a close eye on the marshal's conduct, there is little doubt that his reports to Napoleon on Brune's strong republican leanings were a factor that led to Brune's dismissal the following month.

SPAIN 1808-1809

As an Imperial Aide he travelled with Savary to Madrid in April 1808. He was involved in the intrigue that led to Ferdinand V's journey to Bayonne, and later his overthrow and imprisonment that sparked the Dos Mayos uprising and the war in Spain.

In Paris when things started to go seriously wrong in Spain, Reille now a *comte de l'Empire*, on 28 June 1808 was sent to Perpignan to organize a force to help Duhesme cut off in Catalonia. He reached his destination with commendable speed on 3 July. He found the troops there comprised a heterogeneous body of 7,000 to 8,000 men formed from provisional battalions drawn from the Southern Alps and Piedmont. Taking less than a quarter of his available troops he set out on 5 July for his first objective, the relief of the beleaguered garrison of Figueras. After successfully concluding this, the rest of his division joined him and he turned his attention on the fortress of Rosas. Here the powerful garrison supported by a British naval squadron soon forced him to abandon all hope of seizing the stronghold when he reached it on 12 July.

Retiring inland he linked up with Duhesme before Gerona on 24 July. The investment of the city was carried out with such dilatoriness by Duhesme, that the besiegers soon became the besieged and on 16 August they abandoned the task. The remaining siege guns and stores were destroyed and both commanders returned to their respective bases, Reille to Figueras with comparative ease, while Duhesme barely survived cutting his way through swarms of *somantenes* to Barcelona. He waited for the arrival of Saint Cyr with a new formation (VII Corps) and then took part in the second invasion of Catalonia. He invested Rosas on 6 November and with help from Pino's Italian division the town fell on 5 December 1808. He remained in northern Catalonia covering communications with France while Saint Cyr pressed ahead, relieved Barcelona and occupied the rest of the province.

Imperial Aide 1809-1810

On 28 March 1809 Napoleon recalled him and he resumed his duties as an imperial aide in Germany. He was present at Aspern-Essling on 21-22 May 1809. At Wagram on 6 July he led the Guard Tirailleurs replacing Roguet*, wounded the previous day. At their head he supported Macdonald's assault against the Austrian center, which won the latter his marshal's baton. In August 1809 he moved to Antwerp where ostensibly he was involved in the compilation of a report on the English landings in the Walcheren Peninsula. Most however suspected he was there to keep an eye on Bernadotte who after his dismissal for handling the Saxon Corps at Wagram so poorly, needed to be watched.

Spain 1810-1813

On 29 May 1810 Reille returned to Spain as governor of Navarre. He became so embroiled against Mina's guerillas that large elements of IX Corps originally sent to support Massena's operations in Portugal were tied up lending him support. In January 1811 when news leaked that Napoleon proposed to annex the provinces north of the Ebro into metropolitan France guerilla action intensified further. French strength in the province increased to over 38,000 troops but they achieved little of note. An instance was on 14 June 1811 when he caught Mina near Sanguenza and inflicted a sharp reverse, but the guerillas merely melted away into the mountains to re-emerge once the French had moved on.

By September 1811 after receiving fresh drafts from France and Italy the situation in Aragon and Navarre had stabilized sufficiently to allow Reille to joined Suchet's Army of Aragon as commander of its II Corps. His forces comprising a newly formed 4th Division led by himself and an Italian division under Severoli. Given the task to keep the peace in Aragon and protect Suchet's communications as the main force moved into southern Catalonia and Valencia, he was gradually drawn south as the campaign dragged on. In the end he joined up with Suchet's main army to keep up the drive on Valencia. Hardly had he arrived on the banks of the Guadaluviar, when on 25 December 1811 his troops were thrown into the assault across the river that led to the final encirclement of city.

The bulk of Reille's forces in Valencia, the military situation deteriorated in Aragon and western Catalonia. To pacify north-east Spain Napoleon ordered the formation of a new army, the Ebro, which on 25 January 1812 Reille was recalled to head. Returning to the Ebro with his two divisions he was joined by Palombini's Italians and scattered detachments under Ferino. Based at Lerida with a field force of 20,000 men excluding garrison troops, he set about the subjection of inland Catalonia and Aragon. His generals met little success, for the guerilla leader Eroles ran rings around them, while Mina in Navarre did the same, forcing men to be drawn off to help in operations against him. Finally in April 1812 when Palombini moved west to support Marmont against Wellington his command was virtually non existent.

In the year since it was first intended to incorporate the northern provinces into France, Reille failed to bring peace. Large forces had been put in motion, toilsome marches made over mountains in treacherous weather, bands of insurgents had been dispersed but the country was not conquered. Isolated garrisons were still cut off by the enemy once the main columns had passed on. Communications were no safer than in the previous year. The situation persisted in Aragon throughout the summer of 1812 as dramatic events unfurled elsewhere in the Peninsula.

The Vitoria Campaign June 1813

Appointed commander of the Army of Portugal on 29 November 1812, Reille replaced Souham who had incurred the Imperial wrath for failing to pursue Wellington vigorously enough after the recapture of Burgos. Going into winter quarters his troops had the hopeless task of occupying the provinces of Zamora, Leon, Salamanca, Palencia and Valladolid, with his headquarters at the last named city. His 50,000 men in eight divisions widely scattered across a front of over 200 miles courted disaster. In February 1813, 200 veterans from each regiment were recalled to their depots in France to form the cadres of regiments lost in Russia. In May the army was stripped further by two divisions under Foy*, sent north to help put down the unrest in the Biscay provinces.

At the start of the campaign Wellington with some 52,000 British and 29,000 Portuguese, well equipped, fully recovered from the privations of the previous year and eager for action, faced Reille with only some 17,000 men immediately to hand. When the axe fell, it fell hard. In mid May Wellington set his columns in motion. Graham set out on a long painful 200 mile flank march across the Douro, through the Tras os Montes to Braganca in the farmost north-east corner of Portugal. At the same time from Ciudad Rodrigo Wellington with Hill made for Salamanca. On 26 May 1813 Graham struck across the border completely surprising Reille's cavalry outposts on the Esla. Within four days he crossed the river outflanking the Douro line. A rapid advance along the river's north bank secured Toro and a junction with Wellington and the rest of the army on 3 June.

Bad luck rather than poor dispositions were largely to blame for Reille being outmaneuvered on the Douro. Only the week before Graham struck, Digeon's cavalry had made deep reconnaissances beyond the Esla and saw no sign of the coming blow. Even had they come across Graham's column, which numbered some 55,000 men, it would have taken Reille a week to gather his army together. Daricau's infantry division at Zamora, and Digeon's cavalry were too weak to dispute a flanking movement of such strength, even had they plenty of warning.

In the three weeks that followed Reille tried to hold successive river lines as Jourdan tried to gather in the other French armies of the Centre and the South behind him to form a more solid line of defence. Valladolid fell on 3 June, a defensive line was formed behind the Pisuerga, but that was turned a week later. On 13 June with Jourdan he abandoned Burgos. Gathering in the divisions of Lamartinière and Sarrut, Jourdan hoped the Ebro would form a line for the armies to rally behind. When Reille reached the river on 16 June and took up positions behind it at Frias, Espejo and Puente Lara he soon heard Graham had crossed further upstream at Rocamonde and San Martin de Lines.

Continually flanked, Reille strongly urged the armies give up trying to keep communications open via Bayonne and instead they retire down the Ebro to take up the line from Pamplona and Saragossa joining up with Clausel's and Suchet's armies. Such a concentration would then match Wellington's army, and could hurl it back as it had from Burgos the previous year. The scheme had merit, but Joseph refused to accept it as Napoleon insisted communications with Bayonne be kept open. Another factor influencing the decision was the danger that the French forces in the Biscay provinces would be cut off.

The retreat ended at Vitoria on 20 June 1813 as King Joseph and Marshal Jourdan tried to gather in the remnants of the three French armies, some 50,000 men with 150 cannon. Behind the river Zadorra near Vitoria they intended to make a stand and win time for the huge trains accompanying the armies to get clear towards the French frontier. There was no unified command exercised over the armies that day as the respective leaders Clausel, d'Erlon and Reille made their dispositions as they thought fit. It was a recipe for disaster, through no real fault of theirs but that of Joseph and Jourdan who should have exercised some overall leadership.

Reille on the French right had some 14,000 men to keep open the road to Bayonne that ran parallel to the upper reaches of the Zadorra. His third division, Maucune's, was detached to escort the baggage trains streaming north from Vitoria. To do this he placed Sarrut's division with a brigade of Mermet's light horse in an advanced position a mile and more before the river on a ridge, above the village of Aranguiz flanking the main Bilbao road. Along the road came Longa's Spanish division that Reille mistakenly thought was a diversion, since the rest of Graham's great flanking column was concealed by the approaching Spanish till too late. Faced by three divisions against that of Sarrut, Reille gave up his position on the heights and pulled back to the Zadorra. Here joined by his other division under Lamartinière he hastily established *têtes de pont* covering the bridges that included the villages of Abechuco, Gamorra Mayor and Menor while the remnants of his two infantry divisions formed up behind the river.

Tactically the move was a mistake as Graham's artillery from the heights dominated the road to Bayonne cutting the route to France. The Anglo Portuguese division stormed the ridge and soon took Gamorra Mayor, while another force occupied Abechuco. A desperate struggle followed as the French in vain tried to retake these positions. Graham then approached the Zadorra bridges, but was halted by well placed artillery and the fighting qualities of the two divisions that stood their ground with skill and courage for most of the afternoon.

It was the threat of being taken in the rear after collapse of Army of the Centre led by d'Erlon' to Reille's left that forced his troops to give ground. Whilst the two other French armies fled in the direction of Pamplona, his two divisions withdrew intact and made their way towards the Bidassoa. Foy and Maucune rejoined him on the Bidassoa bring the remnants of his army up to 22,000 men, enough to pose Graham's forces with a major problem. In another moment of misplaced judgement on 1 July he allowed Foy to destroy the bridge over the river at Béhobie as Graham made a half hearted attempt to secure it. The move was foolish considering the forces he had available to defend the position. The loss of an important *tête de pont* at the Spanish end of the bridge also later made it more difficult to launch a counter offensive. The simple fact was that Reille, like most of the other French generals was totally demoralized by the recent disaster at Vitoria and had lost confidence both in himself and in his troops.

THE PYRENEES AND SOUTHERN FRANCE 1813-1814

Soult's reorganization of the broken armies in July resulted in Reille taking command of the Lieutenancy of the Right, numbering 17,235 men formed from his remaining three divisions of the Army of Portugal led by Foy, Maucune and Lamartinière. Moved from the Lower Bidassoa, Reille's force marched to Saint Pied-de-Port where with d'Erlon they formed Soult's counteroffensive through the Pyrenees aimed at relieving the garrisons of Pamplona and San Sebastian. On 25 July his long winding column was first held up by the stubborn defence of Cole's 4th Division before Roncesvalles. Unable to deploy in the narrow defiles, Reille's troops were shot away by the long lines of opposing troops. The advance resumed when a fog enveloped the embattled divisions and Cole withdrew, fearing he would be overwhelmed, enabling the French to occupy the pass.

The next day the French columns were checked at Sorauren where the Allies occupied the last series of ridges before the descend to Pamplona. Maucune failed to deploy sufficient skirmishers to explore ahead. His men staggered up the slopes, and as they reached the crest met crashing salvoes from Cole's division that sent his men reeling back down the slopes. Lamartinière battalions to Maucune's left were more successful, but exhausted by the climb were unable to hold the crest when Byng counterattacked. The offensive spent Reille, retired with his columns to their original positions on the Bidassoa.

On 31 August Reille led Soult's offensive over the Bidassoa to relieve San Sebastian. While Clausel attacked at Vera, Reille's two divisions led by Maucune and Lamartinière crossed the fords and made for the ridge of San Marcial defended by Longa's Spanish division. His troops made progress establishing a foothold on the opposite bank. Then as they tried to outflank the Spanish positions on the mountainside his troops became disordered and were repulsed. Forced back to the river he called on Foy's and Vilatte's divisions to lend support. As the attack was about to resume, Soult received news that d'Erlon had also made little progress on his sector and that San Sebastian was about to fall. The reason for the offensive lost, Soult ordered Reille to withdraw.

The demoralized French army went on the defensive and in September Reille took charge of the six miles of the tidal estuary of the Bidassoa from Biriatou to the sea. For the task he had only two divisions, those of Maucune and Boyer (late Lamartinière's), for the third (Foy) had been placed under d'Erlon beyond the Nive at Saint Jean Pied du Port. Maucune's 4,000 men covered the entire riverline while Boyer's 6,000 formed the reserve several miles to the rear at Urrogne and Bordagain. It was the most weakly defended sector of Soult's twenty five mile front, as the marshal rightly felt the barrier of the lower Bidassoa provided adequate protection.

Much to Reille's discomfort for this very reason, Wellington used the element of surprise to attack that part of the French line considered the most impregnable. Against this thin line Wellington let loose some 24,000 men at dawn on 7 October. The 5th Division concealed behind the deserted town of Fuenterabia burst forward across the sands and made for the fords that were less than three feet deep at low water to take Hendaye. At the same time the 1st Division concealed behind Irun struck across the river at Béhobie. The force thrown against the thin French line on the north bank was so overwhelming that any serious resistance soon collapsed. In two hours Wellington turned the line of the Bidassoa and Soult's army was streaming back to new positions before the

Nivelle. The marshal blamed everybody but himself for the lost battle. Heads were to roll, including Reille's, who was accused in reports to Paris of failing to take reasonable precautions and exercising simple powers of observation. He had taken all measures available to him at the time to reinforce the unfortunate Maucune. The real problem was Reille simply had too few troops available to cover such a long sector. His supporting division, Boyer's, was placed at Soult's orders too far from the front line to lend support when a sudden strike came. For this he was not responsible. Soult instead concentrated his wrath on Maucune, who certainly had been negligent and ought to have detected earlier that something was afoot on the opposite bank. Maucune was sent to the rear in disgrace, but the affair also left Reille's future in doubt especially after his poor showing earlier in the campaign at Sorauren.

From this point on Reille's relationship with Soult was uneasy. His reputation remained intact when the next month the French fell back from their positions along the Nivelle. This time Wellington fell on Clausel's divisions holding the mountainous Lower Rhun in front of the river on 9 November. Supporting attacks were made on d'Erlon's divisions holding the river itself. While Wellington rolled up the French line, Reille succeeded with some skill in retreating in good order, destroying the defences of Saint Jean de Luz and taking up another defensive position between there and Bayonne.

Wellington found himself facing strong French positions defending the area between the Bay of Biscay and the rivers Nive and Adour leading northward toward Bayonne. He realized the danger of being caught in a triangular salient unless he could force a crossing of the Nive. Despite bad weather Hill led five divisions over the Nive near Ustaritz on 9 December, while the rest of Wellington's army launched diversionary attacks towards the Adour and Bayonne. Split by the Nive the Allies were vulnerable to a counterblow as Soult operating on interior lines was in a position to deploy a superior mass to either the left or right of the river. In the end Soult achieved neither. The next day Reille advanced from his positions before Bayonne. His divisions led by Boyer and Leval fell on Hope's 5th Division driving it back to Barouillet. Here they allowed themselves to become bogged down in fighting around the chateau rather than push vigorously ahead. It gave time for elements of the 1st Division to form and bring the French advance to a halt. Clausel the previous night had recrossed the Nive with three divisions and fell on the pickets of the Light Division before Arcangues, but the attack slow to develop was held. By the afternoon of the 11th, the bulk of Wellington's forces now on the west bank of the Nive and with the arrival of his 4th and 7th Divisions all opportunity for the French to crush

Hope's forces passed.

The deteriorating weather ended military operations for 1813 as the French withdrew to Bayonne and formed up behind the Adour. Reille had the task to hold Bayonne and the lower Adour with the divisions of Leval, Abbé, Taupin and Maransin. It became increasingly more difficult when Leval's and Boyer's divisions were taken in January 1814 to join Napoleon forces in the Champagne. Maransin's force then moved upriver and away from his control leaving him with only Abbé's division. Increasingly isolated, he feared being trapped in Bayonne where he had no control over the garrison of 8,000 troops under Thouvenot, though junior to him, was appointed by Imperial Decree and so not subject to his authority. He entered into angry correspondence with Soult, stating a double command of the fortress was ruinous and undermined his own position. Soult was unmovable, so Reille resigned on 15 February 1814, handed over his remaining battalions to Thouvenot and departed for Dax, after sending a letter of complaint to Paris. This gross act of insubordination was overlooked by the War Minister Clarke who ordered Reille to return, and Soult to accept him back with the field army. The question of Thouvenot's authority in Bayonne was not negotiable due to the imperial warrant.

The incident showed neither of the protagonists in good light. Soult for his remote, lofty high-handed attitude, and Reille for irresponsibly precipitating a leadership crisis within the army at a time of national crisis. By 27 February he was back at headquarters in command of the divisions of Rouget and Taupin for the Battle of Orthez. His influence on the day's events was minimal since he had barely taken up his post. The relations between him and Soult from then on were extremely unpleasant. He fought at Tarbes on 20 March 1814 when an attempt to trap Soult with his back to the Pyrenees was foiled, allowing the French to retire on Toulouse. At Toulouse on 10 April he fought in the last major battle of the Peninsular War. Again nominally in command of Taupin's and Maransin's divisions, any resemblance to a corps structure had ceased to exist in Soult's army and again his influence on the events was minimal.

THE RESTORATION AND NAPOLEON'S RETURN 1814-1815

Peace saw him acknowledge the Bourbons. On 27 June 1814 he became a *Chevalier de Saint Louis* and received the appointment of Infantry Inspector General of the 14th and 15th Military Divisions at Caen and Rouen. During this period affairs of the heart also came to the fore when he married Massena's daughter Victoria. On 14 February 1815 he received the Grand Cross of the Legion of Honor. He adroitly managed not to compromise himself when the news of Napoleon's landing at Antibes broke and remained at his post at Caen. Summoned to Paris after Napoleon's arrival, there was no suspicion as to his loyalty. His past experience against Wellington stood him in good stead and on 30 March 1815 he was made commander of the II Corps of the Army of the North.

THE WATERLOO CAMPAIGN

Immediately sent to Valenciennes to prepare for the coming campaign, the men and material he had to deal with were good. In division commanders the calibre of Bachelu, Foy, Girard and Piré he had some of the best in the army. His first movements of the Waterloo Campaign were on the 11 June when his troops left Avesnes for Mauberge. The movement had two purposes, to act as a screen for d'Erlon's I Corps on its flank march from Lille and to form part of the army's left wing at Solre-sur-Sambre. On 14 June 1815 he received his final orders from Napoleon that he was to lead the left wing as it advanced into Belgium. That night his infantry edged towards the frontier without causing suspicion, whilst his cavalry watched the enemy pickets in the distance. His bivouacs were well placed so the glare from fires were screened from observation by the low hills and woods.

On 15 June Bachelu's division at the head of the column secured the crossing over the Sambre at Marchienne after a short sharp struggle. Then cutting across country he pushed Steinmetz's Prussians up the Charleroi-Brussels Road towards Gosselies. By 4.00 P.M. the Prussians were expelled from the village and the advance resumed towards Frasnes where the last formation of Dutch troops before the Quatre Bras crossroads was forming up.

The advance's momentum was lost when Ney arrived at Gosselies to take command of the left wing. Time was taken briefing Ney with the latest situation. Foy and Jérôme's divisions were at Gosselies, Bachelu was up the Brussels Road near Frasnes with Piré's and Lefebvre-Desnoettes's cavalry beyond the village ready to envelop a Nassauer battalion deploying ahead, whilst Girard was snapping at the Prussians' heels as they fell back on Wangenies. No doubt he was put out by the arrival of Ney, seeing another taking the glory for the day's gains and the time lost briefing the marshal proved fatal. When the advance resumed, it was soon called to a halt around 6.30 P.M when a Nassauer battery lending support kept the French cavalry at a respectful distance. There were only 4,000 Nassau troops about Quatre Bras. Another hour would have brought Bachelu's highly charged troops to Quatre Bras where with sufficient artillery support they could have secured the crossroad before nightfall. As a result the opportunity to split the Allied armies on the first day of the campaign was lost.

The aggression Reille showed on the campaign's first day was in complete contrast to that shown on

16 June. No orders were given to resume the march at first light and his divisions only got under way around 10.00 A.M in the morning. Then seeing a brigade of Perponcher's Dutch division skilfully deployed before him, using every inch of the terrain to give the impression of greater numbers, he warned Ney not to attack until the whole corps had mustered. Mindful of past experiences in Spain, Ney was aware that Wellington was the master of the art of defence. Wellington's widely spread flanks prevented effective reconnaissance while at the same time Ney was also convinced that the main body of troops would be hidden by the woods and dips in terrain. In spite of Napoleon's orders that Quatre Bras was to be taken, Reille's divisions did not clear Frasnes till 2.00 P.M whilst Wellington continued to strengthen his position till the numerical advantage was irretrievably lost. Once the battle opened Reille directed the assaults by his divisions vigorously. His men fought gallantly but with the delays, Girard's division detached to Ligny and d'Erlon's corps not appearing at all, the task was simply too much. He suffered over 3,800 casualties and by dusk, Wellington still held Quatre Bras with his troops driven back to their start positions.

From a positive point of view, whilst Reille's troops failed to take Quatre Bras they had kept the Allies apart and left Wellington in a very exposed position. There was the distinct possibility the next day of Napoleon bearing down on Wellington's left from Ligny whilst he and d'Erlon pinned him down with a frontal assault. Inexplicably there was no communication between the two army wings. Ney on the other hand never gave the orders to prepare for an early morning assault, nor did Reille or d'Erlon think to prepare for such a contingency. Napoleon came relentlessly down the road from Ligny while Reille's troops sat watching Wellington's outposts not realizing until too late that the enemy had slipped away. Next to Ney and d'Erlon, Reille must bear responsibility for a tactical blunder that went a long way to losing Napoleon the campaign.

After receiving a furious admonishment from Napoleon, he set off up the Brussels road following d'Erlon's corps. His troops bivouacked around Genappe that evening. Early in the morning of 18 June they broke camp and reached their final deployment positions before Mont Saint Jean stretching left from the Charleroi road to the Nivelle road opposite the chateau of Hougoumont. The sodden terrain delayed the battle's start so Napoleon in buoyant mood held an impromptu council of war over breakfast with his senior commanders. Reille was asked by Napoleon to give his views on the steadfastness of English infantry and Wellington as a commander. In a pessimistic response complemented both. The Emperor in a curt reply blandly dismissed his answer as typical of generals who had fought in Spain and suffered defeat at Wellington's hands.

Reille's orders were to mount a diversionary attack against the Allied left in order to draw troops away from Wellington's center before d'Erlon launched the sledge-hammer blow against it. The crude yet effective plan soon went seriously wrong after the battle commenced at 11.30 A.M. Jérôme's division led an attack against Hougoumont though no orders had specifically been given that it should be taken. Soon Jérôme's whole division was embroiled in a titanic struggle for the chateau plus one of Foy's brigades. As the hours passed over 12,000 French were committed unnecessarily against Allies that never numbered more than 2,000 men. Wellington hardly altered his dispositions and when d'Erlon's attack against the center went in around 1.30 P.M it was beaten back with fearful losses.

The Hougoumont assaults were pointless, it was likened to staking lead against gold. The chateau was of no strategic value except in that it acted as a breakwater to disrupt the smooth flow of Napoleon's assaults. The French could still flow around it to reach their main objective, the Allied line further back. Reille showed marked tactical ineptitude in allowing Napoleon's young

QUATRE BRAS, SKETCHED THREE WEEKS AFTER THE BATTLE, WITH THE GRAVES OF THOSE WHO FELL ON 16TH JUNE.

brother to persist with this folly and clearly lost sight of his objective. Round 6.00 P.M he led an assault ordered by Ney against the Allied right center to the west of La Haye Sainte. Without sufficient cavalry support, at the head of Bachelu's division and Foy's remaining brigade in support his troops were cut to pieces as they reached the crest. Of the 5,000 men that set out, he lost a quarter of his strength within a few minutes before the rest fled back down the slopes. The attack did however distract the Allies and d'Erlon was able to occupy La Haye Sainte precipitating another crisis further along the line.

The remnants of his corps supported the final assault of the Old Guard and when that failed his formations finally broke. Lost during the confusion of the retreat he rejoined Napoleon at Philippeville the next day. He set about reorganizing the broken units as they arrived and by nightfall he had rallied over 10,000 men before handing them over to Soult. He resumed command of his corps when he reached Laon on 25 June. It was in a pitiful state; after returns were submitted only 9,140 men remained compared to the 25,065 at the start of the campaign ten days earlier. He reached Paris on 29 June after passing through Soissons (26th), Villers-Cotterets (27th) and Nanteuil (28th). On 4 July after the conclusion of the Armistice with Wellington and Blucher, he marched out of Paris with his troops and retired behind the Loire south of Orleans. The army disbanded, he was placed on the non-active list on 4 August 1815.

POST NAPOLEONIC CAREER

Reille was included in the general amnesty granted by Louis XVIII on 30 December 1818 and though restored to the active list he remained unattached. He became a Peer of France on 5 March 1819. His reconciliation to the Bourbons appeared complete when in 1820 he was appointed to the honorary post of Gentleman to the Royal Chamber. During this time he served on many consultative committees concerned with military affairs culminating with his appointment as a member of the Council of War on 27 February 1828. Thus by the time of the July Revolution of 1830 he had moved from being a Bonapartist to being a committed Bourbon. Tipped to become a *maréchal* his fortunes took a tumble as he was stripped of all his posts in August 1830. Anxious to keep a low profile he retired to his estate, the Château of Coudreaux, which he had purchased from Michel Ney's destitute family. By August 1836 he had worked himself back into favour when appointed president of various policy making committees covering infantry and cavalry affairs.

Aged sixty-five he passed to the 1st Section of the General Staff on 31 December 1840.

Little was heard of Reille for several years. Then suddenly, Louis Philippe anxious to bolster a flagging and unpopular monarchy, decided to rekindle the Napoleonic Legend by creating a few Marshals of France from generals of note who had served Napoleon. There not being too many candidates about who were not either in their dotage or senile, Reille received the rank on 17 September 1847. He survived the overthrow of Louis Philippe and with Louis Napoleon installed as head of state he was elected a senator on 1 January 1852, representing the right. Further awards came on 13 June 1852 when he received the *Medalle Militaire*. An active supporter of Napoleon III till his end, he saw the Second Empire reach its zenith and died in Paris on 4 March 1860.

ASSESSMENT - A FINE IMPERIAL AIDE, A LIMITED CORPS COMMANDER

Intelligent and brave, Reille was the classic Napoleonic general. A loyal Bonapartist in his earlier years, his intelligence, zeal and bravery merited his appointment as an imperial aide where he showed true devotion. His years in Spain tempered his enthusiasm. The circumstances of Massena's dismissal in May 1811, which he felt unfair was the first sign of disillusionment he showed towards Napoleon. Later in 1814, war weariness coupled with a growing dislike of Soult finally led to him abandoning his post. It was an act of gross insubordination, inexcusable in a time of his country's greatest need. It showed a dark side to his character where status and self interest were paramount. Napoleon's return from Elba saw him as a compliant but by no means an enthusiastic supporter. The presence of any member of the Bonaparte family was enough to give him the jitters. It showed at first at Quatre Bras when Jérôme more or less acted on his own initiative and got away with it. Then at Waterloo he failed lamentably because he was overawed by Jérôme and failed to exercise proper authority, allowing himself again to lose sight of the given objective.

In a tactical situation on a battlefield where the objectives were clear and not too much initiative was required he was decisive. At times he was considered impetuous, too anxious to come to grips with the enemy he was easily distracted and often lost sight his objective. Strategy was not one of his strong points. Overall he was a fine aggressive division commander, but an indifferent leader of formations any larger.

ROGNIAT - JOSEPH, BARON (1776 - 1840)

BACKGROUND AND EARLY CAREER

Joseph Rogniat, the engineer general and military theorist, was born at Saint-Priest (Isère) on 9 November 1776. The son of a Royal Notary, he studied at the Lyons Oratory with a view to entering his father's profession. Patriotism gained the better of him when he later enrolled at the Metz Military College where he gained a commission as a *sous lieutenant du génie* on 28 September 1794.

Posted to the Army of the Rhine, a scarcity of capable engineer officers made promotion swift, *lieutenant* on 5 March 1795 and then *capitaine* on 20 April after the capture of Landau. In November 1796 he distinguished himself at the defence of Kehl. He moved to the Army of England in January 1798 serving on its General Staff, coming into contact with General Bonaparte for the first time. He did not join the expedition to Egypt, instead moving to the Army of Mayence in July 1798, then Switzerland and ultimately the Army of Germany, serving in Delmas's 6th Division with Lecourbe's left wing. Present throughout the 1800 Campaign in Germany he fought at Neubourg, Engen on 3 May, Moeskirsh on 5 May and Biberach on 9 May. When Grandjean became division commander after Delmas's dismissal, Rogniat on 27 June 1800 became his aide de camp. Promotion to *chef be bataillon* followed on 9 July 1800. During Moreau's Winter Campaign he commanded the engineers of the Centre Corps. He fought at Hohenlinden on 3 December and used his bridging train to good effect during the pursuit of the Austrian down the Danube valley.

With Europe at peace after the Peace of Lunéville, Rogniat moved to Belle-Isle off the coast of Brittany on 28 February 1801 as chief of engineers to improve the defences of the island. That task completed, a move to Lyons followed on 15 October 1801 as Assistant Director of Fortifications. He moved to Bayonne in August 1803 and from there to the camp at Brest in December as Augereau's engineer commander.

THE GRANDE ARMÉE 1805-1807

With the break up of the camp on 23 August 1805 and the start of the long march to the Rhine, Rogniat headed the engineers with VII Corps. In the Austrian Campaign he was with Augereau during the march into Voralberg and was present at the capitulation of Jellacic's force at Feldkirch in November. The remainder of the campaign was spent in the vicinity of Ulm covering Napoleon's communications.

In the Prussian Campaign of 1806, he led the small engineer detachment attached to Murat's Cavalry Reserve. He fought at Jena on 14 October 1806 and took part in the pursuit of the Prussians. In the Polish Campaign he was present at Pultusk on 26 December 1806 and fought at Eylau on 8 February 1807. In March 1807 he joined X Corps under Lefebvre to take Danzig. He served throughout a most difficult siege in appalling weather that enhanced the reputation of the engineer corps. After the capitulation of the port on 26 May 1807 he shared in the promotions in his case to *major commandant du génie*. He moved to Swedish Pomerania in July 1807 and served at the siege of Stralsund.

SPAIN - THE SIEGE OF SARAGOSSA 1808-1809

On 19 February 1808 he was promoted *colonel du génie* and joined the General Staff of the Army of Spain. He was detached to Saragossa as deputy commander of engineers during the first unsuccessful siege of the city in September 1808. When Napoleon's full blooded invasion of Spain came in November 1808 Rogniat was with Lannes's II Corps. He fought at Tudela on 23 November before being recalled to help at Madrid if the city had to be taken by storm. Present at the capture of Madrid on 3-4 December he was soon asked to join the second siege of Saragossa, which was proving as difficult as the first. Working under Lacoste, who commanded the engineers, he brought a new lease of energy to the troops as they opened new works. The main walls being breached on 20 January made little difference, as each house in the narrow streets had been turned into a small fortress. Lacoste died on 2 February 1809, shot through the head while peering from a window at the next houses to be attacked. Rogniat took charge of the engineers. A murderous month followed as his men inched their way through the city house by house by blowing down the outside walls of buildings so the infantry could charge into the open shell to deal with the dazed defenders.

It was a fitting testament to his patience and skill as an engineer that he saved the lives of hundreds of French troops. At the same time as the Spanish perimeter gradually shrank he sapped the morale of the defenders till it finally snapped. On 20 February 1809, with Lannes barely in control of a third of the city, the Junta capitulated. Rogniat for his part gained a well deserved promotion to *général de brigade* on 6 March 1809.

THE DANUBE CAMPAIGN 1809

He left Spain with Lannes and served under him with the Army of Germany as engineer commander of II Corps. He continued at his post when Oudinot took over the corps on 28 April 1809. He was of invaluable help to Bertrand during the construction of the pontoons over the Danube before Aspern-Essling. He fought at Aspern Essling on 21-22 May and at Wagram on 5-6 July 1809. On 3 December 1809 he was made *baron de l'Empire*.

THE ARMY OF ARAGON 1810-1812

Rogniat returned to Spain in December 1809 as commander of the Army of Spain's siege train. The post meant little, since there were several armies scattered about Spain and no centralized command, let alone one for a siege train. In April 1810 Suchet secured his services as commander of his engineers and siege train with the Army of Aragon. The first of Suchet's objectives was to clear the Lower Ebro and capture Tortosa. To take the city free navigation of the Ebro was vital, since the country devoid of provisions was inhospitable. Rogniat's direction of the siege train at Lerida from 25 April 1810 till the fortress's fall a month later was a great lift to the Army's morale. His help in the capture of the fortress of Mequinenza on the highest navigable point on the Ebro was vital to Suchet's advance on Tortosa. The feat was more remarkable in that the fortress was on a spur 500 feet above the river, out of range of any siege guns unless placed on the spur. From 15 May he patiently constructed a five mile road up the spur in order for the siege guns to line up against the walls. Once in place the walls soon crumbled and on 13 June the garrison of 1,000 men capitulated.

On 16 December 1810 he started siege operations against Tortosa. The city's defences though not strong, were well manned by a garrison of over 8,000 men. The skill exercised by Rogniat and his colleague Valée who directed the siege guns soon resulted in a breach to the walls and the city's surrender on 2 January 1811.

The siege of Tarragona from 17 May till 28 June 1811 was the crowning point to his and Suchet's careers. Napoleon declared it was inside the city's walls Suchet would find his marshal's baton. Tarragona was a formidable objective. Apart from a garrison numbering some 12,000 men, it was easily supplied by sea. Rogniat nullified the seaward advantage by seizing outlying positions and by skilful placement of guns made the harbor approaches dangerous. The bravery of his engineers became a legend during the siege, and his men suffered over 400 casualties as they worked their way closer to the walls. On 21 June he breached the walls and Habert's division poured through the gap. The citadel proved another major obstacle, which finally fell on 28 June 1811. Suchet secured his marshal's baton, and for Rogniat his reward was a spell of leave in France and promotion to *général de division* on 9 July 1811.

Suchet in October 1811 commenced his invasion of Valencia. The first problem he encountered was the capture of Saguntum twenty miles from the city. Lack of a properly equipped siege train resulted in his men making little impression against the walls. Rogniat returned from France on 11 October and immediately made a detailed study of the enemy defences. His arrival had a considerable effect on morale since he had never failed to reduce a fortress in Spain. Five days later the siege train arrived and, with Valée, he started to work at reducing Saguntum's defences. A diversion came outside the town on 25 October 1811 when he was at Suchet's side when the marshal defeated a relieving army under Blake. The morale of the Saguntum garrison collapsed as a result and the next day the town surrendered before Rogniat effected a breach. He moved on, directed operations at Murviedro and was present at the siege of Valencia that lasted from 1-14 January 1813.

GERMANY AND FRANCE 1813-1814

His work under Suchet established his reputation as one of the foremost engineer officers in the French Army. The campaign in Russia had taken its toll amongst the engineers in the Grande Armée. Eblé was dead and Chasseloup, the senior surviving officer, was a broken man. On 9 January 1813 the order went out for Rogniat's recall from Spain to head the engineers of the Grande Armée in Saxony. He was with Napoleon throughout the Spring and fought at Lutzen and Bautzen. In the Autumn he was present at Dresden, Leipzig and Hanau. He did his work well and whilst not involved in any set piece siege work the engineers generally never let the army down. The destruction of the bridges over the Elster on 18 October 1813 that resulted in a large part of the army being trapped in Leipzig was an unfortunate exception that in no way reflected any poor handling of that arm. His skills were later shown in January 1814 when the Allies surrounded Metz, as he played a key role in its defence till after Napoleon's abdication.

THE RESTORATION AND NAPOLEON'S RETURN

The Bourbons retained Rogniat's services and he was given the task to examine the organization of the Engineer Corps. On 1 June 1814 he received the title chevalier de Saint Louis. When Napoleon landed a Fréjus, he was on his way to Grenoble to report on the defences in the region. News of the landing saw him diverted to join Artois's forces at Lyons but when he neared the city, d'Artois had already fled. Exercising caution, rather than join Napoleon's triumphant progress he sided with the remnants of the Bourbon forces under the Duc d'Angoulême at Nîmes on 17 March. When Grouchy moved against Angoulême, rather than face the prospect of civil war, on 7 April he offered his services to Napoleon. On 20 April 1815 he became a member of the Commission of Defence, a body set up by Napoleon to ready the frontier fortresses for an invasion of France by the Allies. His appointment as commander of engineers with the Army of the North followed on 5 May 1815. The appointment was opposed by Davout who considered his loyalty suspect but, because there were few engineers of his experience around, it went through.

THE WATERLOO CAMPAIGN

Present at both Ligny and Waterloo, there is nothing to suggest that he did not perform his duty throughout the campaign. Like most other commanders he had a sense of foreboding as to the outcome, yet whilst Napoleon remained successful he remained a quiescent follower. After Waterloo however he made the mistake of leaving his post and on 20 June returned to Paris. Davout suspected something sinister in the move, immediately removed him from his post and ordered him to rejoin the Army.

LATER CAREER

Rogniat remained with the Army on the Loire till September 1815 then moved to La Rochelle to disband the large engineer detachment at the port. Recognized as no avid Bonapartist andthere an acute shortage of engineer officers of his standing, the Bourbons retained his services. On 16 February 1816 he became president of the Committee of Fortifications, a body set up to study the strategic consequences of losing key fortresses along the Rhine after the Treaty of Vienna. Then in later years it assumed the important responsibility for the construction and maintenance of the frontier fortresses.

An unpleasant task faced him in April 1816 when he served as a member of the Council of War that tried Drouot. The general conducted a brilliant defence, which resulted in a sentence that raised no major outcry, and Rogniat's career was not affected in future years by the unfortunate incident. In June 1817 he headed a body that supervised the demarcation of the Swiss frontier along the Rhine and the Vosges. At this time he also gained a reputation as a fine military writer and theorist. His papers on firepower and fortifications became standard texts for many military colleges. In 1819 he also wrote a critical account of Napoleon's handling of the Waterloo Campaign. It highlighted

Napoleon's declining powers as a leader rather than the shortcomings of his subordinates as the real cause for the campaign's failure. It elicited a suitably terse response from Saint Helena. He received the Grand Cross of the Legion of Honor on 20 October 1820 and sat on many committees that debated the role of the peace time army in France. Held in high regard in civilian life, it was recognized when he was admitted to the Academy of Sciences on 23 November 1829, a rare honor for a soldier.

His stature as an engineer and professional soldier was such that it remained virtually unaffected by the July Revolution. He had previously been called to serve on the Superior Council of War in July 1828 and then was appointed Senior Inspector General of Engineers in January 1830. The post was abolished by the new government but they had such confidence in his ability they appointed him president of a new Committee of Engineers in October 1831, which in effect was a different name for the same post. He held the post till his death, active till the end lending his weight to policy decisions on military matters. On 19 November he became a Peer of France. He died in Paris on 8 May 1840.

CAREER ASSESSMENT

Rogniat was one of the finest engineer generals in the French army. His direction of siege operations under Suchet was outstanding. If Suchet owed a debt to anyone for his marshal's baton it was to him. The difficult task Suchet faced in conducting siege operations in Catalonia and Valencia is not generally recognized, the campaigns against Wellington largely catching the attention of writers. The size of his army, the problems keeping communications open while confronting fortified towns and cities with an army that barely outnumbered the enemy required men of exceptional talent serving him. Rogniat in charge of the siege train was a vital link in the chain to complete any such plan. His eye for the land and ability to observe the weak point in the enemy's defences was exceptional. The tenacity and patience he displayed as he effected breach after breach were to his credit. As leader he was an inspiration to all. Often as a siege bogged down, his arrival at the scene soon resulted in renewed enthusiasm to complete the work. His ability to put theory into practice, his true professionalism as a soldier enabled him to survive and continue an outstanding career for a further twenty-five years after Napoleon's fall.

ROGUET - FRANÇOIS, COMTE (1770 - 1846)

BACKGROUND AND EARLY CAREER

The son of a locksmith, the "Old Grumbler" François Roguet was born in Toulouse (Haute Garonne) on 12 November 1770. He enlisted as a volunteer in the Guyenne Infantry Regiment (21st Line) on 3 May 1789. By January 1791 he had risen to *caporal* when disillusioned with promotion prospects he deserted. He emerged at the year end in Toulouse as an *adjudant sous-officer* with the Haute Garonne Volunteers.

The outbreak of war in April 1792 saw him with the Army of the Alps. He then moved to the Army of Italy after his promotion on 5 April 1793 to *adjudant major-capitaine*. He was to remain in Italy for the next seven years returning to his old regiment the 21st Line with the *Amalgame* of January 1794. During an attack on the old fortress of Savona on 23 June 1795 a ball passed through his left leg that put him out of action for several months.

The arrival of General Bonaparte as army commander saw him with Joubert's division serving in the 32nd Line. Very active in Bonaparte's opening campaign, he fought at Montenotte on 11 April, Millesimo on 13 April and was at the crossing of the Lodi on 10 May 1796. In a reorganization that followed after the French entered Milan, he moved to the 33rd Line still with Joubert's division. He fought at Castiglione on 5 August. Promoted acting *chef de bataillon* on 25 December 1796, he distinguished himself at Rivoli on 14 January 1797.

In the Second Italian Campaign he received a thigh wound at Pastrengo on 26 March 1799. Campaigning extensively in Northern Italy, he gained promotion to *chef de brigade* of the 33rd Line on 11 June 1799. Present at Novi when Suvorov defeated Joubert on 15 August he retired with the French forces to the Ligurian Alps. During the Marengo Campaign the next year he avoided being trapped in Genoa and served on the Var front under Suchet.

THE GRANDE ARMÉE 1805-1807

He was promoted *général de brigade* on 29 August 1803 and moved to the Camp of Montreuil joining Ney's staff. He got on well with Ney*, as their characters were very similar. When VI Corps was set in motion towards the Danube, Roguet led a brigade with Loison's 2nd Division. He played a key role at Elchingen on 14 October 1805. In the struggle his troops seized the bridge over the Danube before the town while Dupont and Mahler, approaching along the north bank turned the Austrian left. The battle forced the demoralized Austrians back on Ulm, resulting in their final encirclement and surrender a week later. The remainder of the campaign he spent in the Tyrol, a notable exploit he achieved being on 5 November 1805 the capture of the fortress at Scharnitz.

In the Prussian Campaign he led the same formation serving under Marchand who had replaced Loison. He fought at Jena on 14 October 1806 and took part in the investment of Magdeburg from 22 October till 6 November. At the end of December he moved to Gardanne's 2nd Division replacing Marcognet as head of its 1st Brigade. He was shot in the foot at Güttstadt on 5 June 1807 and the Russians took him prisoner. Freed after Tilsit he returned to Paris where he took a post with the 1st Military Division on 10 September 1807.

SPAIN 1808

On 17 March 1808 he became a *baron de l'Empire*. He moved to the 24th Military Division in Holland on 18 May 1808, becoming garrison commander of Cadzand. A return to Paris followed on 23 August where he joined Sebastiani's division, with Lefebvre's IV Corps, preparing for Spain as a brigade commander (the 28th and 32nd Line regiments). His contribution during Napoleon's brief period in Spain was with Lefebvre against Blake's Army of the Left in Galicia that culminated at Zornoza. The Spanish general advancing from the direction of Bilbao tried to sweep around behind the French armies and fall on Napoleon's communications as they massed before the Ebro. Blake realized his mistake too late and caught in a double envelopment, made a stand at the Burgos-Vitoria-Bilbao crossroads near Durango on 31 October. Lefebvre drew up his corps and Roguet leading his brigade smashed its way through the Spanish center. The Spanish losses though not great, around 1,000 men; the victory did contain a threat and enabled Napoleon to launch his hammer blow across the Ebro a week later. Roguet's force was with Lefebvre covering Napoleon's left as the army moved towards Madrid. On 16 November 1808 at Valladolid he was detached. He remained there for two months with a dual role of watching for any movements by Moore in Portugal and the Spanish under Blake to once more threaten communications with France.

THE YOUNG GUARD ON THE DANUBE

Later at Valladolid on 16 January 1809 when Napoleon passed through the city and was so impressed by the bearing of Roguet's troops that the order was issued for him to join the Imperial Guard. Napoleon that day had signed an order creating the tirailleur grenadier and chasseur regiments that were to form the Young Guard. It was Roguet's task to return to Paris and to recruit and train some 3,200 of the strongest and best trained Guard conscripts formed from line and light regiments. Assigned excellent officers and NCO's from their parent regiments, they were to become after the Old Guard the finest infantry in the army. He took up his post in Paris on 19 February 1809. Within a month the regiments were ready and Napoleon reviewing them on 19 March was pleased with their appearance and alertness. On 2 April his troops started their march to the Danube forming the tirailleur brigade of Curial's Young Guard Division. At Strasbourg on 5 April he received news of his formal appointment in the Imperial Guard as lieutenant colonel of grenadiers.

At Aspern-Essling on 22 May 1809 he came to the fore during the defence of Essling on the second day of the battle. Boudet's exhausted division after two days of fighting collapsed under the assaults of Rosenberg's IV Corps, causing a great gap to open on the French right. To answer the crisis, Napoleon ordered Mouton*, the hero of Landshut, with Roguet's Tirailleur Chasseurs and Grenadiers, and the Fusilier Chasseurs to retake Essling with the bayonet. Roguet with Mouton quickly brushed aside the Austrian infantry and had nearly retaken the whole village when Rosenberg ploughing in further reinforcements began to close in behind them. Napoleon saw the threat and ordered another aide General Rapp to lead the Guard Fusilier Grenadiers forward to extract the tirailleurs and retire from Essling. Once engaged, Rapp instead agreed with Mouton they go on the offensive, and with Roguet's regiment in support they retook the village. The remainder of the day Roguet endured a withering cannonade till evening when his men covered the army's withdrawal to the Isle of Lobau. He lost over a quarter of his strength that day, but on their debut the Young Guard established a reputation equal to their senior regiments for the way they drove back a stronger enemy from a defensive position.

At Wagram on 6 July 1809 the army was not faced by the same crisis and Roguet formed part of Napoleon's reserve at Raasdorf. Only at the battle's end did he support Macdonald's final push on Süssenbrünn that broke the Austrian center and helped seal victory.

RETURN TO SPAIN 1810-1812

After his return from Austria Napoleon still had unfinished business in Spain. He ordered Roguet and Dumoustier there with the two Young Guard divisions, each command comprising two regiments of Guard conscripts and two of tirailleurs. On 16 December 1809 Roguet set off from Chartres with the 1st Young Guard Division to join Bessières at Burgos to await Napoleon's arrival. They reached their destination on 6 March 1810 and as time passed Napoleon never came. With a new young bride, Napoleon lost his taste for campaigning. The Guard at Burgos instead of becoming Napoleon's elite strike force, was formed into a separate corps under

Dorsenne as part of Bessières's Army of the North. It had a dual role, to act as a central reserve and to guard the rear of the Imperial armies.

It achieved neither as it was dragged into a vicious guerilla war. Roguet went through two years of total frustration as he joined operations against the guerilla leader Mina that ranged as far afield as the Biscay coast to the provinces of Soria, Burgos and Navarre. More and more troops poured in to contain the guerillas and by Autumn of 1810 over 38,000 men under Drouet*, Reille* and Roguet were in the hunt. The most far reaching consequence of these operations was that Massena never received the support in Portugal he so desperately needed. Mina and other guerilla leaders like Longa conducted brilliant campaigns. The provinces were in a continual state of unrest. Roguet was often faced by small groups of guerillas that united, scattered, then joined forces again. His troops had neither time to prepare for these attacks nor the means to resist them. The generals, obliged to disperse their units, dissipated their efforts. Roguet's officers were on the road twenty four hours out of twenty four.

Bessières, cold and formal, much to Roguet's distaste ordered a systematic reign of terror, seizing hostages and arresting magistrates and priests. Dorsenne, haughty and hard, used the same tactics when he took over. The fear, more often imaginary than real, inspired by the guerillas's movements and the fluidity of their numbers imposed constant changes in any plans of operation against them so that a coherent master plan was impossible. Whilst his men became hardened veterans, discipline often broke down as atrocities were committed by both sides. Furious at the needless killing, on one occasion Roguet ordered a court martial and had three culprits shot as an example to his men. During this period he had successes, at Yuanguas near Soria on 6 September and more notably at Belorado on 10 November 1810 when Mina lost over 800 men as he tried to cut the road to Bayonne.

News of Bessières's recall from Spain arrived on 24 July 1811 and with it came Roguet's long overdue promotion to *général de division*. He had deserved it since Aspern-Essling, and with also came the appointment as *lieutenant colonel* of the 1st Foot Grenadiers of the Imperial Guard. The level of activity in Northern Spain increased as Dorsenne took charge. Like his predecessor he did not possess much ability but he at least tried to implement a plan that would give full support to Marmont's Army of Portugal. Expeditions ranged far and wide into the provinces of Leon and Galicia. In August Roguet marched into the Cantabrian sierras with a large force to deal with Longa who threatened Leon. This mass of rugged country, over fifty miles broad covered by passes and ravines, made operations difficult and led to no tangible results. He then joined the Guard in a march against Abadia in the hope of bringing the Army of Galicia to battle. They pursued it as far as Astorga till it melted in to the sierras. In September he moved to Zamora on the plains of Leon where he kept communications open with Marmont gathering crucial provisions for the Army of Portugal arrayed before Ciudad Rodrigo.

THE RUSSIAN CAMPAIGN OF 1812

On 23 February 1812 whilst in the middle of operations against insurgents around Valladolid, the Young Guard were recalled for the coming campaign against Russia. His men now fit and able veterans aged between 25 and 30, formed the nucleus of the 2nd Young Guard division placed under his command when he arrived in Paris on 12 April 1812. They formed some of the finest troops in the Grande Armée and numbered some 4,070 officers and men. Late arriving from Spain, they caught up with the rest of the Guard at Vilna on 8 July 1812 having completed a four month 2,000 mile march from Burgos losing only 63 men en route. He fought at Borodino on 7 September where Lanabère's brigade (1st Tirailleurs and 1st Voltigeurs) distinguished themselves during the storming of the Great Redoubt.

It was during the retreat from Moscow that Roguet became one of the campaign's heroes as he doggedly kept the remnants of the army together. The lead role he played started at Krasnoe on 15 November. Napoleon after occupying the city, realized that not only did Kutuzov threaten to envelop him, but he could also cut the army in two by placing himself between the main army and the corps of Eugene, Davout and Ney. That night as the Russian campfires glowed, visible to the south and south east of the city, Napoleon carried out a daring plan. At 9.00 P.M he ordered Roguet to fall on Ozhrovski's camp some two miles from Krasnoe. He formed four battalions of fusiliers into three columns and advanced silently towards the enemy. At midnight in a cold so intense that frozen sentries had fallen asleep, the Fusilier Grenadiers and Chasseurs fell on the Russians around their fires driving them into the darkness in great disorder. Roguet lost near 300 men, but the action had a profound effect on Kutuzov. The Russian started to exercise more caution. The enveloping manoeuvre around Krasnoe stopped, as Tormassov's corps ordered to cut the road 20 miles further on at Liady was recalled. The Grande Armée was a few valuable days respite gathering in stragglers and reorganizing itself.

Kutuzov again tried to cut the army in two on 17 November, but was foiled by a force of 5,000 Guard, which included Roguet's division that occupied the Lossima gorge seven miles east of the city. Throughout the day they held back Kutuzov's attacks as first Eugene's and then Davout's corps passed by. Finally after the loss of another 760 men and unable to hold any longer to save Ney's force that never appeared, they withdrew. Filing through Krasnoe they left the blazing city with its 6,000 stragglers to the Russians, but the army had survived. When Roguet reached the Beresina his division was down to 1,500 men. Crossing the river on 27 November he was in the fore in the assault on the Brodnia defile, which pushed back Tshitchagov and opened Napoleon's line of retreat. He struggled manfully to keep his division intact and by the time he reached Konigsberg in East Prussia, though down to 315 officers and men, it was still a recognizable formation.

At forty-two, prematurely old and grey, "Old Man Roguet" as he was known to his men had repeatedly in the campaign proven his toughness and ability as a natural leader. Neither the rigors of the climate nor the misfortunes of war had shaken his confidence in the future as it had other generals of the Guard. Wearing rags for boots he had marched at the head of his division, ate gruel and drunk melting snow. He had slept in the open and was happy to share a campfire with his men. War wise, he never suffered even a head cold during the retreat, whilst his servants and aides either froze to death or lost fingers and toes. Each night he stood watch at the bivouacs, and at dawn forced his benumbed men to their feet. During the day he urged them on, and gave away personal items of clothing to keep them warm. He lost everything he possessed and hadn't a sou. Prince Eugene gave him a pelisse and a cantiniere lent him a 1,000 francs for equipment and a new uniform.

GERMANY 1813

Roguet's determination in Russia made him a natural choice to lead the remnants of the Guard. His numbers swelled with the arrival at Stettin of two battalions of voltigeurs, numbering 1,355 officers and men from Spain on 13 December. By 13 January 1813 when his appointment was confirmed he had organized 4,157 officers and men into a small division at Posen. The old formations were formed into single battalion regiments with companies of 100 men. The rest of the men, over 750 Old Guard Grenadiers and Chasseurs were sent to Mainz to form the cadres for the new formations being gathered.

His division formed the backbone of Eugene's army and performed well during the long retreat to the Elbe. On 28 April summoned from Merseberg to join Napoleon's main army as it moved towards Leipzig he fell in with Dumoustier's mixed Guard division at Naumberg. A quick reorganization followed whereby he took Dumoustier's chasseur and grenadier regiments in exchange for his voltigeurs and tirailleurs to form an Old Guard

Division. He led the formation at Lutzen on 2 May, but held in reserve never fired a shot all day. With Napoleon at Bautzen on 21-22 May the same happened except that when he took up a position on high ground two miles north of the town he suffered casualties from long range Prussian artillery fire. When the army advanced, held in reserve he was unable to retaliate.

The expansion and reorganization of the Guard during the Armistice in resulted in Roguet switching commands with Friant'. On 15 August 1813 he took over the 4th Young Guard Division made up of the 7th-10th Tirailleurs. The resumption of hostilities saw him in the epic four day one hundred and twenty five mile march from the Katzbach to the Elbe to save Dresden. Present at the battle for the city on 26 August, a spent musket ball struck his side badly bruising him as he led an assault on the Ziegel Gate. At Leipzig from 16-18 October his troops held in reserve were not heavily engaged in the struggle for the city. He followed Bertrand's IV Corps out of Leipzig on 18 October and performed valuable service as the army's vanguard during the retreat to the Rhine. At Hanau on 30 October he engaged advance elements of Wrede's Bavarians before the next day Napoleon, ably supported by Drouot', contemptuously brushed the Bavarian army aside.

THE DEFENCE OF BELGIUM

Back in France another reorganization of the Young Guard followed with Roguet moving to Metz to form the 6th Young Guard Division comprising the 9th-12th Tirailleurs. At the same time on 13 November 1813 a revolt broke out in Holland and France's north western defences lay open. Ordered to take his new command to Belgium, he found that on his arrival at Metz the cadres barely existed. Nevertheless, he gathered up whatever troops were at hand and marched to Belgium with orders to defend Louvain and Antwerp. A period of mixed fortunes followed. He first failed to capture Breda on 20 December 1813, was defeated at Turnhout on 10 January 1814 and again at Mannheim four days later. He fell back to Antwerp and was besieged by a strong British and Prussian force under Graham and Bulow, which was later joined by Saxons under the Prince of Saxe-Weimar. He held the city for two months till 27 March 1814 when he broke out, dodged the Allied forces and rejoined Maison's main force at Courtrai. Here with Maison they defeated Thielemann and Saxe-Weimar on 31 March 1814. When orders to cease hostilities reached him on 15 April he was at Lille where his division was still a very potent force, numbering over 6,400 men.

The irony of the campaign in Belgium was that Napoleon was deprived of one of his ablest Guard commanders as well as two of his Young Guard divisions during the crucial campaign in France.

Though these troops contributed for a time towards the containment of the enemy on the northern front, the special talents of the Guard, particularly its amazing mobility, was wasted in this theater when separated from the Emperor.

THE RESTORATION AND NAPOLEON'S RETURN

With Napoleon gone Roguet saw little alternative but acknowledge the Bourbons and in turn on 8 July 1814 received the title *Chevalier de Saint Louis*. He served under the Bourbons as a colonel on the staff of Royal Guard. Napoleon's return caused him a crisis of conscience. At Metz under Oudinot, he obeyed his superior's orders to march against the former emperor, but after three days it became very apparent the troops would defect. Oudinot fled, while he prudently led the grenadiers towards Paris ensuring they avoided any contact with Napoleon's triumphal progress. Napoleon understood his predicament and on 8 April 1815 made him Friant's deputy of the Imperial Guard Foot Grenadiers.

THE WATERLOO CAMPAIGN

Roguet had difficulty bringing the four guard grenadier regiments up to strength. Many veterans had retired, cadres were scattered throughout line regiments, whilst many in the Royal Guard regiment were of dubious quality. To sort all this out was time consuming, and by the time orders were received to concentrate on the Belgian border the 4th Regiment only had a single battalion. For the campaign in Belgium command of the grenadiers was split, Friant retaining overall command, while Roguet took charge of the 3rd and 4th Regiments often referred to as the Middle Guard.

At Ligny on 16 June he joined the final assault of the day by the Guard. Amidst the thunderstorm that engulfed the field he led the first echelon made up of the 2nd, 3rd and 4th Grenadiers that penetrated the west side of Ligny. A second column comprising the 1st Grenadiers and 1st Chasseurs led by Friant approached the village from the east. Faced by the cuirassiers of Milhaud, the Guard heavy cavalry in support plus the divisions of Pécheux and Vichery, the Prussian center collapsed. Pouring through the village and out the other side, his men repelled a last desperate counter attack by Prussian cavalry which tried to stem the tide.

At Waterloo Roguet remained in reserve with his troops throughout the day until the evening when Napoleon committed the Guard in the final assault on Mont Saint Jean. Friant taking charge of his 3rd and 4th Grenadiers led them against the ridge while he remained with Napoleon near La Haye Sainte. He was left with the three 2nd battalions of the 1st Grenadiers, the 1st and 2nd Chasseurs under Napoleon's direct control, with which the Emperor intended to follow to give Wellington the final coup de grace.

Then as Napoleon saw the Guard's assaults fail and that he was faced with ruin, he realized the best he could do was make the three remaining Guard battalions form a human dyke behind which the army could rally. For this purpose he ordered the three battalions into square joining Roguet in the center one, formed by the 1st Grenadiers with their right resting on the Charleroi road about a hundred yards from La Haye Sainte. The squares withstood cavalry charges by Vivian's cavalry with ease as Wellington's infantry and two British Royal Horse Artillery troops closed. The guns opened with grape at 60 yards and the squares poorly placed were unable to reply as fire poured into them. Napoleon ordered Roguet and the other squares to withdraw, and then with a few aides and mounted chasseurs left the field.

Roguet's three battalions were attacked in turn by English dragoons, Brunswick lancers, infantry under Maitland and Mitchell, fell back step by step. Reduced in numbers, too small to form a square three ranks deep they formed themselves into two row deep triangles. Then with bayonets fixed, they slowly cut their way through the throng of fugitives and enemy that barred their path. At each step they stumbled over bodies or fell pierced by bullets. Every fifty yards they halted to reform their ranks and repulse a fresh charge of cavalry or a new attack from infantry. Enemies and fugitives continued to surround the three battalions of Guard as they withdrew. Their retreat though slow, was carried out with precision and order. It was amidst this tumult as they neared La Belle Alliance that Cambronne at the head of the 1st Chasseurs when asked to surrender supposedly gave the curt reply,"Merde!". It was later cleaned up into the immortal phrase,"La Garde meurt, elle ne se rend pas". After La Belle Alliance, the dwindling squares were submerged by the crowd of assailants, and as they dissolved the survivors mingled with the remnants of the Armée du Nord.

Somehow through the chaos Roguet escaped unscathed. The next morning after he reached Charleroi he restored order amongst the remnants of the Guard infantry as they crossed the Sambre during the day. News of Napoleon's abdication caused near mutiny amongst his men. Officers who read the proclamation were threatened. The men suspicious, resentful and bitter, accused them of treason as they broke ranks and approached them with bayonets bared. Generals too tried to calm them, but were booed and jeered. Roguet outraged, with sword in hand showed great courage and stormed up to the undisciplined soldiery. Ignoring the muskets levelled at him he threatened to run the next man through who dared not to resume the march. It was the language soldiers understood and the march

to Paris soon resumed. By the time he reached Soissons he had rallied some 6,000 Guard infantry and took over from the wounded Friant as commander. At the Council of War he was one of the officers who advocated a final battle before Paris. Then after the army retired to the Loire, he presided over the emotional disbandment of the Guard regiments in the ensuing weeks.

LATER LIFE AND CAREER

Suspended on 16 October 1815, Roguet's recall did not come till 30 December 1818 with Saint Cyr's army reforms. A liberal and therefore suspect, he remained unattached until retired on 1 January 1825. The accession of Louis Philippe resulted in his recall on 4 August 1830 as commander of the camp at Saint Omer and the 7th and 16th Military Divisions. He received the Grand Cross of the Legion of Honor on 21 March 1831 and became a Peer of France on 19 November 1831.

Just as his prestige reached its peak, he was suddenly caught in the middle of a maelstrom of civil and political unrest and found it impossible to adapt to the changing role expected of the army when faced by civil unrest. His experience at Lyons as commander of the 7th Military Division in November 1831 was disastrous. The city for months before had been a hot bed of republican intrigue and unrest, but whether sympathetic or not he chose to ignore the dangers. Ordered to restore authority after an outbreak of violence he led 3,500 troops into the city. Taken by surprise he suffered a humiliating defeat as his men were halted by barricades and showers of missiles that rained down from roof tops. Suffering some 350 casualties he withdrew after he heard a further 750 of his men joined the mob.

The military establishment was stunned, and with memories of events at the barricades in Paris the previous year fresh in their minds, retribution was swift. The prospect of failure at the hands of a left wing rabble of workers was unacceptable and Soult' as Minister for War descended on the city with a force of 18,000 that soon took control. Roguet, humiliated and in sympathy with some of the workers' demands, was removed on 8 December 1831. Military honor satisfied, Roguet became a convenient government scapegoat for their failings in Lyons. His career could have ended there and then, but Soult understood the old veteran and knew he was no coward. He secured Roguet a post as an Inspector General and for several years Roguet held various appointments with the Infantry Inspectorate. He also served on several committees studying proposed infantry and cavalry reforms till his return to the Camp of Saint Omer in 1834. Placed on half pay on 28 August 1836 he retired on 15 August 1839.

He lived quietly in Paris and wrote a comprehensive volume of memoirs on his wartime experiences. They were published in 1864 by his son, who rose to become an eminent soldier during the Second Empire. Roguet himself died in Paris on 4 December 1846.

AN ASSESSMENT - THE "OLD GRUMBLER"

As a soldier Roguet never lost his common touch nor forgot his origins. A man feared for his temper, his bark was often far greater than his bite. He also showed uncommon generosity, kindness and concern towards his men as witnessed during the retreat from Moscow. His name rarely gained much prominence during Napoleon's triumphs, for the Emperor's aides were often more adept at stealing the limelight. It was in times of greatest adversity, at Aspern Essling, the retreat from Moscow and then in the aftermath of Waterloo that his indomitable spirit came to the fore. For this reason above all others, Roguet ranks as one of Napoleon's finest division commanders to take the field at Waterloo.

NAPOLEON IS PERSUADED TO LEAVE THE BATTLEFIELD. ROGUET AND THE OLD GUARD FORM THE REARGUARD

BACKGROUND AND EARLY CAREER

Of noble birth and a native of Lorraine, the cavalry general Roussel D'Hurbal was born at Neufchateau (Vosges) on 7 September 1763. His family's German origins resulted in him entering Austrian service as an officer cadet with the Kaunitz Infantry Regiment on 1 January 1782. A gifted horseman on his promotion to *2me lieutenant* on 8 February 1785 he gained a transfer to the Latour Chevaulégers, one of the finest cavalry regiments of the Austrian army. He spent time in the Austrian Netherlands during the insurrection in Belgium during the Winter of 1789-90, with the Latour Chevaulégers gaining promotion to *lieutenant* on 13 October 1789. The outbreak of war against France saw him with the Austrian forces on the Rhine from 1793 till 1797. Although very talented as with many emigré officers at the time he found advancement slow in a foreign army. It was not till after distinguishing himself at Aldenhoven on 2 March 1793 that he was promoted *captain 2nd class* on 20 April 1793 and then had to wait till 1 March 1797 before gaining the full rank.

Roussel served in Germany from 1799 till 1801 against the French armies. The savaging the Austrian army received on all fronts resulted in a shake up amongst the officer class and with reforms introduced by Archduke Charles his prospects improved. He was promoted *major* on 1 March 1802 and then *lieutenant colonel* on 2 September 1804 when he moved to the Latour Chevaulégers. He took part in the Campaign of 1805 and received the Order of Maria Theresa on 11 October for his part in the capture of Munich. At Ulm he escaped the encirclement when on 15 October Archduke Ferdinand broke out north with the force of 6,000 cavalry.

Roussel became *colonel* of the Moritz Liechtenstein Cuirassiers on 1 January 1807. During the Campaign of 1809 he served with Archduke Charles's army. At Aspern-Essling on 22 May he distinguished himself against the French cavalry in spite of being badly concussed by a sabre blow early in the day. Promoted *major general* on 23 May he led a cuirassier brigade comprising of the Erzherzog Franz Cuirassiers. At Wagram on 5-6 July he again emerged with great credit when Hessen Homberg's cavalry valiantly tried to stem the French advance between Sussenbrünn and Aderklaa.

SERVICE WITH FRANCE 1811-1814

In November 1809 Roussel received the title a Commander of the Order of Maria Theresa, in recognition of his part played in the campaign. Peace between Austria and France however had an adverse effect on his career. In terms of the Treaty of Schönbrunn the Austrian army was reduced in size, resulting in him finding himself without a command in October 1810 and placed on half pay. An admirer of Napoleon and anxious to return to France during this period of detente, on 1 April 1811 he resigned from the Austrian Army. In France the French army accepted him on 31 July 1811 with the rank of *général de brigade*. He was a useful find since at the time the French cavalry was going through a period of reorganization with six dragoon regiments converting to Chevauléger lanciers and his experience with light cavalry and dragoons proved useful. Posted to Germany on 3 August 1811 he took charge of the newly raised 9th (Polish) Chevauléger Lanciers at Hamburg. A brief period then followed with the 8th (Polish) Lancers before he joined the general staff of Davout's I Corps on 1 May 1812.

ROUSSEL D'HURBAL - NICOLAS FRANÇOIS, VICOMTE (1763 - 1849)

THE RUSSIAN CAMPAIGN

Before the start of the campaign his experience with the Poles and his fluency in German singled him out for command of the 15th Light Cavalry Brigade with Bruyère's 1st Light Cavalry Division. The formation made up of the 6th and 8th (Polish) Uhlan Regiments and the 2nd Combined Prussian Hussar Regiment was soon in action outside Vilna on 28 June when it engaged the Russian rearguard. After the march resumed, he won another sharp action deeper in Russia at Kosny on 5 July as Bagration's forces slipped past Davout's encircling movement.

As Napoleon drove forward into the gap between the Dnieper and Dvina rivers trying to draw Barclay de Tolly into a major action near Vitebsk, Roussel's horsemen were often used as bait. On 25 July Murat leading the cavalry encountered strong opposition outside Ostrovno on the banks of the Dvina, which at first indicated the Russians had decided to make a stand. Then as the French army massed the Russians disengaged and continued the retreat towards Smolensk. In the series of manoeuvres that followed Roussel played an important part. A well laid trap by Ostermann to fall on Murat's infantry was overturned by Roussel's brigade when it fell on the flank of the Ingremannland Dragoons and took over 200 prisoners. That not enough, they then overran two battalions of Russian infantry as they tried to drive in the French right. The next day the Russians fell on the 8th Légère and 84th Line and before long the French infantry were in difficulty. To avert a crisis, Murat darted forward with Roussel to harangue the 8th Uhlans. Inspired by the Gascon's exhortations and the sight of the advancing Russians they burst into an uncontrolled charge. Caught in front of this avalanche of horse that they merely wanted to use as a feint, Murat and Roussel were caught in the front rank of a melee that forced the enemy back to their original positions. The action accidentally conceived and poorly controlled, luckily had the right result. Murat was flushed with a success that had been a long time coming, and Roussel suddenly thrown into the limelight prospects of promotion beckoned.

He fought at Smolensk on 17 August when under Bruyère's orders around 1.00 P.M. he chased a Russian battery off the plateau that overlooked the Smolensk bridges. A dominant position secured, a sixteen gun battery was set up that rained shot and shell on the Russians posted on the north bank of the Dnieper. In response the Russians erected a twenty four gun battery and a devastating artillery duel continued throughout the afternoon. Roussel's capture of the position and the effect of the fire severely disrupted Russian movement through the city and had an important influence on the outcome of the battle.

At Borodino Roussel took part in the great cavalry struggle that developed during the struggle for the Bagration Fleches. Shot in the leg, there is no record that he took any further part in the campaign. An act of generosity on Napoleon's part as he handed over the army to Murat at Smorgoni on 5 December 1812 was to approve Roussel's promotion to *général de division*.

THE CAMPAIGN IN GERMANY 1813

On 13 February 1813 Roussel joined the cadres of Sebastiani's II Cavalry Corps gathering at Luneberg on the lower Elbe. Formed from the shattered remnants of regiments returned from Russia and depot squadrons called from France, he took command of a hopelessly understrenght 2nd Light Cavalry Division on 19 April numbering barely 300 men. While Napoleon fought at Lutzen on 2 May his cavalry were still operating on the lower Elbe with Eugene Beauharnais's Army of the Elbe. He missed Bautzen on 21-22 May when a series of confused orders resulted in a large portion of the army under Ney failing to reach the battlefield including all Sebastiani's cavalry.

During the Armistice his division was built up to a fine light formation comprising regiments of chasseurs à cheval, lancers and hussars in 19 squadrons numbering over 3,200 officers and men. The outbreak of hostilities saw him with Macdonald's Army of the Bober. Blucher

FRENCH CUIRASSIERS CHARGE AUSTRIAN
GIVEN ROUSSEL D'HURBALS'S CAREER, HE COULD BE ON EITHER SIDE

immediately went on the offensive and forced Macdonald to abandon his positions on the Katzbach and fell back behind the Bober. Roussel's cavalry covered the retreat and at Hainau on 19 August disrupted the Prussian cavalry as they tried to ford the river. A week later, on 26 August a series of misfortunes resulted in him being partly to blame for Macdonald's defeat on the Katzbach. His division was slow to cross the river and fan out in search of Blucher's army and failed to spot the enemy when a violent thunderstorm reduced visibility to a few hundred yards. As the Prussians appeared his commanders failed to appreciate the size of the forces they were about to engage. A few charges against exposed Prussian infantry who could not easily defend themselves as rain had soaked their powder gave them a false sense of superiority. Suddenly when 20,000 Allied horsemen flooded onto the field the hopelessly outnumbered French cavalry was swamped. The infantry divisions under Gérard* (XI Corps) and Lauriston (V Corps), unnerved at the sight of their cavalry fleeing before them in turn were unable to defend themselves and collapsed. The Army of the Bober in the ensuing chaos lost 15,000 men and 36 artillery pieces. Roussel received a sabre blow to the head that put him out of action for the remainder of the campaign. Macdonald took full responsibility for the disaster, which went a long way to explain how Roussel while recovering from his wounds received news of his award as a *baron de l'Empire* on 28 September 1813.

THE 1814 CAMPAIGN IN FRANCE

On 17 January 1814 Roussel returned to duty as head of a new cavalry depot formed at Versailles. He did not remain there for long, for experienced generals were at a premium and on 19 February he joined VI Cavalry Corps under Kellermann*. The 6th Heavy Cavalry Division was formed from veteran dragoon regiments recalled from Spain and was an excellent command. Almost immediately he was attached to Gérard's II Corps where he fought at Troyes on 23 February. He then joined Napoleon's main army and charged the Russians at Craonne on 7 March before rejoining Kellermann in action at Laon on 9-10 March. Thereafter he was detached under Belliard where he fought at Fère Champenoise on 25 March. Here he was partly responsible for a major set back when his cavalry failed to cover Marmont's retirement and as a result Pacthod's infantry division was destroyed by Allied horsemen the next day near Sezanne. The reverse could have irreparably damaged his career, but any complaints from Marmont soon meant little coming from a defector.

Blucher drove directly onto Paris where Roussel's division joined Marmont's forces in defence of the city. He fought on the plain before Saint Denis on 30 March and then after Paris's fall joined Marmont's VI Corps at Essonnes to await events. Exhausted and disillusioned with the war, he fell an easy victim to Marmont's entreaties that the only way to peace was for the army to abandon Napoleon. On 5 April he crossed the enemy lines and then marched his troops, much to their disgust, to Evreux to disband.

The Bourbons soon found a place for him in the new army when on 1 June 1814 he became Cavalry Inspector General for the 6th and 19th Military Divisions. He received the title *Chevalier de Saint Louis* on 19 July 1814. On 11 March 1815 he received orders to join the comte d'Artois's forces gathering at Lyons to bar Napoleon's return. As he moved out with his troops from Besançon it soon became apparent that they had no intention to fight for the King so he discreetly abandoned them and returned to his home to await events. The move probably saved his career since his past association with Marmont did not make him one of Napoleon's favorites.

Summoned by Napoleon on 8 April he took charge of the 2nd Cavalry Division based at Metz. The appointment was strange since Metz was the depot for cuirassiers and carabiniers, and he had never led cuirassiers while in French service. His record also showed his loyalty was suspect and having shown himself to be a classic waverer, he should never have been entrusted with such a prestigious command. Comparisons were made with men like Trelliard, the hardened dragoon veteran from Spain who the previous year operating under Kellermann in France had done valuable work and would have made a better choice. Somehow the appointment went through and that fine general was destined to remain at his post on Belle-Isle-en-Mer during the Hundred Days.

Despite protests as to his suitability, he retained his command when on 3 June the division became 12th Cavalry Division of III Cavalry Corps. His formation outside the guard cavalry was the finest in the army. It comprised the two magnificent carabinier regiments and the 1st and 2nd Cuirassiers, formed into twelve squadrons numbering 1,638 officers and men. Skepticism increased further the next day on news of Kellermann's appointment as corps commander. Roussel was relieved, for Kellermann another arch waverer was one of few friends he had.

THE WATERLOO CAMPAIGN

Delayed crossing the Sambre at Charleroi, Roussel failed to make an appearance the next day at Quatre Bras on 16 June. Sorely missed, Guiton's brigade of L'Héritier's division led by Kellermann charged the positions on its own and nearly gained a spectacular success. His formation arrived for Waterloo completely fresh and was posted with the rest of Kellermann's corps to the left of the Charleroi road behind Reille's troops. He joined the second series of great cavalry charges against Wellington's line around 5.30 P.M. after Milhaud's corps had been repulsed. The mistakes of the previous charges had not been learnt. Still the cavalry was not backed by infantry, nor was it closely supported by adequate horse artillery. He followed L'Héritier, though his division formed Napoleon's last reserve of cavalry. The task was hopeless, the slopes littered with the debris of Milhaud's corps prevented his men moving at little greater than a trot. Like those before them, his men suffered the same cruel fate. He tried to keep back the carabiniers by concealing them in a dip. Ney saw them and forced Kellermann to order forward this last reserve. Had they been available to support the final assault by the Guard the result of the battle may have been very different. The dogged determination showed by his men was revealed later by the returns that showed only some 700 officers and men of his division survived the battle out of 1,700 at the start of the day.

He remained with the army as it fell back to Paris, taking over command of the corps at Clichy when Kellermann left to join the armistice negotiations with the Allies. After peace and the army's disbandment, as with most generals who served Napoleon he was suspended on 1 August 1815. The Ordonnance of 9 September 1815 resulted in his retirement.

THE BOURBON GENERAL

Roussel's sudden recall on 25 June 1816 as Cavalry Inspector General for the 9th Military Division based at Montpellier raised a few eyebrows. It was barely a year since Waterloo and some of his fellow generals were still awaiting trial, whilst others faced imprisonment and even death. A reactionary at heart, he found it easy to embrace the old order. Many prestigious posts followed, culminating in his appointment as a Gentleman of the King's Chamber on 22 April 1822 and then that of *vicomte* on 17 August 1822. At the same time his military career also prospered. On 12 February 1823 he was made commander of a cuirassier division with the Army of the Pyrenees that served in Spain. After his return from that theater he shared in the awards,

commander of the Order of Saint Louis on 19 November 1823 followed by the Spanish Grand Cross of Saint Ferdinand. He moved to Corsica on 30 December 1823 as governor and commander of the 17th Military Division.

The liberal idealism that accompanied the early days of Louis Philippe's reign resulted in Roussel's recall from Corsica. On 30 December 1830 he found himself without employment and then on 7 February 1831 placed in the Reserve Section of the General Staff. He retired from the army on 13 September 1832. The passage of time caused his past associations with the Bourbons to be forgotten. Anxious to recall glories of the past on 26 April 1846 Louis Philippe had the tottering and lame veteran wheeled out to become a Grand Officer of the Legion of Honor. He died in Paris on 1 December 1849, lamented by few.

AN ASSESSMENT

No fervent Bonapartist, he showed a natural reluctance to rally to the Emperor at the beginning of the Hundred Days. This led to criticism when commands were handed out, but on the other hand allowances were often made for men like him as he was a very fine horse soldier and such men could not easily be passed over. As an emigré he rose to the equivalent of *général de brigade* in the Austrian army, which was a fine feat in arguably the best equipped and trained cavalry in Europe. He was no different from many other commanders who respected Napoleon, but were in no way blindly devoted to him, especially in the bad times. His behavior during the Hundred Days was understandable, as a former emigré he had more to lose by opposing Napoleon's return. Once committed the bravery he showed at Waterloo was equal to that of his peers. He chose to follow Napoleon not out of love or loyalty but because he was first a Frenchman and dreaded the prospect of Allied armies once more marching on the soil of France. He felt no ties to the Allies, he had served the Hapsburgs for over twenty years and then they had discarded him; as a professional soldier that hurt. As time passed after Napoleon's fall he played skilfully to the middle of the road and in turn had a long distinguished career under the Bourbons.

BACKGROUND AND EARLY CAREER

Born in Besançon on 4 November 1774, Ruty came from a middle class background. Educated at the College de Besançon he entered the Chalons Artillery School gaining a commission on 1 September 1792. He joined the 2nd Foot Artillery Regiment and saw his first active service with the Army of the North. Shortly after his promotion to *lieutenant* on 6 November 1793 he was wounded by a shell splinter in the leg at Comines. Promoted *capitaine* on 22 February 1796 he moved to the Army of the Rhine-Moselle. Present at the siege of Kehl on the Rhine he was badly disfigured by a musket ball that smashed his jaw in December 1796.

His first contact with Napoleon came when he joined the Army of the Orient and served in Egypt from 1798 till 1801. Promoted *chef de bataillon* on 29 July 1798, he served in Kléber's division. He fought at the Battle of the Pyramids on 21 July. After a period at Gizeh, in February 1799 he joined the expedition to Syria as commander of Kléber's artillery. He fought at Mount Tabor on 16 April and was at the siege of Acre. On his return to Egypt Napoleon appointed him director of the army's artillery park. He fought at Aboukir on 25 July when Napoleon routed the Turkish landing before his return to France. When Kléber reorganized the army after Napoleon's departure he returned to Kléber's old division headed by Verdier as its artillery commander. He spent time at Damietta where near the port on 1 November 1799 he took part in the rout of 7,000 Janissaries at Lesbeh.

Soon after his return to France he was promoted *chef de brigade* on 5 December 1801 and then moved to Strasbourg as commander of the 4th Foot Artillery Regiment. A period followed at Perpignan as director of artillery from January 1802 till May 1804 when he moved to Grenoble as artillery commander.

THE GRANDE ARMÉE 1805-1807

September 1805 saw his call to the Grande Armée as director of the artillery park with Ney's VI Corps. He served in that capacity with VI Corps in the campaigns in Austria, Prussia and Poland. On 8 January 1807 he was promoted *général de brigade* and moved to Murat's Cavalry Reserve as artillery commander. He fought at Friedland on 14 June 1807.

SPAIN AND PORTUGAL 1808-1813

He spent time with the Berlin garrison where after falling seriously ill he was forced to give up his command and on 29 January 1808 took a quieter posting as commandant of the artillery school at Toulouse. He shared in the awards that followed two years of successful campaigns, receiving a pension of 10,000 francs per annum drawn on Westphalia. The creation of the Imperial nobility saw him receive on 11 August 1808 the title *baron de l'Empire*. Called to Bayonne on 24 September 1808 to join Napoleon's staff for the invasion of Spain, no sooner was he there when he found himself posted on 9 November to Gouvion Saint Cyr's VII Corps as artillery commander. He served in Catalonia till a recurrence of ill health ended in his recall to France on 10 August 1809.

He returned to Spain in January 1810 with the grandiose title of commander of the artillery and siege train of the Army of Spain. In that capacity he served with the Army of Portugal at the siege of Ciudad Rodrigo from 6 June till 10 July 1810 and from there moved to the seigeworks at Almeida from 24 July till 28 August 1810. A move followed on 18 November 1810 to Soult's Army of the South in Andalusia. He took part in the campaign to relieve Badajoz and fought at Albuera on 16 May 1811 when Soult's hopes to relieve the fortress were dashed by Beresford. On 1 September 1811 Soult appointed him the army's artillery commander. Being very innovative, while in the lines before Cadiz he invented a new mortar that in the years to come became standard equipment adopted by the army for siege warfare and later carried his name.

RUTY - CHARLES ETIENNE FRANÇOIS, BARON (1774 - 1828)

GERMANY AND FRANCE 1813-1814

Recalled to France with Soult, he was promoted *général de division* on 10 January 1813. He joined the Grande Armée in Saxony on 24 April as artillery commander with Oudinot's XII Corps. Present at Bautzen on 20-21 May his guns did deadly work against Miloradovich's positions on the extreme right of the French line during the battle's second day. Oudinot's corps forced the Spree and had the enemy on the southern flank streaming back several miles by the end of the day. The indifferent handling by Ney of his forces against the Allied right flank and a shortage of cavalry to follow up prevented Napoleon gaining an overwhelming victory.

During the Armistice under Sorbier, Ruty became the Grande Armée's artillery chief of staff. He fought at Leipzig on 16-18 October 1813 and struggled manfully to bring back to France as many pieces as possible after the defeat. For much of the time during in the campaign in France he headed the artillery as Sorbier's health broke down, remaining at Napoleon's side throughout and loyal to till the end.

THE RESTORATION AND WATERLOO

His future appeared secure under the Bourbons when on 21 June 1814, recognizing his skills as a gunner, Dupont as Minister for War appointed him to the Committee for War. This was followed by him becoming a *Chevalier de Saint Louis* on 4 July and then a Grand Officer of the Legion of Honor on 5 August 1814.

Fearing the dangers to France Napoleon's return presented, Ruty chose to side with the Bourbons and remained with the Duc de Berry's forces in Paris till the last moment. His devotion was such that he was quite prepared to turn his guns on Napoleon's troops had he not been thwarted by his gunners. Surprisingly Napoleon made little of the incident in light of his fine service the previous year and on 27 April 1815 appointed Ruty the Army of the North's artillery commander. Ruty accepted without qualm or conscience, much to the dismay of many Bonapartist officers who knew his loyalty was suspect.

With Napoleon throughout the campaign he fought at Ligny and Waterloo, where his failings as a commander were soon exposed. A very capable desk soldier, administrator and organizer Ruty would have been at home as Artillery Chief of Staff rather than its commander. A step from a corps artillery command to that of an army was a big one and clearly incidents arose that revealed he was not up to the task. It was a sign of the times that great artillerymen like Laroboisère and Sorbier were no more. The first had died of exhaustion after the Russian campaign and the other worn out, preferred politics to campaigning. Ruty's reputation and limited experience prevented him from gaining the respect accorded the rank held by his predecessors. He also faced serious handicap in that his loyalty was suspect. Men like Drouot and Desvaux from the Guard had the Emperor's ear when it came to artillery matters. In fact on campaign Napoleon rarely conferred with him, tending to treat him more like a clerk. So ineffectual was Ruty that had there been no one appointed as artillery commander it would have made little difference at all. The interests of individual commanders prevailed and Napoleon's failure to have a strong person at this key post to maintain a tight control contributed to the poor performance of the artillery at Waterloo.

A fundamental principle of warfare, that infantry and cavalry must be supported by guns when attacking an enemy in a strong defensive position was ignored. Not only did the cavalry assaults fail for lack of artillery support but also the final attack by the Old Guard. Ruty was too much of a light-weight in the heat of battle to make such things happen. He was Lallemand's superior but could not or overlooked to order the former to use his four Guard horse batteries to support Ney's cavalry charges. Instead they remained with d'Erlon's guns bombarding the Allied line from a safe distance. Had these guns advanced with Ney and at the end of the day with the Guard, the result of the battle may have been very different.

LATER CAREER AND ASSESSMENT

Experienced artillery generals at a premium and Ruty not considered a threat, the Bourbons did not suspend him after their return. He spent time with the artillery inspectorate till made director general of powders and saltpeters in November 1817, a post responsible for the quality of ammunition procured by the army. A solid reliable professional, he kept clear of politics within the army and in time was rewarded, a Peer of France on 5 March 1819 and a commander of the Order of Saint Louis on 23 May 1825. He died in Paris after a brief illness on 24 April 1828.

Napoleon overrated Ruty and this was a factor that worked against him at Waterloo. Bravery and organizational ability he possessed, but he lacked the respect and will required to use the tactical and strategic skills he no doubt possessed to their best advantage. He was certainly no Drouot and was not on a level with great men like Senarmont, Laroboisère and Sorbier who served the Grande Armée's artillery so well.

SIMMER - FRANÇOIS MARTIN VALENTIN (1776 - 1847)

EARLY CAREER

Simmer was born in the village of Rodemack near Metz (Moselle) on 7 August 1776. As the war clouds gathered on 3 November 1791 he enlisted with the 4th Battalion of the Moselle Volunteers. He rose steadily through the ranks showing promise as a horseman, that secured him a commission as a *sous lieutenant* on 21 September 1792 with the 7th Cavalry Regiment (Royal Etranger). During this period he served with the Army of the North in Belgium and Holland. At the Siege of Maestricht on 2 March 1793 a musket ball broke his right arm. Promoted *lieutenant* on 18 April 1794, another wound this time in the leg at the Siege of Tourcoing on 18 May 1794 put him out of action for several months. After his recovery he returned to the Army of the North where the following year he joined the large detachment under Duhesme* sent to the Vendée to reinforce the Army of the Coast of Cherbourg. He had hardly arrived when shot in the shoulder on 31 August 1795, he was again laid low for several months. Next emerging with the Army of the Rhine he gained promotion on 5 October 1797 to *capitaine*. A spell followed with the Army of Switzerland before returning in 1798 to the Army of the North. On its disbandment in early 1799 he passed to the Army of Gallo-Batavia serving against the Anglo-Russian forces that landed in Holland in September 1799. When Augereau took charge of the army in January 1800 he joined his staff and served in Germany. In the Hohenlinden Campaign he joined Augereau's march through Swabia and Franconia fighting at Berg Eberach on 3 December 1800. After the Peace of Lunéville the Army of Gallo-Batavia's days were numbered but he remained with it till its final disbandment in October 1801.

THE GRANDE ARMÉE 1805-1807

Augereau took an interest in Simmer's well being and appointed him to his staff when he took charge of the Camp of Brest in January 1804. He was with Augereau when the formations at Brest became VII Corps of the Grande Armée on 30 August 1805. With Augereau throughout the Austrian Campaign, he took part in the advance into Bavaria and the Tyrol where the marshal defeated Jellacich's corps at Feldkirch in November.

In the Prussian Campaign of 1806 he fought at Jena on 14 October, distinguishing himself in the struggle for the Isserstadt Woods and later in the day in the capture of Kötschau. The campaign in Poland saw him in action at Augereau's victory at Kolozomb on 24 December and then two days later at the indecisive struggle at Golymin. On 8 February 1807 he was

with VII Corps when it blundered through a heavy snow storm into the teeth of the Russian guns that ended with its destruction at Eylau. The remnants of VII Corps disbanded, Simmer was transferred to the Grande Armée's general staff serving Berthier's deputy Le Camus. The move was opportune, for under Augereau his career had stagnated. Having remained a captain for nearly ten years he was promoted *chef d'escadron* on 12 February 1807. In the renewed campaign he fought at Heilsberg on 10 June and at Friedland on 14 June 1807, suffering wounds in both battles. On 19 March 1808 he shared in the awards distributed after two years of successful campaigning, being granted an annuity of 2,000 francs a year drawn on the revenues of Westphalia.

PORTUGAL 1808

In May 1808 he joined Junot's staff with the Army of Portugal in Lisbon. When the rebellion broke out he served at the siege of Evora from 30 July to 3 August 1808, then at Vimiero on 21 August. After the French defeat he was evacuated from Portugal with the rest of the army returning to France in November 1808.

THE DANUBE CAMPAIGN

Simmer served with the Army of Germany during the campaign against Austria on Tharreau's 1st Division staff with Lannes's II Corps. Present at Aspern-Essling on 21-22 May 1809 he was wounded in the knee. After his recovery he fought at Wagram on 5-6 July. In the awards after the campaign he received another annuity of 4,000 francs drawn on Rome. Promoted *adjudant commandant* on 24 August 1809 he became Tharreau's chief of staff.

After the withdrawal of the French armies from Austria he spent a period in garrison on the Elbe. Recalled to France in May 1810 his division spent two months at Nantes before being disbanded. He served for a time on Berthier's general staff and on 2 September 1810 received the title *baron de l'Empire*. Portugal and France 1810-1811

Simmer secured a posting with Junot's VIII Corps staff serving with Massena's Army of Portugal. He fought at Bussaco on 27 September 1810 and soon after was invalided back to France due to ill health. In January 1811 he took up an appointment as chief of staff with the 3rd Military Division at Metz. In March he moved to the 19th Military Division at Lyons leading a *column mobile* to hunt down deserters.

Again his career appeared stagnant till a vacant post as chief of staff with Compans's 5th Division of Davout's I Corps of Observation of the Elbe came up on 30 September 1811. It was a move his career needed as he had spent several years on the staff of commanders where he had made little progress.

THE RUSSIAN CAMPAIGN

In the Russian Campaign Simmer was very active. He fought at Saltanovka near Mohilev on 23 July 1812 when Davout just failed to envelop Bagration's army. On 17 August he took part in the capture of Smolensk. He distinguished himself during the storming of the Schevardino redoubt by Compans's division on 5 September 1812, but hit in the arm by a musket ball was forced to miss Borodino two days later.

On 8 October 1812 he gained promotion to *général de brigade*. During the retreat from Moscow he led the remnants of I Corps's 2nd Division after Ricard was wounded at Krasnoe on 18 November. He was lucky to survive the action unscathed having three horses killed underneath him during the day. Wounded at the crossing of the Beresina on 28 November he was forced to relinquish his command. Returning to duty on 15 December he joined Eugene's IV Corps serving on his staff as the remnants of the Grande Armée made its way through Poland.

SAXONY 1813

At Berlin on 12 February 1813 he took charge of a brigade with Charpentier's 36th Division of Macdonald's XI Corps. He served at Lutzen 2 May, Bischofswerda on 12 May and at Bautzen 20 May 1813. Recognized for the part he played in retreat from Russia and Poland he was made a Commander of the Legion of Honor on 4 May 1813.

During the Armistice he worked feverishly to prepare his brigade for the coming campaign. Virtually as the opening shots were fired when Macdonald moved against Blucher's Army of Silesia he was severely wounded at Goldberg on 23 August and took no further part in the Autumn Campaign.

FRANCE 1814

In January 1814 Simmer rejoined Macdonald's corps at Chalons-sur-Marne and was with him as it fell back to Meaux on 4 February. He fought at Mormant on 17 February and Ferté-sur-Aube on 28 February. With Macdonald when he evacuated Troyes on 4 March, he then took part in the defence of Nogent-sur-Seine on 17 March and fought with him at Provins the same day. His final action of the campaign was at Saint Dizier on 26 March when Napoleon cut east and fell on the Allies's communications, routing Winzingerode.

THE RESTORATION AND NAPOLEON'S RETURN

The Bourbons retained his services and on 23 June 1814 appointed him troop commander for the Department of Puy-de-Dome at Clermont-Ferrand. He received the title *chevalier de Saint-Louis* on 19 July 1814. When news reached him of Napoleon's return he was ordered by comte d'Artois to join him at Lyons with the 72nd Line. Arriving at the city he found chaos with news that the 5th and 7th Line had already defected. It left him with the stark choice to either flee or defect. Unlike Macdonald who headed the royal troops at Lyons, he chose the later course and joined Napoleon. His action secured his promotion to *général de division* on 21st April 1815 and command of 19th Infantry Division of Lobau's VI Corps.

THE WATERLOO CAMPAIGN

His division formed from the Paris garrison, left the city in the first week of June marching by stages to join the Army of the North on the Belgian border. His formation with the rest of VI Corps were the last to pass through Charleroi and on 16 June were late arriving on the Ligny battlefield. As part of Napoleon's reserve they did not see action during the day, only advancing after the Guard broke the Prussian line.

At Waterloo on 18 June in the early stages of the battle he remained in reserve behind La Belle Alliance. Around 1.30 P.M. when it became apparent the Prussians advancing from Saint Lambert threatened the French right he deployed with VI Corps before the Bois de Paris. From 4.30 P.M with Jeanin's 20th Infantry Division, they fought a stubborn and skilful delaying action against Bulow's corps faced by over four times their number. Gradually forced back to Plançenoit they rallied when reinforced by elements of the Guard under Duhesme. From there they grimly held the ground to the left of the village till after the final attacks by the Guard against Wellington's center failed. The courage shown by his and Jeanin's divisions fighting alongside the Guard went a long way to keeping the jaws of the Allied vice open and saving a large portion of the army when all appeared lost.

He remained with the remnants of his division when it reached Paris on 29 June. With an armistice declared on 4 July he withdrew from the city and retired with his troops behind the Loire. On 1 August 1815, the disbandment of the Army of the North completed, the new government overturned his promotion and he reverted to his rank of *général de brigade*. He was then dismissed and placed on the non active list without pay on 1 September 1815.

POLITICAL CAREER

Exiled from Paris in February 1816 Simmer was fortunate not to be brought to trial due his defection at Lyons, his friendship with Macdonald helped. He was reinstated as a result of the Army Reforms of 30 December 1818 but remained unattached, and held no post till his retirement from the army on 17 March 1825.

He took up a political career and after the elections of June 1828 represented Clermont-Ferrand as a leftist deputy. In June 1830 he was re-elected. The

July Revolution resulted in his recall to the army on 27 February 1831 and restoration of his rank to *général de division*, though he remained unattached. As the demands of leftist colleagues in the Chamber became more radical he often found his military career compromised. He took a more Centrist stance on many issues, which displeased those who voted for him and was a major reason why he lost his seat in the elections of 1834.

Undaunted, he moved to nearby Riom and gained election there in November 1837. He did not stand for re-election in 1839 and retired from the Chamber on 2 March 1839. His military career came to an end when at sixty-five was he was placed in the Reserve on 8 August 1841. In July 1842 he decided to stand once more as a deputy but failed in the election. He lived his remaining years near Clermont-Ferrand where he died on 30 July 1847.

ASSESSMENT - A SOLID RELIABLE PROFESSIONAL

Simmer was a classic example of the general officer that formed the backbone of the French army in Napoleon's day. He was the solid, reliable, uncontroversial professional. He saw his duty to his country, rather than to an individual hence being able to change allegiance without much controversy or difficulty. Whilst he only led a division for one campaign, at Waterloo he certainly proved worthy of the rank. His earlier career was dogged by his persistence in remaining a staff officer, which more often than not was with commanders who were past their best. After a comparatively short time as a *général de brigade,* first with Davout and then Macdonald, he soon showed he merited the rank and was genuinely unlucky not to have been promoted sooner.

THE DEFECTION OF THE 5TH AND 7TH LINE INFANTRY PERSUADED SIMMER TO GO OVER TO NAPOLEON

SOULT - JEAN DE DIEU, DUKE OF DALMATIA (1769 - 1851)

BACKGROUND AND EARLY CAREER

Jean de Dieu "Nicolas" Soult was born on 29 March 1769 - the same year as Napoleon and Wellington - in the small village of Saint-Armans-Labastide (Tarn) that nestled in the Black Mountains of Southern France. His family came from a long line of small merchants and local officials from Languedoc. As the village notary his father was never well off and when he died before Jean was ten the family fell on hard times. Jean's mother a women of great spirit tried to continue the notary business till he as the eldest son could take over.

Jean meanwhile, a physically tough thick-set youngster took over many of the family duties. It made him strong willed, tough and independent but also had a detrimental effect on his schooling. That it resulted in him only receiving a very basic education was a fact he regretted in later life. His mother also found him a handful. He had become the leader of a gang of village ruffians who were involved in all sorts of mischief. Partly to check his high spirits and to train him to take over his father's business, he was successively apprenticed to two notaries in nearby villages. Neither move was a success. In the first village there was an even rowdier element than in Saint Armans, while in the second his employer was such a tyrant he ran away. On his return home he found the bailiffs in the house and took an instant decision to become a soldier. The prospect of becoming a village notary never attracted him, so he enlisted as a volunteer on 16 April 1785. For this he received ten louis, which he immediately sent to his mother to help pay off the family debts.

Soult served for two years with the Royal Infantry Regiment (the 23rd Line) stationed at St-Jean d'Angély near Rochefort (Charente). Not one to accept discipline easily his impatience and unruly behavior led to him being disciplined several times, on one occasion it nearly led to him appearing before a court-martial. In spite of such misdemeanors he showed talent and on 13 June 1787 was promoted *caporal*. He soon became bored and frustrated, and with little hope of ever gaining a commission he tried to obtain a discharge but was refused. Then as the Revolution gathered momentum and discipline within the army started to deteriorate, he returned home on indefinite leave in June 1790 and set up a bakery in Saint Armans. The business not a success,

and opportunities now open to all within the army, he returned to the 23rd Line stationed at Schelestadt on 31 March 1791 as a *caporal fourier* (quartermaster corporal).

From 1 July 1791 when he was promoted *sergent* Soult's rise was meteoric. In January 1792 he became a drill instructor with the 1st Battalion of the Haut Rhin National Guard with the rank of *sous lieutenant*. It was a post normally reserved for exceptional young men promoted from the ranks, who were given the task to train recruits for line regiments. He soon saw that the officers were inexperienced old men who showed little interest and within a few months Soult was effectively unofficial commander of the battalion with only the rank of acting *adjudant major*.

THE ARMY OF THE MOSELLE 1793-1794

The outbreak of war saw him on the Rhine where he emerged with a reputation as an excellent regimental officer covering Custine's withdrawal from Mainz. Later at Wissembourg on 29 March 1793 he led two battalions against the Austrians causing heavy losses. The *Amalgame* of 1793 saw his unit merge with his old regiment (the 23rd Demi-brigade) and its transfer to Hoche's Army of the Moselle in November 1793. Wanting to widen his experience he applied for a staff post and with Hoche's agreement joined Taponier's division staff at Zweibrücken. As a staff officer he was still leading from the front and on 8 December 1793 led a mixed force against the Austrians at Niederbronn, taking some 300 prisoners.

In January 1794 he moved to the army's advanceguard as General Lefebvre's chief of staff. It was an association Soult treasured and played an important part in developing his career, giving him a good grounding in staff work. Promoted acting *adjudant général chef de bataillon* on 7 February 1794, the rank was barely confirmed before another followed on 14 May 1794 as *adjudant général chef de brigade*. Present at Fleurus on 26 June 1794, he was at Lefebvre's side in the thick of the action having three horses killed under him during the day.

THE ARMY OF THE SAMBRE-MEUSE 1794-1799

Soult continued as Lefebvre's chief of staff when on 28 June 1794 the advanceguard became part of the Army of the Sambre-Meuse. His association with Lefebvre ended briefly after his promotion to *général de brigade* on 11 October 1794. An unsettling time followed as he was moved from division to division to fill gaps in the army's command structure. He first headed a demi-brigade of Hatry's division. That was followed with charge of a brigade in Desjardin's division at the siege of Luxembourg from 13 April till 5 June 1795. After the siege he moved to Poncet's division till his return on 6 November 1795 to

Lefebvre and the army's advanceguard.

During Jourdan's offensive into Southern Germany 1796 he distinguished himself on 4 June at Altenkirchen. At the head of a mixed force he drove back a strong force of Austrian cavalry and at the end of the day was largely responsible for the capture of over 4,000 prisoners. At Herbron on 15 June he found himself cut off by a superior force but managed to hold out till Kléber came to his rescue. Then together the two generals fought their way through the enemy to rejoin the rest of the army. During Jourdan's retreat to the Rhine after the defeat at Neresheim on 11 August he again came to the fore when on 29 August his brigade separated from Lefebvre, stubbornly resisted encirclement at Freiberg and Würzburg.

After the campaign he showed his fine grasp of strategy when he wrote a report largely blaming the government for the campaign's failure by not providing the army with an adequate commissariat. His comments helped deflect much of the blame from Jourdan who was severely criticized for advancing on too wide a front. It was the first significant instance of Soult showing he was his own man and not afraid to express his opinion.

In the new year Soult moved to Championnet's division near Solingen. The renewal of hostilities saw him heading the division's advanceguard during the Sambre-Meuse's last offensive. The army led by Hoche, with Soult and Ney* in the van, drove the enemy from Giessen and Steinberg on 21 April 1797 till the armistice declared soon after stopped the run of successes. Soult then spent a quiet period at Solingen before joining the Army of England in January 1798 as a brigade commander at Ghent. The army was little more than a diversion to distract attention away from Bonaparte's expedition to Egypt and was soon disbanded. Soult found himself in September 1798 with the Army of Mayence attached to Saint Cyr's division.

Lefebvre secured his return to the advanceguard division on 9 February 1799. He fought at Pfullendorf on 21 March taking over the division when Lefebvre left the field wounded. Stockach on 25 March was the first occasion he handled a sizeable force in the field. Heavily outnumbered by the Austrians, he led the division sword in hand from the front in a violent attack against their flank. After the initial impact the enemy stood their ground. Then the Archduke Charles launched a vigorous counter-attack causing the French to give way. Jourdan broke off the battle and started to retire to the Rhine. It was a serious reverse for the French, for which Soult was in no way to blame. His initial attack almost succeeded, and during the retreat he provided a stout rearguard.

MASSENA - THE SWITZERLAND CAMPAIGN 1799

On 10 April 1799 Soult achieved his immediate ambition, a command with Massena's Army of Switzerland as head of its 4th Division. His rank of *général de division* was confirmed by the Directory 21 April 1799. The first task he undertook was to pacify the turbulent cantons of Schwyz, Uri and Unterwald in southern Switzerland. The situation there was serious, Frenchmen had been murdered and an Austrian army not far away at Engadine. Moving quickly, on 8 May 1799 he occupied Schwyz and set up his headquarters at Altdorf. Massena gave him a completely free hand to re-establish French authority in this desolate mountain region. It was a military cum political task for which he had become well suited, and using a mixture of firmness and conciliation as he had done in the Rhineland he was successful. The neighboring cantons of Uri and Unterwald were more of a problem. Soult was forced to wage a short vicious campaign for two weeks around the Saint Gothard Pass, which ended with a rebel defeat at Airolo.

In June Soult returned to Central Switzerland to help Massena in the Zurich area where the French were threatened by two Austrian armies under Archduke Charles and Hötze. Present at the first battle for Zurich on 2 June he held the right of the French line in front of the Zurich Beg. His force of some 7,000 men were engaged in fierce hand to hand fighting against twice their number of Austrians and suffered heavy losses. In the battle he used the novel tactic of mounting field guns on boats on Lake Zurich to enfilade the enemy positions. He held his positions but elsewhere Massena was unable to stem the Allied advance and had to evacuate the city. In Massena's reports to the Directory Soult's bravery, innovativeness and military skill received the highest praise.

In the operations that ended in the Second Battle of Zurich Massena showed he was at the time the finest general in Europe, but Soult's profile also rose in his wake. The Army of Switzerland starved of money and resources, held a tight defensive line along Lake Zurich that then stretched along the left bank of the Linth covered by Soult. Korkasov's Russo-Austrian army facing Massena had to be defeated before the arrival of Suvorov's forces fresh from their triumph over Joubert in Italy. Massena agreed to Soult's plan to take the offensive by crossing the Linth near the eastern end of the lake. The river with its marshy banks strongly defended by Hötze's Austrians, made a formidable obstacle. After a detailed reconnaissance Soult realized bridging the river was out of the question. To him a surprise amphibious operation was the only solution. On the night of 24-25 September he gathered a dozen large boats, several small skiffs and three low draft barges each capable of carrying a cannon on their bows. In addition he selected 160 swimmers led by his brother Pierre* that swam across the river in darkness and seized some defences without firing a shot. A small bridgehead established the rest of the division was ferried across. The plan achieved complete surprise and before the Austrians realized what was up they had lost control of the entire bank. Hötze died in the struggle and over 8,000 prisoners were taken. The river defences breached, Massena rolled up their line and by the end of the day Zurich had fallen with the remnants of Korkasov's army streaming back to the Rhine.

On 3 October 1799 Massena ordered Soult to Southern Switzerland with three divisions, the 2nd (Loison), 3rd (Mortier) and 4th (Gazan) to fall on Suvorov's columns marching from Italy. In a series of manoeuvres in the Saint Gotthard region he tried to bring the Russian to battle. Apart from a success at Glavus where he picked up 3,000 prisoners the Russians slipped away after suffering great hardship. He then returned north where his divisions were needed to secure the left bank of the Rhine between Schaffhausen and Constance. The campaign then drew to a close with the arrival of the winter snows.

Massena with Soult as his principal lieutenant had achieved great things. They had driven out with heavy losses a 70,000 strong army that threatened to invade France through Switzerland. One of its commanders was killed and the other Korkasov in disgrace. Suvorov with a second army had turned back and resigned his command. Within a few months Russia had left the Second Coalition leaving Austria alone to face a resurgent France. Massena considered France's foremost commander, in turn lavished praise on the ingenuity of Soult. He referred to Soult as a man who showed imagination and flair for daring enterprises, as well as a fine organizer of men who possessed a mastery of all arms.

THE ARMY OF ITALY - THE SIEGE OF GENOA

Other generals were after Soult's talents, and Moreau for one wanted him to join the new Army of Germany forming under him, which included elements from the Army of Switzerland. Soult however decided to remain with Massena who had moved to Italy, and on 13 December 1799 accepted an appointment with the Army of Italy. In early February 1800 he arrived in Genoa and was horrified by the state of the army that barely numbered 36,000 fit men. Placed in charge of the army's central sector he set up headquarters at Acqui. His first task was to restore his troops' morale, faced by 120,000 Austrians in Italy. His men covered the two main passes over the Apennines. The one led down from the upper Bormida and descended on Savona and Finale while the other the Bocchetta wound its way down to Genoa. With only 12,000 men he had an enormous area to cover from the city itself west to Savona.

On 5 April Melas opened the offensive with 50,000 Austrians heading up the Bormida. Another 10,000 under Hohenzollern poured over the Bocchetta, while a further 15,000 scaled the Trebbia. Soult and Suchet who was in charge of the other wing were driven back. Within a day the Army of Italy was split in two. He retired to Savona and from there to the safety of Genoa, while Suchet made his way to Borghetto. The city surrounded, the siege entered its second stage with French offensive and defensive operations mounted to protect its immediate environs. On 11 April Soult seized Mount Hermette but failed in an attack on Monte Inurea a week later. A renewed Austrian offensive drove him from Sassello and Mount Hermette on 20 April, which made French control of the port virtually untenable. On 30 April he led a valiant counterstroke at the head of the 73rd and 106th Demi-brigades, driving the enemy from the plain of the Two Brothers. From there he led the 3rd Demi-brigade in a separate assault on Fort Quezzi causing the enemy to loose over 4,000 men. Coupled with the arrival of a brig that broke the naval blockade and discharged valuable supplies, the recapture of these strongpoints gave the beleaguered garrison a new lease of life.

His luck ran out on 13 May after he pressed Massena to sanction an attack on Monte Cretto that would loosen the Austrian stranglehold and secure further supplies. Gazan led a diversionary attack from Fort Diamond, which failed when a thunderstorm drenched the field and impeded the attack. The enemy alerted turned on Soult's small force, which after taking Monte Cretto was driven back into the valley. Soult as a point of honor since he had advised the foray rallied the 3rd Demi-brigade and led it back against the enemy. The attack might have been successful had he not been cut down by a musket ball in the leg. His men tried to carry him away, but the enemy saw their prize, dashed forward and took him and his brother Pierre prisoner, while his disconsolate troops fell back to Genoa.

Soult robbed of his possessions on the battlefield, Prince Hohenzollern took care of him for a few days at his headquarters. Then he was passed to the rear and stayed at the Bishop's Palace in Alessandria where he had a better chance to recover from his wound. As he lay in bed Pierre gave him a running commentary of the Battle of Marengo, while watching its progress through a field glass from the window. In the confusion after the Austrian defeat he considered plans of escape but his wound became gangrenous and he nearly lost his leg. In the end realizing hostilities would soon end he gave his parole. With freedom of movement his health improved, in particular when his wife Louise arrived to care for him. Exchanged in August 1800 he completed his recovery in the mountain spa town of Acqui. The wound left him with a marked limp for the rest of his life. It also had a sobering effect, he became more cautious and from that time on led his men from the front on fewer occasions with such bravado.

YEARS OF PEACE (1800 - 1804)

After nine hectic years of battle, the tempo of Soult's life changed as he entered a period without a major combat command. On 1 September 1800 he moved to Piedmont as military governor where his brief was to re-establish French rule in the region. In February 1801 he became Murat's *lieutenant général* with the Army of Observation of the Midi based at Florence. He later moved to Taranto when the army occupied southern Italy in September 1801. In this inhospitable region he had three major tasks, the first of which was to build up good relations with the southern Italians and suppress the banditry. The second was to bring Taranto and other smaller ports in Apulia to a state of full defence so they could act as supply bases for the French army in Egypt. Thirdly he was to turn Taranto into a base for further overseas activities, in effect making it into an Adriatic and Eastern Mediterranean Gibraltar from where French designs could always be a threat.

He got on well with Murat, who in turn spoke highly of him to Napoleon. The First Consul was also intrigued by Soult and wanted to bring the young general into his inner circle. Not having shared in Napoleon's earlier successes, their paths had never crossed and by 1802 they had still never met. Soult in Switzerland and Germany had not been involved in the tempestuous politics of post Brumaire Paris where the years of the Consulate were riddled with conspiracy and treason. Many ambitious soldiers foremost among them Moreau, believed they had as much right to rule France as the First Consul. Soult on the other hand was among the few young generals who had given proof of their military skill, had real administrative gifts and most important of all was uncommitted politically. With such credentials and backed by the recommendations of Massena and Murat on 1 March 1802 he was appointed by Napoleon *colonel général* of the Chasseurs of the Consular Guard before they had even met.

Egypt lost and the army in Southern Italy being wound down, Soult returned to Paris and took up his new post on 1 June 1802. In the next eighteen months he became fully committed to Napoleon's cause and on 28 August 1803 became general in chief of the military camp at Saint Omer. From there in December he moved to the camp at Boulogne where his superb organizational skills were put to work to prepare for the invasion of England. His troops were to form the largest corps, numbering some 46,000 men, which later became IV Corps of the Grande Armée. His 1st Division forming the *camp droit* in Boulogne was led by Saint Hilaire and occupied the fisherman's part of the old town. The 2nd, quartered in the old town formed the *camp gauche* and was led by the tough foulmouthed Vandamme*. At Ambleteuse Legrand was stationed with the 3rd Division, while his colleague Suchet from his Genoa days led the 4th at Wimereux, and Margaron the corps cavalry. It was an impressive concentration of military talent.

THE MARCH TO THE DANUBE

Soult's troops were in superb condition when the Grande Armée broke camp on 26 August 1805 for the march to the Rhine. In the twenty nine day march to Speyer it is believed he did not lose a single man. Once across the river the situation changed, his men living off the land as it was the speediest way to move, stragglers increased. Morale remained good, as it stemmed from the discipline and trust built up between officers and men at Boulogne. His troops were the first to cross the Danube when Vandamme found the bridge intact at Donauworth on 7 October. After detaching Saint Hilaire to cover the crossing he headed for Augsburg, then passed through the Swabian Alps to Memmingen, which he reached on 14 October. He took over 5,000 prisoners en route and effectively cut the Austrians' southern escape route from Ulm.

Present at the Capitulation of Ulm a week later he then passed through Munich on 26 October. Operating under Murat with the Cavalry Reserve and Lannes's II Corps they narrowly missed bringing Kutuzov to battle. The pursuit was hectic and disorganized, Murat a brilliant cavalryman was not the man to lead combined arms. Personal rivalries between the marshals began to show, Soult never liked Lannes and his relationship with Murat also began to sour. After crossing to the north bank of the Danube his troops engaged Bagration's rearguard at Hollabrünn on 16 November before they took post near Austerlitz at the end of the month.

AUSTERLITZ 2 DECEMBER 1805

In the battle, Napoleon planned for IV Corps to bear the brunt of the Allied attack as little heavy fighting had fallen on Soult's troops since the start of the campaign. They occupied the center of the French line behind the Goldbach stream. Vandamme's and Saint Hilaire's divisions faced the Pratzen Heights while his third division, Legrand's held the two villages of Zokolnitz and Telnitz on the right of the French line. Legrand was joined by Davout's divisions as they formed up on his right during the day.

In the early hours of the battle Legrand bore the brunt of the fighting. His steadfastness coupled with the rigid discipline and fine training his troops received at Boulogne paid. Forced out of the villages before Friant's* division arrived, his line buckled but held and played a vital part in holding the French line before the main blow fell. It was on the Pratzen Heights where the battle reached a decision. At 7.30 A.M. Soult started to move against the enemy. His assault was one of the most memorable feats of arms by any formation of the French Army. The lower slopes of the Pratzen were concealed by a thick mist and smoke of French campfires as the two divisions, Saint Hilaire's on the right and Vandamme's to the left moved forward unseen. The elements hid the advance till the last moment when the battalion columns preceded by skirmishers reached the crest and emerged into brilliant sunshine. There they fell on the exposed flank of a huge Russian column moving across their front to support the attack on the French right.

Kutuzov saw the danger and hurriedly changed the direction of the main Russian thrust against the French right. His columns reinforced by reserves turned on Soult's forces on the plateau. Soult personally took control as a strong Russian counter-attack developed. Saint Hilaire's division under great pressure considered a partial withdrawal. The crisis passed when Saint Hilaire led a fierce bayonet charge that Soult supported by personally directing a battery of 12-pounder cannon on the enemy. Vandamme on the left, hotly engaged held his own and consolidated his position. At midday after Napoleon joined Soult at the little chapel of Saint Anton on the highest point of the plateau another crisis occurred. The final Allied reserve, the Russian Imperial Guard launched a last desperate assault. Their forces including the Chevalier Guard fell with devastating effect on one of Vandamme's brigades that had become separated. The 4th Line, which Napoleon's brother Joseph had once commanded during the Boulogne days lost its eagle. Reinforcements arrived from Drouet's* division of Bernadotte's I Corps to stabilize the situation and together with Bessières's Guard cavalry they broke the last Russian counter-attack.

The Allied center split and their army broken in two, Napoleon held a brief council of war. He then directed Soult to launch Saint Hilaire's and Vandamme's divisions upon the flank and rear of the main army, which was still heavily engaged against Davout's corps and Legrand's division. The Russian and Austrian troops made desperate attempts to cut through the ranks of encircling French, but hemmed in were doomed. Driven back against the Satschan Pond the defeat became a disaster, their casualties mounting to some 15,000 killed and wounded with a further 12,000 prisoners. The French lost some 7,000 men. It was Soult's handling of his three divisions, one in a critical defensive role, the other two leading the great attack on the Pratzen that made him the main architect of Napoleon's finest victory. Napoleon concurred, and not one to easily ascribe such glory

to others enthusiastically referred to Soult as the "first tactician in Europe". It was undoubtedly Soult's finest hour.

THE PRUSSIAN AND POLISH CAMPAIGNS

After the Treaty of Pressburg Soult remained in Austria and southern Germany with IV Corps for six months to ensure the peace conditions were fulfilled. The outbreak of war against Prussia in October 1806 saw his corps form the right wing of the Grande Armée's march into Saxony. At Jena on 14 October 1806 he crossed the Saale and led the attack against the Prussian left whilst the corps of Lannes and Augereau advanced on Jena from the south. At the head of Saint Hilaire's troops he drove a large force under Holtzendorf from Lobstadt. A potentially fine enveloping movement was at hand when he had to halt and commit IV Corps to the main battle around Vierzehnheiligen after Ney in a rash move found himself cut off.

In the pursuit across Prussia Soult's forces moved with exemplary speed. On 6 November he took part in the storming of Lubeck where Legrand's division in particular distinguished itself. The next day with Bernadotte he received the surrender of Blucher's forces at Schwartau. He rested his troops a week at Berlin before continuing the march into Poland via Frankfurt-on-Oder and Posen. After crossing the Vistula he spent January in winter quarters north of Warsaw with his headquarters at the small town of Passnitz. His 22,000 men still the largest corps in the army suffered terribly first from mud, then snow and ice. Under such conditions far from home morale collapsed, a foretaste of 1812 which Napoleon failed to heed.

At the beginning of February Napoleon forced to resume the offensive, Soult formed part of the army's main thrust. The objective was Konigsberg capital of East Prussia. A major action soon developed at Jonkowo near Allenstein on 3 February 1807 when Soult tried to pin the enemy by cutting through Bergfried to launch a holding attack on Bennigsen's front. He met strong resistance from a Russian division before the town and it was not till darkness that he was able to cross the Alle. Forewarned by the speed of the French concentration and with Soult in possession of the town Bennigsen withdrew during the night. Soult dogged the Russian retirement fighting at Hof on 6 February. Then the next day after a confused struggle lasting several hours he secured the small town of Eylau.

It was the terrible weather that was largely responsible for the fateful Battle of Eylau on 8 February 1807. Neither side really wanted to fight in such conditions, but both commanders were prepared too simply for the shelter Eylau provided. From there it escalated as both Napoleon and Bennigsen saw a chance to overwhelm the other. Soult's three divisions first in line held the French center. In the early hours Saint Hilaire suffered heavily as his division bore the full fury of the Russian artillery. Augereau then launched his corps into the battle to relieve the pressure on Saint Hilaire. Disaster struck when caught in a blinding snowstorm Augereau's VII Corps veered off line into the teeth of the Russian artillery. Seventy guns in minutes reduced Augereau's formations to a shambles and left a gaping hole in the French line. Soult's other two divisions exhausted by the previous day's fighting, could do little to help. Also faced by Russian reserves of unknown number, any movement by them endangered the whole French position before the arrival of Ney's and Bernadotte's forces. In the end Murat's cavalry saved the day when they checked the Russian advance against the French center, while Davout's threat to the Russian left drew further forces from Soult's threatened sector.

In the battle Napoleon and Murat, severely shaken by the destruction to the army, considered retreat. Such a move meant abandoning the wounded and would have had a disastrous effect on morale. Soult was the one who kept his nerve in the crisis and made the celebrated remark by reminding all that the Russians were the same position and that, "French bullets were not made of cotton wool". He urged the army stand and wait the arrival of Ney and Bernadotte. They did and in the morning the news came in of the Russian withdrawal.

The losses at Eylau were horrendous, the offensive was abandoned and Soult retired with the rest of the army behind the Passarge. By June with new drafts from France he had rebuilt his strength to 30,000 men. In the new offensive with Murat he fought Bennigsen in an inconclusive struggle at Heilsberg on 10 June. Then directed to march on Konigsberg with Murat and Davout he missed Friedland on 14 June 1807. News of the French triumph destroyed the morale of Konigsberg's defenders, and on 16 June Soult entered the city taking a huge quantity of stores as well as being left to care for thousands of enemy sick and wounded.

After the Treaty of Tilsit Soult became military governor of Old Prussia, the region that stretched from Konigsberg to the Oder. His headquarters first at Konigsberg, he then moved to Ebling and later Stettin. He was later appointed military governor general of Berlin in October 1807. In both posts he went a long way to help make the French occupation as tolerable as possible. It was not forgotten. Remembering his consideration, after Waterloo the Prussians allowed him to settle in the Rhineland when he was exiled from France.

It was the time for rewards and Soult was showered with orders and decorations, which included the Order of Saint Hubert from Bavaria and the Seraphin of Sweden. Other titles and annuities came his way, including one of 45,000 francs a year drawn on the

Grand Duchy of Warsaw. In all his official titles and awards gave him an annual income of over 300,000 francs. When compared to an average soldier's annual pay of 250 francs it was a considerable sum. The most important honor he received was that of Duke of Dalmatia on 29 June 1808. Whilst flattered at being made a duke, he was disappointed it since he had no claim to that remote outpost of the Empire. More significantly it didn't commemorate any of his feats of arms in particular Austerlitz.

NORTHERN SPAIN AND CORUNNA 1808-09

On 3 November 1808 Napoleon appointed Soult as commander of II Corps of the Grande Armée in Northern Spain, replacing Bessières who had proven woefully inadequate. When he took over II Corps numbered some 18,000 men made up of three infantry divisions led respectively by Mouton*, Bonnet and Merle, plus light cavalry under his former aide de camp Franceschi and the attached dragoons of Milhaud*.

Napoleon had sharply criticized his brother Joseph for the abandonment of Burgos and regarded its immediate recapture essential to his reconquest of Spain. Soult wasted little time and within a few hours of taking over from Bessières his cavalry was in close contact with Belvedere's Army of Estremadura. Near the village of Gamonal on 10 November he brushed aside Belvedere who lost 3,000 men and the next day he entered Burgos. He then cut north and defeated Blake's Army of the Center at Reynosa on 12 November while Victor followed up with another crushing blow at Espinosa. Blake effectively destroyed fled along the mule tracks into the mountain region of the Asturias and Leon while Soult marched on and occupied Santander on 17 November. He left a division to hold the port and then returned to Reynosa. Here he took on the dual role of covering Napoleon's communications while keeping an eye on Romana's army in Galicia and Baird's British force that had landed at Corunna.

In early December with Napoleon and the main army at Madrid, Soult found the strategic situation had changed in Northern Spain. His corps scattered, he had only some 11,000 men immediately available to face the combined forces of Romana and Moore bearing down on him. In real danger of being overrun he acted without authority and ordered up some of Napoleon's reinforcements from Burgos. It was a fortuitous move, for when Moore's cavalry overturned his horse at Sahagun on 20 December he had men available to contain the threat.

Moore, later threatened by Napoleon's hurried advance from Madrid, on 25 December gave orders for the retreat to Corunna. Soult, now forming the French van, pushed his men to the limit trying to catch Moore in order to bring him to battle. It was during this march through atrocious weather conditions that Soult's title of the Duke of Dalmatia was turned by the wry humor of the hard pressed British soldiery into that of the "Duke of Damnation".

Soult fought Moore's rearguard at Astorga on 1 January 1809, again at Lugo on 6-8 January and finally brought him to battle before Corunna 16 January. Reduced to fewer than 12,000 infantry, 20 guns and 3,200 cavalry of limited use in that terrain, he had little hope of success against Moore's 16,000 men in defensive positions. He felt duty bound to try and, after suffering 1,500 casualties against some 800 English, failed to breach the line. Moore mortally wounded died that night as Hope pressed on with the army's embarkation. He had almost finished by the morning of 18 January when Soult's guns came into action from the cliffs overlooking the bay. In the confusion that resulted four transports ran aground and the evacuation was delayed till Royal Navy warships silenced the French guns. On the night of 19 January all remaining supplies were put to the torch and Soult entered the port the next day. He followed up his success by capturing the Spanish naval base and arsenal at Ferrol on 27 January after the English tried to dispute the port with a naval squadron.

Soult acquitted himself reasonably well with his first major independent command. In spite of not delivering a final coup de grace at Corunna, he had humiliated Moore's army and caused it to arrive home in a terrible state. Such were the fortunes of war few would realize that within four months Soult would also lead his army back from Portugal in equal disarray hounded by a new expeditionary force under Wellesley.

THE INVASION OF NORTHERN PORTUGAL– MARCH 1809

Immediately after Corunna Soult received very precise orders from Berthier for the next stage of the campaign - the conquest of Portugal. He was to play the lead role by advancing through Northern Portugal to Oporto and then Lisbon. For the task his forces were increased to four infantry divisions commanded respectively by Merle, Mermet, Delaborde and Heudelet, two heavy cavalry divisions under Lorge and Lahoussaye, and light cavalry under Franceschi. These additions increased II Corps's strength to about 30,000 men after the losses of the recent campaign.

The timetable Berthier supplied was hopeless, being based on the assumption that the march to Lisbon would be a mere promenade with Soult expected to reach Oporto on 5 February and Lisbon ten days later. No provision was made for the fact that Soult needed to rest his men and make provision for Ney's corps to occupy the territory he vacated. It was not till mid February that his men moved up to the River Minho, the frontier between Spain and Portugal. The river in flood and without a pontoon train he had to cut inland. The march dogged by guerillas, he defeated

La Romana at Monterey on 4 March and again at Verin the next day before crossing the Minho and seizing Chaves on 12 March. He then headed southwest over narrow mountain roads towards the old cathedral city of Braga. He fought at Carvalho on 17 March and before Braga three days later. The stubborn resistance he faced from recently raised Portuguese units well provided with artillery and helped by swarms of armed militia surprised him.

When Soult reached Braga he found the city deserted and Freire the governor who was prepared to negotiate, murdered. It took days to restore the confidence of the populace and bring the city back to some form of normality. The episode shook him, as he had seriously underestimated the scale of Portuguese resistance. His lack of sound military intelligence in the face of a hostile population threatened to be as serious in Portugal as it was in Spain. The hearts and minds of the people had to be won. To do this when his troops resumed their march on Oporto they inundated the countryside with proclamations of goodwill. He sent a message ahead of the army urging the bishop of Oporto who was the city's tyrannical unofficial ruler to surrender in order to calm the populace. When he received no reply he sent a similar message with General Foy* who spoke Portuguese to treat with the city authorities. It did not stop Foy being roughly handled and immediately thrown into prison.

With that, all hope of a peaceful settlement went and he was left with no alternative to launching a full-scale assault on Oporto on 29 March with Merle's, Mermet's and Delaborde's divisions. They fought their way through the city street by street till they reached the pontoon bridge over the Douro. Maddened by the futility of street fighting when his soldiers came across the bodies of badly mutilated French prisoners it caused them to run amuck. The city sacked end to end a wave of atrocities and reprisals followed before Soult restored order. Apart from the civilian deaths the victory was overwhelming, the Portuguese lost some 10,000 men, nearly 200 guns and vast supplies of ammunition. In addition over 30 English ships loaded with supplies were captured in the harbor. By comparison Soult's losses numbered no more than 80 killed and 300 wounded.

KING "NICHOLAS" - THE ARGENTON CONSPIRACY

Soult abandoned the immediate military option to continue the advance on Lisbon. Resistance had proven far stronger than expected, and he simply did not have the troops available. Rumors abounded of a second English landing at Lisbon to reinforce the troops still there. His communications with Spain were nonexistent and he had no news of Victor's I Corps that was to advance from Western Spain in support. He instead remained at Oporto to consolidate both his military and political position in Northern Portugal. Oporto was Portugal's second largest city and in some respects more important than Lisbon itself. In Soult's favour there was a decidedly hostile feeling within the city towards the royal house of Braganza he wanted to exploit. The influential trading families also showed an intense dislike towards English merchants who for so long dominated the port wine trade in Oporto. Prepared to first exploit these weaknesses to the full, only after the arrival of reinforcements did he plan to resume his march on Lisbon.

He soon won support by confiscating English businesses and handing them over to locals. The ill will generated by Junot's occupation two years before gradually dimmed in peoples' minds. With the feared Bishop of Oporto a fugitive in Lisbon and the Braganzas fugitives in Brazil a political vacuum developed. There was a move by leading citizens to accept a measure of French protection rather than missives from the ramshackle government in Lisbon. More important the simple people also began to accept the French. They were treated with consideration and freedom of religion was respected. A Portuguese National Guard was set up and law and order returned to the countryside where people had fallen prey to the excesses of guerilla bands. A French party

was formed that advocated detaching Portugal from its British connections and aligning itself more closely to France.

Soult's critics suspected he planned to make himself head of state. The possibility of him even being offered a crown entered peoples' minds. He was a realist and did not discourage the rumors, Napoleon had encouraged him to use whatever means to detach Portugal from England and bind her closer to France. If the acceptance of a Portuguese crown was involved, either for a member of the Bonaparte family or even himself, he was prepared to go along with it provided it had the Emperor's blessing. His generals unfortunately read much more into it. They feared that with Soult about to carve himself a kingdom, it would leave them in the lurch far away from prospects of further glory in a remote outpost of the Empire. Such feelings were understandable, for Soult was a dictatorial man by nature, and one his biggest failings was an inability to level or communicate freely with his generals.

A far more serious affair that undermined Soult's position at Oporto was the d'Argenton conspiracy. As Wellesley moved up from Lisbon he knew Soult's every move largely due a French traitor, an ex-Jacobin called d'Argenton who served as a captain in the 18th

SOULT'S TROOPS IN ACTION AT OPORTO

Dragoons. This officer had apparently crossed the lines, met Wellington and claimed that the French forces in the Peninsular were seething with discontent. He also claimed there was a plot to cooperate with the British in a march of mutinous officers and men from the French armies in Spain and Portugal to the Pyrenees. Here they would join with other disaffected French armies in an attempt to move on Paris and dethrone the Emperor who was at the time in Austria. D'Argenton even suggested that Soult would join the insurrection if it gained support. Another variation of d'Argenton's story was that Soult's army contained two factions. The first he claimed, which included Soult was violently anti Napoleon while the other headed by Loison and other senior officers wanted to take action against Soult because he planned to make himself king. D'Argenton for his part hoped in return for a large sum of money Wellington would help the Portuguese to back Soult's royal ambitions.

Arrested on 8 May, d'Argenton's allegations greatly exaggerated and distorted came to Soult's attention at a critical time. Wellesley's army had pushed his outposts on the Vouga back to Oporto. For a vital few days he lost confidence in himself, becoming suspicious of those generals he disliked and distrusted. Yet there was no shred of evidence to indicate that either he or any of his generals were implicated in any of the plots mentioned. He had fallen prey to possibly one of the finest pieces of disinformation of the Napoleonic Wars.

THE RETREAT FROM OPORTO - MAY 1809

The brief campaign that followed was a disaster for Soult. Despite his careful preparations to defend the lower reaches of the Douro, Wellesley fooled him and stormed across the river above the city on 12 May. Soult's veterans were the victims of a magnificent surprise attack. So precipitate was their retreat that he was forced to abandon his sick and wounded as well as much of his artillery. His position became more desperate when Loison covering communications with Spain allowed Almarante to fall to Silveira's and Beresford's forces. Hemmed in against the Sierra Catalina with the enemy occupying both ends of the road at Oporto and Amarante his only possible line of retreat was northward across the mountains towards Galicia. A shepherd fortunately told him of a rocky track through the Catalina known only to locals. Immediately he ordered his troops to destroy the remainder of the artillery, burn the baggage and abandon the wounded who could not make their way on foot. With only their muskets his long column scrambled their way over the mountains and joined Loison's force at Guimaraes.

Wellesley continued to dog his retreat by taking Braga while Silveira blocked the alternate route via Chaves. Again Soult cut northward over narrow mountain tracks. Porto Novo on the Rio Cavado was seized from a strong force of Portuguese in a daring night attack. After crossing the river the army struggled on pursued by Wellesley and the Portuguese. On 19 May he crossed the Spanish frontier near Orense, some three months after he had first entered Portugal. His army reduced to 19,000 strong had left some 6,000 men behind, killed, wounded and prisoners. He was without guns, ammunition or stores. His men half dead with exhaustion were in tatters, many without muskets. The army so easily could have met the same fate as Dupont at Baylen but through Soult's determination and skill it survived. For all his failings in allowing himself to be outmaneuvered on the Douro, Soult's retreat was an achievement ranking with Moore's to Corunna four months before and revealed in him a toughness few expected.

WESTERN SPAIN AND THE TALAVERA CAMPAIGN.

Ney should have been able to help Soult's troops as they poured back into Galicia. Unfortunately Soult's high-handed overbearing manner demanding provisions and support alienated the touchy Ney. Little or no help was offered, which resulted in mutual recriminations between the two. The most damning consequence of their failure to co-operate resulted in the French withdrawal from Galicia. Soult started the rot by setting up headquarters at Zamora in Old Castile to prepare his corps for a new struggle against Romana's army. It left Ney isolated and denuded of troops in Galicia to the extent that by July he had evacuated the province.

In spite of these misfortunes and wranglings Napoleon still considered Soult the most able of the marshals in Spain. Orders soon came from him via King Joseph for Soult to take charge of the three corps in western Spain. He had little difficulty dealing with Mortier in command of V Corps, but Ney with VI Corps continued to be the main thorn in his side and little was achieved. At the same time he tried to convince Joseph that the occupation of territory was not paramount to wining wars. More important was the necessity for the three corps in Western Spain work together to deal with Wellesley.

In July 1809 Soult managed to concentrate the three corps at Salamanca ready to re-enter Portugal via Ciudad Rodrigo. Wellesley forestalled his plan when he made a lunge across the border with the intent of joining Cuesta's Army of Estremadura in a march on Madrid. Outnumbered and outflanked Victor's I Corps centered on Placensia fell back. Joseph panicked when Madrid appeared threatened by Cuesta's army cutting behind Victor. As a result Victor was ordered to fall back on the capital rather than hold his original position. The move prevented Soult offering immediate help as the center of operations moved away from him. Joseph changed tactics when

Victor's and Sebastiani's forces united and on 24 July turned on Cuesta routing a force of 6,000 Spanish at Santa Ollalla. Joseph's confidence renewed by the success as the Spanish fell back to join Wellesley at Talavera, he took the fateful decision to engage the two armies before Soult's arrival. The result was the Battle of Talavera on 28 July 1809, a battle fought by the French at quite the wrong time and in the wrong place.

After a brutal struggle Joseph's army was repulsed and retired behind the Alberche. Wellesley for his troubles received the title Viscount Wellington of Talavera, but still remained in a very vulnerable position. In the days that followed he nearly came to disaster as Soult's three corps tried to cut him off from Portugal. Mortier scattered a Portuguese force under Robert Wilson in the Banos Pass and entered Placensia forcing Wellington to abandon his sick and wounded. Realizing that Soult was between him and Portugal and his way down the Tagus valley was blocked, Wellington crossed to the left bank of the river at Arzobispo. Soult followed and seized the crossing from a force of Spaniards on 8 August. With skill Wellington moved his army away from the Tagus by difficult mountain roads to Truxillo. Ney on the right bank was sent to intercept him, but let Soult down when he failed to find the ford at Almaraz. Had Ney crossed the river, Wellington boxed in on both sides had little chance of escape.

Wellington overcame the most exacting crisis of his Peninsular career, reached Badajoz and finally crossed into Portugal. Soult wanted to continue the campaign by marching along the north bank of the Tagus to Lisbon but Joseph vetoed it. The King was again preoccupied with the safety of Madrid when Victor was detached to face another threat from La Mancha. Surprisingly Napoleon supported Joseph, he had become wary of poorly prepared marches into Portugal and perhaps had in mind leading an invasion himself the following year. Soult was disappointed and it certainly soured his relationship with Joseph, which at the best of times was never good since the days when Joseph served under him as a *colonel* at the Camp of Boulogne. From a military standpoint a great opportunity was lost. Wellington had been totally out maneuvered, his army forced south of the Tagus Lisbon lay wide open. There were no defences of note for Soult to overcome like the Lines of Torres Vedras that were Massena's undoing two years later. Such an opportunity never presented itself again. For the next two years Wellington remained obstinately and wisely in Portugal consolidating his position.

KING JOSEPH'S MAJOR GÉNÉRAL

For the second time in six months Soult's fortunes reached a low ebb. His advice was ignored and his well laid plans that promised so much had come to nothing. Napoleon also turned on him in a series of critical letters to him and Joseph. When it appeared he was totally out of favour a letter arrived from Schönbrunn dated 20 September 1809 appointing him King Joseph's *Major Général* replacing Jourdan. The defeats required a shake up in the command structure. Furthermore Jourdan had shown signs of subscribing to too many of Joseph's "Spanish nationalist" ideas and to stiffen his brother's resolve Napoleon needed a man of Soult's proven loyalty and ability.

Three main tasks faced the French when Soult took up the post. The first was to complete the conquest of Spain, the second to reconquer Portugal and lastly to drive out the English. Hardly had he taken over in Madrid when another crisis loomed. In La Mancha Areizaga with some 60,000 men of the largest Spanish army still operating, in late October made a confident thrust through the passes of the Sierra Morena towards Toledo and Madrid. The French forces widely dispersed,, Victor's I Corps in La Mancha fell back before the Spanish advance. Soult's old II Corps led by Heudelet was on the lower Tagus between Almaraz and Alcantara covering Wellington still licking his wounds on the Guadiana. Mortier's V Corps moving up from Talavera to Toledo was the only major force between the Tagus and Madrid. A decisive move by Areizaga across the river at Aranjuez could have led to the fall of the capital. Areizaga then lost his nerve as rumors of Victor about to fall on his flank caused him to dither for three days at Aranjuez before he fell back to Ocana.

The respite gave Soult time to concentrate Joseph's forces and go on the offensive. On 18 October his cavalry led by Milhaud crossed the Tagus at Aranjuez and drove in the Spanish outposts. Areizaga rather than retire decided to give battle. He felt confident as Soult had only 32,000 men with him and the Tagus to his back. The advantage in numbers he possessed was soon thrown away by some foolish dispositions. His left flank 15,000 strong was placed behind a deep ravine, the Rivin de la Canda, the only physical feature in the area, which effectively prevented them from taking any part in the battle. The battle itself began with a heavy French artillery bombardment. Mortier leading the infantry attacked the Spanish frontally. The Spanish infantry committed, Milhaud's cavalry followed with an onslaught against Freire's cavalry on the Spanish right. Soon overwhelmed Milhaud then literally rolled up the Spanish line. It was a disaster for the Spaniards who lost some 5,000 killed and wounded, and over 20,000 men captured. The French losses were a little over 1,700 killed and wounded mainly from Mortier's infantry. The victory was probably the finest French success of the Peninsular War largely due to Soult's meticulous planning.

THE CONQUEST OF ANDALUSIA

After much advice to the contrary Soult convinced Joseph the conquest of Andalusia was paramount before any further moves against Portugal. The merits of the move were debatable. Oman in his work sharply criticized Soult for the great harm the conquest of Andalusia later did to the French cause in the Peninsula. Without the benefit of hindsight, at the time it appeared correct since after the return of the French armies from the Danube huge reinforcements were pouring over the Pyrenees. Andalusia was the richest province of Spain and the armies in the south were broken after Ocana.

The invasion of Andalusia started on 10 January 1810 when the French marched through the passes of the Sierra Morena. Their march was little more than a progress with the Junta in Seville fleeing to Cadiz. In less than three weeks Joseph and Soult accompanied by a glittering array of generals on 1 February entered the ancient capital in triumph. Soult make a great mistake not urging Joseph to head straight for Cadiz instead of wasting time savouring a triumphant entry into Seville. Cadiz was poorly defended and in a state of great confusion, and a determined French attack could easily have succeeded. By the time Victor arrived before the city and summoned it to surrender on 5 February it was too late. The defences had been hurriedly repaired and English and Spanish reinforcements had arrived. Too strongly defended for an immediate assault to succeed Victor settled down to a siege. Cadiz never fell and the bulk of Victor's corps was to spend the next two years before the city.

At this time in March 1810 French fortunes were at their height in Spain. Except for Cadiz and the inaccessible reaches of the Sierra Nevada open resistance in Andalusia had ceased. The Andalusians who had spent centuries longer than the rest of Spain under Moorish rule, seemed prepared to accept a ruler like Joseph who appeared to display overwhelming power. They turned out *en masse* to welcome him and he in turn was equally captivated by them. Soult was not as easily convinced, these were the people who slaughtered the French at Baylen eighteen months before.

RULER OF ANDALUSIA 1810-1812

From May 1810 when King Joseph departed for Madrid until the French left Southern Spain in August 1812 Soult reigned over Andalusia as viceroy in everything but name. His military and civil independence from Joseph was established by imperial decree when Napoleon made him commander in chief of the Army of the South on 14 July 1810. The order confirmed his authority over the three army corps of Mortier, Victor and Sebastiani numbering on paper some 100,000 men. Guerilla activity continued intermittently in Southern Spain but never gripped the country as in the northern reaches of Galicia, Castile and Leon. Units of Sebastiani's IV Corps engaged them in Grenada and the Sierra Morena but never cleared the region, however they did manage to contain the problem and large parts of Andalusia remained peaceful. The transport of supplies was rarely interrupted and couriers could travel virtually without escort. It testified to Soult's abilities not only as a fine field commander but also as an astute politician and organizer of civil affairs. Much of the success was due to the way he exercised control over his senior officers who had strict orders to maintain good relations with civil authorities and the population in general. Compared to the rest of Spain the economy and trade thrived. Within a year French power appeared stable and looked permanent. The intellectuals and middle classes, who generally had some qualified admiration for the ideals of the French Revolution as did the nobility in Seville and Cordoba, went over to the French in considerable numbers.

One problem was Soult's relations with Joseph, which became increasingly strained. Joseph was jealous of Soult and resented Napoleon's favorable treatment of him. Soult was Joseph's representative, yet the Emperor treated Soult as if he was his own representative ruling a large slice of his brother's kingdom while ignoring directives from Madrid. Soult took advantage of the situation and aggravated the situation further by referring direct to Paris and ignoring Joseph when he needed advice.

MILITARY OPERATIONS WESTERN SPAIN

In Southern Spain Soult's military operations covered three main areas, Estremadura in the west to the Portuguese frontier, the siege of Cadiz and operations in the mountainous regions of Grenada and Murcia. In the Summer and Autumn of 1810 operations in Estremadura were mainly the responsibility of Mortier's V Corps supported by Reynier's II Corps of the Army of the Center on the Tagus. The continual complaint Soult received from his generals was a shortage of manpower, including Mortier who was generally the least troublesome of his commanders. Generally Mortier did well. He attacked and defeated Romana's Spaniards at Villa Garcia, forced a small maritime invasion force to re-embark at Moguer near the Gulf of Cadiz, and operated with success between the Rio Tinto and the Portuguese border. The position looked so secure that for a while in the southwest Soult was able to rely on locally recruited Spanish troops to defend Seville when he later took to the field to support Massena in Portugal.

In January 1811 Soult belatedly started to support Massena in Portugal by marching on Badajoz, which dominated the southern route to Lisbon. His plans at first went well and after a short siege from 11-22 January he achieved his first objective when the fortress of Olivenza fell. His communications through Estremadura secure he then arrived before Badajoz on 26 January with Mortier's V Corps. The fortress proved a tougher nut to crack than expected and with only Girard's' and Gazan's divisions available he was barely able to maintain a blockade. Fortune then smiled on him at Gebora on 19 February when with Mortier they defeated a relief force under Mendizabel inflicting some 2,000 casualties and taking a further 5,000 Spanish prisoner. The garrison at Badajoz clung on grimly till its surrender on 11 March 1811 when the inhabitants urged the governor Imaz to do so after the walls were breached.

The success was intended to relieve some pressure on the Army of Portugal but it all happened too late. Massena four days earlier had ordered a retreat from the Lines of Torres Vedras. Generally criticized for not offering Massena help sooner, Soult's presence would have made little difference in terms of men and materials since after detaching men to secure his communications in Estremadura he only had a handful of troops available. What was of greater significance, and Soult never appreciated it to the full, was that his mere presence on the border with a large force was enough to severely restrict the scope of Wellington's operations. Problems arose in Andalusia that required Soult to return to Seville after leaving a strong garrison at Badajoz. An English landing near Algeciras menaced Victor's forces before Cadiz while another Spanish army under Ballasteros threatened Seville. After a chase both forces made their way to the safety of Gibraltar.

At the end of April 1811 the French garrison at Badajoz also came under threat. With Massena forced back to the Portuguese border around Ciudad Rodrigo, Beresford's Anglo-Portuguese army became bolder and laid siege to Badajoz. Soult left Seville on 10 May and quickly gathered his forces in Estremadura to fall on Beresford and raise the siege. The plan fell apart at Albuera on 16 May 1811. Beresford with 20,000 men left the siegeworks and set up position at the village of Albuera to block the French advance. Unknown to Soult another Spanish army of the same size under Blake had joined Beresford. With some 24,000 men at his disposal Soult felt confident of success. In the morning he made a strong diversionary attack against the Allied left flank and the village of Albuera. His main attack fell on the Allied right, where concealed by olive groves he achieved complete surprise with three quarters of his army supported by Latour Maubourg's cavalry.

It was only after Soult was fully committed and his troops came up against the Walloon Guards did they realize they faced two armies. His generals, in particular Girard, fought magnificently. The Spanish offered unusually strong resistance but eventually gave way. It was then the devastating fire, first from Colborne's brigade before it was overwhelmed by French cavalry and then from Cole's division, that brought the French to a halt. Soult's reserves already committed in order to gain a decision, he fell back to the lower ground to regroup and renew the battle

the next day. The losses on both sides made it the bloodiest battle of the Peninsula War with over 8,000 French and 6,000 Allied casualties. When news arrived that night of Massena's defeat by Wellington at Fuentes d'Onoro he abandoned his plans and fell back to Llerena. Beresford's army also in no condition to follow up the success remained fixed to the high ground and then returned to the siegeworks.

Soult's critics made much of his defeat. The comment that he was a great planner, but lacked flexibility to amend them when faced by the unexpected was valid. Faced by such an overwhelming enemy he should never have fought the battle in the first place. Why had his intelligence failed? In spite of the huge disadvantage he possessed in numbers, the sheer brilliance of the plan nearly allowed him to win. Had Godinot on his right flank not prematurely broken off the action enabling Beresford to pour reinforcements south to counter the threat to his right, the result may have been very different.

On 16 June Soult and Marmont with the Army of Portugal united at Merida to relieve Badajoz. It was a good object lesson in what could be achieved when the army commanders co-operated. Faced by a combined force of 60,000 men Wellington with only 44,000 to face them decided to abandon the siege and retired on 19 June. From there the French operation lost its momentum as both marshals were wary of being drawn into Portugal. Their intelligence was poor, and they were unaware of the political as well as military problems Wellington faced with the Portuguese, Spanish and his government at home. Soult and Marmont soon started to quarrel over their differing military priorities. Soult believed in the complete subjugation of Southern Spain came before all else. Marmont felt Portugal was the main objective and that the defence and provisioning of Badajoz was Soult's responsibility while he dealt with Wellington. A compromise was reached when control of d'Erlon's old IX Corps returned to Soult in exchange for him accepting responsibility for Estremadura and Badajoz, while Marmont was to be responsible for operations north of the Tagus.

OPERATIONS IN ANDALUSIA AND SOUTH-EAST SPAIN 1811-1812

Soult began not only to face more serious military problems in Andalusia and south-east Spain but also problems with his commanders. The capture of Cadiz remained Soult's most urgent problem. Graham's amphibious excursion from Cadiz threatened the French rear areas and led to Victor suffering a costly defeat at Barrosa on 3 March 1811. At times when it appeared the siege was about to succeed, Soult had either to draw forces away to support Massena in Portugal or succor Badajoz. Apart from tying up most of I Corps it soured his dealings with Victor. The marshal became a liability, his outbursts were bad for morale and in the end he departed for France on health grounds in February 1812.

Before Victor's departure Sebastiani, Soult's lieutenant in Grenada and Murcia, had also been increasingly troublesome. The previous year IV Corps had occupied as much territory as possible along the coast between Malaga and Cartagena in an effort to link up with Suchet's forces in Valencia. The progress then stopped as Sebastiani incessantly complained of lack of men and support. Tired of such bickering Soult accepted his resignation in May 1811 and the more amenable Leval took over, bringing for a while more stability to the region.

Suchet's occupation of Valencia further east had the adverse effect of pushing the defeated Spanish armies into Murcia and Granada, which resulted in an increase of activity in the area. Soult headed an expedition into the region in August 1811, relieved the fortress of Niebla, then scattered the insurgents at Venta del Bahul on 9 August and at Baeza the next day.

Badajoz, Salamanca - The End in Andalusia

With the fall of Ciudad Rodrigo in January 1812 Wellington turned his attention to Badajoz and on 16 March invested the fortress. Soult received no help from Marmont as Napoleon had ordered him to persist with operations north of the Tagus and was before Almeida. On his own Soult gathered a force together to relieve the fortress but his approach hastened Wellington to launch a costly assault on 6 April and after a bloody struggle through the streets the ordeal of Badajoz came to an end.

The fall of Badajoz enabled Wellington to break out of the confines of Portugal and go on the offensive. Communications between the armies of Portugal and the South were split which caused the simmering dispute over the operational control of d'Erlon's troops to come to a head when Joseph ordered the 12,000 men to join Marmont. Soult was grievously offended since Hill's forces almost immediately began to move into Estremadura. He also sensed that this undermining of his authority was the start of a wider plan secretly negotiated by Joseph and the Cortes in Cadiz to evacuate Andalusia. Soult threatened to resign, but Joseph wisely did nothing as he needed him since Wellington was once more on the move.

The whole strategic position changed in Spain after Marmont's defeat at Salamanca on 22 July 1812. As Joseph evacuated Madrid, Soult refused to consider the evacuation of Andalusia and impertinently offered Andalusia as the rallying point for the French armies. Joseph ignored the advice and ordered him comply with the evacuation or resign. Soult's bluff called, for he knew the province was untenable, he immediately set in motion its evacuation. The outlying garrisons went first, Niebla on 12 August, Ronda and Medina Sidonia three days later. On 25 August

1812 the siegeworks at Cadiz were abandoned. He pulled out of Seville on 27 August and with him came convoys of civilians fearing for the future. Wagons were piled high with booty, Soult in particular had developed a taste for art treasures that were to adorn his homes in France. He fell back to Cordoba where he was joined by d'Erlon on 30 August who had retired from Estremadura. After a pause at Granada which they evacuated on 16 September the Army of the South some 45,000 strong linked up with Suchet's Army of Valencia at the end of the month.

Out of the turmoil of defeat there resulted a meeting between King Joseph, Jourdan, Suchet and Soult at Fuente la Higuera. More by accident than design three French armies were close at hand in Eastern Spain to turn on Wellington's overextended army. A plan thrashed out, Suchet remained in Valencia while Joseph's Army of the Center under d'Erlon, and Soult's Army of the South began to march north to link with the Army of Portugal. Wellington in the north faced the Army of Portugal at Burgos where he became bogged down in the siege of its citadel. Also faced by a revitalized French army outside the city and Soult approaching Madrid from the south, he raised the siege on 21 October and fell back to the Douro. Hill left to shield Madrid from the south could not hold Joseph's combined armies of some 65,000 men. Soult advancing via Albacete cleared the Tagus on 30 October and on 2 November re-entered the capital with Joseph. Hurled back from the Douro and the Tagus the two halves of Wellington's army had to march for their lives. Harried by Soult, after a series of punishing marches they made it to the line of the Tormes behind Salamanca, the original position where Wellington had confronted Marmont the previous July. With both sides exhausted and neither prepared to move, Soult established himself at Toledo for the winter months.

The Saxon Interlude - Spring 1813

The defeat in Russia was the reason for Soult's recall by Napoleon on 3 January 1813. Napoleon was woefully short of marshals of the first rank, as Davout was occupied on the lower Elbe, Massena was sick and in disgrace, while Ney was still recovering from his experiences in Russia. The reputation that preceded Soult meant Napoleon needed him to help rebuild the armies and hold himself ready for any special role he may be given. The order took time to transmit and was followed by Soult making careful arrangements before handing over, which meant he did not leave Spain till 6 March 1813.

Initially Soult was to assist the ailing Berthier with a view to succeeding him. He also took on a special task to reorganize the Old Guard infantry led by Roguet'. Then with Bessières's death on 2 May he took over his duties as senior officer within the Imperial Guard. He fought at Lutzen where his direction of the attacks by the Young Guard made an important contribution to the victory. At Bautzen he helped Bertrand˙ prepare the assaults by his old IV Corps against the Allied center where Blucher held the high ground behind the Spree. The attack was meticulously planned, with the first day spent building an elaborate earthwork on the Spree's east bank. Behind it the sappers then built pontoon bridges in order to cross the river. From their cover he planned for IV Corps to burst out and smash through the Allied center in a repeat of their storming of the Pratzen at Austerlitz. The circumstances were very similar, but Ney's and Lauriston's corps to his left were late joining the battle. The delay enabled Blucher to pour reinforcements into this threatened sector, which slowed the advance. Then when Ney began to envelop his right later in the day Blucher was able to conduct an orderly withdrawal.

THE PYRENEES AND SOUTHERN FRANCE 1813-1814

News of Joseph's defeat at Vitoria meant Soult's immediate return to the Spanish theater, replacing him as commander of all forces on 11 July 1813. He reorganized the defeated armies into a single Army of Spain made up of nine infantry and three cavalry divisions totalling some 85,000 men into three lieutenancies or wings. Each of the wings on the French side of the Pyrenees were composed of three infantry divisions and supporting cavalry. Reille˙ commander of the right wing was based at Vera covering the lower Bidassoa, d'Erlon took the center and Clauzel the left. Gazan who originally succeeded Soult as commander of the Army of the South resumed his post as chief of staff.

Soult did not believe in a static defence behind the Bidassoa and instead intended to march through the Pyrenees to relieve Pamplona. Once he succeeded with this surprise manoeuvre he planned to march westward, attack the main body of Wellington's army from the rear, relieve San Sebastian and then either press on to the Ebro or join up with Suchet in Catalonia. He first ordered Reille to lead the bulk of his forces quickly and quietly from the lower Bidassoa along the French side of the Pyrenees to join Clauzel's near Saint Jean Pied-de-Port. Thus the greater part of his army took position on the mountainous frontier ready to pass through the narrow defiles to Pamplona. Clauzel and Reille took the lead while d'Erlon on their right was to give full support. The concentration along narrow roads and in pouring rain went unobserved. It was a brilliant achievement considering Soult had taken over the demoralized army only thirteen days before.

The series of battles that followed, the Col de Maja, Roncesvalles and the First and Second Sorauren from 25-30 July was the last do or die effort to wrest the initiative from Wellington. As with so many other of Soult's plans the concept was brilliant but he failed in its execution. After suffering some 16,000 casualties his forces were driven back over the mountains. The relief of San Sebastian then became Soult's most immediate concern. On 31 August he made the mistake of launching Reille and Clauzel across the Bidassoa in two separate uncoordinated attacks rather than concentrate his forces. The planning was hurried and unworthy of Soult but the relief of the fortress had become an obsession and a matter of honor. He suffered some 3,800 casualties before forced to retire, including his division commanders Vandermaesen killed and Lamartiniere mortally wounded. The Allied lost half the number and on 8 September San Sebastian surrendered. The futility of Soult's offensive showed when only 1,300 Frenchmen marched out of the fortress to lay down their arms.

It took Wellington a month after San Sebastian's fall to feel confident enough to launch an offensive across the Bidassoa. When it came on 7 October Soult's careful preparations to meet it collapsed. Expecting Wellington to outflank the Bidassoa by an advance through the mountains via Saint Jean Pied-de-Port Soult set up extensive defence works in that region to secure his left. Wellington fooled him by making a personal appearance before Pied-de-Port, then to distract him further directed a diversion against Clauzel's forces. When night fell he left the scene and next morning appeared before Reille's sector on the lower reaches of the river where he directed his main thrust. Maucune's division was taken completely by surprise as a heavy artillery bombardment preceded the Allied attack. Two infantry columns poured across the river at low tide and the few troops posted at the low tide fords were overrun. Maucune still held the heights with his 3rd and 15th Line regiments but lost his nerve and without waiting for the rest of his division to arrive, beat a hasty retreat. Had Maucune held on Soult could have mounted a vigorous counter-attack and placed the Allied army in great difficulty. The onrushing tide would have split the army and made it hard for them to withdraw. Maucune instead allowed Wellington to consolidate his position and Soult's reserves when they arrived, unable to match them in numbers, opted to retire to the next set of defensive positions along the Nivelle.

Wellington needed to clear the Pyrenees before the onset of winter launched his offensive to clear the Nivelle valley on 10 November. Expecting an Allied attack along the coast Soult placed a full third of his 60,000 men along the narrow coastal strip at Saint Jean-de-Luz with the rest of the army scattered in defensive positions along the valley up to the Pyrenees. The main blow by five divisions fell on Clauzel's sector holding the mountainous Petit Rhun in front of the river. Supporting attacks by another three divisions were made on d'Erlon's positions holding the river-line itself. By midday the Light Division had taken the three forts on the Petit Rhun while Beresford with three divisions seized the strategic bridge at Amotz. A French counter-attack by Foy from Saint Jean Pied-de-Port failed to stop Hill's advance on the far right. A breakthrough achieved, Wellington wheeled left and rolled up the French line. Reille's divisions on the French right when threatened hurriedly destroyed their defences before Saint Jean-de-Luz and fell back towards Bayonne.

Again Soult had suffered defeat in spite of careful preparations. His problem was that a passive defence was no substitute for the first class soldiers that he lacked. Morale had collapsed, the generals were blaming each other and the three wings in the battle had operated with little or no coordination. Defeatism was rife, and it was sheer willpower on Soult's part that kept the army going.

He established a new strong position defending the area between the Bay of Biscay and the rivers Adour and Nive leading northward to Bayonne. In danger of being cooped up in a triangular salient Hill forced the Nive at Ustaritz on 9 December, while the rest of the Allied army launched distracting attacks towards the Adour and Bayonne. With Hill with five divisions on the west bank of the Nive, linked only by a pontoon bridge Wellington's army split and very vulnerable. Soult saw his chance and on 10 December started to move eight divisions from Bayonne to fall on the forces west of the Nive. He gained a measure of surprise before the Light Division held its ground on the ridge behind Bassussary. Desultory fighting continued for the next two days as neither side sought a major battle. Further west along the coastal strip before Bayonne Reille led a diversion against Barrouillet that broke through the Portuguese line. A desperate struggle around the *mairie* followed till the arrival of Allied reinforcements around 2.00 PM from St Jean-de-Luz, whereupon Soult called off the attack. The struggle around Bayonne settled down to intermittent skirmishing for the next two days as both sides maneuvered for position.

On 13 December Hill found himself isolated on the east bank of the Nive with 14,000 men after a sudden storm had washed away the pontoons. Knowing it would take at least twelve hours for aid to reach Hill's force, Soult sent out four divisions from Bayonne supported by the two already facing Hill to storm the ridge of Saint Pierre d'Irube. The French attack was delayed due to bridge congestion while crossing the Adour. Then rather than wait till the whole force was assembled, each division went in as it arrived. Hill held a strong position, the battle was hard and at one stage a English unit fled the field believing all was lost. The gap was soon plugged and the struggle continued until Abbé's attacks petered out faced by an unbroken line while the other French divisions to the rear refused to advance. The near mutiny coincided with Wellington's arrival at the head of the first reserves. Soult also hearing Clauzel's

attacks before Bayonne had also failed saw no alternative but to withdraw leaving some 3,000 casualties on the field in some of the fiercest fighting of the campaign.

Winter conditions caused the armies to remain static from mid-December 1813 to February 1814. Morale and discipline amongst the French deteriorated further with frequent quarrels amongst the senior officers. Reille resigned his command after one such confrontation with Soult over the authority exercised by Thouvenot, the young governor of Bayonne. A sense of duty and the imminent prospect of renewed hostilities made the disillusioned general return a week later. The army also visibly shank before Soult. The defeats in Germany caused Krause's German brigade to defect and to avoid further problems he ordered the remaining German and Spanish troops in French service to be disarmed. The worst blow came when he was ordered to release the two infantry divisions of Leval, Brayer and Trelliard's dragoons to join Napoleon in Eastern France.

When Wellington breached the line of the Adour between Bayonne and the sea Soult left Bayonne to fend for itself and retired eastwards to a strong position at Orthez. Here his outnumbered forces were again defeated on 27 February 1814. Faced with a difficult decision whether to retire north to Bordeaux or east towards Toulouse he opted for the latter. The country was more fertile compared to the sandy area south of Bordeaux where supply would be difficult. Moving towards Toulouse also meant drawing closer to Suchet with the hope of the two forces joining to deal with Wellington. Wellington was forced to split his army and detach Beresford to take Bordeaux. The numerical advantage he had against Soult disappeared as he was committed to follow him, since he could not allow the French to threaten his flank if he made for Bordeaux. Soult made another stand at Tarbes on 20 March before he withdrew to the outskirts of Toulouse.

The battle before Toulouse on 10 April 1814 was the last bloody action of the campaign. News of Napoleon's abdication on 6 April failed to reach the armies in time. Soult's 40,000 troops put up a fine struggle against Wellington's 50,000 men present on the day. Fighting for the redoubts on the Calvinet Heights, which dominated the city from the east was particularly heavy. At the end of the day Soult pulled back to Toulouse with French losses some 3,000 against 4,000 suffered by the Allies. Apart from the enemy's presence the city had become untenable due to the influence of the pro-Bourbon populace and Soult retired to Castres where he heard of Napoleon's fate.

Toulouse was reckoned to be one of his finest battles. With his ragged weary army he faced three Allied armies, the British, the Spanish and the Portuguese, led by the Coalition's most able commander, Wellington. Overall his defensive campaign in southern France was possibly his finest. Whilst his positions were turned and he was defeated time and time again he managed to keep his army together. His contribution was significant to the French cause in that he prevented Wellington from realizing his dream of marching through France, crossing the Loire and then entering Paris from the south before the Allies.

THE RESTORATION AND MINISTER FOR WAR

Except for Davout, Soult was the last marshal to recognize Louis XVIII as King of France on 19 April 1814. His army being the last to submit caused the Bourbons to categorize him as an irreconcilable and thus unsuited for command. As a result Suchet took command of his army on 22 April 1814 and he returned to his home near Paris. As the weeks passed he felt a need to be active once more, in particular to help in the country's reconstruction, and within it the Army. He put out feelers for a posting and on 21 June 1814 was made commander of the 13th Military Division at Rennes in Western France.

Dupont of Baylen fame proved an inept Minister for War in the intervening months and a strong man respected by the Army and politicians alike was needed. No Bourbon available the choice was narrowed down to serving marshals. Soult was amongst the three recognized by the Bourbons as being the most able, which included Massena and Davout. Massena was old and tired while Davout flatly refused to serve in any capacity. Suchet also entered the race as a late candidate but Soult's organizational skills and reputation for hard work resulted in his securing the post on 3 December 1814.

The tasks Soult faced were near impossible, dealing with the Army's problems of past and present regimes. Appointments were a major concern that alienated both Royalists and veterans of the Grande Armée alike. After years of sacrifice and exile the Royalists felt indignant that more commands did not come their way. Likewise Napoleon's former followers were equally aggrieved when they lost theirs. The cafes of Paris became a hotbed of unemployed malcontent officers, a problem that led Soult to a crisis he so easily could have avoided.

On his desk in the Ministry lay the portfolio of General Exelmans' then an inspector general of cavalry based in Paris. Exelmans had been one of Murat's senior aides and unfortunately a letter written by him offering his services to help Murat make a bid for the crown of a united Italy had been intercepted. Dupont while minister wisely contented himself with simply reprimanding the general and letting matters rest. When Soult succeeded Dupont, Exelmans continued to make noises. Not prepared to accept such behavior from a prominent officer Soult relieved him of his command, placed him on half-pay and ordered him to leave Paris. Given Exelmans's prominent position in the army and his sympathy towards disaf-

fected officers, it was a reasonable step. Exelmans refused to leave Paris using his wife who was expecting a baby as an excuse. The King urged Soult to be calm, but the matter soon got out of hand as comte d'Artois and his son de Berry raged against Exelmans at dinner parties demanding his arrest. Left with no alternative Soult had him arrested in front of his distressed wife and taken under escort to Soissons. The matter appeared over till Exelmans escaped and returned home demanding of Soult that he be allowed to face a court martial.

Soult not prepared to duck a challenge, Exelmans was brought before a military tribunal at Lille in January 1815. Charges included defying the King, disobeying orders, of spying and offering his sword to "the general who ruled Naples". They were in most cases justified. Exelmans conducted a skilful and dignified defence, was respectful towards the King and made much of the fact that France was not at war with Murat. D'Erlon who chaired the tribunal gave him a sympathetic hearing and he was unanimously acquitted. There was uproar from both sides. A secret Bonapartist sympathizer d'Erlon had old scores to settle with Soult and was prepared to go to any lengths to cause Soult and the government embarrassment. Soult considered resignation but the King with whom he had a good rapport was very supportive and urged him to stay. Old Imperial officers, who had previously welcomed his arrival at the Ministry for War, on the other hand now began to fear and distrust him.

When Soult's fall came a few weeks later, it came suddenly. News of Napoleon's landing reached Paris on 5 March and within six days he was toppled. In February he had ordered the movement of large numbers of troops to Lyons as the first stage of an Allied move to overthrow Murat in Naples. Fortuitously their presence in the region played into Napoleon's hands. Soult ordered the troops to intercept him but as news of the defection of regiment after regiment arrived Soult was accused of being party to a wider conspiracy. When a military revolt with d'Erlon implicated broke out in the northern garrisons, he with the King were among the few that kept their heads. Mortier was sent to deal with it and in a few days the ringleaders were arrested. Soult then gave orders for a reserve army of some 100,000 men to gather south of Paris at Melun. He hoped the recall of half-pay officers and old regimental loyalties would win them over to the King and provide a disciplined army to face Napoleon. In the end it did, not for the King but Napoleon!

An outcast to both Bourbons and Bonapartists alike, by 11 March 1815 criticism had reached such a peak even the King's ministers demanded Soult's departure. When he resigned Louis was devastated, for apart from being a conscientious and hard working minister Soult's behavior had always during this time been honorable. Even his harshest critics among them Houssaye and Thiers, acknowledged he made a good Minister for War in at a very difficult time for France. The Exelmans affair was unfortunate, and the movement of the troops to the east was sheer coincidence. Napoleon had the final say from Saint Helena where he referred to Soult in O'Meara's work as being in no way privy to his plans to return.

NAPOLEON'S MAJOR GÉNÉRAL

Soult retired to his country home Villeneuve l'Etang near Saint Cloud to await events. Twice Napoleon summoned him for a meeting but he declined. It was his colleague Clauzel who finally persuaded him to see the Emperor. When they met they spoke together frankly. Soult never tried to deny that he had issued orders for Napoleon's arrest on news of his landing as he considered it his duty to do so. He also insisted he wanted to live quietly at his home as a private citizen. Napoleon on the other hand needed him, as he was a fine political prize. Having served the Bourbons, Soult's acceptance of an important post also signalled a reconciliation between the old and new orders. Foremost as a realist Napoleon admired his military and administrative talents besides recognizing he was possibly the most hard working and conscientious of all his marshals.

It took till 9 May 1815, a month after his first talks with Napoleon, to return to the fold and accept the post of *Major Général* of the Army of the North. Immediately questions as to his loyalty and suitability for the post arose. The events of Waterloo later built a myth which compared the missing Berthier's gifts as a chief of staff to Soult's failings. Closer analysis makes such claims appear unjustified and at times just an attempt by Bonapartist writers to shift attention away from Napoleon's own failings as a leader during the campaign. Soult had extensive experience as a chief of staff, which went back to his days under Lefebvre. He also served with success under King Joseph in the campaign that led to the victory at Ocana and during the invasion of Andalusia. He understudied the ailing and overworked Berthier in the Spring Campaign of 1813, being earmarked as a possible successor. His attention to detail and administrative skills after Berthier and Davout were foremost amongst the marshalate.

Napoleon found reason to criticize him for a certain lack of care in exercising his clerical duties. A case in point involved a rebuke he received for failing to send important messages in triplicate or quadruplicate. That frankly was not really his fault, since without as much as a second thought such practices had been the norm under Berthier and as such should have automatically continued. His subordinates with years of staff experience behind them should have ensured such basic precautions were taken.

Soult's biggest problem was with Napoleon himself. He considered himself a chief of staff of the Prussian mould, preferring to fight the battle himself and keep his commander-in-chief informed. It worked well under the gallant old Lefebvre and with the militarily incapable like Joseph. He had problems taking such a role under the jealous Berthier. A genius like Napoleon required a "note taker" of the Berthier mould, rather than a person who would try to second guess him. Napoleon should not be criticized for Soult's selection but rather for failing to make the best use of his talents. One attribute Soult possessed which Napoleon didn't use to his advantage was Soult's good knowledge of the countryside in Belgium from his days with the Army of the Sambre-Meuse. Napoleon on the other hand had never campaigned in this area. More important, Soult had the greatest experience of the marshals in fighting the British and Napoleon's principal opponent, Wellington.

Davout supported Soult's appointment, but there were several scores senior officers wished to settle, several of which had arisen in his short time as Minister for War. The arrest and trial of Exelmans was the most celebrated. The peppery Vandamme refused to acknowledge his appointment until Davout confirmed it in writing. Reille's and d'Erlon's animosity towards him went back to Spain. More recently to the revolt of the northern garrisons, it was Soult that had threatened to have those senior officers of I Corps arrested and shot who joined he revolt. It was not a promising start.

Contrary to popular belief, Soult and his staff generally performed well. Houssaye summed it up in his work on the campaign when he went as far to say that, "only brilliant staff work could have achieved the rapid and secret concentration of such large forces in northern France in June." The final comment on Soult's merits as *Major Général* come from Napoleon himself. Apart from the barbed references to his failure to use enough staff officers to carry messages, the Emperor who spared Grouchy and Ney of no criticism did not lay any of the campaign's misfortunes at Soult's door. Indeed at Saint Helena he told Doctor O'Meara the author of *Napoleon in Exile* that Soult in fact, "made an excellent *Major Général*."

THE WATERLOO CAMPAIGN

Soult was made a Peer of France on 2 June 1815. Three days later the final preparations for the campaign were underway when he issued orders for the Army of the North with its five army corps, the Cavalry Reserve and the Imperial Guard to move up to the border. It was a complicated operation as National Guardsmen secretly moved to positions vacated by regular troops. He soon ran into trouble for not sending instructions to Grouchy's cavalry formations to move sooner than 11 June. The resultant rush to the border nearly caused the concentration to go wrong with many regiments arriving exhausted before the campaign had even begun.

As the army crossed into Belgium things again began to go wrong with Soult and his staff receiving criticism as vital messages went astray. Vandamme never received his marching orders because he went missing and the messenger fell from his horse. The result was several hours delay as a great traffic jam occurred when Lobau's VI Corps following III Corps crashed into his troops still trying to get moving. The failure of d'Erlon corps to appear at Ligny on 16 June was another reason for his critics to question his ability as *Major Général*. Whilst the censure d'Erlon and Ney received for the affair was justified, the criticism apportioned to Soult generally was not. The sequence of events that caused d'Erlon not to fall on Blucher's flank rests principally with him and to a lesser extent Ney.

Around 3.00 p.m. Napoleon's aide de la Bedoyère came across the head of d'Erlon's column heading for Quatre Bras and acting on Napoleon's orders ordered it to head for Wagnelée to fall on Blucher's flank. Ney at Quatre Bras, expecting d'Erlon imminent arrival as the battle reached a critical stage, became increasingly agitated. He then learnt of d'Erlon's new movements around 4.00 P.M. just as Soult's aide Forbin-Janson arrived. This unfortunate officer carried messages from Napoleon urging Ney to press Wellington with more vigor and also ordering him to detach d'Erlon to Ligny. Ney by now in a blinding rage as it appeared d'Erlon had disobeyed his orders directed his abuse at the flustered staff officer who only managed to blurt out the first part of his message before galloping off. Forbin-Janson only realized later he had forgotten to deliver the vital message concerning d'Erlon's movements.

Ney was still unaware that it was Napoleon who had ordered d'Erlon to Ligny where the main battle was being fought. Aware it was his duty to take Quatre Bras he immediately sent messages ordering d'Erlon's recall. D'Erlon a capable and reliable commander, inexplicably for reasons only known to himself when he received Ney's message around 6.00 P.M. allowed Ney's orders to override those of his commander-in-chief Napoleon. About to fall on Blucher, d'Erlon turned his troops about and 20,000 troops ended the day wandering between two battles without firing a shot all day. Their presence at either one would have proven decisive.

Ney was at fault for not giving Forbin-Janson a hearing, while d'Erlon faced by a terrible dilemma was more guilty since he didn't comply with Napoleon's order. Because the unfortunate Forbin-Janson happened to belong to Soult's staff, here was the reason for his critics to place the blame on him. An area where Soult can be more fairly criticized is his failure during the course of the day to keep Ney better informed of Napoleon's and the main army's movements. Had Ney been made aware earlier that Ligny was the scene of the main battle and not Quatre Bras, he in all probability would have kept his head and behaved quite differently. A past master of holding actions and rearguard struggles Ney could have changed his plan to suit the circumstances and not interfered with d'Erlon's movements.

Before Waterloo Soult with Reille took the cautious line of warning Napoleon against attacking British infantry in defensive positions. They advocated a battle of manoeuvre rather than a headlong drive against Wellington's center. An overconfident Napoleon rounded on them for expressing such doubts, ridiculing them unfairly because Wellington had beaten them in Spain. In spite of reservations Soult did his duty and fine staff work helped the army to be in position by 9.00 AM after the previous night's heavy storms.

Soult was at Napoleon's side throughout the day and at the end of the battle when all was lost it was he who kept his head. Napoleon in despair hoped to die at the head of a forlorn charge by a battalion of the Old Guard, while several of his aides rushed around with little regard for his safety trying also to get themselves killed. Soult took charge, laid his hand firmly on Napoleon's bridle and with General Gourgand led him away from the field.

He took command of the army after Napoleon left for Paris, succeeded in stemming the rout and by the time he made contact with Grouchy at Laon some semblance of order had returned. The news of Napoleon's abdication on 24 June came as a devastating blow. Up till then as a master of the fighting withdrawal he hoped the enemy advance could be stemmed in the forests of Compiègne or by holding the Oise crossings. "Without an Emperor to serve", he insisted in resigning his command on "health grounds" and handed over to Grouchy on 26 June 1815. It was an abdication of responsibility that was possibly the most shameful episode of his life. Once in Paris he then had the temerity to attend the House of Peers and make speech supporting the immediate return of the Bourbons as the best hope for the country. Many Peers had already advocated the same course, and his intervention was construed by critics as a feeble attempt to build bridges and lessen the blow expected to fall on him with the King's return. Rather than face the consequences, he quit Paris on 3 July and took refuge with friends in Lozère before finding his way to Saint Armans.

EXILE AND RETURN

Soult was included at the head of the second category of personages listed by the Ordonnances of 24 July 1815, which meant banishment. The order not published till the following December gave him time to put his affairs in order and he quit France on 12 January 1816. A place of exile did not prove a problem as his wife was from the Duchy of Berg. His fair treatment of the Rhinelanders during his time with the Army of the Sambre-Meuse coupled with his sympathetic behavior towards the Prussians as governor of Berlin led him being allowed to settle near Dusseldorf.

He attempted to write his memoirs but soon realized with his limited education writing was difficult. Also as a guest of Prussia and fearful of losing his property in France he was weary of making political judgements. He continued writing at intervals, even when he resumed an active political career but only managed to complete his early life and career as far as the Revolutionary Wars. They were later published after his death by his son Hector but aroused little interest.

The Royal Family had always felt Soult had acted honorably during the events leading to the King's flight. At the request of comte d'Artois and the duc d'Angoulême he was allowed to return to France on 26 May 1819. His marshal's baton was restored on 6 June 1820, but in the next ten years he took no active part in political or military affairs. Instead he used his considerable wealth, largely amassed in Spain, to help fund various business enterprises.

POLITICIAN AND ELDER STATESMAN

By 1830 and the Bourbons seemingly secure, Now sixty Soult was quite prepared to retire and bask in the glory of his reputation as a patriot soldier and the "grand old man" of Napoleon's marshals. The July Revolution suddenly changed all this and brought back his profile into the public eye. As a liberal Royalist, he could easily identify with Louis Philippe and decided to offer his services. Over the years he had nurtured a good friendship with the new monarch and in the first shaky months acted as an advisor and close confident. The government also needed a new Minister for War, Gérard appeared not up to the task and on 17 November 1830 Soult succeeded him. He was to remain as minister for his second term till 18 July 1834.

He plunged into his work with the same customary energy he had used sixteen years before with a daily routine that started at 5.00 A.M.. Reform was his objective as the army was in poor shape, with low morale and poorly led. He needed a spiritually united army after the disturbances and mutinies of 1830. He had to take unpopular measures including special powers in November 1831 when rioting in Lyons supported by mutinous elements of the city's garrison got out of hand. At the same time he had to face the problem of Belgium that dragged on for over a year. In the end Gérard with the Army of the North re-entered the country and laid siege to Antwerp, which capitulated after a month in December 1832, finally securing Dutch recognition of Belgian independence.

Promotion, pay, recruiting, the general staff, military justice and prisons, drill and service regulations were shaken up during his second and third ministries. He did not always succeed. Close to his heart was a plan to form a properly trained reserve, so reducing the length of conscription for regulars. The army was continually short of manpower, with the government and opposition never really grasping the problem. In the decades that followed France lagged behind in this area and paid the ultimate price against Prussia in 1870.

SOULT IN HIS LATER YEARS

Soult took on added responsibilities when he succeeded Casimer Périer as President of the Council of Ministers in October 1832. He faced further problems including Spain when Don Carlos tried to seize power from the young Queen Isabella. It provoked a ministerial crisis with Thiers favouring armed intervention, while the King and Soult were firmly opposed. He clashed over the Lyons riots of 1834, which he suppressed with a firm hand then prevaricated for weeks over the fate of the ringleaders. As time passed he was increasingly out of step with his fellow ministers. In the end they plotted his fall and forced his resignation over the government's Algeria policy. Algeria had been a continual problem since 1830 and as in Spain he was never able to master the intricacies of guerilla warfare or the possibility of negotiating with them. He wanted to go out and put the countryside to the sword himself, but he was too old. He firmly advocated the conquest and settlement of the territory as solely a military problem and nothing else. When his colleagues set the trap by insisting on the appointment of a civilian governor, he felt betrayed and resigned in protest on 18 July 1834.

Louis Philippe genuinely missed him and repeatedly asked him to return to office. He re-entered public life when he travelled to England for Queen Victoria's coronation on 25 April 1838. The tour was a magnificent success. He renewed acquaintance with Wellington, toured naval bases and army camps, visited factories. At Birmingham during a speech he made reference to the interests of France and Britain being mutually dependent on each other and there never a need for the two countries ever to go to war again with each other. In spite of such generous words even in Soult's time as head of the government and in the years that followed the two countries continued to have problems. They included Mehmet Ali in Egypt, the Ottoman Turks, and at the end of the century the colonial rivalries that culminated in the incident at Fashoda. But most important of all, they were all solved by negotiation. Soult can with some justification lay claim to being the one who laid the foundations of the *Entente Cordiale*.

Louis Philippe secured his recall as head of government when he became Minister for Foreign Affairs and President of the Council in place of the Duc of Montebello from 12 May 1839 to 1 March 1840. A short period out of office followed till 29 October 1840 when he returned as President of the Council in place of Thiers and also took up his old post as Minister for War. He was to serve the King faithfully for a further seven years. The government brought tranquillity to France at home. As the King's closest supporter he was weary of Bonapartism and when the famous column of the Grande Armée was completed at Boulogne for "political reasons" he was not at its dedication. Louis Bonaparte, languishing for much of Soult's time in office in the fortress of Ham, was considered by Soult as a dangerous upstart who was not worthy of his uncle's name. He saw however the need for France to have a sense of pride in Napoleon's accomplishments and sanctioned the return of his remains to the Invalides.

As Minister for War his works included further modernization of the army including a reorganization of the General Staff. After fierce debates the defences of Paris were improved by the building of fortresses around the city. The war in Algeria continued and took up much of his time, but the problem remained insoluble. Aged 75 he finally begged the King to allow him to retire from the War Ministry in 10 November 1845 and continued as President of the Council in name only till September 1847. As a parting gesture Louis Philippe took the exceptional step of making him on 26 September 1847 Marshal General, a rank only previously held by Turenne, Villars and Saxe.

After his departure from office scandal frequently rocked the government. When the monarchy fell in February 1848 Soult was at the King's side but was able to offer little help. In the political turmoil that followed his sympathies lay with the strong-willed General Cavaignac as President of the Second Republic rather than Louis Bonaparte. He died on 12 December 1851, a few days after the coup d'état, which secured the Prince President the presidency for life as a preliminary to becoming Emperor Napoleon III.

AN ASSESSMENT - THE MALIGNED MARSHAL

Soult was possibly the most misunderstood and unfairly treated of Napoleon's marshals by writers and critics. Called the "Maligned Marshal" by Alan Hayman in his biography on Soult, there is no better way to describe this controversial figure who attracted criticism like bees to a honey pot. He was a brilliant soldier, one of the chief architects of the Grande Armée and one of the creators of the modern French army. He was a subtle politician, a fine civilian administrator, an art admirer and art plunderer. He was intensely ambitious, otherwise he would never have succeeded in the vicious promotion scrambles that characterized the armies of Revolutionary France. Aloof by temperament, he was immensely determined and physically very strong. He was sometimes over-deferential to his superiors while being harsh to his immediate subordinates, a trait common to those at the time who had risen rapidly beyond their station. He was respected by his men, but certainly not loved by them.

Never having fought a campaign with Napoleon till after he received his marshal's baton, it is correct to say he owed Napoleon little or nothing and as a result he was his own man. He was no Bonapartist, but rather a French patriot who with a clear conscience was able to serve successive governments of France as and when they existed and changed.

SOULT - PIERRE BENOÎT, BARON (1770 - 1839)

BACKGROUND AND EARLY CAREER

The younger brother of Jean de Dieu "Nicolas" Soult, Pierre Soult was born in the village of Saint Armans Labastide (Tarn), southern France on 19 July 1770. His father, a struggling local notary, died before he was seven placing the family in straitened circumstances. His mother, a determined woman, wanted her sons to take over her late husband's business and apprenticed them to notaries in a nearby village. Neither had the aptitude or inclination to follow their father's career. When Jean left to join the army, it was only a matter of time before Pierre, easily influenced by his strong willed brother would follow.

On 28 September 1788 he enlisted with the Regiment de Touraine (33rd Line). In possession of a basic education he gained promotion to *caporal* in the old army on 24 March 1791. On the outbreak of war in April 1792 he saw service with Burcy's brigade of the Army of the Moselle. He was wounded near on Saverne on 24 October 1793. In January 1794 he joined his brother as an unofficial aide, when Jean was making a name for himself as chief of staff with Lefebvre's advance guard of the Army of the Moselle. It was the start of an association, which lasted for the next twenty years, the success of his career becoming very dependent on fortunes of his elder brother. He fought at Fleurus on 24 June 1794 and took part in operations that cleared the left bank of the Rhine.

AIDE DE CAMP TO JEAN SOULT

For three years Pierre campaigned at his brother's side with the Army of the Sambre-Meuse, first serving with him in Hatry's division. A period then followed from April to June 1795 at the Siege of Luxembourg when Jean was attached to Desjardin's command. In September 1795 he distinguished himself when the army crossed the Rhine and made its way to the Lahn. The crossing was fraught with difficulty, there being long defiles to pass with the river banks steep and the left bank strengthened by redoubts. Poncet the division commander maneuvered to cross near Diez on 21 September, while Pierre at the head of a small detachment of the 66th Demi-brigade found a ford the enemy believed impassable. The small party dashed across, established a bridgehead on the left bank while the rest of the brigade crossed and then stormed into Diez.

THE CROSSING OF THE LINTH

304

The incident had other benefits for Pierre when General Klein commanding the cavalry decided to cross taking advantage of the narrow ford. Not one to heed warnings of its dangers, Klein wandered off the line and was swept away. Pierre plunged into the waters and saved the spluttering dripping veteran. The incident did not go unnoticed and Pierre later received a flattering letter from the Directory commending him for his bravery. On 21 April 1796 he received a commission as *sous lieutenant*.

In 1796 back with Lefebvre's division he took part in Jourdan's campaign deep into Germany. He served at Altenkirchen on 4 June and fought at Wurzburg on 18 August before the army had to make a hurried retreat to the Rhine. The following Spring he fought at Pfullendorf on 21 March and at Stockach 25 March 1797. At Herborn on 21 April 1797 his daring came to note when at the head of thirty hussars he surprised an enemy column and took 320 Austrians prisoner. For his efforts he gained immediate promotion to *lieutenant*.

In the period of peace that followed he spent a short period with the 6th Chasseurs à Cheval before rejoining his brother with the Army of England in January 1798. In April 1798 he moved to Germany when Jean took command of a division with the Army of Mayence. On 28 October 1798 he was promoted *capitaine*. He took part in the Swiss Campaign of 1799 and was cut down and taken prisoner on 5 June while defending the entrenchments before Zurich. Jean secured his exchange on 29 June and he returned to duties after promotion to *chef d'escadron*. In the second battle for Zurich on 25 September 1799 he again distinguished himself during his brother's audacious crossing of the Linth. In charge of a unit of 160 men he forded the river in darkness and fell on the enemy outposts. Complete surprise achieved, before the enemy recovered they lost control of the river line as his brother crossed with the main force.

On 13 December 1799 he moved to Massena's Army of Italy. Present at the siege of Genoa he was in charge of a small cavalry force that supported the attack on Monte Creto on 13 May 1800. The assault failed, and when he tried to carry his wounded brother to safety both were captured. Taken to Alessandria they were imprisoned in the Bishops's Palace. Here from a window with a field glass they watched in the distance as Napoleon overwhelmed the Austrians at Marengo. After the battle they gained their parole on 17 June 1800. Pierre joined Jean, who on becoming military governor of Piedmont faced the difficult task of bringing order to the region. At the end of the year he moved to Taranto on his brother's appointment as lieutenant general of the Army of Observation of the Midi.

When Jean returned to France he remained in Italy gaining promotion to *chef de brigade* with the 25th Chasseurs à Cheval on 30 December 1802. He served in Mermet's dragoon division of Massena's Army of Italy during campaign against Austria in 1805, being present at the crossing of the Adige and in action at Caldiero on 30 October 1805.

THE GRANDE ARMÉE 1806-1807

Pierre rejoined Jean, now head of IV Corps of the Grande Armée as an aide de camp for the campaigns in Prussia and Poland. He fought at Jena on 14 October and joined the pursuit of the Prussian army, which ended with Blucher's capitulation at Lubeck on 6 November 1806. In Poland he fought at Eylau on 8 February 1807 when IV Corps resisted Russian attacks all day and Napoleon received his first serious reverse. At Heilsberg on 10 June he was bayoneted in the stomach, the wound forcing him to miss the remainder of the campaign.

THE PENINSULA WAR 1808-1814

On 11 July 1807 he received promotion to *général de brigade*. He served on Berthier's staff during Napoleon's invasion of Spain till posted on 28 November 1808 to Santander as military governor. In June 1809 he rejoined his brother as cavalry commander for II Corps after Franchesci's capture by guerilas. The move was not a success, the cavalry lost a certain of its sharpness and Franchesci a fine leader was sorely missed. Pierre did not merit the post, for there were others more capable and experienced available. The whole affair smelled of nepotism, yet there were few generals who happy to stake their careers with such a difficult and demanding man as Jean Soult.

After Talavera during his brother's pursuit of Wellington's army, Pierre had some success. His cavalry plus Houssaye's dragoons crossed the Tagus at Arzobispo on 8 August and surprised the Allied rearguard taking 1,400 prisoners and recovering fourteen of the seventeen French guns lost at Talavera. He retained his command when Jean became King Joseph's chief of staff and Heudelet took over II Corps. In July 1810 he joined Massena's invasion of Portugal serving under Reynier who replaced Heudelet in April 1810. He fought at Bussaco on 27 September 1810 and received a shell splinter in the leg. He soon recovered, and under Montbrun his cavalry were the first to discover the Lines of Torres Vedras on 11 October as the army neared Lisbon. He spent the next six months before the lines and then helped cover Massena's retreat to the Spanish border. As Reynier's corps fell back he fought at Sabugal on 3 April but was unable to prevent the Light Division giving it a severe mauling.

In the summer of 1811 he moved to Andalusia where for a time he led the 3rd Dragoon Division of Sebastiani's IV Corps while Milhaud' was on leave.

Co-operating with Latour Maubourg's cavalry he caught Freire's Army of Murcia at Las Vertientes on 10 August 1811 when it threatened Granada. In the action his horsemen put 2,200 enemy horse to flight before forcing Freire's 15,000 infantry to take to the hills, removing the threat from eastern Andalusia for the rest of the year. In January 1812 he led a raid into Murcia to draw Spanish troops opposing Suchet's advance on Valencia. The venture proved a fiasco. His 800 horsemen encountered little resistance and on 25 January entered Murcia with Pierre setting up quarters in the archbishop's palace. Having failed to post adequate pickets a force of cavalry under La Carrera swept into the city the next day and in the chaos nearly captured him. Pierre unnerved by the encounter and convinced a huge force was nearby, gave up the plan and started back to Granada.

On 7 February 1812 he became commander of the 3rd Cavalry Division of the Army of the South. Ordered north, he spent several months in Estremadura lending support to forces under d'Erlon* being hard pressed by Hill. In July 1812 he was detached to Marmont's Army of Portugal operating against Wellington. While seizing the bridge at Alba de Tormes on 23 July before the Battle of Salamanca he was shot in the arm. The success proved vital when after Marmont's defeat it secured the French a line of retreat. He covered d'Erlon's withdrawal from Estremadura in August. In the Autumn at the head of the 2nd Dragoon Division he took part in the offensive that drove Wellington from Madrid to Salamanca. In the pursuit he tried to outflank Wellington and fall on his retreating columns by crossing the Tormes further north. But seeing the move the Allies slipped away and fell back to Ciudad Rodrigo. A great opportunity lost, much of the blame fell on Pierre as throughout the brief campaign his critics felt he showed little vigor during the pursuit. Speculation reached such a peak that commanders believed his brother had such little faith in him that he had deliberately held Pierre back lest he make a fool of himself.

As the armies went into winter quarters, Pierre at Toledo set about reorganizing his cavalry. In spite of the criticism of the previous campaign, on 3 March 1813 he received promotion to *général de division* and was placed in command of all light cavalry with the Army of the South. He took part in the events that led to the Battle of Vitoria, fought on 21 June 1813. Throughout the day he did little, acting with the same indecision as so many other generals. His division stood in reserve a mile behind Conroux's and Darricau's formations on the slopes near Arinez as the French line collapsed. He then fled the field with his horsemen, giving little regard for the infantry left behind. As a result out of the chaos of retreat he managed to bring his division back to France reasonably intact. On his brother's recall to

head the armies in Spain he retained command of the light cavalry. The country unsuited for cavalry operations he found opportunities limited and only played a minor role in the remainder of the campaign. From 9-13 December he took part in the offensive against Wellington on the Nive at Saint Pierre de Irube. Then once he made contact with the enemy he found the terrain prevented him deploying. His cavalry then had little influence on the battle's outcome, which ended in failure when the army withdrew to Bayonne and fell behind the Adour.

He was present at Orthez on 27 February 1814, but with his brother fighting a defensive infantry action his cavalry remained inactive and merely covered the retreat. Falling back on Toulouse he fought a successful action at Tarbes on 20 March where he mauled Somerset's cavalry. At Toulouse on 10 April 1814 his cavalry kept communications open with Carcassone while the last major battle of the war raged around the city. It was at Carcassone on 12 April 1814 that he learnt of Napoleon's abdication and the end of hostilities.

THE RESTORATION AND NAPOLEON'S RETURN

His division disbanded, on 16 June 1814 Pierre found himself unattached. He received the title *Chevalier de Saint Louis* on 13 August 1814. Recalled on 30 December 1814 he became Cavalry Inspector General for the 9th, 10th, 11th and 20th Military Divisions. He received the news of Napoleon's return with dismay, dreading the prospect of France being faced by civil war. On 6 March 1815 he joined the Duc d'Angoulême's forces as cavalry commander at Bordeaux and remained loyal as the situation deteriorated. Moving across country he was at Montpellier when he realized the Bourbon cause was lost. He decided to return home to Saint Armans, but the garrison arrested him on 3 April as he tried to leave. When he heard Jean had submitted to Napoleon he decided to follow suit and swear allegiance. He entered politics and was elected Deputy representing Castres (Tarn) on 16 May 1815. Davout suspected his loyalty and did not recommend him for a command with the new armies being formed. However, Jean as the Army of the North's chief of staff interceded on his behalf and on 7 June 1815 Pierre took command of the 4th Cavalry Division with Pajol's I Cavalry Corps. His command was one of the most colorful in the army, being made up of four hussar regiments.

THE WATERLOO CAMPAIGN

As the campaign opened on 15 June Soult performed well in charge of advance elements of Pajol's cavalry that burst across the border in the early hours and headed towards Charleroi. Reaching the Sambre around 8.30 A.M., Pajol sent him forward but heavy fire from troops in skirmish order concealed behind hedges and in the southernmost houses along its banks drove him back. While waiting for Vandamme's infantry he tried in vain to find a ford by which he might cross to the other side. A detachment of Guard marines and engineers arrived and after a stiff firefight cleared the barricade on the bridge over the river. The way clear his hussars charged over it into the town. Pushing up the road to Fleurus he kept the Prussians on the run before being halted by a strong force holed up in the woods behind Gilly. Urged on by Napoleon after a delay arose due to Vandamme and Grouchy arguing over who should lead the assault he worked his way around the woods and by dusk had reached Lambusart. The Prussians in danger of being outflanked fell back that night to positions behind Ligny.

Only lightly engaged at Ligny on 16 June, Soult's part was mainly confined to cavalry feints on the French right that prevented Thielemann's III Corps supporting Blucher's center when Napoleon's main blow fell on it at the end of the day. The next day following the retreating Prussians one of his regiments captured a battery retiring along the Namur road. This information he passed on up through Pajol to Grouchy and, interpreted as a Prussian retreat on Liege, had a major outcome on the campaign. It was what Grouchy wanted to hear. Grouchy dallied despite Gérard and Vandamme urging him during the day that the Prussians were retiring via Gembloux to Wavre. Soult wandering towards Liege failed to determine the Prussian movements and it was only by evening it became clear that his assumptions were wrong.

Present at Wavre on 18 June his regiments seized the bridge over the Dyle at Limale in the late afternoon. After the news of Napoleon's defeat his cavalry continued to probe aggressively against Thielemann's corps winning valuable time for the rest of Grouchy's force to make its way towards safety via Namur. He rejoined Grouchy at Philippeville and continued to cover the retreat to Paris. He was with the army when it retired to the Loire and took part in its disbandment. On 1 August 1815 he was placed on the non active list.

LATER LIFE AND CAREER

During the period of Bourbon rule he returned to his home at Saint Armans and retired from the army on 16 February 1825. Recalled on 6 September 1830 after the overthrow of Charles X, he headed the troops in the Departments of Tarn and Aveyron. He then moved to the Army General Staff on 7 February 1831 after his brother became Minister for War. On 21 March 1831 he received the Grand Cross of the Legion of Honor. Unattached for a while in 1833 he then moved to the 10th Military Division at Toulouse as its commander on 13 September 1833. Unattached from 28 August 1836 he passed to the Reserve Section on 5 August 1839. He continued to lead a fairly public life during the July Monarchy and was involved in several business ventures with his brother that brought certain wealth to the people in his home region. He died at his home in Tarbes (Hautes Pyrenees) on 7 May 1843.

As a leader Pierre Soult owed much to his brother. Whilst a brave officer and a fine regimental leader, he lacked the talent for anything greater. Was it not for the backing of his brother it is very unlikely he would ever have risen to general's rank. He was fortunate Jean tolerated his frequent failings, which was more than Napoleon did for his kith and kin. The favoritism shown by his brother certainly ranked with some of the more extreme cases of nepotism practised by the Bonaparte family.

STROLZ - JEAN BAPTISTE ALEXANDRE, CHEVALIER (1771 - 1841)

BACKGROUND AND EARLY CAREER

The son of a road inspector, the cavalry general Jean Strolz was born in Belfort, Eastern France on 6 August 1771. His first active service was as an unofficial aide to General Ferrier with the Army of the Rhine where he distinguished himself during a raid on Porentruy in August 1792. After Ferrier's removal, he joined the 1st Chasseurs à Cheval as a trooper on 8 April 1793, serving in the Army of the Moselle. He fought at Fleurus on 26 June 1794 and after the battle passed to the Army of the Sambre-Meuse, joining General Kléber's staff on 7 August 1794. Promoted acting *sous lieutenant* on 22 September with the 16th Chasseurs à Cheval, it took over a year for his rank to be confirmed on 25 December 1795, not an unusual occurrence in the army at the time. Thereafter his progress was steady, with promotion to lieutenant on 25 December 1796. He declined an offer to follow Kléber to Egypt and instead on 20 May 1798 joined the veteran General Hatry as his aide de camp. With Hatry,Hatry he was promoted *capitaine* on 23 June 1798 and spent time with him in Holland. Further promotion to *chef d'escadron* came on 21 January 1799 with a move the same month to the Army of Italy. By April 1800, Hatry was too old for active campaigning and took a quieter posting in France, which affected Strolz's prospects. Hatry, recognizing his fine qualities, wrote to Moreau recommending him, which resulted in his appointment as one of his aides on 7 April 1800. He took part in the Spring Campaign in Germany and being German speaking was with Moreau's party during negotiations that led to the Armistice of Parsdorf concluded on 15 July 1800. In the Winter Campaign he was at Moreau's side in the events that led to the Battle of Hohenlinden on 3 December 1800.

ITALY AND NAPLES 1801-1809

Strolz rejoined the 16th Chasseurs à Cheval on 24 August 1801 and stationed in Northern Italy remained with them till 29 October 1803 when he transferred to the 19th Chasseurs à Cheval. On the outbreak of war against Austria in 1805 he served on Massena's general staff during the campaign in Italy. He distinguished himself at Veronette on 29 October 1805 and was promoted *colonel* on the battlefield.

In February 1806 he joined the invasion of the Kingdom of Naples. When Joseph Bonaparte later became King of Naples he joined the influx of French courtiers and officials who followed by accepting secondment to Neapolitan Army on 25 July 1806 as head the Neapolitan Guard Chevaulégers. On 11 November 1807 he became provincial commander of the troops in Basilicate. A more active field command followed when on 14 March 1808 he headed a mixed brigade with Maurice Mathieu's division during the reoccupation of Calabria. After the brief campaign he was made a Commander of the Order of the Two Sicilies on 19 May 1808. Whilst in Neapolitan service he developed a good rapport with King Joseph, and on Joseph's departure for Spain became his first equerry and senior aide de camp.

SERVICE IN SPAIN 1808-1813

Strolz left Naples for Spain on 8 June 1808 joining Joseph at Vitoria at the end of August. He helped restore Joseph's flagging morale after fleeing Madrid following Dupont's disaster at Baylen. He resumed his duties under Joseph, now in the service of Spain, and remained at Joseph's side throughout the first difficult year of his reign. Always anxious to resume a more active military career he secured a brigade command (10th and 26th Chasseurs à Cheval) in Merlin's light cavalry division as part of Sebastiani's IV Corps during the Talavera Campaign.

At Talavera he took advantage of Anson's poor handling of Wellesley's cavalry and on 28 July 1809 was one of the few to shine during the French defeat. Anson's cavalry made up of the 1st Light Dragoons of the King's German Legion and the 23rd Light Dragoons had rashly attacked Villatte's division deployed in the open, which suspiciously made no move to retire or form square. The simple reason for the French not moving was a deep gully that lay before their lines hidden from view. Anson's horsemen plunged into it and became disordered as they tried to scramble out the other side. Strolz with his brigade on Villatte's flank then proceeded to fall on the English cavalry causing them heavy loss. He had another success when he helped Sebastiani's forces rout Venegas's advance on Madrid at Almonacid on 11 August 1809.

On 17 October 1809 he became a Knight of the Royal Order of Spain, which was followed by another award when he became a Commander of the same order on 22 December 1809. He later received the Spanish Grand Cross on 18 June 1810.

Joining Joseph's invasion of Andalusia, he was on 15 February 1811 promoted *lieutenant général* in the Spanish Army. In the four years that followed he however never held another field command. As Joseph's senior aide and equerry his main function was to act in an advisory capacity on military and ceremonial matters. He remained at Joseph's side throughout this turbulent period, travelling the length and breadth of Spain. With Joseph's expulsion from Spain after the defeat at Vitoria he returned to French service with the rank of *général de brigade* on 23 January 1814. No field command offered, he spent his time as Joseph's aide during the final days of the Empire. After protests from Joseph his rank of *lieutenant général* was recognized in the French Army on 19 February 1814. He served at Joseph's side during the defence of Paris on 30 March 1814.

THE RESTORATION AND NAPOLEON'S RETURN

The Bourbons recognized his rank of *lieutenant général* but it did not prevent him being placed on half pay on 10 July 1814. They made a conciliatory gesture by granting him the title *Chevalier de Saint Louis* on 1 November 1814. A close confidant of Joseph not surprisingly he rallied to Napoleon on his return and on 26 March took the post as commander of the Strasbourg garrison. Then as the Army of the North gathered he received late orders to head the 9th Cavalry Division of Exelmans's II Cavalry Corps on 7 June 1815.

THE WATERLOO CAMPAIGN

In the campaign he soon came into his own and dispelled doubts as to his ability to lead a division. He distinguished himself at Ligny on 16 June with a well timed charge against Hobe's brigade as it tried to fall on the French right. At Wavre on 18 June Grouchy held him in reserve to support the infantry once they had forced the Dyle and as a result he saw little action. During the retreat to Namur he did fine work keeping the Prussian horse at bay, particularly when it became a race to secure the bridge over the Sambre at that town. He rejoined the rest of the army when it rallied at Laon and was with the rearguard as it fell back to Paris. He distinguished himself at Roquencourt near Versailles on 1 July where his division was part of a well laid ambush prepared by Exelmans. In the action they caught Sohr's cavalry brigade on its own as the Prussians tried to approach

Paris from the south. The resultant loss of five to six hundred Prussian horse had a sobering effect on an overconfident Blucher. Whilst it was a welcome fill up for French morale, it had in the end little impact on the campaign's outcome as the Armistice followed two days later.

LATER LIFE AND CAREER ASSESSMENT

Strolz remained with the army and took part in its disbandment till dismissed on 25 July 1815. Not considered an extreme Bonapartist he secured his recall on 1 April 1820, but remained a while unattached till on 31 January 1821 when he took the post as troop commander for the Department of Le Finistère. He remained at Brest for several years and only moved after the July Revolution when on 1 September 1830 he joined the Gendarmerie as an inspector general. The Gendarmerie were going through a period of change and he spent much time on serving on commissions examining and implementing new service regulations.

Also interested in politics, he gained election to the Chamber of Deputies as representative for Belfort on 5 July 1831. He continued his military career and on 5 June 1832 became Gendarmerie Inspector General for the 3rd, 6th and 16th Military Divisions. His work was highly regarded and he became a Grand Officer of the Legion of Honor on 18 April 1834. As a deputy he also gained a reputation for hard work and honesty and gained re-election in June 1834. Plagued by ill health he relinquished his active military duties on 6 August 1836. Finding it increasingly difficult to attend the long sittings of the Chamber he did not offer himself for re-election in August 1837. He was admitted to the Reserve Section of the General Staff on 15 August 1839. Highly regarded by Louis Philippe he was to be made a Peer of France but before he took up his seat he died in Paris on 27 October 1841.

Strolz was a very accomplished soldier, but the eight years he spent in the service of Naples and Spain did little to help to his career. Joseph Bonaparte's martial abilities were non-existent and anyone who formed part of his military entourage was invariably tarred with the same brush, with few prospects once they left the service of Spain. When he gained the equivalent general rank in the French army, his service record particularly in the Waterloo Campaign soon showed he merited the rank.

SUBERVIE - JACQUES GERVAIS, BARON (1776 - 1856)

Subervie was born in Lectoure (Gers) on 1 September 1776. He first came to notice when elected *lieutenant* by his men with the 2nd Battalion of the Gers Volunteers on 20 June 1792. He served in the Army of the Pyrenees and gained promotion to *capitaine* on 21 September 1793. The conclusion of peace with Spain brought the army's disbandment and he found himself without a post and placed on half pay on 12 October 1795. Anxious to return to active duty and a personal friend of Jean Lannes (the later marshal had joined the Gers Volunteers on the same day) he secured a post as his aide on 4 September 1797 with the Army of Italy. With Lannes joining the Army of the Orient Subervie followed and left for Egypt with him in May 1798. He took part in the capture of Malta on 12 June 1798 and while on the island caught dysentery and had to be left behind. Vaubois, commander of the island's garrison, appointed him as his aide and Subervie remained with him till Malta's capitulation on 5 September 1800.

He spent time in Portugal as Lannes's aide from November 1801 till the Spring of 1804 when Lannes was French ambassador to Lisbon. On 18 April 1803 he gained promotion to *chef d'escadron*. Returning to France he moved to the encampment at Ambleteuse where Lannes was preparing for the invasion of England. With Lannes during the Austrian Campaign he was at the Capitulation of Ulm and served at Austerlitz on 2 December 1805. Anxious for a line command he left Lannes on his promotion to *colonel* and took charge of the 10th Chasseurs à Cheval from Auguste Colbert on 27 December 1805.

PRUSSIA AND POLAND 1806-1807

During the Prussian Campaign his regiment operated with Colbert's light brigade of Ney's VI Corps of the Grande Armée. He fought at Jena on 14 October 1806 and took part in the pursuit of the broken Prussian Army. During the brief siege of Magdeburg that lasted till 11 November 1806 his horsemen formed the cavalry screen that protected the siegeworks. He moved on into Poland with his regiment and on 6 December took part in a sharp action covering Ney's crossing of the Vistula at Thorn on 6 December.

Involved in Ney's overambitious drive towards Konigsberg he was with Colbert at Schippenbeil on 21 January 1807 when the Russians fell on the over-extended advanceguard and his regiment suffered a severe mauling as a result. The confused events prior to the battle at Eylau on 8 February when Ney failed to make an appearance till late in the day appeared to dent Subervie's career prospects for a short period. He was following the Prussian rearguard when Lestocq managed to give his cavalry the slip before Eylau. The Prussians able to reach the battlefield uninterrupted fell on Davout's flank and nearly won the battle. Ney left on the other side of the field wondering where the Prussians had disappeared, had a few sharp words for Subervie after Napoleon had rounded on him. The frustrations of an inconclusive campaign were put right at Friedland on 14 June 1807. Under Colbert he joined Grouchy's cavalry on the French left, where late in the day they routed Uvarov's cavalry and harried as far as the Niemen.

SPAIN 1808-1812

In the Spring of 1808, Subervie with the 10th Chasseurs à Cheval joined Lasalle's division in Spain. After Dos Mayos Lasalle was anxious to regain favour with Napoleon and conducted a brilliant but brutal campaign against Cuesta's Army of Castile. The Spaniard tried to raise the populace around Valladolid and sever communications with France. Countering this Subervie's chasseurs played a key role. On 6 June they seized the bridge over the Pisuerga at Torquemada on the Valladolid-Burgos highway. They then harried the enemy as they fell back on Valladolid till Cuesta unwisely decided he had sufficient troops at hand to bring them back over the river and make a stand at Cabezon. Lasalle at the head of the cavalry with Subervie at his side destroyed the enemy force of 5,000 within minutes. The raw Spanish troops exposed were swept aside and each unit systematically broken. The only escape route was a single bridge that became clogged by fleeing men, which caused the enemy to panic further and be ripped to pieces. For the loss of barely fifty men they secured Valladolid, one of the most important towns of Northern Spain.

At Medina del Rio Seco 14 July 1808 he formed part of Lasalle's reserve that poured through the gap in Blake's line after Bessières's attacks had pinned the infantry. Blake lost over 3,000 men in his ill fated venture to retake Valladolid, and only Bessières calling off the pursuit early prevented an overwhelming victory.

When Napoleon took charge in Spain and crossed the Ebro in force, Subervie under Lasalle joined the renewed offensive. He was with Lasalle in the charge at Gamonal on 8 November 1808 that smashed through the center of Belvedere's Army of Estremadura and opened the way to Burgos. In

January 1809 he was with Victor's forces as they fanned out towards the Portuguese frontier while Napoleon moved north to trap Moore's army. In Estremadura Cuesta rebuilt a new army of 15,000 at Badajoz and Medina and decided to go on the offensive. Moving up the Tagus he pushed back Lasalle's outposts, which included Subervie, before occupying Almaraz. On 15 March Victor's preparations complete, he crossed the Tagus and with Subervie in the van marched via Talavera and Arzobispo to seek battle with Cuesta.

Exercising uncharacteristic caution Cuesta withdrew. His subordinate Henestrosa in charge of the cavalry drew Subervie's regiment into a well laid trap as he fell back over the Sierra Guadeloupe via the pass of Santa Cruz. At Berrocal on 20 March the Spanish Royal Carabineers and Guard Cavalry fell on the 10th Chasseurs à Cheval and after a sharp skirmish drew off. The next day undeterred by the incident Subervie continued to push forward recklessly far ahead of the rest of Lasalle's division. At the small village of Miajades a small body of Spanish cavalry presented itself on the road, while hidden on each side were the Infante and Almanza regiments of La Romana's expeditionary force from Denmark. Subervie charged the body to his front and suddenly saw himself enveloped and surrounded by the two regiments placed in ambush. In a furious melee that followed he lost over 150 men before they cut their way out when the sight of Lasalle's division following up caused Henestrosa to order a prompt retreat, which was accomplished without loss. Subervie left tearing his hair out at the sight of his broken regiment, received a severe reprimand from Lasalle for not exercising greater care.

Subervie was able to exact revenge when on 29 March 1809 Victor faced Cuesta at Medellin on the banks of the Guadiana. Operating with Lasalle and the 2nd Hussars on the French left, they suffered heavily as they tried to overcome the Spanish guns with a bold charge. Cuesta then advanced at the head of one of the finest combined Spanish infantry and cavalry assaults of the War and drove them back further. Charging again the 2nd Hussars managed to drive off the Andalusian Lancers. The movement uncovered Albuquerque's infantry, who unprepared to receive cavalry, offered brief resistance before they gave way. The humiliations of the previous week and the bombardment they suffered earlier in the day was reason for Subervie's men to give no quarter. The Spanish suffered 10,000 casualties, three quarters being put to the sword with only 1,850 prisoners taken, compared to estimates of 1,000 French casualties.

After Lasalle's departure for the Danube Subervie served under Paris who took over the division. He fought at Ocana on 19 November 1809 when south of Madrid Soult crushed Areizaga's threat to the city.

In the largest cavalry engagement of the Peninsula War his regiment was amongst the 8,000 horse that confronted each other on the open plain. The Spanish cavalry, unnerved after suffering 400 casualties at the hands of enemy horse the previous day, were in no mood to face the French again so soon. When the two armies met Soult skilfully kept his cavalry out of sight concealed by olive groves on the French left. Then with the Spanish infantry committed, Subervie joined the great charge from the olive groves led by Milhaud* that broke up Freire's shaky cavalry and then proceeded to roll up the Spanish line. It was possibly the most decisive battle of the Peninsular War with the Spanish suffering some 4,000 killed and wounded and a further 14,000 prisoners opposed to French losses of 2,000.

In January 1810 Subervie joined Sebastiani's IV Corps for the invasion of Andalusia. With the main force to occupy Granada province he entered the city on 5 February. The subjugation of the valleys of the Sierra Nevada and the coastal area followed. On 1 March he was at the rout of 2,000 cavalry under Freire that escaped Ocana. He joined Sebastiani's march on Malaga via Antequera and on 5 February outside the city helped disperse several thousand irregulars.

The insurgents took to the hills and he spent much of his next two years in Andalusia trying to track them in the Sierras de Ronda, Nevada and along the eastern coast. He also kept a watch on the movements of the Army of Murcia on the eastern border. A rare success came at Baza on 4 November 1810 when with a mixed force of 3,300 under Milhaud he took part in the defeat of 8,000 foot and 1,000 cavalry of Blake's Army of Murcia when it threatened the city of Granada. Promoted *général de brigade* on 6 August 1811 he served on Suchet's staff during the Army of Catalonia's invasion of Valencia. He fought at Saguntum on 25 October and was present at the fall of Valencia itself on 8 January 1812.

THE RUSSIAN CAMPAIGN

Recalled to France on 22 February 1812, he was then despatched to Poland on 28 March to head a light cavalry brigade of Prussian and Polish troops for the coming campaign against Russia. In May he renewed his association with Sebastiani when the latter became his division commander heading the 2nd Light Cavalry Division of Montbrun's II Cavalry Corps, with his formation becoming the 16th Light Cavalry Brigade. Part of the Napoleon's vanguard as the French and their allies poured across the Niemen, his troops had an active time against cossacks. On 5 July 1812 his 10th Polish Hussars won a sharp action on the banks of the Disna.

Present at Borodino on 7 September 1812 Subervie's men suffered severely from a cruel bombardment by the Russians as they waited in the open to join the assaults on the Great Redoubt. Montbrun died while waiting for orders and Subervie was laid low by a splinter that tore through his thigh. Medical facilities in Moscow totally inadequate he was fortunate to be evacuated to Vilna soon after, so missing the horrors of the retreat, which almost certainly would have killed him. As it was, the wound did not heal and in February 1813 the Ministry of War placed him on indefinite leave.

GERMANY AND FRANCE 1813-1814

On 11 August 1813 Subervie returned to duty as acting commander of the 9th Light Cavalry Division of V Cavalry Corps till 15 October when Piré took over and he received a brigade command under him. The division itself was formed from depot squadrons of hussar and chasseur à Cheval regiments, which offered very unpromising material. As a result it spent most of its time away from the main spheres of operations keeping lines of communications open and carrying escort duties. Present at Leipzig, he was with the cavalry engaged before Wachau, though records indicate the division did little that day to halt the Allied advance.

He remained with Piré's division during the campaign in France, and built a sound reputation as he fought at Saint Die on 10 January, Brienne on 29 January, Champaubert on 10 February and Montereau on 18 February. Detached he took part in the defence of Paris on 30 March, where caught in a melee by Prussian uhlans he was left for dead after taking three lance wounds. From his hospital bed he received news of his promotion to général de division on 3 April 1814 in one of Napoleon's last acts as Emperor. The Bourbons, not to be outdone, annulled it five days later. He became a *Chevalier de Saint Louis* on 19 July 1814 followed by confirmation of his promotion on 23 July 1814. The totally inconsistent treatment continued when on 1 September he was placed on half pay.

THE WATERLOO CAMPAIGN

With such shabby treatment meted out to him it was not surprising he rallied to Napoleon's cause on his return. He renewed an old association with Pajol when he took command of the 5th Cavalry Division of I Cavalry Corps on 3 June 1815. In the advance on Charleroi on 15 June he was not heavily engaged during the day as the brunt of the corps' action at Charleroi and Gilly fell on Pierre Soult's division. At Ligny on 16 June, detached from Pajol he moved to the French left to support the assaults of Vandamme's III Corps. A breakthrough not achieved in this sector he spent the day much of the day involved in artillery duels and cavalry feints that failed to uncover the Prussian right.

The next day his division was attached to Lobau's VI Corps, which had no cavalry. Ideally placed on the French left his force moved up the Quatre Bras road in search of Wellington's army. Given the slip, with Colbert's lancers they caught up with Vivian and Vandeleur's squadrons covering the rearguard around 3.00 PM. He tried to charge in the midst of a blinding thunderstorm outside Genappe, but a few English horse batteries for a while kept his men at bay. Once the English guns moved off he spent the rest of the day in appalling conditions literally prodding the backs of the rearguard. In the evening as he approached Mont Saint Jean he was brought to a halt by cannon fire from the ridge that unmasked Wellington's position.

During the battle the next day Subervie's division formed part of Lobau's reserve positioned behind La Belle Alliance. Around 1.00 P.M., with the Prussians observed in the distance emerging from Chapelle Saint Lambert, Napoleon ordered his and Domon's divisions towards the Bois de Paris to delay their advance while VI Corps followed. As the Prussians emerged from the woods around 4.30 P.M. his cavalry in a brilliantly timed charge overwhelmed Prince William of Prussia's cavalry of Bulow's IV Corps. The Prussian infantry paralyzed, it won Napoleon valuable time before their main weight fell on Lobau's infantry. Throughout the struggle as the Prussians gradually turned the French right Subervie and Domon harried the enemy advance. Weight of numbers finally forced Lobau back on Plancenoit with Subervie able to cover the French withdrawal and leave the field reasonably intact at the end of the day. He rejoined Pajol at Laon on 25 June and from there retired by stages to Paris. After the Armistice he moved to behind the Loire where he remained with his troops as they disbanded. On 7 August his work complete he was placed on the non active list.

LATER CAREER AND ASSESSMENT

A partial reconciliation with the Bourbons came with his recall in July 1820, but it was followed by a lengthy period unattached. A scare supposedly caused by the influence former Bonapartist generals held in the army resulted in his compulsory retirement on 16 February 1825. He found an outlet in politics and with his colleague Pajol actively conspired to bring down the monarchy in the July Revolution. On 30 July 1830 when the success of the revolt was in the balance he was made commander of troops in Paris and kept order in the city when Pajol led the insurgents against the Bourbon troops at Rambouillet. On 4 August, with order restored and Louis Philippe on the throne, he relinquished his post.

On 13 August 1830 a grateful government appointed him to head an enquiry to examine the plight of officers who lost their posts with the fall of the Bourbons. It was the start of a very distinguished career that lasted for a further twenty three years. On 3 September recalled to the Army General Staff he became a cavalry inspector general for the 1st and 16th Military Divisions. His leftist political leanings soon caused embarrassment and when elected deputy for Lectoure in July 1831 he found himself unattached. Recall came with the post of inspector general for the 12th Arrondissement in 1834, which coincided with his re-election as deputy. A fairly influential figure in the Chamber he was a member of numerous bodies set up to reform the cavalry. As the years passed he served as inspector general in the 8th, 3rd and 10th Arrondissements till his retirement on 2 September 1841.

A shock came when the radical lawyer Salvandy defeated him in the elections of September 1842. He regained his seat on 1 August 1846 and actively conspired against Guizot's government. With the abdication of Louis Philippe he joined the provisional government on 25 February 1848 as its Minister for War till replaced by Cavaignac after the April elections. Many honorary positions and awards came his way as he took up the role of retired soldier and elder statesman. On 4 March he took the post of Grand Chancellor of the Legion of Honor, followed by receipt of the Grand Cross of the Legion of Honor

on 11 December 1848. The gradual decline into radical republicanism resulted in his retirement from the Army on 8 June 1848. He also gave up his duties as Grand Chancellor as he found on his re-election as a member of the Legislative Assembly in May 1849 the post too demanding.

Louis Napoleon as President of France and then Emperor gave his career fresh impetus when he recalled him from retirement on 1 January 1853. Placed in the Reserve Section of the General Staff Napoleon III frequently referred to him for advise on military matters. With Flahaut he had the rare distinction of serving both the First and Second Empires as division commanders. He died at his home Chateau de Paranchère, Ligueux, Gironde on 10 March 1856.

Subervie was another solid reliable professional that made up the backbone of the Imperial Army. The frequent reverses in Spain made promotion slow but once he attained general rank his performance in France in 1814 showed he was worthy of it. At Waterloo his endeavors certainly dispelled any doubts that he deserved a divisional command. After Waterloo his career slightly blighted by politics was long and distinguished. He showed his professionalism to the end.

ONE OF SUBERVIE'S LANCERS

TESTE - FRANÇOIS ANTOINE, BARON (1775 - 1862)

BACKGROUND AND EARLY CAREER

Son of an advocate François Teste was born on 19 November 1775 in the small town of Bagnols-sur-Céze (Gard), north of Nîmes in the Rhone Valley. He enlisted as a grenadier in the Gard National Guard at Nîmes in August 1792. Educated and intelligent with undoubted leadership qualities he rose rapidly through the ranks to become *sergent* on 1 April 1793. His first active service was in South-east France with the forces sent to reduce the Royalist stronghold at Jalés. Elected *lieutenant colonel* of a conscript battalion from Pont Saint-Esprit (later to become the *4e bataillon de la Montagne*) on 4 September 1793 his unit served with the Army of the Pyrenees Orientales. His budding career came to a halt on 30 May 1794 when reverses on that front resulted in purges that ended in his suspension for incompetence by the Representative Dartigoeyte. His suspension lifted on 17 March 1796, it took him till 11 September 1798 before he was reinstated with the rank of *chef de bataillon* on the regimental staff of 49th Line with the Army of the Danube.

A period 1799-1800 followed with the Army of Switzerland as senior *chef de bataillon* of the 87th Line. He soon came to the notice of General Chabran and on 22 November 1799 became his aide de camp. A move followed to Chalons-sur-Saone where he helped Chabran to form the 1st Reserve Division made up of fourteen depot battalions from regiments posted with the Army of the Orient. On 14 February 1800 he became the division's chief of staff and then with the official formation of the Army of the Reserve on 26 April his formation became its 5th Division. He was present during the march into Switzerland and the crossing of the Alps via the Little Saint Bernard Pass. The passage through the mountains blocked by the Austrians at Fort Bard he took part in the siege of the fortress from 26 May till 1 June 1800. The siege over, Chabran's division moved into the Po valley under Duhesme's* command, with Teste at division headquarters at Casale watching the river crossings for a possible breakout by Melas's army when Napoleon fought at Marengo. After the battle and the Convention of Alessandria Teste remained in Italy gaining promotion to colonel of the 5th Line on 9 August 1800. A reorganization of the armies in Italy followed after the Treaty of Lunéville with the result that his regiment passed to Murat's Army du Midi gathered at Florence to counter the threat from Naples. He served with the army in Southern Italy till its disbandment in 1802.

ITALY AND DALMATIA 1805-1809

The next four years saw him with the Army of Italy and with the outbreak of war against Austria in September 1805 his regiment served with Molitor's 3rd Division. He distinguished himself in the struggle for San Pietro on 30 October during the crossing of the Adige, while Massena triumphed at nearby Caldiero. Massena promoted him acting *général de brigade* on 4 November 1805. His brigade was comprised of his old regiment the 5th Line and the 23rd, forming part of the occupation force that moved into Dalmatia in April 1806. His rank was later confirmed on 28 July 1806. In September-October 1806 helped defend Ragusa (Dubrovnik) when Marmont held out against a Russian force led by Seniavin from Corfu supported by Montenegrins that disputed the French occupation of the Dalmatian coast. In June 1807 he became commander of the garrison at Split and then the following month he rejoined the Army of Italy. Varying periods followed as commandant at Brescia, Verona and Treviso as part of Souham's division.

THE AUSTRIAN CAMPAIGN OF 1809

War imminent against Austria, Teste joined Grenier's 3rd Division as commander of its 3rd Brigade (1st Line and 8th Légère). He struck up a good rapport with this solid veteran of Italian campaigns whose experience went back to leading a corps under Moreau in Germany in 1800. At the Battle of Sacile on 16 April with the 1st Line he led the final assault against the Austrian strong point in the village of Porcia. He left the field when his foot was broken by a shell splinter earlier in the day but returned when Barbou's division on his left gave way and managed to rally it. Barely mobile he insisted on retaining his command and when Eugene reorganized the army Pacthod become his new commander while Grenier returned to a corps command. He fought at Raab on 14 June where in the battle's early stages his troops were held in reserve till the struggle reached its climax. Then as Archduke John committed his last reserves and drove back Durutte's and Severoli's divisions before the Szabadhegy heights Teste's troops joined the battle. With Prince Eugene at his side they brought Jellacic's advance to a halt while Durutte and Severoli rallied. Together the three divisions then hurled the Austrians from the heights. Eugene now in control of the key position and Grouchy's cavalry successful on the right the battle was won.

At Wagram on 6 July 1809 he joined Macdonald's great assault against the Austrian center on the battle's second day. As Macdonald led a massive battalion square formation against the enemy, Teste's brigade in the van suffered a withering fire as Archduke Charles committed his last reserves to stem the oncoming French. The attack faltered, but as Pacthod moved up with the rest of the division supported by the Guard artillery it was enough to cause Charles to realize the battle was lost and order a retreat.

Awards followed, Napoleon's birthday on 15 August 1809 brought Teste an annual pension of 4,000 francs draw from Rome. He returned to Italy in the Autumn and the following year after his recall to France due to ill health in October he became a *baron de l'Empire* on 21 November 1810.

GERMANY AND RUSSIA 1811-1812

His return to duty on 2 April 1811 saw him posted to the Corps of Observation of the Elbe as commander of the 2nd Brigade of Compans's 5th Division. Hardly had he arrived at his post when he found himself reassigned as military governor to the fortress of Kustrin on the Elbe. As the build up of troops in Prussia and Poland gathered momentum he rejoined Compans's division on 13 September 1811. Part of Davout's I Corps for the campaign in Russia, his brigade was a fine formation made up of five battalions of the 57th Line numbering some 3,150 officers and men. The march into Russia immediately proved costly in men and equipment as well as frustrating as the Russians avoided battle. Teste's first task was to protect the supply convoys as a massive traffic jam developed around Vilna accompanied by rumors the enemy would evade the Grande Armée's columns and strike back at Napoleon's communications.

He rejoined Compans after the fall of Vilna and was part of Davout's center column that tried to envelop Bagration's army. The plan failed and when he joined the action at Mohilev on 8 July the Georgian had given them the slip. Present at Smolensk on 17 August during the great struggle before the city his brigade remained in reserve. At Viazma on 28 August Murat ordered him forward, but the move proved too reckless and Davout, amidst threats and recriminations from the King of Naples, countermanded the order. He fought for the first time in the campaign on 5 September when his men stormed the Shevardino position two miles before Borodino. What initially started as a minor action turned into a major struggle as Compans first led forward Duppelin's brigade to probe the village defences. Driven back the battle became a trial of strength as the divisions of Morand* and Friant* were drawn into against the Russian center. They too were also brought to a halt, and it was Teste who broke the deadlock at the head of a furious charge by the 57th and 61st Line that swept through the village and secured the redoubt.

At Borodino two days later he played a key role in the assaults on the Bagration fleches. At 7.00 A.M. Davout opened the battle with a bombardment by over 100 cannon. Compans's division then led the way with Teste's brigade in the fore. Supported by a further 30 guns that moved forward with the infantry columns that unlimbered and fired as they neared the enemy, his men cleared the woods of skirmishers at the foot of the slopes. At the head of a furious charge that burst from the woods his men broke into the westernmost fleche. The Russians rallied and the 2nd Grenadier Division supported by elements from the 27th Division drove him back him back as Morand's attack on the second fleche also faltered. Casualties mounted, Compans fell wounded and Teste took over the division. The situation became more confused when Davout was stunned after his horse fell on him, reducing him to the role of a dazed spectator. Around 10.A.M. with the 24th Légère of Ledru's division in support, Teste leading the 5th Division retook the fleche. Shot in the arm he also had to leave the field while his men grimly held onto their gains for the remainder of the day.

Sent to Viazma to recover he took the post as the city's military governor. Here he helped maintain the vast depot and kept communications open with Moscow. He rejoined Compans on 3 November when the army fell back through the city during the retreat from Moscow. Engaged in a vicious struggle outside the city he helped thwart an attempt by Miloradovich to cut between I and IV Corps and trap Davout's force. His brigade reduced to a bare few hundred men after they pushed aside the Russians played little further part in the campaign. When it reached Konigsberg in East Prussia and returns were submitted on 5 January 1813, only 167 officers and men remained.

GERMANY 1813

Teste returned to France and his solid performance in Russia resulted in his promotion to *général de division* on 13 February 1813. Responsibility followed for bringing to readiness the 4th Division of the Corps of Observation of the Rhine formed from conscripts. On 12 March his formation, renumbered the 23rd Infantry Division, was allocated to Marmont's VI Corps. The formation was not ready when Marmont moved his corps from Mainz into Saxony and it was decided to leave it in Westphalia to help maintain order. When a cossack thrust across the Elbe threatened Cassel he moved on the city to steady the nerves of Napoleon's brother Jérôme*. The threat removed he took the appointment as military governor of Magdeburg on 18 April with his division moving into the fortress to form its garrison. The line of the Elbe secured in Westphalia by these movements, Napoleon was able to concentrate against the Allies in Saxony.

During the Armistice he built his division up to its full strength of around 10,000 men with 14 battalions and 14 cannon before it passed to Vandamme's[*] I Corps on 30 July. The renewal of hostilities saw him detached from Vandamme with a brigade in action at Dresden on 26 August. With Dumoustier's division on the French right they recaptured Lobtau and the next day with Pajol's cavalry in support they trapped Meszko's division at Pennrich. On 28 August he left his troops at Dresden and made a long round about journey to rejoin his 2nd Brigade led by Quiot, which had crossed the Elbe at Pirna with Vandamme's force. He fought at Kulm on 29 August when the Allies halted Vandamme's advance. Then the next day when Kleist's Prussians emerged in Vandamme's rear, Teste with Vandamme was one of the 10,000 prisoners taken.

THE RESTORATION AND NAPOLEON'S RETURN

He returned to France with Vandamme in June 1814. Realizing there was little alternative he took a more conciliatory view than Vandamme and acknowledged the Bourbons. For his pains he was made a *Chevalier de Saint Louis* on 8 July 1814 and on 31 August 1814 became commander of troops in the Department of Pas de Calais. Napoleon on his return retained him at his post where he had a tough time in an area that teemed with Bourbon sympathizers. The situation was aggravated by Louis XVIII issuing a proclamation from Ghent vowing to cross the border and re-establish a military presence in the northern departments. Royalist agents worked to subvert the populace whilst many government officials and prefects were openly anti-Bonapartist. Vandamme joined him as overall commander for the region and helped eliminate the Royalist influence in Pas de Calais and Oise. Showing a bold military presence Teste operated several mobile columns passing through the countryside appealing to the populace's sense of patriotism rather than enforce harsh measures. The tactics soon worked, Bourbon activity decreased and on 23 April 1815 he was recalled to Paris. Despatched to Laon he took command of the 21st Infantry Division with VI Corps of the Army of the North.

THE WATERLOO CAMPAIGN

During the first three days of the campaign his role was very much that of spectator. He missed Ligny, as his division last on VI Corps's route de marche through Charleroi, only reached the battlefield late in the day. His formation fresh, the following day it was attached to Grouchy's wing and joined Pajol's cavalry in search of the Prussians. He could have fallen on the Prussian rearguard at Gembloux and pinned their retreat had Exelmans,

who first sighted them, called on his help, but instead the enemy was allowed to slip away. At Wavre on 18 June he showed his true metal after Pajol's cavalry seized the bridge over the Dyle at Limale. His division though the smallest in the army numbering less than 3,000 men, secured the French position on the opposite bank when it drove Stupenagel's brigade from the village itself. Then in the gathering darkness as Gérard's division joined the struggle they drove the Prussians from the high ground north of the village and by 11.00 P.M. were in possession of the road to Mont Saint Jean.

The next day, unaware of the fate of the rest of the army he joined the renewed attacks on Wavre by moving against the village of Bierges. A tough struggle followed, where he lost his fine brigade commander General Penne. Around 9.00 A.M. the 31st Prussian regiment retired and the way lay open to the Wavre-Brussels road. An hour later news of Waterloo broke and his was the first infantry formation sent to cover Grouchy's retreat to Namur and secure a crossing over the Sambre. The next day he took position outside the city till III and IV Corps passed to safety. Then occupying the antiquated fortress till nightfall, he covered the passage of Grouchy's forces as they retired up the Meuse valley into France.

Out of the ruins of defeat, his heroic defence of this gateway to France in the dying stages of the campaign propelled him from being a reliable career soldier to the status of a national hero. For the defence of the time-worn broken-down fortifications of Namur he could barely muster 2,000 men and 8 field guns. Undismayed his small force boldly manned the old ramparts and secured the gates. Hardly were they in position when the Prussians of Pirch I's corps delivered their first assault. Met by a hail of grape-shot and musketry they reeled back from the glacis leaving the slopes littered with dead and wounded. Without delay Pirch I ordered a second assault, while Teste's men short of ammunition, held their fire till the last moment and again mowed down the Prussians on the glacis. After losing over 1,500 men Pirch I decided to break off the action.

Teste felt he had given Grouchy sufficient time to clear Namur and rather than face a third attack in failing light with little ammunition he began to retire. The withdrawal noticed by the Prussians they dashed forward, burst into Namur and pressed on to seize the bridge over the Sambre. Teste however had foreseen the danger by placing a detachment of engineers in loop-holed houses covering the gate to the bridge. This fire supported by a battery of guns on the opposite bank checked the Prussians and enabled his rear-guard to pass through the "Gate of France" and over the bridge prepared for demolition. When all were clear the sappers set alight bundles of straw. The flames spread from the gate to the houses along the river and prevented any further pursuit. Teste's spirited resistance helped Grouchy's force reach Dinant unmolested and gain a few days respite.

He remained with the army after it rallied at Laon. Then with the conclusion of the Armistice on 4 July he retired with his troops to the Loire. In the days of uncertainty that followed Davout made him responsible for the defence of the river's left bank at Orleans. The disbandment of the army complete, he was placed on the non active list on 3 August 1815.

LATER CAREER

On 30 December 1818 he was recalled but several frustrating years followed unattached on the General Staff. He eventually became an Inspector General of Infantry for the 13th Military Division at Rennes on 7 May 1828. A true professional, the fall of Charles X did not affect his prospects and on 3 August 1830 he took command of troops in 14th Military division with headquarters at Rouen. A move to the General Staff in Paris in February 1831 helped him the following August secure command of the Army of the North's 4th Division for the invasion of Belgium. After the brief campaign he returned to Rouen in his last active post as commander in January 1832.

A grateful Belgian government later made him a Commander of the Order of Leopold on 15 October 1839, which was followed by his elevation as a Peer of France on 7 November 1839. He passed to the Reserve on 12 November 1843, and then in the early days of the Second Republic when politicians were anxious to cleanse themselves of relics from the past, he was retired on 8 June 1848. A fact that did not help, was his younger brother had been a minister in one of Louis Philippe's governments, had gained a certain notoriety due to his involvement in several financial scandals that had resulted in his trial and imprisonment. A brief revival in fortunes came after Louis Napoleon's election as President of France when on 14 December 1849 he received the Grand Cross of the Legion of Honor. There was little to justify the award apart from it being another cynical attempt by Louis Napoleon to remind people of past glories by wheeling out forgotten old generals. Tired and still affected by his brother's scandals he turned down invitations to resume an active life and lived in quiet retirement. He died at Angoulème on 8 December 1862 just after his eighty-seventh birthday, living the longest of Napoleon's generals in the Waterloo Campaign.

AN ASSESSMENT

Promotion came relatively late for Teste, as to many who served in theaters away from Napoleon. His advancement was also hindered by his close association with generals Grenier and Durutte, both for many years Republican sympathizers and friends of the disgraced Moreau. Under the eye of Napoleon for the first time at Wagram he deserved a divisional command at this stage, but had to wait till after Borodino. In 1815 when Davout recommended his appointment to head a division of Lobau's VI Corps, which was made up of some of the finest line regiments in the army, it showed the importance attached to the post. His coolheadedness covering Grouchy's retreat after Waterloo was one of the finest performances of the campaign and once more showed he was a commander of class. A sound professional, he epitomized the younger generation of aggressive commander that rose to prominence in the latter part of Napoleon's career.

VANDAMME - DOMINIQUE JOSEPH RENÉ (1770 - 1830)

The rough, tough, hardfighting Vandamme, the *bête noir* of many an unfortunate marshal, was born in the market town of Cassel (Nord) on 5 November 1770. The son of a surgeon he fell under the patronage of the Duc de Biron, who later commanded the Army of the North. He enrolled as a pupil at the Ecole de Militaire de Paris. He was an intelligent student, but his rebellious and quarrelsome nature however ended with him failing to gain a commission. Undeterred, he enlisted as a private with the Martinique Colonial Regiment joining its depot battalion at Lorient on 27 July 1788. His background, education and forceful personality enabled him to rise through the ranks and by the time he arrived in Martinique at the end of March 1789 he was a *sergent*.

With events in France rapidly gathering momentum and garrison duty in the colonies not exactly to his liking he deserted in April 1790. He found his way back to France and next appeared in July 1790 as grenadier captain with the ragtag Cassel National Guard. The unit later disbanded, on 7th June 1791 he enlisted as a *fusilier* with the Brie Infantry Regiment (later the 24th Line). After a short period he moved on obtaining an honorable discharge on health grounds on 25th August 1792. He almost immediately set about raising a volunteer company of light infantry known as the Chasseurs de Vandamme with himself as *capitaine* and commanding officer, which was incorporated into the Army of the North on 13 September 1792. It later became the Chasseurs du Mont-de-Cassel with him still its commanding officer serving during Dumouriez's campaign in Holland. When the unit amalgamated with others to form the light battalion of Mont-de-Cassel his men on 1st August 1793 elected him their *colonel*.

THE ARMY OF THE NORTH 1793-1794

The Army of the North having been driven from Belgium into Vandamme's home district, he had a unique opportunity as he knew the countryside well. The army's commander Houchard to break the deadlock and relieve Dunkirk gave Vandamme on 5 September 1794 command of a column sent to its relief. With this advanceguard of 4,000 men he played a key role in drawing Freytag's covering army away from the port and in its subsequent defeat at Hondschoote on 8 September. Then turning on the Duke of York's exposed force around Dunkirk he pursued it across the Belgian frontier breaking contact once he reached Furnes on 12 September.

As a colonel he had commanded a division sized formation with great success. Matters were soon put right on 27 September 1793 when he gained promotion to *général de brigade* and made commander of the camp at Cassel to prepare for a new offensive into Belgium. Hardly had he settled in his new post when on 8 October he moved to Dunkirk and replaced Souham as garrison commander. In the renewed offensive he advanced up the coast recapturing Furnes on 22 October before laying siege to Nieuport.

At his young age he lacked the experience required and badly underestimated the difficulty of the task at hand. His preparations totally inadequate, without a siege train and adequate means of transport he tried to reduce the walls with just field artillery. When the enemy breached the dykes on 29 October and flooded the area his troops fell back to Furnes in great disarray. He lost his command as a result and the Representatives ordered his arrest on 1 November 1793. For a time his life was in very real danger since Houchard, who he admired, had recently run afoul of the Deputies, been arrested, tried and executed. He managed to talk his way out and on 11 November was reinstated and given command of a brigade with Ferrand's division.

When Moreau replaced Ferrand at Cassel on 14 April 1794 a very active year followed for Vandamme. On 28 April he relieved Werwicq. Two days later after taking Menin he was again in trouble and nearly lost his command for allowing the enemy to break out of the town. He fought at Courtrai on 11 May with mixed results and then received praise for driving back Clairfayt's assaults from Tourcoing on 18 May. From 4 to 18 June he covered the French forces that completed the siege at Ypres and then advancing into Belgium he took Bruges on 29 June and Ostend the next day.

The reduction of Nieuport had again proved a problem for the Army of the North. Not prepared to allow the second siege to jeopardize his career Vandamme arrived before the lines on 4 July and, taking control of the operations amidst great protests, successfully concluded the operation on 19 July 1794. He then crossed into Holland and on 28 July completed the capture of the Isle of Cadzand. After that he took time to reduce Venlo on the Maas, which held out till 26 October 1794. Then before winter set in on 9 November 1794 he established a bridgehead over the Rhine at Buderich.

On 24 December 1794 he took over Moreau's division on the latter's promotion as the army's deputy commander. Campaigning on the Lower Rhine region he captured Arnhem on 17 January 1795. He then served with Macdonald's right wing clearing the Duke of York's forces from the north-eastern provinces of Holland and Hanover, concluding the campaign with the occupation of Bremen on 14 April 1795.

Recalled to Paris by the Directory, after a brief period of leave he rejoined the Army of the North in Holland on 7 June 1795. Ordered by Moreau to head a large detachment of reinforcements to join the Army of the Sambre-Meuse he raised objections. Moreau, keen to be rid of a former confidant who had become a rather quarrelsome high handed subordinate, dismissed him for insubordination and looting. He undoubtedly was guilty of the second offence, but there were many others far worse than him who were not brought to book. It left a stain on his career that others were to profit from whenever they needed to settle old scores.

ARMY OF THE RHINE 1795-1798

Vandamme returned to his home at Cassel. However the Directory, in need of hard headed generals not afraid to put La Vendée to fire, he was called to join the Army of the West at rennes on 29 September 1795. Hardly had he arrived when he was called to the Army of the Rhine-Moselle joining Gouvion Saint Cyr's 11th Division on 22 November 1795. Then with the army undergoing continual reorganizations he passed to Duhesme's 7th Division on 6 May 1796. With Duhesme* during Moreau's advance deep into Germany, he avoided the controversy that surrounded Duhesme when his division failed to appear at Neresheim on 11 August and Moreau suffered a sound defeat at the hands of Archduke Charles. He took over the division on Duhesme's suspension and won fresh laurels as he covered the retreat at Freiburg on 24 August and Biberach on 20 October. The army once more behind the Rhine, he then conducted a stubborn defence of Kehl during November, maintaining a tenuous foothold on the right bank.

Ill health in December 1796 forced him to take leave till the following April. Duhesme meanwhile rehabilitated, Vandamme resumed his post under him, heading the division's advanceguard when Moreau's army crossed the Rhine on 20 April 1797. With brigades led by Davout and Lecourbe in support, he crossed at Diersheim and established a bridgehead that enabled the rest of the army to pour across. A rapid advance followed as news of General Bonaparte's spectacular advances in Italy spurred the Army of the Rhine into action. When news of the Armistice of Leoben concluded by Bonaparte on 23 April 1797 reached him five days later he had reached Gengenbach.

In 1798, with attention focused on an invasion of England, Vandamme moved to the Army of England and had his first taste of serving Bonaparte. Neither his demeanor nor reputation impressed Bonaparte, with the result he never received an invitation to join the expedition to Egypt and instead found himself posted to Cherbourg as local commander.

As relations deteriorated with Austria, in September 1798 he moved to the Army of Mayence as commander of a brigade with Saint Suzanne's division. When Jourdan took charge of the army he persuaded Vandamme to accept promotion to *général de division* on 5 February 1799 and gave him command of the army's advanceguard. When hostilities commenced on 12 March 1799 his force served with Saint Cyr's left wing of the army of the Danube. Crossing the Rhine he advanced through the Black Forest and reached Freudenstadt on 5 March. Continually at loggerheads with Saint Cyr, the latter replaced him with Lefebvre, who a more senior commander had just returned to duty after a spell of illness. Vandamme joined Jordan's staff till an independent command was found in order to get him from under his chief's feet. Detached with a small force of two infantry regiments and three squadrons of cavalry he was given the task to maintain contact with Bernadotte's Army of the Lower Rhine. Reports filtered through of Archduke Charles threatening to counter the French offensive by cutting between the armies and crossing the Rhine at Kehl.

Vandamme pushed ahead trying to locate the enemy to no avail with the result that his 3,000 men were sorely missed when on 21 March 1799 Charles defeated Jourdan at Ostrach. At the time he was at Friedingen on the Danube's left bank far from his original objectives. He rejoined Jourdan at Stockach on 25 March and took part in Saint Cyr's assaults against the Austrian right. With Soult* in support they drove the enemy back and took over 3,000 prisoners, but when Jourdan's assaults faltered elsewhere there was little option but to retire. As the army retreated to the Rhine he was in charge of the rearguard. At Muhlingen on 28 March he turned on the overconfident Austrians and nearly captured Charles in the sharp action that followed, and in the process took a further 1,000 prisoners before joining the rest of the army at Moeskirch.

THE ARMY OF BATAVIA 1799

Critical of Jourdan's handling of the campaign Vandamme soon ran foul of his chief, who keen to be rid of his troublesome subordinate brought charges of extortion, looting and embezzlement against him. Recalled by the Council of War on 27 April 1799 to face charges he spent three months in Paris waiting for his case to be heard. Fortunately for him the military situation had deteriorated on all fronts. Rather than let a talented but at times insubordinate general languish in Paris the Directory posted him to Brune's Army of Batavia as commander of its 1st Division. Arriving in Holland on 6 September he was soon in action at Zyp four days later. He then fought at Bergen on 19 September inflicting over 4,000 casualties against the Anglo-Russian force and was largely responsible for halting the Allied advance in Northern Holland. He was with Brune at Alkmaar on 2 October and Castricum on 6 October where he again distinguished himself. When the Allies abandoned their plans he joined the negotiations on 21 October that ended in their evacuation from Holland. His part in this important campaign was crucial and went a long way to securing Brune his marshal's baton in 1804.

THE ARMY OF THE RHINE 1800

After a period of leave at Cassel Vandamme was recalled on 26 January 1800. He then spent a short period in Paris before joining Moreau's Army of the Rhine at Basel on 30 March. He fought at Moeskirsh on 5 May, and Memmingen five days later. He was again in trouble when he criticized Moreau for not pursuing the campaign vigorously enough after the army's initial successes. Quick to take advantage, Moreau brought charges of looting and other various administrative irregularities against him and on 23 May 1800 had him recalled to France. After a hearing on 17 August 1800 he was cleared, but the stigma stuck and he remained unattached. Desperate, he secured an interview with the First Consul and convinced him of his unfair treatment with the result received command of a division with the Army of the Reserve on 6 September 1800.

THE AUSTERLITZ CAMPAIGN

He remained with the Army of the Reserve till 19 September 1801 when he was appointed commander of the 16th Military Division at Lille. Then with the build up of forces for the invasion of England he moved to the Camp of Saint Omer on 30 August 1803 where he took over a division there. He spent the next two years bringing his troops to a high state of efficiency and with the formation of the Grande Armée on 25 August 1805 his division became the 2nd Division of Soult's II Corps.

Marching across France he crossed the Rhine at Speyer on 25 September. Then pushing on through Baden and Württemberg he reached the Danube at Donauworth on 6 October. Temporarily foiled by the Austrians, who had burnt the bridge over the river he seized another at nearby Munster that enabled II Corps to pour across. Forming the advanceguard he reached Augsburg on 9 October, cutting Mack's communications with Vienna. Pressing on through Landsberg, he then cut west through Memmingen and Biberach reaching Ulm from the south-west on 16 October. With the encirclement of Mack's Austrians completed it resulted in their capitulation on 20 October 1805.

At Austerlitz on 2 December 1805 he played a leading role in the defeat of the combined Austrian and Russian armies. Concealed by the morning mist he surprised Miloradovich's division and drove it from the Pratzen Heights, splitting the Allied front in two. Counter attacked by Kutuzov's guard infantry he drove them off, but his flank was left in the air due to Bernadotte failing to come up in support and one of his brigades was badly mauled by enemy cavalry. Bessières's timely intervention with the Guard cavalry saved the situation and Vandamme then wheeling southwards took part in the final destruction of the Allies caught against the Satschen Pond.

THE PRUSSIAN CAMPAIGN - THE OCCUPATION OF SILESIA

Continually at loggerheads with Soult, he was replaced by Leval on 16 July 1806 and attached temporarily to the Grande Armée's general staff. Energetic leaders needed to exploit the Prussian defeat at Jena, on 20 October he replaced Mahler as head of the 3rd Division of Ney's III Corps. Engaged in the pursuit he helped bring the siege at Magdeburg to a speedy conclusion on 11 November 1806 after terrorizing the populace into submission by threats of what would happen if the French took the city by storm.

On 27 November 1806 he took command of IX Corps made up of Bavarian and Württemberg troops under Jérôme Bonaparte*. His task was twofold, to occupy Silesia and teach Napoleon's youngest brother the arts of generalship and warfare. The Prussians humiliated by Jena and the disgraceful submission of fortresses of the likes of Magdeburg, Custrin and Stettin, were prepared to sell themselves dearly to restore their country's honor. The defences of cities like Breslau and Glogau, built by Frederick the Great to preserve the greatest conquest of his reign, were in good order and their magazines well provisioned. For Vandamme the task was daunting since he had little experience of siege warfare or any knowledge of engineering. Undeterred, his lively instinct for war soon made short work of Glogau, the first fortress he decided to crack. He adopted the tactics that had succeeded at Magdeburg by terrorizing the inhabitants into persuading the garrison to surrender. He first placed a battery of cannon and heavy mortars before the walls and after some threats followed up with a bombardment. The place capitulated on 2 December with great quantities of stores, which enabled Vandamme to continue his remorseless campaign with renewed vigor.

Breslau, the capital of Silesia, thirty miles up the Oder was the next target. A city of 60,000 people defended by a garrison of 6,000 men meant a lengthy siege, which Vandamme wished to avoid. As at Glogau he intended to intimidate the population into submission. He selected the suburb of Saint Nicholas to set up his batteries and from there rained down incendiary devices into the city's interior. The bombardment did not however achieve the same results as the garrison commander was made of sterner stuff. Goaded on by Jérôme Vandamme attempted a rash assault against a weak point on the outer walls on the night of 22-23 December. In darkness the Württembergers started to paddle silently across the moat on rafts. Then suddenly after days of cloud the skies cleared and revealed a full moon with Vandamme's men crossing visible to all - the attempt failed.

Meanwhile the Prince of Anhalt-Pless who commanded all troops in Silesia raised a relieving force of some 12,000 men and marched on the Breslau renewing the hopes of the populace. Nothing worked better in Vandamme's favour than to settle in open country the question of Breslau's fate. On 30 December with a mixed force of Bavarians and the 13th French Line he scattered the relieving force at Kleinberg. All hope of relief lost, morale in the city plummeted. At the same time a deep frost froze the moat and the outer walls became vulnerable. The populace fearful of the consequences of their wealthy city being taken by storm, the governor was forced to treat with Vandamme. On 7 January 1807 after a siege that lasted barely a month Breslau capitulated.

This conquest was not only brilliant, but particularly useful in the resources it procured for the French army. It assured the French of Silesia, the richest province of Prussia and one of the richest in Europe. Napoleon congratulated Vandamme, and after him Jérôme who he now felt had won his spurs. As an independent commander, it was Vandamme's finest hour.

There was still unfinished business in Silesia. Napoleon wanted all the fortresses that lay near the Bohemian border taken and destroyed, with the intention to leave Prussia's Bohemian border with Austria weak and unprotected. One by one XI Corps spent the remainder of the campaign reducing these fortresses. Schweidnitz held out from 10 January to 16 February, Neisse from 23 February till 16 June and Glatz from 21 June till 28 June 1807 when news of an armistice after Friedland caused the garrison to surrender.

Europe at peace after the Treaty of Tilsit, Vandamme returned to his former post as commander of the 16th Military Division at Lille on 11 November 1807. He later moved to the Camp of Boulogne as commander on 16 August 1808. Many awards for past campaigns came his way, including the Grand Cross of Württemberg on 29 February 1808. The introduction of the Imperial Nobility resulted in the former hard-bitten republican receiving the title *comte d'Unsebourg* on 19 March 1809.

THE WÜRTTEMBERG CONTINGENT - THE DANUBE CAMPAIGN 1809

The deteriorating situation with Austria saw his appointment as commander of the Württemberg contingent with the Army of Germany on 20 March 1809. The move was undoubtedly a calculated one, another semi-independent command in charge of the Württemberg troops whom he respected. It suited both him and Napoleon. Napoleon needed a man of his talents, but not to cause unnecessary discord with any of his marshals because Vandamme found it difficult to accept anyone's authority.

At the beginning of the campaign against Austria the Württembergers were centered around Donauworth before Napoleon on his arrival on 18 April sent them forward to Ingolstadt. From there Vandamme linked up with Lefebvre's VII Corps, outflanked and drove in Hiller's left wing of the Austrian army at Abensberg on 20 April. He followed the Austrians as they fell back to Landshut and started to probe east along the Isar. In the right position at the right time Napoleon ordered him at midnight of 21-22 April to form the army's advance guard to relieve the hard pressed Davout at Eckmühl.

His Württembergers put up a fine display. Leading the way they slogged twenty-three miles in darkness up a muddy single lane road. Then at dawn after four hours marching at the double they drove out the Austrian outposts at Ergoldsbach before arriving around mid-day on the main battlefield. They hit the Austrian left centered on Lindach clearing them from the village and from there stormed across the Gross Laber into Eckmühl.

Vandamme joined the pursuit of Charles's forces and on 4 May he entered Linz. He had orders to remain at the city, a key crossing point on the Danube, and reorganize the government of the province. Across the river the Austrians held the northern suburbs and felt reasonably secure as they had broken the bridge. Vandamme demanded from Austrian garrison commander that the city's surrender included the north bank. That demand refused, he proceeded to bombard the Austrians across the river. A bold amphibious crossing of the river followed the next day and with the entire city in Württemberg hands a vital bridgehead was established on the north bank.

The possibility of an Austrian attempt to sever Napoleon's communications by crossing the Danube at Linz was a very real threat. Vandamme took precautions, repaired the bridge and strengthened the defences on the north bank. It paid off when Kollowrat in Bohemia assembled his corps and 20,000 men soon marched on the city. The Austrians appeared before Linz on 17 May and Kollowrat immediately hurled his columns against the bridgehead. Vandamme's men held on, and when Bernadotte arrived with elements of the Saxon Corps, Kollowrat retired. The Austrians fell back to Freystadt and there remained inactive with all ideas of an offensive abandoned.

On 1 June Vandamme became commander of VIII Corps of the Grande Armée when his Württemberg contingent received corps status as recognition of their fine contribution in the campaign. In June the bulk of his corps moved to Vienna where they formed the city's garrison and covered the river crossings to the west. Anxious to be in action he left his troops and accompanied Napoleon's staff on 6 July during the battle of Wagram, receiving a shoulder wound. After hostilities ended he supervised the occupation of Styria during August before returning to Vienna as head of the garrison till the army's withdrawal in November.

On his return to France he replaced Saint Suzanne as commander of the Camp of Boulogne on 7 February 1810. He was soon in trouble when within twenty-four hours of his arrival he evicted the mayor from his home in order to use it as his headquarters. The incident prompted Napoleon's famous remark, "If there were two Vandammes one would hang the other." On 30 August 1811 he moved to Cherbourg as commander of the 14th Military Division.

JÉRÔME BONAPARTE AND THE RUSSIAN CAMPAIGN

On 21 February 1812 he was appointed deputy commander of the VIII (Westphalian) Corps of the Grand Army under Jérôme Bonaparte. The uneasy relationship that existed in 1807 soon boiled over. Hardly had the army crossed the Niemen when charges of insubordination were brought against him by Jérôme, which resulted in his dismissal on 3 July 1812. In the weeks that followed he was sorely missed as Junot who had taken over from him proved totally inept. The latter's bumbling leadership caused VIII Corps to miss reaching Valutina in time to cut off the Russians retreating from Smolensk. Such an opportunity for glory would never have passed the naturally belligerent Vandamme.

GERMANY AND KULM 1813

Back in France he was without employment till 17 March 1813 when he took command of two divisions at Wesel in Holland with the task of restoring order in the 32nd Military Division after Carra Saint Cyr had abandoned Hamburg. Operating under Davout, whilst there was little love lost between the two, there was accord and Vandamme did as ordered. He advanced on Hamburg from the south-west and took Harburg on 29 April. Turning his attention to the is-

VANDAMME CAPTURED AT THE BATTLE OF KULM

lands at the mouth of the Elbe he captured Wilhelmsburg on 9 May before he entering the city on 31 May 1813.

The reorganization of the French armies during the Armistice resulted in him replacing Davout at the head of I Corps on 1 July 1813 and moving the corps to Dresden. His command at the start of the campaign comprised the divisions of Phillipon, Dumonceau, Teste and the cavalry of Corbineau, numbering some 36,700 men of all arms.

When the campaign started, the Army of Bohemia advanced on Dresden. Napoleon seeing an opportunity ordered Vandamme to cross the Elbe upstream and take Pirna. The plan was then for Vandamme to work his way to the rear of the Allies, ready to sever their line of retreat at Peterswalde. On 26 August while the main Allied army tried to storm Dresden he crossed the river at Zittau, drove back Ostermann's outposts and seized the Pirna Plateau. Gradually he pushed Ostermann back and reached Peterswalde on 28 August. Contemptuous of the Allies, overestimating their plight, and more concerned at being the first to Prague he ignored the dangers of his force becoming increasingly isolated. The additional prospects of Austria knocked out of the war and a marshal's baton within his grasp clouded his judgement. The next day he pressed on to the outskirts of Teplitz where Ostermann turned on him after being reinforced by the Grand Duke Constantine's Russian Guard. At first forced to give a little ground, Vandamme fell back to Kulm to wait for support from Saint Cyr's XIV Corps.

The next day the 30 September the Allies realizing the threat to their communications as they fell back from Dresden massed over 50,000 men to keep the route open. Vandamme confidently held the enemy's assaults until beset in the rear by a Prussian Corps under Kleist. This general was trying to evade Saint Cyr's pursuing troops, had swung eastward from his original line of retreat in the hope of slipping past Vandamme's rear. Instead, he blundered into Vandamme's main body at the very height of the struggle with Ostermann. Boxed in by 64,000 Allies, in the furious action that followed Vandamme swung his corps around and tried to stampede their way to safety through Kleist's equally surprised Prussians. Over half his men escaped, but Vandamme with generals Haxo[*] and Quiot was among the 13,000 prisoners and 48 guns lost, whilst another 6,000 were killed or wounded in the battle.

Vandamme taken to Prague, a dramatic interview followed when brought before Czar Alexander. He nearly ran the Grand Duke Constantine through with his sword when the latter demanded his sword before entering the Czar's presence. The Czar then joined the fracas and in turn demanded his sword. Muskets levelled and swords bared, Vandamme's sarcasm was not lost on him when he said it was easy

to disarm him while he was alone, but it would be more noble and dangerous if tried on the battlefield. Infuriated by the slight the Czar responded by saying he refused to accept a sword from a murderer, thief and brigand. Vandamme countered that such accusations were irrelevant from a man who had the blood of his murdered father on his hands. The interview promptly ended as the guards bundled him into a carriage. He spent a period imprisoned in Moscow, before moving to Viazma only to return to France in July 1814.

THE RESTORATION AND NAPOLEON'S RETURN

Placed on the non active list on 2 September 1814 he sought an interview with Louis XVIII and offered to serve the Bourbons. The King refused to see him and on 22 September 1814 ordered him from Paris and exiled him to Cassel. Not surprisingly he rallied to Napoleon on his return, was recalled and appointed *commandant superieur* at Dunkirk on 9 April 1815 where he immediately set about effectively putting down potential royalist opposition. Earmarked for command of III Corps of the Army of the North comprising the infantry divisions of Habert, Berthezene, Lefol and the cavalry of Domon, he only managed to take up his post in the first week of June.

THE WATERLOO CAMPAIGN

On 11 June his corps moved from Rocroi to Beaumont with orders to screen the rest of the army as it concentrated near the Belgian border. The campaign started badly for him. Because Napoleon's orders to cross the border at 3.00 A.M. on the morning of 15 June did not reach him. He had inexplicably left his headquarters and spent the night at a nearby chateau. As a result Lobau's VI Corps closing up began to collide with his troops and a terrible snarl up ensued. The delay meant that there was no infantry support when Domon's cavalry tried to seize the bridges over the Sambre at Charleroi. Later in the day after he had crossed the river he did not know been placed under Grouchy's command when ordered to support an assault on Pirch's brigade at Gilly. His old faults once more came to the fore as he treated Grouchy with contempt, considering him a mere, "commander of cavalry". For two hours he wasted time arguing how the attack should go in. The impasse was resolved when Napoleon arrived, who exasperated by the delays ordered an artillery barrage followed by an infantry assault, which soon had the Prussians falling back to Fleurus. Later ordered by Grouchy to push on to Fleurus Vandamme flatly refused as he had already ordered his men to make camp and would only take orders from the Emperor.

At Ligny the next day he led the assaults against the Prussian right centered in the Saint Armand hamlets. By 5.00 P.M. his troops had carried the twin villages of Saint Armand and routed a Prussian attempt to envelop the French left but could make no further progress due to heavy fire from Brye. Then just as Blucher launched a furious counter attack, d'Erlon's corps appeared in the distance, his men panicked after mistaking them for Wellingtons columns and lost their hard won gains. Only when Duhesme intervened with the Young Guard was the situation stabilized and his men were able to gain new heart and hurl the enemy back.

The next day he showed little energy under Grouchy. Following the Prussians he took all day to reach Gembloux, a distance of only seven miles. Hearing the opening cannonade at Waterloo on 18 June, with he and Gérard exchanged harsh words with Grouchy, urging him to march to the sound of the guns. Then when he heard the Prussians were at Wavre he was suddenly jarred into action. He urged his men forward and in the late afternoon attacked the town in a most reckless fashion without proper reconnaissance or artillery preparation. Advancing through the narrow streets in dense columns his men quickly cleared the east bank but lost heavily as they tried to rush the bridges over the Dyle. Caught by a plunging fire from the Prussian batteries and sharpshooters from the heights on the opposite bank, his troops were caught in a cauldron and could neither advance nor retreat till dusk.

On the morning of 19 June he crossed the river unopposed and occupied Wavre, the Prussians having fallen back to La Bavette on the Brussels road. Following up he drove them from that village, when around midday he heard the news of Waterloo. He immediately withdrew through Wavre and by evening had reached Gembloux where the rest of Grouchy's force had concentrated. With Gérard wounded the previous day he took on the added responsibility of leading IV Corps.

On 20 June he severely compromised Grouchy's retreat when he could not be found as he had decided to spend the night in a nearby house without informing anyone where he was. He missed Grouchy's orders issued during the night and as a result failed to post III Corps astride the Gembloux road to cover the withdrawal of the rest of the army wing. His divisional commanders acting on their initiative began to fall back towards Namur without warning IV Corps. The move resulted in IV Corps' flank being uncovered and mauled by enemy cavalry the next morning until Grouchy at the head of a brigade of hussars drove them off.

Having safely passed through Namur he fell back on Laon, reaching the city on 25 June where he rejoined the rest of the army. In command of the rearguard he reached the outskirts of Paris on 28 June. Adopting a hard line he doggedly insisted on a battle before Paris, convinced that the widely spread Allies could be defeated in detail. The politicians in the end won the day and he retired with the rest of the army to the Loire on 4 July after the Armistice was signed.

LATER LIFE AND ASSESSMENT OF CAREER

Exiled by the Ordonnance of 24 July 1815 Vandamme quit his command on 7 August. He spent time with friends at Olivet and then Vierzun in the hope that the hue and cry would die down. Returning to Cassel in November he was declared an exile and forced to leave the country on 12 January 1816. He moved to Ghent in Belgium till 16 May 1816 when he was expelled from the Kingdom of the Netherlands. He then settled in the United States living near Philadelphia where the locals surprisingly found him well mannered and accommodating. He returned to Ghent in June 1819 and from there was allowed to return to France the following December. Reconciled with the Bourbons his rank and privileges were restored on 1 April 1820, but he never held any post up till his retirement on 1 January 1825. Ill health dogged his last years and he died of throat cancer at Cassel on 15 July 1830.

Undoubtedly one of Napoleon's foremost generals, the rough violent hardfighting Vandamme was his own worst enemy. His career was marred by the fact that he showed neither respect nor willingness to serve under anyone but Napoleon himself. As a young man his rebellious nature and uncontrollable temper first prevented him gaining a commission. The lesson never sank in, and during the Revolutionary Wars he fell out successively with Moreau, Saint Cyr and Jourdan. This resulted in him failing to attain higher commands, which would have placed him high amongst the candidates included in the first batch of marshals created in 1804. He had all the attributes required, intelligence, bravery and total devotion to Napoleon.

From 1805 onwards a baton was within his grasp but again his temper, lack of tact and often outright insubordination was to continue landing him in trouble, resulting in him being given semi-independent commands that were largely side shows. Command of I Corps in 1813 was his first real opportunity for many years to show his real worth. His failure at Kulm was unlucky. The risk he took was in keeping with Napoleon's style of warfare and his orders were clear. The Allied army was beaten and demoralized, and the appearance of Kleist in his rear was as equally unexpected for the Prussian as it was for Vandamme.

At times his behavior during the Waterloo Campaign was inexcusable, but it is easy to apportion blame on a person after a campaign is lost. Napoleon in his memoirs certainly did not place much importance on his behavior affecting the outcome of the campaign. With victory such indiscretions would have been ignored and, after the tenacity he displayed at Ligny,.with Drouot and Gérard he would have received the marshal's baton he so coveted.

VICHERY - LOUIS JOSEPH, BARON (1767-1831)

EARLY CAREER

Vichery was born in the village of Frévent (Pas de Calais) on 23 September 1767. He enlisted in the *Corps de Montréal* on 5 May 1781, remaining with them till 1 January 1787 when he obtained an honorable discharge after reaching the rank of *sergent*. In January 1790 he emerged as a quartermaster sergeant with the rebel forces in Belgium. Then when their bid for independence collapsed, he left and appeared the following year with the Paris National Guard. In June 1792 with the rank of *sergent* he joined the 1st Battalion of the Belgian Volunteers attached to the Army of the North. The turmoil that army was going through at the time ensured his rapid promotion, *sergent-major* on 1 July 1792, *sous lieutenant* on 10 November and *adjudant-major* on 1 March 1793. He became aide de camp to *général de brigade* Dumonceau on 16 October and was present at the siege of Lille in November 1793. In April 1794 he joined Souham's division staff and saw extensive service in Belgium and Holland.

With the establishment of the Batavian Republic he decided to enter their service and on 6 June 1796 joined the staff of Dumonceau's division. In July 1797 he was promoted *lieutenant colonel*. In October 1797 he joined the Batavian fleet that sailed from Texel to join French fleet at Brest and was then to land troops off Ireland. The fleet was beaten by the Royal Navy off Camperdown on 11 October and Vichery was fortunate to return to Texel unscathed.

During the Anglo-Russian invasion of Holland in 1799 he served in Daendel's division and fought at Zyp on 10 September, Bergen on 19 September, Alkmaar on 2 October and Castricum on 6 October 1799. In the Winter of 1800-1801 under Dumonceau's command he took part in Augereau's operations on the River Main that supported Moreau's left during the French advance into Bavaria. In June 1803 he was with Dumonceau during Mortier's occupation of Hanover after the breakdown of the Treaty of Amiens. A period of garrison duty followed in Hanover during which he was promoted to *colonel* on 9 December 1803. He returned to Holland in March 1804 and based at the Camp of Utrecht made preparations for the invasion of England.

THE GRANDE ARMÉE 1805-1807

With the Grande Armée officially coming into being on 30 August 1805 Vichery took the post as chief of staff with Dumonceau's 3rd Division of Marmont's II Corps. He remained with Dumonceau throughout the Austrian Campaign and was prominent during the capture of the bridges over the Danube at Ingolstadt on 9 October.

On 24 September 1806 he became a *brigadier general* in the Dutch Army. In the Prussian Campaign he continued to serve under Dumonceau when elements of the Dutch Army with Mortier's VIII Corps helped reduce the Prussian fortresses in Western Germany, and then occupied the territories of Prussia's ally Hesse-Cassel.

SERVICE UNDER LOUIS BONAPARTE

On 20 March 1807 Vichery took up the appointment as senior aide de camp to Louis Bonaparte, King of Holland. Ever anxious to exercise his sovereignty Louis decided to copy some of the traditions of the French Army. He created Marshals, which resulted in some harsh words from Napoleon. Then to add insult to injury Louis on 6 April 1807 appointed Vichery *General-Major* of the Dutch Army, the equivalent post held by Berthier in the French Army. The whole affair was a farce since in Holland the post carried little weight as the Dutch Army had no real autonomy or muscle to act on its own. It made Vichery look ridiculous and gave him an elevated status beyond his station that proved to be an albatross around his neck. Few French commanders would readily take orders from a Dutch Marshal let alone its *Général Major*. In spite of such handicaps he proved an able administrator and was able to meet the continual demands of Napoleon for more troops to support the operations of the French armies.

The real test came on 29 July 1809 when Lord Chatham landed in the Scheldt estuary with a force of 40,000 British troops that soon occupied Walcheren. The French forces in Belgium led by Charmberlhac and 8,000 Dutch troops under Vichery hastily blocked the approaches to Antwerp. After Louis's arrival he did sterling work in the region, pulling together the scattered Dutch forces that managed to keep the English pinned to the fever ridden Walcheren countryside until the arrival of French reinforcements.

Napoleon had little confidence in Louis, and on 16 August 1809 replaced him with the discredited Bernadotte as commander of the Army of Brabant. Vichery, offended by the slight against Louis, resigned his post and accompanied him back to Amsterdam. The situation between Louis and Napoleon deteriorated further. When Louis abdicated in July 1810 and Holland was incorporated into metropolitan France, Vichery's post of *Général Major* ceased to exist. Closely aligned to the fortunes of Louis Bonaparte now in disgrace, his career in Holland effectively ended. Without a post he served for a short period with Molitor's division in the Corps of Observation of Holland, assisting with the merger of Dutch troops into the French Army.

PORTUGAL AND SPAIN 1810-1813

On 20 September 1810 he accepted command of a provisional brigade made up of replacement battalions on the move to Spain in a division under Claperède. He caught up with his command at Salamanca on 11 November 1810 and while there was formally readmitted into the French Army with the rank of *général de brigade*. During the winter his troops based at Celorico and Trancoso tried to keep open the links between Massena's Army of Portugal and its base at Almeida. The following Spring, after Massena's retreat from Portugal he took part in the campaign to relieve Almeida. He fought at Fuentes d'Onoro on 5 May 1811 where his brigade suffered heavily in the street fighting for Fuentes itself, losing over 400 men before forced to withdraw.

The following month he joined Soult's army in Estramadura. His brigade was disbanded and its battalions allocated to existing regiments within the army. The influx of reinforcements resulted in a reorganization of the army and Vichery received command of the 1st Brigade of Claperède's 2nd Division of d'Erlon's' V Corps. After a period in Estramadura he moved to Granada in February 1812 as commander of the 2nd Brigade of Leval's 4th Division of the Army of the South. He had some success against Ballasteros and for a while was instrumental in keeping the Spaniard in check after wining a sharp action against him at Alcanin on 4 June 1812.

Marmont's defeat at Salamanca on 24 July 1812 spelt ruin for Soult's survival in the Southern Spain, so Vichery joined the withdrawal from Andalusia. In the Autumn he took part in Soult's renewed offensive that recovered Madrid. The Winter of 1812-1813 saw him based at Toledo in Central Spain, where engaged against guerillas he was wounded in an action at Siguenza on 3 February 1813.

DAVOUT - HAMBURG 1813-1814

He returned to France to recover, and on 30 May 1813 was promoted *général de division*. Sent to Hamburg, he joined Davout's XIII Corps on the Lower Elbe on 28 July 1813 as commander of the 50th Infantry Division. He was with Davout when operations against Wegesack's Swedish Corps began on 17 August. Over five days as the advance gathered momentum they occupied Schwerin and a large part of Western Mecklemburg. The news of Oudinot's failure before Berlin foiled the plan. Vichery with the rest of Davout's forces were left exposed on the right bank of the Elbe and had little choice but to withdraw to Hamburg. Here he remained locked up in the city with Davout's forces till their surrender on 11 May 1814. During the siege he worked well under Davout and this undoubtedly went a long way to securing him a command for the Waterloo Campaign.

THE WATERLOO CAMPAIGN

Vichery's return to France saw his appointment as commander at Dunkirk on 8 September. He was made a Commander of the Legion of Honor on 28 September 1814. Well out of the way when Napoleon returned he was not faced with any agonizing decision over his loyalty to the King. He simply waited at Dunkirk for events to develop and on 31 March 1815 received orders from Davout to report to Metz to join Gérard's IV Corps a commander of the 13th Infantry Division.

On 10 June Vichery's preparations complete, his troops left Metz to concentrate on the Belgian border. It was a long march and with little respite they crossed into Belgium on the morning of 15 June. The advance was soon thrown into confusion by Bourmont's defection at the head of the advance guard. Unaware of the incident, his troops crashed into the rear of Bourmont's column. A huge traffic jam developed till Gérard resolved the command issue. Valuable hours lost, Charleroi as an objective was abandoned. Instead Vichery cut across country with IV Corps and at dusk reached the Sambre at Chatelet. He crossed the river the next morning, but with the rest of Gérard's corps was late at Ligny. It delayed the start of the battle, which later was to prove so crucial to the campaign's outcome.

His division was held in reserve while Pécheux's led the initial assaults against Ligny. After two unsuccessful attempts, Vichery was dragged into the cauldron that engulfed the village. His troops drew in further Prussian reserves but the task proved too much and he made little progress. The intervention of the Old Guard around 8.00 P.M. was crucial and with his troops in support they finally broke the Prussian hold over the village.

He fought at Wavre on 18 June, where he unsuccessfully tried to force the Dyle at Bierges. It was not till evening that he crossed at Limale after Pajol's cavalry had secured the bridge. Firmly placed on the Limale Plateau he joined the assault on Wavre the next day. Then as the enemy were falling back from the outskirts, he heard of Napoleon's fate at Waterloo. Fast to react, under Grouchy's orders he disengaged, withdrew his division through Wavre and recrossed the Dyle. He reached Gembloux that evening where he rejoined the rest of Grouchy's wing. The next day on 20 June he reached Namur and from there crossed into France.

During the retreat to Paris he fell under Vandamme's command as Gérard had fallen at Wavre. On 2 July he fought a sharp action at Clichy, which coupled with Exelmans's success south of Paris the same day, was enough to give many French generals some hope the Allies could be forced to the conference table. Vichery however thought otherwise, and felt that after Napoleon's abdication continued resistance was futile. When it appeared the hard-liners might win the day and prevent the Armistice being signed, he quit his command on 3 July 1815. He however returned to his post after the armistice went through the next day and helped with the disbandment of the Army. On 28 July he was removed from his post and forcibly retired by the Ordonnance of 4 September 1815.

LATER CAREER AND ASSESSMENT

Recognized as a voice for reconciliation he was allowed to reside in Paris, which was unusual for a general who had taken part in the campaign in Belgium. A partial reconciliation with the Bourbons came on 4 May 1817 when he was made a baron. He took no part in military affairs till after the July Revolution when on 13 August 1830 he was asked to head a government commission that examined the plight of unemployed officers.

On 13 February 1831 he was formally recalled and on 13 May 1831 received the appointment of Infantry Inspector General for the 4th and 12th Military Divisions. He spent 1832 as Inspector General for the 12th Military Division at Nantes. Ill health forced him to give up his post on 14 November 1832. He retired on 1 January 1833 but did not survive long, dying of a stroke in Paris on 22 February 1833.

Vichery was an interesting case, for he certainly had highs and lows to his career which were no real fault of his own. He possessed a certain stoicism that enabled him to overcome the bad times. His loss of rank when he re-entered French service appeared harsh and unnecessary. Yet he soon proved in Spain he was able commander. Davout recognized his talent and was not let down by him in 1813. Ligny and Wavre also showed him as a tenacious and stubborn leader. In conclusion, he was no fanatical Bonapartist, but another capable and reliable professional, typical of many that formed the backbone of the officer class in the French Army at the time.

THE EVENING OF LIGNY, GNEISENAU GIVING ORDERS FOR THE BEATEN PRUSSIAN ARMY TO RETREAT

327

WATIER - PIERRE
(1770 - 1846)

EARLY CAREER

The cuirassier general Pierre Watier was born in Laon (Aisne) on 4 September 1770. His first active service was with an independent cavalry squadron that he joined as a *sous lieutenant* at Arras on 3 September 1792. He campaigned with the Army of the North for a short period before on 2 April 1793 his unit merged with the 12th Chasseurs à Cheval. Promoted *lieutenant* on 26 May 1793 he then transferred to the 16th Chasseurs à Cheval.

The army at the time going through a period of change and the wholesale dismissal of officers was not uncommon. Promotion was comparatively easy provided one showed the necessary revolutionary zeal, which Watier did. He gained promotion to *capitaine* on 14 August 1793, and to *chef d'escadron* on 18 November 1793. He spent from 1795 till 1797 with his regiment in the Army of the Interior. In 1797 he moved to the Army of Batavia in Holland when he joined the 4th Dragoons. He took command of the regiment on his promotion to *chef de brigade* on 4 September 1799. An active period followed. He fought at Zyp on 10 September, Bergen on 1 October and at Castricum on 6 October 1799.

After the campaign he remained in Holland as part of Augereau's newly formed Army of Gallo-Batavia. He served in Barbou's division during Augereau's campaign in Germany in the Winter of 1800. His formation with the advanceguard was at the capture of Schweinfurth on 26 November, fought at Burg Eberach on 3 December and distinguished himself at Nuremburg on 18 December 1800. When the Austrians under Klenau later turned on Augereau, Watier's horsemen played a key role covering the retreat to the Rhine. After the campaign he remained with the Army till its disbandment on 22 October 1801.

THE GRANDE ARMÉE

From June 1803 he was based at the Camp of Bruges with the 4th Dragoons as they prepared for the invasion of England. He gained a certain notoriety when he opposed plans to form dismounted dragoon regiments. He was so incensed by the move, he secured a meeting with Napoleon on 4 August 1804 to argue the case. He earned Napoleon's respect for his forthright manner, but his efforts were in vain. The dismounted dragoons remained without their horses and when the Grande Armée was set in motion on 30 August 1805, two of his squadrons were included with Baraguey d'Hilliers's Foot Dragoon division. He stayed with the remainder of the regiment, which served with Walther's 2nd Dragoon Division as part of Murat's Cavalry Reserve.

With the cavalry screen that concealed the French approach to the Danube, he seized the bridge over the river at Munster on 7 October 1805. The next day moving along the south bank he secured another bridge over the Lech near Rain. While resting by the river, the sound of distant cannon drew him towards Wertingen where he joined a struggle outside the town. It was the first major engagement of the campaign as his regiment joined in the destruction of Auffenberg's force of 7,000 Austrians.

After the capitulation of Ulm he operated with Mortier's VIII Corps on the Danube's northern bank. Separated from his men during the struggle at Durrenstein on 11 November, he tried to cross the river in a small boat with General Graindorge but was captured by the Russians. Exchanged after Austerlitz, on 24 December 1805 he was promoted *général de brigade* and took command of Sebastiani's brigade, who in turn replaced Walther.

PRUSSIA AND POLAND 1806-1807

In June 1806 his reputation as a fine leader of dragoons resulted in his appointment as commander of the Dragoon Depots at Versailles and Saint Germain. Hardly had he taken up the post when on 11 July 1806 he was posted to Bernadotte's I Corps as commander of the 2nd Light Cavalry Brigade (2nd, 4th Hussars and 5th Chasseurs à Cheval) with Tilly's Division. In the Prussian Campaign his brigade formed part of the cavalry screen covering the advance of Bernadotte's Corps. In action at Saalburg on 8 October he repulsed elements of Tauenzien's advance guard. The next day at Schleiz he supported the assaults by Drouet* that won the first major engagement of the campaign. He was compromised by Bernadotte's indecisiveness when with the rest of I Corps he did not appear at either Jena or Auerstadt on 14 October 1806.

Active in the pursuit of the enemy across Prussia, overconfidence nearly caused his undoing at Criwitz near Lubeck on 3 November. Bernadotte ordered him forward to the village after Maison's division had driven the Prussians from it. The enemy had retired in good order but then sent forward two dragoon regiments to counter Watier's light cavalry. Instead of charging them, Watier sent forward an officer to demand their surrender, when against orders some of his men opened fire. The outraged Prussians charged and before Watier could get under way his troops were driven back in confusion. Both he and Bernadotte narrowly escaped after taking refuge in a square formed by one of Pacthod's battalions.

On 1 December 1806 he left Bernadotte to command the 3rd Cavalry Brigade of the Cavalry Reserve, comprising of the 11th Chasseurs à Cheval and the Bavarian Royal Chevaulégers. He passed through East Prussia and Poland, and crossed the

Vistula at Wyszogorod on 16 December. On 24 December he fought at Kolozomb where he covered Augereau's VII Corps as it crossed the Ukra. The following day he overwhelmed a Russian column near Golymin. On 30 December his brigade merged with others to form a new light cavalry division under Lasalle.

Watier's association with Lasalle was not a good one. Lasalle had built up a fine reputation in Prussia, but ran out of luck in Poland. It reflected on Watier where at Eylau on 8 February 1807 he was kept idle on the French left. At Heilsberg on 11 June a badly timed charge ordered by Murat, resulted in the rout of his brigade with the rest of Lasalle's division. He was attached to Grouchy's command on the French left at Friedland on 14 June but his involvement was limited.

SPAIN 1808-1812

In the awards that followed the campaign he received the Bavarian Order of the Lion on 29 June in recognition of the way he led their troops. He remained in Poland till November 1807 and then, recalled to France, led a hussar brigade under Grouchy with Moncey's Observation Corps of the Ocean Coasts based at Bayonne. On 8 January 1808 he crossed into Spain and led his cavalry as the vanguard when the French occupied the main towns in Biscay and Navarre. He then moved to Madrid and from there joined Moncey's expedition against Valencia in June 1808. After its failure, and news of the reverse at Baylen, he covered the French retreat to the Ebro in July.

He was with Moncey's III Corps during Napoleon's invasion of Spain. He fought at Tudela on 23 November and in the pursuit that followed destroyed a Spanish column at Bubierca on 29 November as it fell back on Saragossa. During December he covered the approaches to Saragossa as Moncey invested the city. In January 1809 he showed his talents as a combined arms leader. At the head of a mixed force of two infantry battalions and two cavalry regiments he scoured the southern banks of the Ebro and kept the insurgents from the besieging force. On 26 January 1809 he had a notable success against Francisco Palafox at Alcaniz on the Lower Ebro. Four to five thousand peasants supported by a newly raised regiment of Aragonese volunteers tried to resist his small column, but were routed before the town. Vast quantities of provisions meant for the garrison at Saragossa fell into his hands.

On 16 June 1809 he left Spain for the Army of Germany as commander of the 2nd Brigade with Saint Sulpice's 2nd Cuirassier Division. By the time he reached the Danube the campaign was over. On 12 November 1809 he was made *comte de Sainte Alphonse* for his service in Aragon. A year later after a period of duty in Austria and Germany as the French armies withdrew from the Danube, returned to Bayonne on 4 October 1810. Here with two cavalry regiments from Caffarelli's Division of Reserve of the Army of Spain, he formed the nucleus of the cavalry force that in the following year served with the Army of the North. Once in Spain he had a tough time trying to corner guerilla forces in Galicia followed by the same lack of success in Old Castile.

At the end of April 1811 with Bessières's detachment from the Army of the North he joined Massena's Army of Portugal to help relieve Almeida. He fought at Fuentes de Onoro on 8 May 1811, where his brigade (11th, 12th, 24th Chasseurs and 5th Hussars) operated under Montbrun, who led the cavalry. His horse formed part of Montbrun's great cavalry sweep that tried to work its way around Wellington's right flank as the main battle raged to the north. In action against Arentschildt's four horse regiments he pushed them back from Pozo Bello. Wellington's right exposed, the move threatened to throw his forces into a precipitous retreat, but Houston's 7th Division was able to deploy in square to meet the threat. Watier's cavalry passed through the intervals between the squares and then encountered Craufurd's Light Division. Again they were unable to make much impression and a plan that offered so much fizzled out. The infantry support promised from Mermet's and Marchand's divisions had lagged far behind, and became embroiled in the fighting around Fuentes de Onoro itself.

On 31 July 1811 Watier was promoted *général de division* and appointed commander of the cavalry of the Army of the North. In September 1811 a rare period of cooperation between army commanders occurred when Dorsenne's Army of the North joined Marmont's Army of Portugal to relieve Ciudad Rodrigo. His cavalry was in the fore when that task was accomplished on 23 September 1811. Over enthusiasm soon led him into trouble when he led two brigades from Ciudad on the morning of 25 September to sweep the line of the Azava and probe Wellington's defences. Crossing the river at Carpio, his Berg Lancers reached the ridge beyond, before they were charged by Anson's 14th and 16th Light Dragoons. Driven back two miles to the Azava, the French advance beyond Carpio was called off.

Montbrun made a more successful sally against Wellington's defences further south at El Bodon, which resulted in an English withdrawal to Alfayates on 27 September. Watier followed till halted at Aldea da Ponte by pickets of Cole's 4th Division and Slade's dragoons. He waited till Thiébault's division came up and seeing the village was important decided to attack. An indecisive action followed, Watier skirmished with Slade's dragoons while Thiébault took the village and the Anglo-Portuguese withdrew two miles to their main positions on the ridge behind Alfayates. Both armies then took up defensive positions before they dispersed to their winter quarters, Watier's cavalry returning to their base at Valladolid.

THE RUSSIAN CAMPAIGN 1812

On 9 January 1812 Watier was recalled to France. Posted to Mainz, he took command of the 2nd Light Cavalry Division of Montbrun's II Cavalry Corps preparing for the campaign against Russia. A long march across Germany and Poland followed to join the Grand Army on the Niemen. On his arrival in mid-May his careful preparations came to nought when he had to switch commands with Sebastiani and take over the 2nd Cuirassier Division. In the long run the move suited the temperaments of both commanders since he was more at home with formations and Sebastiani with light. His new division was a fine formation (5th, 8th, 10th Cuirassiers and a company of the 2nd Lanciers) and numbered over 2,500 men at the start of the campaign.

The hardships of the long march took their toll. By the time he reached Borodino on 7 September 1812 his force was down to around 1,200 men and he had not taken part in any major engagement. Deployed in the open opposite the Great Redoubt during the early stages of the battle, he suffered heavily from artillery fire. Montbrun was killed while trying to maneuver the troops from the deluge of shells that fell among them. To fill the gap, Napoleon sent Caulaincourt from Berthier's staff to lead the II Cavalry Corps. Around 1.00 P.M. it moved forward to support Eugene's IV Corps assault the Raevsky Redoubt. With Caulaincourt at their head, Watier's three cuirassier regiments charged. Caulaincourt was killed almost at once as the cuirassiers surged around the sides and into the rear of the redoubt. Watier's men maddened by the slaughter they had suffered earlier in the day, their charge went out of control and outstripped the supporting infantry. From within the redoubt they were driven back by a storm of musketry. Their retirement then ploughed into Defrance, who was following with his carabiniers. Saxon cuirassiers with Lorge's division in the third wave effected a breakthrough by conducting a suicidal charge straight over the top of the breastworks. After regrouping his division, Watier was then involved in the murderous cavalry melee that developed behind the redoubt which lasted till 5.00 P.M. when the Russians drew off exhausted.

After the battle Sebastiani took charge of II Cavalry Corps and together they pushed the Russians to the gates of Moscow. After the occupation of the city, Watier was deployed to the south near Vinkovo with the bulk of the cavalry to watch Kutuzov's movements. During the retreat his division completely disintegrated. By the time it reached the Beresina on 25 November he could barely scrape together a single squadron. Two weeks later when he crossed the Niemen not one of his men was mounted.

GERMANY 1813-1814

On 15 February 1813 he was confirmed as commander of the 2nd Cuirassier Division. Based at Magdeburg, he feverishly tried to rebuild his division from depot squadrons and conscripts rushed forward from France. The shortage of horses was his biggest problem and after a month his division barely mustered 600 men. His plans were disrupted when an emergency arose after Carra Saint Cyr was ignominiously bundled out of Hamburg. On 18 March he left his command and set out with a mixed force for Bremen to lend support. Here he joined Vandamme's and Davout's forces on the lower Elbe, and took part in the operations that led to the recapture of Hamburg. With the renewal of hostilities after the Armistice he remained with Davout and was appointed his cavalry commander on 3 September 1813. After Davout's initial successes at the start of the campaign he covered XIII Corps as it fell back to Hamburg and was present throughout the siege of the city till its surrender on 9 May 1814.

THE RESTORATION AND THE WATERLOO CAMPAIGN

on his return to France he was left unattached and placed on the non-active list on 1 July 1814. He became a *Chevalier de Saint Louis* on 19 July. Unemployed, he found it easy to rally to Napoleon after his return and on 31 March 1815 was given the 2nd Cavalry Division with Reille's II Corps. Heavy cavalry being his arm, in May he transferred to Milhaud's IV Cavalry Corps as commander of the 13th Cavalry (Cuirassier) Division. Hindered by lack of suitable mounts, his division started the campaign considerably understrenght with two of his four regiments only fielding two squadrons.

Present at Ligny on 16 June he was held in reserve throughout the day till the Old Guard broke the Prussian line between Ligny and Sombreffe, when he joined in the pursuit. At Waterloo on 18 June 1815 his troops were positioned to the left of the Charleroi road, behind d'Erlon's Corps. They did not enter the battle till round 3.30 P.M. when Ney, mistakenly sensing Wellington was in retreat, prematurely ordered the cavalry forward to deliver the coup de grace. Watier in the second echelon moved forward to support Delort's charge against the Allies on the ridge. Following Delort, his horses made heavy weather of the climb through the muddy, heavily trodden terrain, and with each salvo the column's advance was brought to a crashing halt. Spurred on by trumpets sounding the charge, the cuirassiers rushed the cannon and overran the crewless batteries one after the other. As they reached the crest they met a wall of bayonets formed by the Allied infantry in square. Lunging and plunging at the infantry whilst suffering from their fire, they were charged by Wellington's cavalry before falling back to reform and repeat the movement.

After four charges, his exhausted men gave up. At the end of the battle he managed to regroup his battered regiments and was able to withdraw his division from the field reasonably intact. He remained with his division during the retreat to Paris and then after the Armistice to the Loire.

LATER LIFE AND ASSESSMENT

On 1 August 1815, his division disbanded, Watier found himself suspended without pay. Recalled after Saint Cyr's Army Reforms of 30 December 1818 he remained unattached and held no post till 21 April 1820 when he became an Inspector General of Cavalry. His good work with the Cavalry Inspectorate resulted in him receiving the title Grand Officer of the Legion of Honor on 1 May 1821.

In the years that followed he held numerous posts with the Cavalry and Gendarmerie Inspectorates. A solid and reliable professional, he kept out of politics which enabled his career to be unaffected by the July Revolution of 1830. He moved to the Army General Staff on 4 September 1831 where he remained till placed in the Reserve on 4 September 1835. He finally retired from the Army on 15 August 1839. He died in Paris on 3 February 1846.

As a general, Watier was neither brilliant nor controversial. He reflected the characteristics of many generals who served Napoleon, that of a solid, brave and loyal professional. As to his loyalty, it was first to his country rather than to any particular dynasty and so enabled him to survive the turbulent period France passed through.

THE ARMY OF THE NORTH JUNE 1815

The size of the French Army that marched into Belgium depends largely upon whether officers are excluded and only rank and file listed as was the custom at the time with the British Army. It is largely for this reason that figures for the French Army vary depending on the source from anything between 115,000 and 124,000 men.

The parade states given in this work are from the returns preserved at the *Archives du service Historique* at Vincennes, contained in carton C15, files 34-35. They were compiled between 10-15 June 1815 and give the parade state for each formation at the start of the campaign. Following Waterloo the Army compiled returns between 23-25 June as the corps commanders tried to ascertain the debilitating effects of the campaign in Belgium and the retreat to France. Understandably the post Waterloo period for the Army was chaotic, and many returns submitted were incomplete, often giving corps or division strengths, but none for the individual units. Alos when information was unavailable, commanders often resorted to the expedient of providing figures from the previous return. That information which is intact is shown and allows one to look at the state of the army after the battle, and to see the effect the campaign had on the formations led by the commanders mentioned in the text.

Where figures are *italicized* estimates have been used in order to reconcile to actual figures given for division or corps totals submitted. this problem is evident amongst the cavalry and artillery returns where the situation was further confused by the arrival of additional ordinance and cavalry squadrons after the campaign was underway, leaving some formations stronger at the campaign's end than at the beginning.

From returns compiled between 10-15 June 1815 the strength of the Army of the North was as follows:

Infantry	175 battalions	88426
Cavalry	164 squadrons	21630
Artillery and Engineers	358 guns	12596
Total		122652

Casualties during the campaign can be estimated and include many who lightly wounded later rejoined the ranks:

15 June
Charleroi and the Sambre crossings	200
The advance to Gilly and Quatre Bras	400

16 June
Ligny	13721
Quatre Bras	4100

17 June
Pursuit to Waterloo	125

18 June
Waterloo and the French retreat	43656
Wavre and the French retreat	2400
Total Losses	64,602

The state of the Army after Soult rallied the remnants and rejoined Grouchy can be summarized from the returns compiled from 23-25 June:

Infantry	38190
Cavalry	16411
Artillery and Engineers	7938
Total after the campaign	62539

The above figure of 62,539 officers and men compiled from returns at Laon would indicate that between the two dates a further 4,500 men joined the army, which is born out in the case of the cavalry, several fourth squadrons having arrived after the campaign's start.

Commander in Chief Emperor Napoleon

Senior Field Commanders	Marshal Michel Ney, Prince de Moskova, Commander of the Left Wing
	Marshal Marquis Emmanuel de Grouchy, Commander of the Right Wing
The Military Household	*Général de Division* comte Bertrand, Grand Marshal of the Palace
	Master of the Horse *Général de Division* Fouler, comte de Relinque
Imperial Aides de Camp	*Général de Division* Lebrun, duc de Plaisance
	Général de Division comte Corbineau
	Général de Division comte Flahaut de la Billarderie
	Général de Division comte Dejean
	Général de Brigade baron Bernard
	Général de Brigade comte de la Bédoyère
	Général de Brigade baron Bussy
The Imperial Staff	Army Chief of Staff: Marshal Jean de Dieu Soult, duc de Dalmatie
	Chief of General staff: *Général de Division* comte Bailly de Monthion
	Deputy Chief of General Staff: *Général de Brigade* baron Gressot
	Deputy Chief of General Staff: *Général de Brigade* baron Couture
	Deputy Chief of General Staff: *Général de Brigade* baron Lebel

Artillery	Commander in Chief of Artillery: *Général de Division* comte Ruty
	Inspector General of Artillery: *Général de Division* baron Neigre
Engineers	Commander in Chief of Engineers: *Général de Division* baron Rogniat
Gendarmerie	Commander in Chief of Gendarmes: *Général de Division* baron Radet

The Imperial Guard
Commander in Chief: *Général de Division* comte Drouot,
commanding in the absence of Marshal Mortier, duc de Treviso
Chief of Staff *Général de Division* comte Drouot

Imperial Guard Artillery Reserve: *Général de Division* baron Desvaux de Saint Maurice
Foot Artillery Commander: *Général de Division* baron Henri Lallemand

	15 June 1815		25 June 1815	
	Officers	Men	Officers	Men
Old Guard Foot Artillery Regiment, 1st Company				
six 12lb cannon, two 6in. howitzers	5	119	2	70
Old Guard Artillery Train 1st company	2	108	1	77
Old Guard Foot Artillery Regiment, 2nd Company				
six 12lb cannon, two 6in. howitzers	5	115	2	71
Old Guard Artillery Train 2nd company	2	106	1	76
Old Guard Foot Artillery Regiment, 3rd Company				
six 12lb cannon, two 6in. howitzers	5	120	2	70
Old Guard Artillery Train 3rd company	2	107	2	76
Old Guard Foot Artillery Regiment, 1st Company				
six 12lb cannon, two 6in. howitzers	5	116	3	73
Old Guard Artillery Train 1st company	2	97	2	79
Guard Artillery Strength (32 guns)	28	888	15	592

Imperial Guard Engineers and Equipment Train: *Général de Division* baron Haxo

	12 June 1815		25 June 1815	
	Officers	Men	Officers	Men
Old Guard Engineers	3	109	3	89
Marines of the Guard	3	104	2	77
Equipment Train of the Guard	8	390	14	359
Ouvriers	17	261	1	16
Engineers and Equipment Train Strength	31	864	20	541

Imperial Guard Division of Grenadiers: *Général de Division* comte Friant
Deputy Commander: *Général de Division* comte Roguet

	Bn	12 June 1815		25 June 1815	
		Officers	Men	Officers	Men
1st Brigade: *Général de Division* comte Friant					
1st (Old Guard) Grenadier Regiment	(2)	41	1239	29	615
2nd (Old Guard) Grenadier Regiment	(2)	36	1055	22	352
2nd Brigade: *Général de Division* comte Roguet					
3rd (Old Guard) Grenadier Regiment	(2)	34	1130	19	182
4th (Old Guard) Grenadier Regiment	(1)	27	493	9	91
	(7)	138	3917	79	1240
Division Artillery and Train					
Old Guard Foot Artillery Regt, 5th company					
Six 6lb cannon, two 5.5in. howitzers		4	114	2	70
Old Guard Artillery Train, 5th company		2	92	2	68
7th Foot Artillery Regt, 11th company					
Six 6lb cannon, two 5.5in. howitzers		4	105	3	63
Line Auxiliary Train company		3	111	2	68
Artisan Company of the Guard		1	84	-	-
		16	506	9	269
Division Strength (16 guns)	(7)	152	44423	88	1509

Imperial Guard Division of Chasseurs: *Général de Division* comte Morand
Deputy Commander: *Général de Division* comte Michel

		12 June 1815		25 June 1815	
	Bn	Officers	Men	Officers	Men
1st Brigade: *Général de Division* comte Morand					
1st (Old Guard) Chasseur Regiment	(2)	36	1271	26	615
2nd (Old Guard) Chasseur Regiment	(2)	32	1231	20	355
2nd Brigade: *Général de Division* comte Michel					
3rd (Old Guard) Chasseur Regiment	(2)	34	1028	6	159
4th (Old Guard) Chasseur Regiment	(2)	30	1041	8	236
	(8)	132	4471	60	1312
Division Artillery and Train					
Old Guard Foot Artillery Regt, 6th company					
Six 6lb cannon, two 5.5in. howitzers		4	98	1	70
Old Guard Artillery Train, 6th company		2	89	1	78
Auxiliary Foot Artillery company					
Six 6lb cannon, two 5.5in. howitzers		4	100	-	-
Line Auxiliary Train company		3	116	1	49
		13	403	3	197
Division Strength (16 guns)	(8)	145	4874	63	1509

Young Guard Division: *Général de Division* comte Duhesme
Deputy Commander: *Général de Division* comte Barrois

		12 June 1815		25 June 1815	
	Bn	Officers	Men	Officers	Men
1st Brigade: *Général de Brigade* Chartrand					
1st Tirailleur Regiment	(2)	26	1083	9	83
1st Voltigeur Regiment	(2)	31	1188	14	182
2nd Brigade: *Général de Division* comte Michel					
3rd Tirailleur Regiment	(2)	28	960	16	148
3rd Voltigeur Regiment	(2)	32	935	24	122
	(8)	117	4166	63	535
Division Artillery and Train					
7th Foot Artillery Regt, 12th company					
Six 6lb cannon, two 5.5in. howitzers		4	107	5	78
6th Train Squadron, 10th company		3	171	2	78
7th Foot Artillery Regt, 13th company					
Six 6lb cannon, two 5.5in. howitzers		3	102	-	-
Line Auxiliary Train company		1	107	1	49
		11	487	8	205
Division Strength (16 guns)	(8)	128	4653	71	740

Imperial Guard Light Cavalry Division: *Général de Division* comte Lefebvre Desnoettes

		12 June 1815		25 June 1815	
	Sqns	Officers	Men	Officers	Men
1st Brigade: *Général de Division* baron Lallemand					
Guard Chasseurs à Cheval Regiment	(5)	59	1138	26	618
2nd Brigade: *Général de Division* baron Colbert					
Guard Chasseurs à Cheval Regiment	(5)	47	833	30	525
		106	1971	56	1143

Division Horse Artillery and Train
Guard Horse Artillery Regt, 1st company

Six 6lb cannon, two 5.5in. howitzers	5	99	-	47
Guard Artillery Train, 7th company	2	79	1	77
Guard Horse Artillery Regt, 2nd company				
Six 6lb cannon, two 5.5in. howitzers	5	96	-	47
Guard Artillery Train, 8th company	2	74	1	74
	14	348	2	245

Division Strength (12 guns)	(10)	120	2319	58	1388

Imperial Guard Heavy Cavalry Division: *Général de Division* comte Guyot

	Sqns	15 June 1815 Officers	Men	25 June 1815 Officers	Men
1st Brigade: *Général de Brigade* Jamin					
Guard Grenadiers à Cheval Regiment	(4)	44	752	30	369
2nd Brigade: *Général de Division* baron Letort					
Guard (Empress) Dragoon Regiment	(4)	51	765	29	310
3rd Brigade: *Capitaine* Dyonnet					
Guard Elite Gendarmes	(1)	4	102	-	-
	(9)	99	1619	59	679

Division Horse Artillery and Train
Guard Horse Artillery Regt, 3rd company

Six 6lb cannon, two 5.5in. howitzers	5	94	-	46
Guard Artillery Train, 9th company	2	77	1	77
Guard Horse Artillery Regt, 4th company				
Six 6lb cannon, two 5.5in. howitzers	4	91	-	47
Guard Artillery Train, 10th company	2	75	1	75
	13	337	2	245

Division Strength (12 guns)	(9)	112	1956	61	924

Summary Imperial Guard	15 June 1815			25 June 1815		
	Guns	Officers	Men	Guns	Officers	Men
Infantry (23 battalions)	48	685	16200	16	307	4187
Cavalry (19 squadrons)	24	106	1400	-	64	738
Artillery (10 foot and 4 Horse Companies)	32	30	1066	18	5	150
Engineers		21	330		16	109
	104	842	18996	34	392	5184

I Army Corps

Commander: *Général de Division* Drouet, comte d'Erlon
Chief of Staff: *Général de Brigade* baron Delacambre

Corps Artillery Reserve: Corps Artillery Commander *Général de Brigade* baron Desales

	10 June 1815 Officers	Men	24 June 1815 Officers	Men
6th Foot Artillery Regt, 11th company				
Six 12lb cannon, two 6in. howitzers	3	84	3	64
1st Train Squadron, 6th company	1	118	2	86
Artillery Strength (8 guns)	4	202	5	150

Corps Engineers: Commander *Général de Brigade* baron Garbé

1st Engineer Regiment, 2nd Battalion, 1st-5th companies	21	330	16	109

1st Infantry Division: Commander *Général de Brigade* baron Quiot de Passage

		10 June 1815		24 June 1815	
	Bn	Officers	Men	Officers	Men
1st Brigade: *Général de Brigade* Quiot de Passage					
54th Line Infantry Regiment	(2)	41	921	25	320
55th Line Infantry Regiment	(2)	45	1103	14	426
2nd Brigade: *Général de Brigade* Bourgeois					
28th Line Infantry Regiment	(2)	42	856	22	278
105th Line Infantry Regiment	(2)	42	941	13	265
	(8)	170	3821	74	1280
Division Artillery and Train					
6th Foot Artillery Regt, 20th company					
Six 6lb cannon, two 5.5in. howitzers		4	81	-	-
1st Train Squadron, 5th company		3	103	-	-
Division Strength (8 guns)	(8)	177	4005	74	1289

2nd Infantry Division: Commander *Général de Division* baron Donzelot

		10 June 1815		24 June 1815	
	Bn	Officers	Men	Officers	Men
1st Brigade: *Général de Brigade* Schmitz					
10th Light Infantry Regiment	(3)	61	1814	33	556
17th Line Infantry Regiment	(2)	42	1015	18	285
2nd Brigade: *Général de Brigade* Aulard					
19th Line Infantry Regiment	(2)	43	989	17	335
51st Line Infantry Regiment	(2)	42	1126	19	289
	(9)	188	4944	87	1465
Division Artillery and Train					
6th Foot Artillery Regt, 10th company					
Six 6lb cannon, two 5.5in. howitzers		3	86	-	-
1st Train Squadron, 9th company		1	95	-	-
Division Strength (8 guns)	(9)	192	5125	87	1465

3rd Infantry Division: Commander *Général de Division* baron Marcognet

		10 June 1815		24 June 1815	
	Bn	Officers	Men	Officers	Men
1st Brigade: *Général de Brigade* Nouguès					
54th Line Infantry Regiment	(2)	41	996	12	158
55th Line Infantry Regiment	(2)	43	845	19	220
2nd Brigade: *Général de Brigade* Bourgeois					
21st Line Infantry Regiment	(2)	40	934	9	78
45th Line Infantry Regiment	(2)	43	960	14	155
	(8)	167	3735	54	611
Division Artillery and Train					
6th Foot Artillery Regt, 19th company					
Six 6lb cannon, two 5.5in. howitzers		4	81	-	-
1st Train Squadron, 2nd company		2	92	-	-
Division Strength (8 guns)	(8)	173	3908	54	611

4th Infantry Division: Commander *Général de Division* baron Durutte

		10 June 1815		24 June 1815	
	Bn	Officers	Men	Officers	Men
1st Brigade: *Général de Brigade* Pégot					
8th Line Infantry Regiment	(2)	40	943	30	33
29th Line Infantry Regiment	(2)	40	1106	13	352
2nd Brigade: *Général de Brigade* Brue					
85th Line Infantry Regiment	(2)	40	591	25	198
95th Line Infantry Regiment	(2)	40	1060	24	239
	(8)	160	3700	92	822

Division Artillery and Train
6th Foot Artillery Regt, 20th company

	Squ	Officers	Men	Officers	Men
Six 6lb cannon, two 5.5in. howitzers		3	81	-	-
1st Train Squadron, 5th company		1	92	=	=
Division Strength (16 guns)	(8)	164	4005	92	822

1st Cavalry Division: Commander *Général de Division* baron Jacquinot

		15 June 1815		25 June 1815	
	Squ	Officers	Men	Officers	Men
1st Brigade: *Général de Brigade* Bruno					
7th Hussar Regiment	(3)	28	411	19	212
3rd Chasseur à Cheval Regiment	(3)	29	336	27	262
2nd Brigade: *Général de Brigade* Gobrecht					
3rd Chevauléger Lancier Regiment	(3)	27	379	12	180
4th Chevauléger Lancier Regiment	(2)	22	274	6	84
	(11)	106	1400	64	738

Division Artillery and Train
1st Horse Artillery Regt, 2nd company

	Squ	Officers	Men	Officers	Men
Four 6lb cannon, two 5.5in. howitzers		3	70	-	-
1st Train Squadron, 4th company		2	83	=	=
Division Strength (6 guns)	(11)	111	1553	64	738

Summary I Army Corps

	15 June 1815		25 June 1815	
	Officers	Men	Officers	Men
Infantry (33 battalions)	685	16200	307	4187
Cavalry (11 squadrons)	106	1400	64	738
Artillery (46 guns)	30	1066	5	150
Engineers (5 companies)	21	330	16	109
	842	18996	392	5184

II Army Corps
Commander: *Général de Division* comte Reille
Chief of Staff: *Général de Brigade* baron Pamphile Lacroix

Corps Artillery Reserve: Corps Artillery Commander *Général de Brigade* baron Pelletier

	10 June 1815		24 June 1815	
	Officers	Men	Officers	Men
6th Foot Artillery Regt, 11th company				
Six 12lb cannon, two 6in. howitzers	4	96	4	77
1st Train Squadron, 6th company	2	114	2	100
Artillery Strength (8 guns)	6	210	6	177

Corps Engineers: Commander: *Général de Brigade* baron de Richemont

	Officers	Men	Officers	Men
1st Engineer Regiment, 1st Battalion, 1st-5th companies	22	409	9	9

5th Infantry Division: Commander *Général de Brigade* baron Bachelu

		10 June 1815		24 June 1815	
	Bn	Officers	Men	Officers	Men
1st Brigade: *Général de Brigade* Husson					
3rd Line Infantry Regiment	(2)	42	1101	21	292
61st Line Infantry Regiment	(2)	41	817	26	206
2nd Brigade: *Général de Brigade* Campi					
72nd Line Infantry Regiment	(2)	42	953	27	335
108th Line Infantry Regiment	(3)	51	1046	22	185
	(9)	186	3917	96	1018

Division Artillery and Train
6th Foot Artillery Regt, 20th company

	Bn	Officers	Men	Officers	Men
Six 6lb cannon, two 5.5in. howitzers		4	86	-	-
1st Train Squadron, 5th company		2	99	=	=
Division Strength (8 guns)	(9)	192	4102	96	1018

337

6th Infantry Division: *Général de Division* Prince Jérôme Bonaparte
Deputy Commander: *Général de Division* comte Guilleminot

	Bn	10 June 1815 Officers	Men	24 June 1815 Officers	Men
1st Brigade: *Général de Brigade* Baudin					
1st Light Infantry Regiment	(3)	64	1824	35	572
2nd Light Infantry Regiment	(3)	94	2247	43	634
2nd Brigade: *Général de Brigade* Soye					
1st Line Infantry Regiment	(3)	59	1736	21	535
2nd Line Infantry Regiment	(3)	65	1730	29	585
	(12)	282	7537	128	2326
Division Artillery and Train					
2nd Foot Artillery Regt, 2nd company					
Six 6lb cannon, two 5.5in. howitzers		4	92	4	64
1st Train Squadron, 10th company		2	102	3	93
Division Strength (8 guns)	(12)	288	7731	135	2483

7th Infantry Division: *Général de Division* comte Girard

	Bn	10 June 1815 Officers	Men	24 June 1815 Officers	Men
1st Brigade: *Général de Brigade* Devilliers					
11th Light Infantry Regiment	(2)	42	913	34	502
82nd Line Infantry Regiment	(2)	40	1110	33	585
2nd Brigade: *Général de Brigade* Piat					
12th Line Infantry Regiment	(3)	51	1141	34	474
4th Line Infantry Regiment	(2)	44	1157	34	630
	(9)	177	4321	136	2191
Division Artillery and Train					
6th Foot Artillery Regt, 19th company					
Six 6lb cannon, two 5.5in. howitzers		3	74	3	80
1st Train Squadron, 1st company		1	58	1	49
5th Train Squadron, 2nd company		1	43	1	30
Division Strength (8 guns)	(9)	182	4496	141	2350

9th Infantry Division: *Général de Division* comte Foy

	Bn	15 June 1815 Officers	Men	25 June 1815 Officers	Men
1st Brigade: *Général de Brigade* Gauthier					
92nd Line Infantry Regiment	(2)	40	998	29	486
93rd Line Infantry Regiment	(3)	59	1427	32	442
2nd Brigade: *Général de Brigade* Jamin					
100th Line Infantry Regiment	(3)	51	1067	10	152
4th Light Infantry Regiment	(3)	61	1573	25	355
	(11)	211	5095	96	1435
Division Artillery and Train					
6th Foot Artillery Regt, 1st company					
Six 6lb cannon, two 5.5in. howitzers		4	84	-	-
1st Train Squadron, 10th company		2	102	2	79
Division Strength (8 guns)	(11)	217	5276	98	1514

2nd Cavalry Division: *Général de Division* baron Piré

	Squ	15 June 1815 Officers	Men	25 June 1815 Officers	Men
1st Brigade: *Général de Brigade* Huber					
1st Chasseur à Cheval Regiment	(4)	40	445	32	278
6th Chasseur à Cheval Regiment	(4)	34	526	29	260
2nd Brigade: *Général de Brigade* Wathiez					
5th Chevauléger Lancier Regiment	(3)	25	387	26	250
6th Chevauléger Lancier Regiment	(4)	34	371	20	200
		133	1729	107	988
Division Artillery and Train					
4th Horse Artillery Regt, 2nd company					
Four 6lb cannon, two 5.5in. howitzers		4	76	-	-
5th Train Squadron, 2nd company		2	81	-	-
Division Strength (6 guns)	(15)	139	1886	107	988

Summary II Army Corps	Officers	Men	Officers	Men
Infantry (39 battalions)	856	20870	456	6970
Cavalry (15 squadrons)	133	1729	107	988
Artillery (46 guns)	35	1102	18	493
Engineers	22	409	9	99
	1046	24110	590	8550

III Army Corps
Commander: *Général de Division* Vandamme, comte d'Unsebourg
Chief of Staff: *Général de Brigade* baron Revest

Corps Artillery Reserve: *Général de Brigade* baron Doguereau

	10 June 1815 Officers	Men	24 June 1815 Officers	Men
2nd Foot Artillery Regt, 1st company				
Six 12lb cannon, two 6in. howitzers	4	95	*11*	*108*
2nd Foot Artillery Regt, 19th company				
Six 12lb cannon, two 6in. howitzers	4	97	*8*	*97*
5th Train Squadron, 6th company	2	104	*9*	*104*
Artillery Strength (8 guns)	10	296	28	309

Corps Engineers: *Général de Brigade* chevalier Nempde-Dupoyet

	10 June 1815		24 June 1815	
1st Engineer Regiment, 2nd Battalion, 1st & 2nd companies	14	288	8	158

8th Infantry Division: *Général de Brigade* baron Lefol

	Bn	15 June 1815 Officers	Men	25 June 1815 Officers	Men
1st Brigade: *Général de Brigade* Billard					
15th Light Infantry Regiment	(3)	62	1676	64	779
23rd Line Infantry Regiment	(3)	62	1152	64	1538
2nd Brigade: *Général de Brigade* Corsin					
37th Line Infantry Regiment	(3)	59	1117	85	884
64th Line Infantry Regiment	(2)	40	891	48	572
	(11)	223	4836	261	1018
Division Artillery and Train					
6th Foot Artillery Regt, 7th company					
Six 6lb cannon, two 5.5in. howitzers		4	83	9	83
1st Train Squadron, 1st company		2	97	7	97
Division Strength (8 guns)	(11)	230	5042	96	1445

10th Infantry Division: *Général de Division* baron Habert

		15 June 1815		25 June 1815	
	Bn	Officers	Men	Officers	Men
1st Brigade: *Général de Brigade* Baudin					
34th Line Infantry Regiment	(3)	55	1384	56	790
88th Line Infantry Regiment	(3)	57	1265	34	666
2nd Brigade: *Général de Brigade* Soye					
22nd Line Infantry Regiment	(3)	55	1406	44	536
70th Line Infantry Regiment	(2)	45	909	31	599
2nd Swiss Line Infantry Regiment	(1)	21	386	14	146
	(12)	233	5350	179	2737
Division Artillery and Train					
2nd Foot Artillery Regt, 18th company					
Six 6lb cannon, two 5.5in. howitzers		4	89	9	89
5th Train Squadron, 4th company		2	92	7	92
Division Strength (8 guns)	(12)	239	5531	195	2918

11th Infantry Division: *Général de Division* baron Berthezène

		10 June 1815		25 June 1815	
	Bn	Officers	Men	Officers	Men
1st Brigade: *Général de Brigade* Dufour					
12th Line Infantry Regiment	(2)	41	1171	38	588
56th Line Infantry Regiment	(2)	42	1234	32	660
2nd Brigade: *Général de Brigade* Logarde					
33rd Line Infantry Regiment	(2)	39	1097	27	663
86th Line Infantry Regiment	(2)	44	870	37	625
	(8)	166	4372	134	2536
Division Artillery and Train					
2nd Foot Artillery Regt, 17th company					
Six 6lb cannon, two 5.5in. howitzers		4	96	9	96
5th Train Squadron, 5th company		2	94	7	94
Division Strength (8 guns)	(8)	172	4562	140	2726

3rd Cavalry Division: *Général de Division* baron Domon

		10 June 1815		25 June 1815	
	Squ	Officers	Men	Officers	Men
1st Brigade: *Général de Brigade* Dommanget					
4th Chasseur à Cheval Regiment	(3)	31	306	40	291
9th Chasseur à Cheval Regiment	(3)	25	337	24	201
2nd Brigade: *Général de Brigade* Vinot					
12th Chasseur à Cheval Regiment	(3)	29	289	33	315
	(9)	85	932	97	807
Division Artillery and Train					
2nd Horse Artillery Regt, 4th company					
Four 6lb cannon, two 5.5in. howitzers		3	74	-	-
5th Train Squadron, 3rd company		3	100	-	-
Division Strength (6 guns)	(9)	91	1106	97	807

Summary III Army Corps

	10 June 1815		25 June 1815	
	Officers	Men	Officers	Men
Infantry (31 battalions)	622	14558	574	9046
Cavalry (9 squadrons)	85	932	97	807
Artillery (38 guns)	34	1021	76	860
Engineers	14	288	8	158
	1046	16799	755	10871

IV Army Corps
Commander: *Général de Division* comte Gérard
Chief of Staff: *Général de Brigade* chevalier Saint Rémy

Corps Artillery Reserve: *Général de Brigade* Baltus de Pouilly

	13 June 1815		24 June 1815	
	Officers	Men	Officers	Men
5th Foot Artillery Regt, 4th company				
Six 12lb cannon, two 6in. howitzers	4	94	4	94
5th Foot Artillery Regt, 5th company				
Six 6lb cannon, two 5.5in. howitzers	4	96	4	96
2nd Train Squadron, 8th company	2	95	2	95
2nd Train Squadron, 10th company	2	98	2	98
Artillery Strength (16 guns)	12	383	12	383

Corps Engineers: *Général de Brigade* du Friche de Valazé

	Officers	Men	Officers	Men
2nd Engineer Regiment, 2nd Battalion, 3rd-5th companies	7	223	14	184
2nd squadron, Engineer Equipment Train	4	125	4	125
1st pontonier Battalion, 4th company	4	64	4	64
	15	412	22	373

12th Infantry Division: *Général de Brigade* baron Pécheu

	Bn	10 June 1815		24 June 1815	
		Officers	Men	Officers	Men
1st Brigade: *Général de Brigade* Rome					
30th Line Infantry Regiment	(3)	54	1086	37	605
96th Line Infantry Regiment	(3)	51	1172	30	664
2nd Brigade: *Général de Brigade* Schaeffer					
63rd Line Infantry Regiment	(3)	53	1112	35	573
6th Light Infantry Regiment	(1)	20	596	44	725
	(10)	178	3966	146	2567
Division Artillery and Train					
5th Foot Artillery Regt, 2nd company					
Six 6lb cannon, two 5.5in. howitzers		3	100	3	100
1st Train Squadron, 1st company		2	97	2	97
Division Strength (8 guns)	(10)	183	4143	151	2764

13th Infantry Division: *Général de Division* baron Vichery

	Bn	13 June 1815		24 June 1815	
		Officers	Men	Officers	Men
1st Brigade: *Général de Brigade* Le Capitaine					
59th Line Infantry Regiment	(2)	42	997	44	748
17th Line Infantry Regiment	(2)	40	934	36	818
2nd Brigade: *Général de Brigade* Desprez					
48th Line Infantry Regiment	(2)	43	834	40	663
69th Line Infantry Regiment	(2)	40	1024	38	730
	(8)	165	3789	158	2959
Division Artillery and Train					
5th Foot Artillery Regt, 1st company					
Six 6lb cannon, two 5.5in. howitzers		4	84	4	84
2nd Train Squadron, 2nd company		4	84	4	84
Division Strength (8 guns)	(12)	173	3959	166	3127

14th Infantry Division: *Général de Division* comte Bourmont

	Bn	13 June 1815 Officers	Men	24 June 1815 Officers	Men
1st Brigade: *Général de Brigade* Hulot					
9th Light Infantry Regiment	(2)	43	1218	40	685
111th Line Infantry Regiment	(2)	43	1023	36	524
2nd Brigade: *Général de Brigade* Toussaint					
44th Line Infantry Regiment	(2)	43	935	31	751
50th Line Infantry Regiment	(2)	40	867	38	730
	(8)	169	4043	145	2690
Division Artillery and Train					
5th Foot Artillery Regt, 3rd company					
Six 6lb cannon, two 5.5in. howitzers		5	174	5	174
2nd Train Squadron, 1st company		5	174	5	174
Division Strength (8 guns)	(8)	179	4291	155	2938

6th Cavalry Division: *Général de Division* baron Maurin

	Squ	13 June 1815 Officers	Men	25 June 1815 Officers	Men
1st Brigade: *Général de Brigade* Vallin					
6th Hussar Regiment	(3)	26	387	30	322
8th Chasseur à Cheval Regiment	(3)	30	373	37	435
2nd Brigade: *Général de Brigade* Berruyer					
6th Dragoon Regiment	(3)	28	205	31	289
16th Dragoon Regiment	(2)	26	301	26	332
	(11)	110	1266	124	1378
Division Artillery and Train					
2nd Horse Artillery Regt, 4th company					
Four 6lb cannon, two 5.5in. howitzers		3	77	3	77
5th Train Squadron, 3rd company		2	74	2	74
Division Strength (6 guns)	(11)	115	1417	129	1529

Summary IV Army Corps

	13 June 1815 Officers	Men	25 June 1815 Officers	Men
Infantry (26 battalions)	512	11798	439	7931
Cavalry (11 squadrons)	110	1266	124	1378
Artillery (38 guns)	40	1247	40	1247
Engineers	15	412	22	373
	677	14723	625	10929

VI Army Corps

Commander: *Général de Division* Mouton, comte Lobau
Chief of Staff: *Général de Brigade* baron Durrieu

Corps Artillery Reserve: *Général de Division* baron Noury

	10 June 1815 Officers	Men	24 June 1815 Officers	Men
8th Foot Artillery Regt, 4th company				
Six 12lb cannon, two 6in. howitzers	3	92	-	-
8th Train Squadron, 5th company	2	127	-	-
Attached from Imperial Guard Corps				
Line Horse Artillery Company				
Four 6lb cannon, two 5.5in. howitzers	3	70	-	-
Line Horse Artillery Train company	2	96	-	-
Artillery Strength (14 guns)	10	385	-	-

Corps Engineers: *Général de Brigade* Sabatier

	10 June 1815 Officers	Men	24 June 1815 Officers	Men
3rd Engineer Regiment, 1st Battalion, 1st-3rd companies	10	286	19	372

19th Infantry Division: *Général de Division* baron Simmer

		15 June 1815		25 June 1815	
	Bn	Officers	Men	Officers	Men
1st Brigade: *Général de Brigade* Bellair					
5th Line Infantry Regiment	(2)	42	910	29	325
11th Line Infantry Regiment	(3)	61	1135	44	575
2nd Brigade: *Général de Brigade* Thévnet					
27th Line Infantry Regiment	(2)	39	782	34	437
84th Line Infantry Regiment	(2)	45	894	25	415
	(9)	187	3721	132	1752
Division Artillery and Train					
8th Foot Artillery Regt, 1st company					
Six 6lb cannon, two 5.5in. howitzers		3	83	-	-
7th Train Squadron, 1st company		3	53	-	-
8th Train Squadron, 4th company		1	37	-	-
3rd Equipment Squadron, 1st company		2	62	-	-
Division Strength (8 guns)	(9)	196	3956	132	1752

20th Infantry Division: *Général de Division* baron Jeanin

		15 June 1815		25 June 1815	
	Bn	Officers	Men	Officers	Men
1st Brigade: *Général de Brigade* Le Capitaine					
5th Light Infantry Regiment	(2)	48	838	27	371
10th Line Infantry Regiment	(2)	56	1375	35	408
2nd Brigade: *Général de Brigade* Desprez					
47th Line Infantry Regiment		(Detached to the Vendée)			
107th Line Infantry Regiment	(2)	44	691	18	253
	(6)	148	2904	80	1032
Division Artillery and Train					
8th Foot Artillery Regt, 2nd company					
Six 6lb cannon, two 5.5in. howitzers		3	84	-	-
8th Train Squadron, 3rd company		2	84	-	-
3rd Equipment Squadron, 1st company		2	84	-	-
Division Strength (8 guns)	(6)	155	3156	80	1032

21st Infantry Division: *Général de Division* comte Teste

		10 June 1815		25 June 1815	
	Bn	Officers	Men	Officers	Men
1st Brigade: *Général de Brigade* Lafitte (absent)					
8th Light Infantry Regiment	(2)	42	896	39	534
40th Line Infantry Regiment		(Detached to the Vendée)			
2nd Brigade: *Général de Brigade* Penne					
65th Line Infantry Regiment	(1)	22	481	22	210
75th Line Infantry Regiment	(2)	42	939	40	482
	(5)	106	2316	101	1226
Division Artillery and Train					
8th Foot Artillery Regt, 3rd company					
Six 6lb cannon, two 5.5in. howitzers		3	91	3	62
6th Train Squadron, 4th company		5	70	2	50
3rd Equipment Squadron, 1st company		1	27	-	-
Division Strength (8 guns)	(5)	112	2505	106	1338

Summary VI Army Corps		10 June 1815		25 June 1815	
		Officers	Men	Officers	Men
Infantry (20 battalions)		441	8941	313	4010
Cavalry		-	-	-	-
Artillery (38 guns)		32	1060	5	112
Engineers		11	286	19	372
		484	10287	337	4494

I Reserve Cavalry Corps
Commander: *Général de Division* comte Pajol

4th Cavalry Division: *Général de Division* baron Soult

		15 June 1815		25 June 1815	
	Squ	Officers	Men	Officers	Men
1st Brigade: *Général de Brigade* Saint Laurent					
1st Hussar Regiment	(4)	36	489	33	507
4th Hussar Regiment	(4)	29	346	32	346
2nd Brigade: *Général de Brigade* Ameil					
5th Hussar Regiment	(4)	29	399	29	399
	(12)	94	1234	94	1252
Division Artillery and Train					
1st Horse Artillery Regt, 1st company					
Four 6lb cannon, two 5.5in. howitzers		2	70	2	70
1st Train Squadron, 3rd company		1	84	1	84
Division Strength (6 guns)	(12)	97	1388	97	1380

5th Cavalry Division: *Général de Division* baron Subervie

		15 June 1815		25 June 1815	
	Squ	Officers	Men	Officers	Men
1st Brigade: *Général de Brigade* Colbert					
1st Chevauléger Lancier Regiment	(4)	40	375	40	375
2nd Chevauléger Lancier Regiment	(4)	41	379	41	379
2nd Brigade: *Général de Brigade* Merlin					
11th Chasseur à Cheval Regiment	(3)	27	458	27	438
	(11)	108	1212	108	1192
Division Artillery and Train					
1st Horse Artillery Regt, 3rd company					
Four 6lb cannon, two 5.5in. howitzers		2	74	2	74
4th Train Squadron, 4th company		2	89	2	91
Division Strength (6 guns)	(11)	112	1375	112	1357

Summary I Reserve Cavalry Corps	Officers	Men	Officers	Men
Cavalry (23 squadrons)	202	2446	202	2444
Artillery (12 guns)	7	317	6	319
	209	2763	248	2763

II Reserve Cavalry Corps
Commander: *Général de Division* comte Exelmans

9th Cavalry Division: *Général de Division* chevalier Strolz

		15 June 1815		25 June 1815	
	Squ	Officers	Men	Officers	Men
1st Brigade: *Général de Brigade* Burthe					
5th Dragoon Regiment	(4)	41	465	40	507
13th Dragoon Regiment	(4)	35	389	34	346
2nd Brigade: *Général de Brigade* Vincent					
15th Dragoon Regiment	(4)	34	381	34	383
20th Dragoon Regiment	(4)	31	316	31	316
	(16)	141	1551	139	316

Division Artillery and Train
1st Horse Artillery Regt, 4th company

Four 6lb cannon, two 5.5in. howitzers		4	66	4	66
6th Train Squadron, 1st company		2	79	2	79
Division Strength (6 guns)	(16)	147	1696	145	1697

10th Cavalry Division: *Général de Division* baron Chastel

		15 June 1815		25 June 1815	
	Squ	Officers	Men	Officers	Men
1st Brigade: *Général de Brigade* Bonnemains					
4th Dragoon Regiment	(3)	35	330	35	309
12th Dragoon Regiment	(3)	30	310	34	291
2nd Brigade: *Général de Brigade* Berton					
14th Dragoon Regiment	(3)	34	339	34	317
17th Dragoon Regiment	(3)	39	287	38	282
	(12)	108	1212	137	1199
Division Artillery and Train					
4th Horse Artillery Regt, 4th company					
Four 6lb cannon, two 5.5in. howitzers		2	60	2	54
8th Train Squadron, 1st company		2	81	2	73
Division Strength (6 guns)	(12)	142	1407	127	1267

Summary II Reserve Cavalry Corps

	15 June 1815		25 June 1815	
	Officers	Men	Officers	Men
Cavalry (27 squadrons)	279	2817	276	2751
Artillery (12 guns)	12	286	12	286
	291	3103	288	3037

III Reserve Cavalry Corps
Commander: *Général de Division* comte Kellermann

11th Cavalry Division: *Général de Division* baron l'Héritier

		15 June 1815		25 June 1815	
	Squ	Officers	Men	Officers	Men
1st Brigade: *Général de Brigade* Picquet					
2nd Dragoon Regiment	(4)	43	550	24	301
7th Dragoon Regiment	(4)	41	476	26	218
2nd Brigade: *Général de Brigade* Guiton					
8th Cuirassier Regiment	(3)	32	427	12	91
11th Cuirassier Regiment	(2)	24	308	14	110
	(13)	140	1761	76	720
Division Artillery and Train					
2nd Horse Artillery Regt, 3rd company					
Four 6lb cannon, two 5.5in. howitzers		3	75	2	40
2nd Train Squadron, 3rd company		2	81	2	40
Division Strength (6 guns)	(13)	145	1917	80	800

12th Cavalry Division: *Général de Division* baron Roussel d'Hurbal

		15 June 1815		25 June 1815	
	Squ	Officers	Men	Officers	Men
1st Brigade: *Général de Brigade* Blanchard					
1st Carabinier Regiment	(3)	32	402	15	121
2nd Carabinier Regiment	(3)	30	383	15	180
2nd Brigade: *Général de Brigade* Donop					
14th Cuirassier Regiment	(2)	21	290	17	166
17th Cuirassier Regiment	(4)	38	442	8	142
	(12)	121	1517	55	609

Division Artillery and Train
4th Horse Artillery Regt, 4th company

	Squ	Officers	Men	Officers	Men
Four 6lb cannon, two 5.5in. howitzers		3	75	2	40
8th Train Squadron, 1st company		2	81	2	40
Division Strength (6 guns)	(12)	126	1670	59	689

Summary III Reserve Cavalry Corps

	15 June 1815		25 June 1815	
	Officers	Men	Officers	Men
Cavalry (27 squadrons)	261	3278	131	1329
Artillery (12 guns)	10	309	8	160
	271	3585	139	1489

IV Reserve Cavalry Corps
Commander: *Général de Division* comte Milhaud

13th Cavalry Division: *Général de Division* Watier, comte de Saint Alphonse

		15 June 1815		25 June 1815	
	Squ	Officers	Men	Officers	Men
1st Brigade: *Général de Brigade* Dubois					
1st Cuirassier Regiment	(4)	43	422	*	*
4th Cuirassier Regiment	(3)	30	284	*	*
2nd Brigade: *Général de Brigade* Travers					
7th Cuirassier Regiment	(2)	32	158	*	*
12th Cuirassier Regiment	(2)	24	234	*	*
	(11)	119	1098	155	1500
Division Artillery and Train					
2nd Horse Artillery Regt, 3rd company					
Four 6lb cannon, two 5.5in. howitzers		3	75	3	75
2nd Train Squadron, 3rd company		2	79	2	79
Division Strength (6 guns)	(11)	124	1252	160	1654

14th Cavalry Division: *Général de Division* baron Delort

		15 June 1815		25 June 1815	
	Squ	Officers	Men	Officers	Men
1st Brigade: *Général de Brigade* Farine					
5th Cuirassier Regiment	(3)	39	479	*	*
10th Cuirassier Regiment	(3)	32	327	*	*
2nd Brigade: *Général de Brigade* Vial					
6th Cuirassier Regiment	(3)	22	263	*	*
9th Cuirassier Regiment	(3)	34	378	*	*
	(12)	127	1447	149	1522
Division Artillery and Train					
4th Horse Artillery Regt, 4th company					
Four 6lb cannon, two 5.5in. howitzers		4	82	4	82
8th Train Squadron, 1st company		1	78	1	78
Division Strength (6 guns)	(12)	132	1607	154	1682

Summary IV Reserve Cavalry Corps

	15 June 1815		25 June 1815	
	Officers	Men	Officers	Men
Cavalry (23 squadrons)	246	2545	304	3022
Artillery (12 guns)	10	314	10	314
	256	2259	314	3336

	Belgium Holland 1793-94	Germany 1796-97	First Italian 1796-97	Egypt Syria 1798-1801	Second Italian–Marengo 1800	Germany Winter 1800	Austria–Ulm Austerlitz 1805	Prussia–Jena Auerstadt 1806	Poland–Eylau Friedland 1807	Spain Portugal 1808-09	Austria–Wagram 1809	Spain Portugal 1812	Russia–Borodino 1812	Germany Spring 1813	Germany Autumn 1813	North and Eastern France 1814	Spain Southern France 1813-14	Quatre Bras	Ligny	Wavre	Waterloo
Bachelu – Gilbert	-	•	-	•	-	-	•	-	•	•	-	•	•	•	-	-	•	-	-	-	•
Bailly de Monthion – François	-	•	-	-	•	-	•	•	•	-	•	•	•	•	-	-	•	-	•	-	•
Barrois – Pierre	•	•	-	•	-	-	•	•	•	•	-	•	•	•	•	•	-	-	•	-	-
Berthezène – Pierre	-	-	•	-	•	-	•	•	•	•	•	•	•	•	•	•	-	-	•	-	-
Bertrand – Henri	-	-	•	•	-	-	•	•	•	-	•	-	•	•	•	•	-	-	•	-	•
Bonaparte – Jérome, Prince	-	-	-	-	-	-	-	•	-	-	•	-	•	•	•	•	-	-	•	-	•
Bourmont – Louis	-	-	-	-	-	-	-	-	-	-	-	=	=	=	=	-	-	-	-	-	-
Chastel – Louis	-	-	•	•	-	•	•	•	•	-	•	-	•	•	•	•	-	-	•	•	-
Colbert Chabanais – Pierre	-	-	-	•	-	-	•	•	•	•	•	•	•	•	•	•	-	-	•	-	•
Corbineau – Jean	•	•	-	-	•	-	•	•	•	-	•	-	•	•	•	•	-	-	•	-	•
Dejean – Pierre	-	-	-	-	-	-	•	-	•	-	•	-	•	•	•	•	-	-	•	-	•
Delort – Jacques	-	-	-	-	-	•	•	-	•	-	•	-	-	•	•	•	-	-	•	-	•
Desvaux de Saint Maurice – Jean	-	-	-	-	-	-	•	•	•	-	•	-	•	•	•	•	-	-	•	-	•
Domon – Jean	•	•	-	-	•	-	•	•	•	-	•	-	•	•	•	•	-	-	•	-	•
Donzelot – François	•	•	-	•	-	-	-	-	-	-	-	-	-	-	-	•	•	-	-	-	•
Drouet – Jean, Comte d'Erlon	•	•	-	•	-	-	•	•	•	•	•	•	-	-	•	•	•	-	-	-	•
Drouot – Antoine	•	•	-	-	•	-	•	-	•	-	•	-	•	•	•	•	-	-	•	-	•
Duhesme – Philibert	•	•	-	-	•	-	-	-	-	•	-	-	-	-	•	-	-	-	•	-	•
Durutte – Pierre	•	•	-	-	•	-	-	-	•	-	•	-	•	•	•	•	-	-	•	-	•
Exelmans – Rémy	•	•	-	-	•	-	•	•	•	-	•	-	•	•	•	•	-	-	•	•	-
Flahaut de la Billarderie – Auguste	-	-	-	•	-	-	•	•	•	-	•	-	•	•	•	•	-	-	•	-	•
Fouler – Albert	•	•	-	•	-	-	•	•	•	-	•	-	•	•	•	•	-	-	•	-	•
Foy – Maximilien	•	•	-	•	-	•	-	-	-	•	-	•	-	-	•	•	•	-	-	-	•
Friant – Louis	•	•	•	•	-	-	•	•	•	-	•	-	•	•	•	•	-	-	-	-	•
Gérard – Maurice	•	•	•	-	-	-	-	-	•	-	•	-	•	•	•	•	-	-	•	-	-
Girard – Jean	-	-	•	•	-	-	•	•	•	•	•	•	-	•	•	•	-	-	•	-	-
Grouchy – Emmanuel	-	-	-	•	-	•	•	•	•	-	•	-	•	-	•	•	-	-	•	•	-
Guilleminot – Armand	•	•	-	-	•	-	•	•	•	-	•	-	•	•	•	•	-	-	•	-	-
Guyot – Claude	-	-	-	•	-	-	•	-	•	-	•	-	•	•	•	•	-	-	•	-	•
Habert – Pierre	•	•	-	•	-	-	•	-	•	-	•	-	•	•	•	-	•	-	•	-	-
Haxo – François	-	-	•	-	•	-	•	-	•	-	•	-	•	•	•	•	-	-	•	-	•
Jacquinot – Charles	•	•	-	-	•	-	•	•	•	-	•	-	•	•	•	•	-	-	-	-	•
Jeanin – Jean	-	-	•	•	-	-	•	•	•	-	-	-	-	-	-	-	-	-	•	-	•
Kellermann – François	-	-	-	-	•	-	•	•	•	-	-	•	-	•	•	•	-	-	•	-	•
Lallemand – François	•	-	•	•	-	•	•	-	-	-	-	-	•	-	-	-	-	-	•	-	•

	Belgium Holland 1793-94	Germany 1796-97	First Italian 1796-97	Egypt Syria 1798-1801	Second Italian–Marengo 1800	Germany Winter 1800	Austria–Ulm Austerlitz 1805	Prussia–Jena Auerstadt 1806	Poland–Eylau Friedland 1807	Spain Portugal 1808-09	Austria–Wagram 1809	Spain Portugal 1812	Russia–Borodino 1812	Germany Spring 1813	Germany Autumn 1813	North and Eastern France 1814	Spain Southern France 1813-14	Quatre Bras	Ligny	Wavre	Waterloo
Lallemand – Henri	-	-	-	•	-	-	•	-	-	-	•	-	•	•	•	•	-	-	•	-	•
Lefebvre-Desnoettes – Charles	•	•	•	•	-	-	•	•	-	-	-	•	•	•	•	•	-	•	•	-	•
Lefol – Etienne	•	•	-	-	•	-	-	•	-	-	-	-	•	•	•	•	-	•	•	-	•
Letort – Louis	•	-	-	-	-	-	-	-	-	-	-	-	•	•	•	•	-	•	•	-	•
L'Héritier – Samuel	-	•	•	-	-	-	-	-	-	•	-	•	•	•	•	•	-	•	•	-	•
Marcognet – Pierre	-	•	•	-	-	-	-	-	-	•	-	•	•	•	•	•	-	•	•	-	•
Maurin – Antoine	•	•	•	-	-	-	-	-	-	•	-	•	•	•	•	•	-	•	•	•	•
Michel – Claude	-	•	•	-	-	-	•	•	•	-	-	-	•	•	•	•	-	•	•	-	•
Milhaud – Edouard	•	-	-	•	-	-	•	•	•	-	-	-	•	•	•	•	-	•	•	-	•
Morand – Charles	•	•	-	•	-	-	•	•	•	-	•	-	•	•	•	•	-	•	•	-	•
Mouton – Georges, Comte Lobau	-	-	-	•	-	-	•	•	•	-	•	-	•	•	•	•	-	•	•	-	•
Neigre – Gabriel	-	•	-	-	-	-	•	-	•	-	•	-	•	•	•	•	-	•	•	-	•
Ney – Michel	•	•	•	-	-	•	•	•	•	-	•	-	•	•	•	•	-	•	•	-	•
Noury – Henri-Marie	•	-	-	•	-	-	•	•	•	-	•	-	•	•	•	-	-	•	•	-	•
Pajol – Claude	•	•	•	-	-	-	•	•	•	-	-	-	•	•	•	•	-	•	•	-	-
Pécheux – Nicolas	•	-	-	-	-	-	•	-	•	-	-	-	•	-	-	•	-	•	•	-	•
Piré – Hippolyte	-	-	-	-	-	-	•	•	•	-	-	-	•	•	•	•	-	•	•	-	•
Radet – Etienne	•	-	-	•	-	-	-	-	-	•	-	•	•	•	•	•	-	•	•	-	•
Reille – Honore	-	-	•	•	-	•	•	•	•	•	-	•	-	-	-	-	•	•	•	-	•
Rogniat – Joseph	-	•	-	-	-	•	•	-	-	•	-	•	•	•	•	•	-	•	•	-	•
Roguet – François	-	-	•	•	-	-	-	-	-	•	-	•	•	•	•	•	-	•	•	-	•
Roussel d'Hurbal – Nicolas	-	-	-	-	-	-	-	-	-	-	-	•	•	•	•	•	-	•	•	-	•
Ruty –Charles	-	-	-	•	-	-	-	-	•	-	•	-	•	•	•	•	-	•	•	-	•
Simmer – François	•	•	-	-	-	-	-	•	•	-	•	-	•	•	•	-	-	•	•	-	•
Soult – Jean	•	•	•	-	-	-	•	•	•	-	-	-	•	-	-	-	•	•	-	-	•
Soult – Pierre	-	-	•	•	-	-	-	-	-	•	-	•	•	•	•	-	•	•	•	•	-
Strolz – Jean	•	•	-	-	•	-	•	•	-	-	-	-	-	-	-	•	-	•	•	-	-
Subervie – Jacques	-	•	•	-	-	-	•	-	-	-	-	•	•	•	•	•	-	•	•	-	•
Teste – François	-	-	-	-	•	-	-	-	-	-	-	-	•	•	•	•	-	•	•	•	-
Vandamme – Dominique	•	•	•	-	-	•	•	•	•	-	•	-	•	•	•	-	-	•	•	-	-
Vichery – Louis	•	•	-	-	-	•	•	•	•	-	•	-	•	•	•	-	-	•	•	•	•
Watier – Pierre	•	-	-	-	-	•	•	•	•	-	•	-	•	•	•	-	-	-	•	-	•

348

Wargames Chart

For the benefit of wargamers, and armed with the foregoing knowledge of individual commanders, the author has assigned a one to ten rating to each general covering various aspects of their leadership and command capabilities. The rating covers their whole careers while they held either a brigade, division, corps or even army command. To put the factors in perspective one can assume that Napoleon at his best would have achieved a perfect score on all counts, though certainly not throughout his career, since at Waterloo, he was well beyond his best.

The ratings are purely the assessment of the author, who has dabbled at Napoleonic wargaming and are not by any means the final word. They are assigned as follows:

Control

The ability of a commander to control the men under his command, the level of discipline within his formation. In particular in battle to restrain them from behaving impetuously by charging without orders or vice versa heading off in willful retreat at any given moment. The commander must be flexible in responding to changes in orders issued by a superior. A high rating being given to an aggressive, decisive man who would act immediately to changing circumstances and carry out a given order to the letter. A low rating would apply to a slow cautious commander, who would invariably fail to respond in time, or at all.

Attack

The ability of the commander to raise his men's morale, instill them with confidence to come to grips with the enemy.

Defence.

The commander to possess the same attributes as with the attack factor but in a defensive position, or where he is in charge of a rearguard action.

Superior Commander

An overall assessment of these men using the previous factors mentioned but applying it to them as independent commanders, who in their careers led a formation greater than a division, i.e. an army corps, an army wing or even an army.

Command and Capabilities Table

Commander	Control	Attack	Defence	Superior Commander
Bachelu	7	7	7	-
Bailly de Monthion	7	7	7	-
Barrois	7	8	7	-
Bethezéne	7	7	7	6
Bertrand	6	6	6	5
Bonaparte, J	5	6	7	4
Bonaparte, N	10	10	10	10
Bourmont	7	7	7	7
Cambronne	7	7	7	-
Chastel	6	7	7	-
Colbert	7	8	7	-
Corbineau	7	7	7	-
Dejean	6	6	6	-
Delort	7	7	7	-
Desvaux de St. Maurice	7	8	7	-
Domon	7	7	8	-
Donzelot	5	5	6	-
Drouet comte d'Erlon	7	6	7	7
Drouot	8	8	8	8
Duhesme	7	8	8	4
Durutte	7	7	8	-
Exelmans	7	8	7	5

Commander	Control	Attack	Defence	Superior Commander
Flahaut de la Billarderie	6	6	6	-
Fouler	6	6	6	-
Foy	7	7	7	6
Friant	7	8	7	-
Gérard	8	7	7	7
Girard	6	7	7	5
Guilleminot	6	7	6	-
Grouchy	8	7	7	6
Guyot	7	7	7	-
Habert	7	7	8	-
Haxo	6	7	7	-
Jacquinot	7	7	7	-
Jeanin	6	6	7	-
Kellermann	8	8	8	7
Lallemand, F.	7	7	7	-
Lallemand, H.	6	7	7	-
Lebrun	7	7	6	5
Lefebvre-Desnoettes	7	7	6	-
Lefol	7	7	7	-
Letort	7	8	7	-
L'Héritier	7	7	7	-
Marcognet	6	5	6	-
Maurin	6	6	6	-
Michel	7	8	8	-
Milhaud	7	7	8	7
Morand	8	8	8	6
Mouton	8	8	8	6
Neigre	7	7	7	-
Noury	7	7	7	-
Ney	8	9	9	6
Pajol	7	8	8	7
Pêcheux	7	7	8	-
Piré	8	7	8	-
Radet	7	-	-	7
Reille	7	7	7	7
Rogniat	6	7	8	-
Roguet	8	8	8	-
Roussel d'Hurbal	7	7	7	-
Ruty	6	6	6	6
Simmer	7	7	7	-
Soult, Jean	9	7	9	8
Soult, Pierre	5	6	5	-
Strolz	6	6	6	-
Subervie	6	6	7	-
Teste	7	7	8	-
Vandamme	7	8	8	7
Vichery	6	6	7	5
Watier	7	7	7	-

BIBLIOGRAPHY

Arnold, J.R.	*Crisis on the Danube*, Arms and Armour Press, London, 1992
Atteridge, A.H.	*Joachim Murat*, Brentano's, New York, 1911
Becke, A.E.	*Napoleon and Waterloo*, Kegan Paul, London, 1936
Bernay, F. de	*Son of Talleyrand*, Collins, London 1956
Bowden, Scott	*Napoleon's Grande Armeé of 1813*, Emperor's Press, Chicago, 1990
Bowden, S. & Tarbox, C.	*Armies of the Danube, 1809*, Emperor's Press, Chicago, 1989
Brett-James, A.	*Europe against Napoleon*, MacMillan, New York, 1970
Brett-James, A.	*1812*, MacMillan, London 1966
Brett-James, A.	*The Hundred Days*, MacMillan, 1964
Britten Austin, F.	*1812 - The March to Moscow*, Greenhill Books, London, 1993
Chandler, D.	*Waterloo*, Osprey Publishing, London 1980
Chandler, D.	*Dictionary of the Napoleonic Wars*, Arms & Armour Press, London, 1979
Chandler, D.	*Napoleon's Marshals*, MacMillan, New York, 1987
Chandler, D.	*The Campaigns of Napoleon*, MacMillan, New York, 1966
Chlapowski,D.	*Memoirs of a Polish Lancer*, Emperor's Press, Chicago, 1992
Connelly, O.	*Napoleon's Satellite Kingdoms*, MacMillan, Toronto, 1965
Connelly, O.	*Blundering to Glory*, Scholarly Resources, Delaware, 1987
Détaille, E.	*L'Armée Française*, reprint 1889 edition, Waxtel & Hasenauer, New York 1992
Duffy, C.	*Austerlitz*, Seeley, London 1977
Duffy, C.	*Borodino*, Seeley, London 1972
Esposito, V.	*A Military History and Atlas of the Napoleonic Wars*, AMS Press, NY, 1978
Elting, J.R.	*Swords around the Throne*, Weidenfeld and Nicolson, London, 1988
Epstein, R.M.	*Prince Eugene at War: 1809*, Empire Press, Arlington, Texas, 1984
Fregosi, P.	*Dreams of Empire*, Hutchinson, London, 1989
Furse, G. A.	*Marengo and Hohenlinden*, Worey Publications, facsimile edition, 1993
Gallaher, J.G.	*The Iron Marshal*, Feffer & Simons, London, 1976
Gates, D.	*The Spanish Ulcer*, George Allen & Unwin, London 1986
Gill, J.	*With Eagles to Glory: Napoleon and his German Allies*, Greenhill Books, London, 1992
Glover, M.	*The Peninsular War 1807-1814*, David & Charles, London, 1974
Gourgand, G.	The Campaign of 1815, Paris, 1818
Griffith, P.	*Military Thought in the French Army 1815-1851*, Manchester University Press, Manchester, and New York, 1989
Hamilton-Williams, D.	*Waterloo - New Perspectives*, Arms and Armour, London 1993
Hayman, P.	*Soult: Napoleon's Maligned Marshal*, Arms & Armour Press, London, 1990
Herold, C.J.	*Bonaparte in Egypt*, Hamish Hamilton, London, 1964
Horricks, R.	*In Flight with the Eagle: A Guide to Napoleon's Elite*, Costello, Tunbridge Wells, 1988
Horricks, R.	*Marshal Ney: The Romance and the Real*, Midas Books, Tunbridge Wells, 1982
Houssaye, H.	*1814*, Paris, 1888, facsimile Worley Publications, Tyne & Wear, 1991
Houssaye, H.	*Waterloo*, Paris, 1893, facsimile Worley Publications, Tyne & Wear, 1991
Humble, R.	*Napoleon's Peninsular Marshals*, Purcell Book Series, London 1973
Johnson, D.	*Napoleon's Cavalry*, Batsford, London 1978
Johnson, D.	*The French Cavalry 1792-1815*, Belmount Publishing, London 1989
Lachouque, H.	*The Anatomy of Glory*, Brown University Press, Providence R.I., 1962
Lachouque, H.	*Waterloo*, Arms & Armour Press, London 1974
Lucas Dubreton, J.	*The Restoration of the July Monarchy*, William Heinemann, London 1929
Lucas Dubreton, J.	*Soldats du Napoleon*, Paris 1948
Mackenzie, N.	*The Escape from Elba: The Fall and Flight of Napoleon 1814-1815*, Oxford University Press, 1882
Manceron, C.	*Austerlitz*, George Allen, London, 1966
Manceron, C.	*Napoleon Recaptures Paris*, George Allen, London 1968
Marbot, J.	*The Memoirs of Baron Marbot*, facsimile, Greenhill Books, London, 1988
Martineau, G.	*Napoleon's Last Journey*, John Murrey, London, 1971
Martineau, G.	*Napoleon Surrenders*, John Murrey, London, 1971
Nafziger, G.	*Lutzen and Bautzen*, Emperor's Press, Chicago, Chicago, 1993
Nafziger, G.	*Napoleon's Invasion of Russia*, Presidio Press, California, 1988

Napier, W.	*A History of the War in the Peninsula and in the South of France 1807-1814*, facsimile, London 1992-1993
Oman, C.	*A History of the Peninsular War*, AMS Press, New York, 1980
Petre, L.	*Napoleon's Conquest of Prussia 1806*, Arms & Armour Press, London 1972
Petre, L.	*Napoleon and the Archduke Charles*, Arms & Armour Press, London 1976
Petre, L.	*Napoleon's Campaign in Poland 1807*, Arms & Armour Press, London 1976
Petre, L.	*Napoleon's Last Campaign in Germany 1813*, Arms & Armour Press, London 1974
Petre, L.	*Napoleon at Bay 1814*, Arms & Armour Press, London 1977
Phipps	*The Armies of the First French Republic*, Oxford University Press, 1926 - 1939
Porch, D.	*Army and Revolution: France 1815-1848*, Routledge & Kegan Paul, London and Boston, 1974
Rogers, H.C.	*Napoleon's Army*, Ian Allen, London, 1974
Schom, A.	*One Hundred Days: Napoleon's Road to Waterloo*, Michael Joseph, London, 1993
Schom, A.	*Trafalgar: Countdown to Battle 1803-1805*, Michael Joseph, London, 1993
Siborne, W.	*History of the Waterloo Campaign*, Greenhill Books, London, 1990
Six, Georges	*Le Dictionnaire biographique des généraux et amiraux de la Révolution et de l'Empire*, George Saffoy, Paris, 1934
Stacton, D.	*The Bonapartes - 200 Years*, Hodder & Stoughton, London 1966
Thiébault, baron P.	*Memoires of Baron Thiébault*, facsimile edition 1994, Worley Publications
Thiers, M. Adolphe	*History of the Consulate and Empire*, London (20 Volumes), 1845 et seq.
Tranie J. & Carmigniani J.	*La Patrie en Danger 1792-1793: Les Campagnes de la Revolution Tome 1*, Lavauzelle, Paris, 1987
Tranie J. & Carmigniani J.	*Les Guerres de l'Ouest 1793-1815*, Lavauzelle, Paris, 1989
Tranie J. & Carmigniani J.	*Napoleon Bonaparte: 1er Campagne d'Italie*, Pygmalion, Paris, 1990
Tranie J. & Carmigniani J.	*Napoleon 2me Campagne d'Italie*, Pygmalion, Paris, 1991
Tranie J. & Carmigniani J.	*Napoleon Bonaparte: La Campagne d'Egypte*, Pygmalion, Paris, 1988
Tranie J. & Carmigniani J.	*Napoleon et Austriche: La Campagne de 1809*, Copernic, Paris, 1979
Tranie J. & Carmigniani J.	*Napoleon et l'Allemagne: Prusse 1806*, Lavauzelle, Paris, 1984
Tranie J. & Carmigniani J.	*Napoleon: 1813 La Campagne de l'Allemagne*, Pygmalion, Paris, 1987
Tranie J. & Carmigniani J.	*Napoleon: 1814 La Campagne de France*, Pygmalion, Paris, 1989
Tulard, J.	*Dictionnaire Napoleon*, Fayard, Paris, 1987
Watson, S.	*By Command of the Emperor: A Life of Marshal Berthier*, Trotman, England, 1988
Wood, E.	*Cavalry in the Waterloo Campaign*, London, 1895